Digital Thread-Based Lifecycle Health Management System for Hot Strip Mill Process

Zhang Chuanfang　Zhang Hanwen　Peng Kaixiang

(Color pictures)

Beijing
Metallurgical Industry Press
2024

Metallurgical Industry Press
39 Songzhuyuan North Alley, Dongcheng District, Beijing 100009, China

图书在版编目(CIP)数据

基于数字孪生的热连轧过程生命周期健康管理系统 = Digital Thread-Based Lifecycle Health Management System for Hot Strip Mill Process:英文/张传放,张瀚文,彭开香著.—北京:冶金工业出版社,2024.3

ISBN 978-7-5024-9743-9

Ⅰ.①基… Ⅱ.①张… ②张… ③彭… Ⅲ.①热轧—英文 Ⅳ.①TG335.11

中国国家版本馆 CIP 数据核字(2024)第 043168 号

Digital Thread-Based Lifecycle Health Management System for Hot Strip Mill Process

出版发行	冶金工业出版社	**电　　话**	(010)64027926
地　　址	北京市东城区嵩祝院北巷 39 号	**邮　　编**	100009
网　　址	www.mip1953.com	**电子信箱**	service@mip1953.com

责任编辑　卢　敏　张佳丽　美术编辑　彭子赫　版式设计　郑小利
责任校对　郑　娟　责任印制　窦　唯
北京建宏印刷有限公司印刷
2024 年 3 月第 1 版,2024 年 3 月第 1 次印刷
710mm×1000mm　1/16;31.5 印张;614 千字;485 页
定价 198.00 元

投稿电话　(010)64027932　投稿信箱　tougao@cnmip.com.cn
营销中心电话　(010)64044283
冶金工业出版社天猫旗舰店　yjgycbs.tmall.com
(本书如有印装质量问题,本社营销中心负责退换)

Preface

The manufacturing industry is the main body of the national economy, the foundation of establishing a country, the tool of revitalizing the country, the foundation of strengthening the country, and the main battlefield of technological innovation. The steel industry is a typical process industry. In 2019, China's crude steel production reached 950 million tons, exceeding half of the global crude steel production and accounting for about 8% of GDP. The phased achievements of supply side reform have accelerated the transformation and upgrading of the industry, bringing development opportunities for intelligent steel manufacturing. At present, the steel industry is generally facing challenges such as "large but not strong" and overcapacity, product homogenization, long R&D cycles for new products, low labor productivity, unstable quality, and overall low profits for enterprises. As the core process of the steel industry, the digital transformation and operational safety management of the hot strip mill process (HSMP) are increasingly receiving attention from enterprises, and are also one of the effective ways for enterprises to improve quality and efficiency.

With the integration and implementation of the new generation of information technology and manufacturing industry, countries around the world have introduced their own advanced manufacturing development strategies, such as Germany's "National Industrial Strategy 2030" and the United States' "Industrial Internet". At the same time, in the context of the grand strategy of "manufacturing power" and "network power", China has introduced the development and implementation strategies of manufacturing countries such as "Internet +". The report of the 20th National Congress of the Communist Party of China clearly stated that "we will promote new industrialization, accelerate the construction of a strong manufacturing country, a strong quality country, a strong aerospace county, a strong transportation country, a strong network country, and a digital China. We will implement industrial infrastructure reconstruction projects and major technological equimment research project, support the development of specialized, refined, and innovative enterprises, and promote the high-end, intelligent, and green development of the manufacturing industry".

Digital twin technology is the cornerstone of digital transformation, which we are currently witnessing in the new industry 4.0 revolution. Digital twin is accessible now more than ever and many reputable and innovative companies such as Tesla and Siemens have adopted it with varying success. As the most important foundational technology in digital manufacturing, digital thread provides a standard framework for key parameters that are always available from design to operation and maintenance stages. The term "digital thread" refers to the set of data created across the life cycle of a given component. It is called a thread for several reasons. First, for the data to be most useful, it must be collected starting at the beginning of a component's life cycle, from raw materials, processing, transit, manufacturing, installation, operation, maintenance/repair, and the end-of-life. The "thread" must be pulled for it to maintain its value, and a break in the thread can severely compromise the entire thread's integrity.

Why is the digital thread so important? Here, a metaphor might be more useful. If you were a pilot and worked with another pilot long term, you might be able to operate sufficiently simply by knowing that the co-pilot had passed their certification and maintained that certification by completing refresher training at the required intervals. However, if you had a true insight into the co-pilot, understood their childhood, personality, fears, weaknesses, strengths, and interests and had a deep understanding of what was happening in their life each time you flew together, you would be able to much better coordinate your duties, understand if there were underlying risks (depression, a lack of sleep the night before, personal tragedies, hidden addiction, etc.) or alternatively understand additional strengths (e. g., dedication to excel in order to receive higher certification). In the second scenario, you would be able to understand at a much deeper level the quality and integrity of the co-pilot, which would allow you to more accurately understand key risks and at the same time be able to create efficiencies by basing maintenance on their actual needs instead of a conservative time table.

This book takes the significant demand for real-time monitoring and reliability maintenance technology for safety and stability under the supply side reform of China's steel industry as the research background, breaking through the limitations of the insufficient connection between HSMP health management and digital twin, laying a technical foundation for achieving efficient and high-quality production of HSMP, and meeting the significant demand for safety and reliability technology by HSMP. Focus on the following content:

(1) Design and Implementation of HSMP digital thread. Significant technological advancements have been driven in areas such as HSMP non-ideal data expansion and feature fusion technology, unified description of multi-source heterogeneous data and multi-stage digital representation of information, analysis of association patterns driven by data and mechanism collaboration, and mining and transformation of implicit knowledge. This has

overcome the difficulties in establishing model data network structure and knowledge modeling, completed the architecture design of the digital thread platform, and achieved comprehensive integration of data sources, this provides theoretical and technical support for the construction of digital thread and the integration and deployment of related algorithms for the deep integration of model data knowledge in the future.

(2) Lifecycle health management of HSMP. Starting from the deep analysis of the three dimensions of "multi-level, plant-wide process, and variable operating conditiuns", using a combination of fcature extraction and information fusion, comprehensively and accurately perceive the abnormal quality information of the entire process; Secondly, starting from in-depth analysis of the complex characteristics of quality anomalies, a model for fault detection, fault diagnosis, fault classification, and operational status evaluation is constructed through methods such as statistical analysis, neural networks, and deep learning. The model deeply reveals the causes of faults and their degree of harm, and can effectively evaluate the operational status under normal operating conditions, providing important algorithmic support for the digital thread platform.

This book will focus on the basic theories and key technologies of HSMP safety, quality anomaly diagnosis, and operational status evaluation. The theories and methods studied are predictive, real-time, and systematic, providing core theories and new technologies for improving the overall operational safety, quality stability, and reliability of the steel industry process. In order to further enrich the connotation and breadth of intelligent manufacturing in the steel industry.

The authors thank the National Key R&D Program of China (No. 2021YFB3301201) for funding the research over the past 3 years. The authors also thank Associate Professor Ma Liang, Associate Professor Zhang Kai, Associate Professor Liang Qiao, Zhang Xueyi, Zhang Chi, Zhang Cuicui, and Wang Yaqi, the postgraduate students from School of Automation and Electrical Engineering, University of Science and Technology Beijing, for the hard work in system design and programming.

Zhang Chuanfang
Zhang Hanwen
Peng Kaixiang

Contents

Chapter 1 Background

With the deep integration and development of information technology and manufacturing technology, the production mode of manufacturing industry is undergoing profound changes, and the development of advanced manufacturing industry is in an important period of strategic opportunities. Major industrialized countries around the world have put forward new national industrial development strategies. For example, Industry 4.0, the National Strategic Plan and New Industrial France strive to develop advanced manufacturing through "re-industrialization". Although these strategies are put forward in different backgrounds, one of their common goals is to realize the interactive integration of the physical world and the information world of manufacturing, thereby driving the intelligent development of the entire manufacturing industry. In order to raise the country's manufacturing power, China introduced "Made in China 2025" in 2015 as a national strategy to promote manufacturing power. Under the wave of transformation and upgrading of manufacturing industry, the structure and function, process flow, regulatory environment and evaluation indicators of traditional manufacturing system are becoming more and more complicated, evolving into a new complex product manufacturing process. The corresponding research on digital representation, factor modeling, production regulation and production line optimization is the key to the future core competitiveness of manufacturing industry. It has aroused great attention and long-term exploration in academia and industry.

As a typical representative of the manufacturing industry, hot strip mill process (HSMP) is acomplex industrial process, which is always characterized by complex operation mechanism, large scale, multiple operation conditions, and high efficiency. There are more than 300 control loops and 15000 process variables in the whole production line, which are coupled with each other to guarantee that the steel thickness is precisely reduced to the desired value. Based on the combination of knowledge and data-based process industry decomposition method, as described in **Fig. 1.1**, the industrial HSMP, in turn, can be divided into six subprocesses (reheating furnaces, roughing mill unit, transfer table & crop shear, finishing mill unit, laminar colling unit, coiler). The steel slab is first heated up to about 1200 ℃ in the reheating furnace.

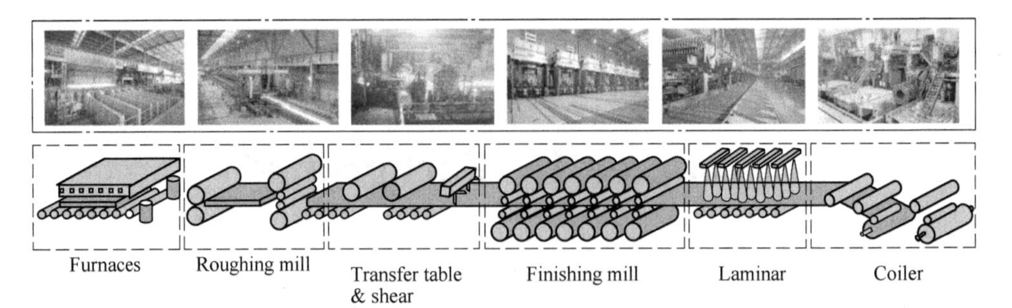

Fig. 1. 1 Schematic layout of HSMP

(Scan the QR code on front of the book for color picture)

During the roughing mill process (RMP), the thickness and the length of hot steel slab are shaped roughly to the expectations. After being passed through the transfer table, the finishing mill process (FMP) gives further precise gauge reduction toward the preset width and thickness. Then, the extremely hot strip steels are cooled to the desired temperature by the pieces of laminar-cooling equipment, and are finally coiled by the coiler for convenient loading. Among these subprocesses, FMP is the key subprocess for ensuring continuous production, stability, and high precision of the final products, which will serve as the focus of subprocess in this book.

The roughing mill group is composed of reversible four-high mill with an edger mill, of which the edger mill is used to improve the width accuracy and the shape of head and tail of middle billet. The hydraulic-press device is located in the upper part of rough roll, which is used to adjust the gap, and control the slab pressure and exit thickness. The strip width is measured by the photoelectric width measuring instrument that is based on the principles of thermal radiation, laser triangulation, and charge coupled device imaging. Moreover, it can be observed that there are seven stands in the FMP, of which each stand has two supporting rolls (in two sides) and two working rolls (in the middle). In each stand, a hydraulic system is equipped to offer the needed rolling and bending forces such that the steel thickness can be reduced; an electromechanical system is also placed to rotate the rolls so that the strip steel can be smoothly moved forward. Meanwhile, the rolling and bending forces can be measured in real time by piezomagnetic and strain gauge sensors. More remarkably, due to the high speed and temperature, the thickness. Due to the lack of real-time monitoring capabilities for the production process, the traditional HSMP is prone to slow production control response, poor quality inspection, reduced production efficiency, and difficulty in meeting the current needs of intelligent factory monitoring system construction, which affects the profitability and sustainable development of steel enterprises. Therefore, the real-time display and online monitoring of equipment operating condition are increasingly urgent. Digital thread technology can accurately simulate and depict the behavior of physical entities in the real world, and the establishment of digital thread for HSMP can make the production process more digital and transparent, which has important practical significance for the production process optimization, cost reduction, efficiency improvement and quality improvement of iron and steel enterprises. In order to meet the needs of intelligent manufacturing for production process monitoring, this book combines DTH technology, fault detection and diagnosis (FDD) technology, operating performance assessment (OPA) technology, remaining useful life (RUL) prediction technology to put forward a digital thread-based lifecycle health management system for HSMP.

1.1 Introduction

1.1.1 Industrial internet

The Industrial internet is a new type of infrastructure, application mode, and industrial ecosystem that deeply integrates the new generation of information and communication technology with the industrial economy. By comprehensively connecting people, machines, things, systems, etc., it constructs a new manufacturing and service system that covers the entire industrial chain and value chain, providing a way for the digital, networked, and intelligent development of industry and even industry, it was an important cornerstone of the Fourth Industrial Revolution.

The industrial internet is not a simple application of the internet inindustry but has richer connotations and extensions. It is based on the network, platform as the center, data as the element, and security as the guarantee. It is not only the infrastructure for industrial digitization, networking, and intelligent transformation, but also an application model for the deep integration of the Internet, big data, artificial intelligence, and the real economy. It is also a new business form and industry that will reshape the enterprise form, supply chain, and industrial chain.

At present, the integrated application of industrial internet has been widely expanded to key industries in the national economy, forming six new models: platform-based design, intelligent manufacturing, networked collaboration, personalized customization, service-oriented extension, and digital management. The role of empowerment, intelligence, and value assignment is constantly emerging, effectively promoting the quality improvement, efficiency enhancement, cost reduction, green, and safe development of the real economy.

The industrial internet includes four major systems: network, platform, data, and security. It is not only the infrastructure for industrial digitization, networking, and intelligent transformation, but also an application model for the deep integration of the internet, big data, artificial intelligence, and the real economy. It is also a new form of business and industry that will reshape enterprise form, supply chain, and industrial chain.

The network system is the foundation. The industrial internet network system includes three parts: network interconnection, data exchange, and identity resolution. Network interconnection enables data transmission between elements, including external and internal networks of enterprises. Typical technologies include traditional industrial buses, industrial Ethernet, as well as innovative time sensitive networks (TSN), deterministic networks, 5G, and other technologies. The enterprise external network is built based on the needs of industrial high-performance, high reliability, high flexibility, and high security networks, used to connect various institutions, upstream and downstream enterprises, users, and products of the enterprise. The enterprise intranet is used to connect personnel, machines, materials, environment, and systems within the enterprise, mainly including information networks and control networks. Currently, the development of internal network technology presents three characteristics: information networks and control networks are moving towards integration, industrial fieldbus is evolving towards industrial Ethernet, and industrial wireless technology is accelerating its development. Data exchange refers to the standardized description and unified modeling of data to achieve mutual understanding of information transmission between elements. Data exchange involves different levels such as data transmission, data semantics and syntax. Among them, typical data transmission technologies include embedded process control unified architecture (OPC UA), message queue telemetry transmission (MQTT), data distribution service (DDS), etc. Data semantic grammar mainly refers to information models, and typical technologies include semantic dictionaries, automation markup languages (AML), instrument markup languages (IML), etc. The identification analysis system realizes the marking, management, and positioning of elements, consisting of identification codes, identification analysis systems, and identification data services. By assigning identification codes to physical resources such as materials, machines, products, and virtual resources such as processes, software, models, and data, it realizes the logical positioning and information query of physical entities and virtual objects, and supports data sharing and sharing across enterprises, regions, and industries. The identification resolution system in China includes five national top-level nodes, international root nodes, secondary nodes, enterprise nodes, and recursive nodes. The national top-level node is the key hub of China's industrial internet identity resolution system, the international root node is the key node for cross-border resolution of various international resolution systems, the secondary node is a node that provides identity resolution public services for specific industries or multiple industries, and the recursive node is a public service node that improves overall service performance and accelerates resolution speed through caching and other technical means. Identification resolution applications can be divided into static identification applications and active identification applications based on the type of carrier. Static identification applications use one-dimensional codes, two-dimensional codes, radio frequency identification codes (RFID), near-field communication identifiers (NFC), and other carriers, and require the use of scanning guns, mobile apps, and other reading and writing terminals to trigger the identification parsing process. Active identification involves embedding identification in chips, communication modules, and terminals, and actively sending parsing requests to parsing nodes through the network.

The platform system is the backbone. The industrial internet platform system includes four levels: edge layer, IaaS, PaaS, and SaaS, which are equivalent to the "operating system" of the industrial internet and have four main functions. One is data aggregation. The multi-source, heterogeneous, and massive data collected at the network level is transmitted to industrial internet platforms, providing a foundation for deep analysis and application. The second is modeling analysis. Provide algorithmic models for big data and artificial intelligence analysis, as well as various simulation tools such as physics and chemistry, combined with technologies such as digital twins and industrial intelligence, to mine and analyze massive data, and achieve data-driven scientific decision-making and intelligent applications. The third is knowledge reuse. Translate industrial experience knowledge into model and knowledge bases on the platform, and facilitate secondary development and repeated calls through industrial microservice components, accelerating the precipitation and popularization of common capabilities. The fourth is application innovation. We provide various industrial apps and cloud software for research and development design, equipment management, enterprise operations, resource scheduling, and other scenarios to help enterprises improve quality and efficiency.

The data system is an essential element. Industrial Internet data has three characteristics. One is the importance. Data is the foundation for achieving digitization, networking, and intelligence. Without data collection, circulation, aggregation, calculation, and analysis, various new models are like passive water, and digital transformation becomes a rootless tree. The second is professionalism. The value of industrial internet data lies in its analysis and utilization, which must rely on industry knowledge and industrial mechanisms. The manufacturing industry is diverse and diverse, and each model and algorithm require long-term accumulation and professional team. Only through deep cultivation and meticulous work can data value be realized. The third is complexity. The data used in the industrial internet comes from various links of "research, production, supply, sales, and service", various elements of "human machine, material, method, and environment", and various systems such as ERP, MES, and PLC. The dimensions and complexity of the industrial internet far exceed that of the consumer internet, and it faces challenges such as difficult collection, diverse formats, and complex analysis.

The security system is the guarantee. The industrial internet security system involves various network security issues such as equipment, control, network, platform, industrial APP, data, etc. Its core task is to ensure the healthy and orderly development of the industrial internet through monitoring and warning, emergency response, detection and evaluation, functional testing, and other means. Compared with traditional internet security, industrial internet security has three major characteristics. One is that it involves a wide range. The industrial internet has broken the relatively closed and trustworthy environment of traditional industries, and network attacks can directly reach the production line. The explosive growth of networked devices and the widespread application of industrial internet platforms have continued to expand the scope of network attacks. Secondly, it has a significant impact. The industrial internet covers real economic fields such as manufacturing and energy, and once network attacks or sabotage occur, security incidents have a serious impact. Thirdly, the foundation of enterprise protection is weak. At present, the safety awareness and protection capabilities of most industrial enterprises in China are still weak, and the overall safety guarantee ability needs to be further improved.

Compared to the consumer internet, the industrial internet has many essential differences. One is that the connection objects are different. The consumer internet mainly connects people, and the scenario is relatively simple. The industrial internet connects people, machines, things, systems, as well as the entire industry chain and value chain, with a far greater number of connections than the consumer internet, making the scenario even more complex. The second is that the technical requirements are different. The industrial internet directly involves industrial production, requiring higher reliability, stronger security, and lower latency of transmission networks. Thirdly, the user attributes are different. The consumer internet is aimed at the public, with strong common user needs but relatively low level of specialization. The industrial internet is oriented towards a thousand industries and must be closely integrated with the technology, knowledge, experience, and pain points of various industries and fields. The above characteristics determine that the diversity, professionalism, and complexity of the industrial internet are more prominent, and determine that the development of the industrial internet is not achieved overnight and requires sustained efforts and long-term efforts.

1.1.2 Digital twin

In 2003, the concept of digital twin was first proposed by Professor Grieves of the University of Michigan in a speech on product lifecycle management (PLM). He believes that it is possible to collect data from physical entities and construct a virtual entity with a true mapping of the entity in the virtual space, forming a continuous channel for data interaction between the entity and the virtual entity, which will closely run through the entire production, manufacturing, and saleslifecycle[1]. In 2010, NASA defined the term "Digital Twin" in its technical report for the first time. It was also NASA that first used the concept of digital twin and applied it to actual scenarios. NASA applied digital twin technology to space vehicles to obtain flight data of physical aircraft. Through a series of simulations of digital twin running on the ground, it successfully predicted possible future situations, providing a powerful basis for staff to make correct decisions[2].

In order to achieve better operation and maintenance of equipment, General Electric Company has developed a virtual model using digital twin technology and applied it in the industrial field[3]. Siemens has applied digital twin technology in automation equipment and computer-aided manufacturing system[4], and designed and manufactured a small industrial personal computer Nanobox PC with digital twin technology in order to carry out a new model of digital industrial manufacturing. Greyce et al. completed data exchange technology based on digital twin technology, where data interacts in real-time between physical and virtual spaces[5]. Glaessgen and Stargel proposed the establishment of a high-fidelity model for aircraft, which is supported by digital twin technology and combined with historical data of the aircraft during flight to improve its operational safety and maintenance reliability[6]. Rios et al. published their first applied paper relying on digital twin technology in the intelligent manufacturing industry, opening up a new path to deeply integrate digital twin technology with the field of intelligent manufacturing, and achieving standardized processes and rules for modeling technology based on digital twin for the first time[7]. Canedo proposed to manage the large-scale Industrial internet of things and conduct in-depth analysis and optimization of the Internet of Things based on the digital twin technology in the context of the performance, communication, or interaction with other devices of the Internet of Things in the operation stage and the Internet of Things devices in operation[8]. Kostenko et al. proposed the application of digital twin technology in knowledge-based diagnostic methods, and achieved its application in dynamic and static diagnosis through research on device data[9]. Cichon and Roßmann proposed a digital twin model, focusing on the structure and functionality of the model, and developed an embeddable digital twin platform to improve the performance of human-computer interaction[10].

Countries all over the world have added digital twin technology to national strategic development goals, and various scientific research institutions and enterprises have become the main force of national Digital transformation. As a key technology, digital twin technology is the top priority of research and innovation. Chinese scholars have also established a certain foundation in their research on digital twin technology[11]. Tao et al. proposed the concept of digital twin workshop and analyzed the management elements of the workshop. They proposed a new model of digital twin control workshop with three stages: physical workshop and information workshop are independent of each other, information workshop assists in physical space, and physical workshop and information workshop interact and integrate[12]. Li studied the optimization design of detection equipment based on digital twin technology by analyzing three models of information, mathematics, and virtual space on the device. He also conducted research on the method of data communication between physical entities and digital twin, and studied the decoupling method in overall multi-objective optimization based on real-time data of the model[13]. Zhuang et al. conducted in-depth research on digital twin technology, proposed new insights into the concept of digital twin, proposed a new architecture of digital twin technology, conceived a blueprint for the future development of digital technology, and considered the possibility of applying this technology to specific industrial fields[14]. Chen et al. applied digital twin technology to the assembly of aircraft parts and proposed a new production control model[15]. Qu et al. proposed the combination of digital twin technology and the process flow of enterprise operation, which can achieve full lifecycle management of products from design, development, operation, and maintenance. Relying on information integration technology, it can visualize the entire production cycle and achieve the visualization of the entire operation state on a computer. Through continuous iteration of control, analysis, and optimization processes, it can optimize the production process and improve production efficiency[16].

In summary, current research on digital twin technology by domestic and foreign experts and scholars mainly focuses on building frameworks and theoretical research. In terms of the application of digital twin technology, it is mainly focused on the aerospace field or specific mechanical equipment. There is relatively little research on the iron and steel industry, especially the HSMP.

1.1.3 Digital thread

The concept of digital thread appeared in the report of global horizons final report: US air force global science and technology vision[17]. As the report pointed out, the idea of digital thread is to extend MBE (model-based enterprise) and digital twin to better cover and link the life cycle. In this report, digital thread is described as "the creation and use of a digital surrogate of a material system that allows dynamic, real-time assessment of the system's current and future capabilities to inform decisions in the Capability Planning and Analysis, Preliminary Design, and Detailed Design, Manufacturing and Sustainment acquisition phases", and the digital surrogate is "a physics-based technical description of the weapon system resulting from the generation, management, and application of data, models, and information from authoritative sources across the system's life cycle" . Before the emergence of the digital thread, the main emphasis in practice was on the vertical integration of systems and platforms. To achieve the goals of rapid deployment, cross-platform integration and modular upgrades, a digital thread based on data storage and analysis, model building and computation was proposed to enable quick equipment building, testing and upgrading in conjunction with technology upgrades and requirements iterations.

Digital thread now is commonly defined as "an extensible, configurable and Agency enterprise-level analytical framework that seamlessly expedites thecontrolled interplay of authoritative data, information, and knowledge in the enterprise data-information-knowledge systems, based on the Digital System Model template, to inform decisionmakers throughout a system's life cycle by providing the capability to access, integrate and transform disparate data into actionable information"[18]. A digital thread can be viewed as a framework that provides a holistic view of a system and full life-cycle traceability based on system information, models and the standards and contextual state in which the system operates. The information in a digital thread contains data from multiple sources, including flat files such as tabular documents, computer models, real-time data streams, data from hardware, etc. [19].

Providing a single, authoritative "source of truth" is an essential function of the digital thread[20]. The ability of a digital thread to provide knowledge of the current state of the system for all types of needs is critical to support decision analysis in a digital environment. As a single, authoritative source of truth, the digital thread also enables all users (stakeholders) who need to know the status of a system to access the same source of truth, thereby reducing errors and unnecessary duplication of information generation and storage, which is shown in **Fig. 1. 2**.

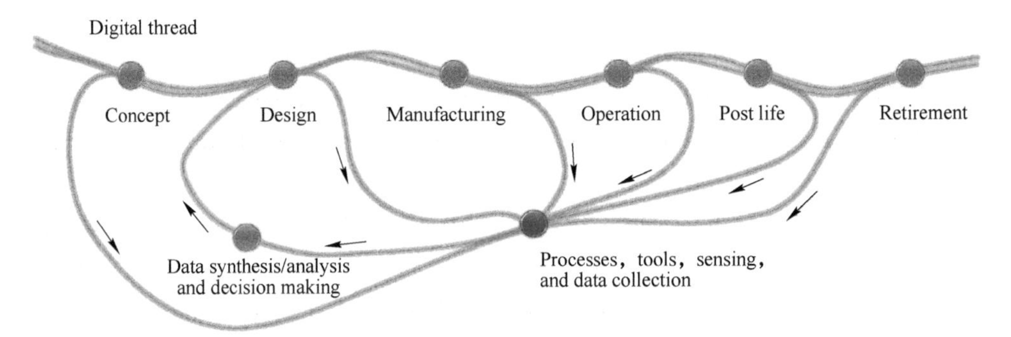

Fig. 1. 2 Illustration of digital thread

Seamlessly connecting all phases of the life cycle is an essential feature of the definition of the digital thread. Digital thread collects, analyzes, processes and presents data and information from each phase and realizes data interaction and decision supports in different phases. For example, data from the operation stage and maintenance stage can be linked to establish fault detection and health management mechanisms better, and knowledge from the operation and maintenance stage can also be fed back to the design stage to achieve higher quality and high-reliability design iterations through key parameter improvements, etc.

As an example, to illustrate the application of digital thread, Northrop Grumman established a digital thread infrastructure to support the material review board (MRB) in making F35 defective product handling decisions. Collect data from all stages of the life cycle, such as design, production and operation, to achieve feedback in the entire life cycle and the entire value chain. Based on these data, combined with the study of automatic algorithms in a 3D environment, fast and accurate structural qualification analysis is achieved to reduce processing time. The MRB process for F35 project structures, supported by digital thread, can reduce processing time by 33%. We will compare and correlate digital thread with digital twin in the next paragraph to provide a deeper understanding of digital thread's connotations and functions and give a complete definition.

How can the paradigm of digital threads and digital twin be characterized and distinguished? From the articles retrieved during the exploration phase, it is clear that digital threading is emerging in the context of Industry 4.0 and adopting specific technologies such as cloud computing, Internet of things, and artificial intelligence. Therefore, it is necessary to collect these key technologies, explore their interrelationships, and compare their contributions to the digital threading paradigm[21]. Liu et al. pointed out that the digital twin is a bridge and link "between the physical world and the information world"[22]. As a result, they can provide services that are closer to real-time, more efficient, and smarter. According to Aheleroff et al.[23], digital twin represents digital copies of physical entities, including bidirectional dynamic mappings between physical objects and their digital models. The digital model has a structure of connected elements and meta-information. However, Julien and Martin criticize this definition as being too limited, since a digital twin may exist before its physical counterpart is produced for optimal design, and after the physical object has been processed to lay the foundation for another digital twin[24]. Furthermore, NASA describes a digital twin as an "integrated multi-physics, multi-scale, probabilistic simulation of a vehicle or system" that reflects its physical counterpart using the best available physical models, sensor updates, etc. Liu et al. believe that the ability to collect and maintain system operation, history, behavior and state information is the core aspect of digital twin[25]. More specifically, Leng et al. interpret the digital twin concept as a result of modeling and simulation techniques aimed at optimizing manufacturing, aviation, healthcare, and other medical fields[26].

Digital threads, on the other hand, cover the entire product or system development lifecycle, including design, engineering, and operations. Kwon et al. viewed digital threads as a combination of model-based definition, fabrication, and inspection. According to the authors, in smart manufacturing, digital threads are essential for connecting and coordinating data across the entire product lifecycle[27]. The goal is to provide insights for better forecasting and decision-making. Similarly, Xiang et al. pointed out the potential for providing information to decision makers through the ability to access, integrate, and transform decentralized data into operational information[28]. As a result, digital threads allow data streams to be connected to create an integrated view that contains all functional views of the product lifecycle. Finally, some retrieved literature also covers the relationship between digital twin and digital threads. One article referred to digital threads as "the backbone of digital twin" . Similarly, Leng et al. classify digital threads as "critical" to creating digital twin solutions for manufacturing because of their ability to continuously collect and link a wide range of relevant data and analytical models throughout the life cycle of physical systems. Duan and Tian also pointed out that digital threading is a "key enabling technology" for digital twin[29]. In short, digital threads describe processes, and digital twin symbolize technology. Digital threads are symbiotic with digital twin to provide data support for decision-making throughout the product life cycle.

1.2 Review of process monitoring and fault diagnosis

Process monitoring and fault diagnosis (PMFD) are mainly for the detection, separation and identification of faults that have occurred in industrial processes, that is, to first determine whether the system has a fault, and then locate the location and type of the fault, and finally determine the size and occurrence time of the fault. From the perspective of quantitative analysis, process monitoring, and fault diagnosis methods can be generally divided into analytical model-based and data-driven methods[30].

1.2.1 Analytical model-based PMFD

The mathematical model of the diagnosed object is established according to the basic principles of mathematics and physics and applied to PMFD. By comparing the measurement information with the prior information derived from the analytical model, residual signals are constructed and analyzed and processed, to realize PMFD of complex industrial processes[31]. Among them, linear system-oriented methods mainly include: (1) adaptive method[32], which estimates actuator efficiency and detects actuator faults by designing a robust adaptive fault observer; (2) Multi-objective optimization method transforms the fault detection filter design problem into a filter design problem[33]; (3) Weighted matrix method[34], which uses the low frequency domain characteristics in the early stage of performance degradation to achieve fault detection to limit the frequency domain range of faults; (4) The finite frequency method[35], combined with generalized KYP lemma, directly describes the finite frequency characteristics of faults and external disturbances. From the research results of linear system fault diagnosis, the research of its theory and method is relatively mature, which provides a good reference for nonlinear system fault diagnosis, but most of the methods and conclusions are not easy to be extended to nonlinear system fault diagnosis. The methods for nonlinear systems mainly include: (1) Adaptive observer method[36], which is suitable for fault detection and isolation of nonlinear systems with fault or linear parameter uncertainty; (2) The sliding mode observer method[37] has good robustness to the uncertainties and external disturbances of the system; (3) Unknown input observer method[38], using the degree of freedom in observer design to realize the decoupling of residual signal from interference and modeling uncertainty; (4) Fuzzy observer methods are mostly focused on affine nonlinear systems, but little research has been done on general nonlinear systems[39]. In general, although the research of nonlinear system fault diagnosis has achieved fruitful results, most of them are aimed at special nonlinear systems, such as Lipschitz nonlinear system, sector bounded nonlinear system, etc., and most of the current research focuses on the design of residual generator.

1.2.2 Data-driven-based PMFD

Data-driven-based PMFD methods use statistical analysis, cluster analysis, spectrum analysis, neural network and other data processing and analysis methods to establish a trouble-free system and the actual system between the statistical distribution law is compared and calculated to achieve fault diagnosis. It has been successfully applied in chemical, pharmaceutical, steel, polymer and other complex industrial production processes, and has achieved good results[40-44].

Among the many data-driven PMFD methods, the most research papers and application cases are based on multivariate statistics. Its main theories include multivariate statistical process monitoring (MSPM) methods with principal component analysis (PCA), independent component analysis (ICA), partial least squares (PLS), gauge variable analysis (CVA) and other core methods, but these methods all assume that process data obey a single Gaussian distribution and stable working conditions. The application range of multivariate statistical process monitoring technology is greatly limited. Therefore, most of the current researches on multivariate statistical methods are aimed at improving the above basic models to adapt to complex conditions such as nonlinear[45], dynamic[46], multimodal[47] and intermittent[48]. However, when small faults are submerged in non-Gaussian, multimodal and nonlinear data, the above methods are difficult to obtain good diagnostic results[49]. At the same time, the current data-driven methods mainly focus on fault detection, and there are few researches on fault identification[50] and root cause diagnosis[51-52], which can provide information support for field operators to quickly and accurately analyze fault variables and locate fault causes, and have important practical significance. In addition, due to the complex characteristics of complex industrial processes such as large-scale, complex and integrated, quantitative analysis of fault degree still has some problems that need to be further solved in specific applications, especially for the fault assessment problem of multi-level process industry and complex and variable working conditions, there is still a lack of effective methods[53].

In general, the analytical model-based method is mainly applied to the control loop or a subsystem, while the data-based method is mainly applied to the equipment or a single process of small-scale continuous process. With the large-scale, complex and integrated development of modern industrial processes, plant-wide PMFD is the frontier hotspot and development direction in this field in the future, and some researches have been carried out at home and abroad.

1.2.3 Fault types in HSMP

A fault can be understood as at least one important variable or feature in the system deviating from the normal range. The generalized fault can be understood as any abnormal phenomenon of the system, which makes the system show the characteristics that people do not expect. In general, the fault has at least two different meanings: first, it refers to the "innate" inherent defects or defects of the system functional components; One refers to the abnormal functional changes that occur during the use or operation of the system. As the main object of process monitoring, most of the dynamic faults refer to the latter category. According to the different parts of the fault can be divided into four categories: sensor or instrument fault, actuator fault, process object fault and controller fault.

(1) Sensor or instrument fault refers to the interruption, constant deviation or large drift of the sensor signal used to detect process variables in the control loop, and the measured information cannot be accurately obtained. The specific performance is that there is a certain difference between the measured value of the object variable and the actual value. In the actual industrial process, due to the interference of the external environment and its own characteristics, the sensor will have many different types of faults, such as: deviation fault, drift fault, accuracy reduction fault and complete fault. The first three types of faults are called "soft faults" and the latter are called "hard

faults" . Because the fault phenomenon is not obvious, it is difficult to detect, which makes the soft fault more harmful than the hard fault to some extent. If there is a position sensor fault in the Hydraulic Gap Control (HGC) of hot tandem rolling, it is usually shown that the HGC resources turn red, accompanied by the HGC position over deviation alarm, the mill level value increases on the HMI screen, the mill level tilt, and the mill cannot be ready to draw steel.

(2) Actuator fault refers to the actuator used to execute the control command in the control loop is stuck, constant deviation or constant gain change and cannot execute the control command correctly. The specific performance is that there is a certain difference between the input command of the actuator and the actual output, such as the valve is stuck. The common actuator fault in hydraulic roll gap control of HSMP, such as hydraulic valve group fault, leads to system leakage.

(3) Process object fault refers to the fault caused by the abnormal occurrence of some source parts or even subsystems in the controlled object. For example, the fault of the baling machine in the last link of the continuous production of HSMP will directly affect the production rhythm and capacity efficiency of the hot rolling production line.

(4) Controller fault refers to the fault caused by an abnormal controller used to execute control commands in the control loop. For example, the electric dust removal power supply equipment is one of the essential equipment of the electric dust removal system, and the performance of the electric dust removal power supply equipment has a great impact on the dust removal efficiency. Under the same other conditions, the dust removal efficiency of electric dust removal depends on the dust driving speed, and the driving speed is increased with the increase of the intensity of the charged electric field and the intensity of the dust collecting electric field. To obtain the highest dust removal efficiency, it is necessary to increase the intensity of the electric field as much as possible, and the intensity of the electric field depends entirely on the power supply device, which is called the power supply.

1.3 Review of root cause diagnosis and propagation path identification

At present, the main research methods of root cause diagnosis and propagation path identification are knowledge based, data based, and joint knowledge and data based.

First, the knowledge-based approach builds a topological model for qualitative analysis by mining the potential correlation and causal information of process industrial processes. In the symbolic directed graph (SDG) model, nodes and directed edges are used to represent variables and the potential causality between variables[54]. The adjacency matrix quantified the correlation strength of variables by cosine similarity and partial correlation coefficient to establish causal topology model. Fault tree method is used to determine the root cause of system fault by reasoning and analyzing the logical relation of events. At present, Petri Net has achieved certain application effect in the fault diagnosis of power grid and electromechanical system.

Second, the data-based method is currently a research hotspot in this field[55]. Based on the large volume of historical process data, this method mines the causal correlation and effective information among variables, and establishes the causal matrix or topology diagram, so as to realize fault tracing. Cross-correlation analysis analyzes the causal correlation between variables from the perspective of time series to achieve fault tracing in industrial processes[56]. Transfer entropy (TE) is the construction of causality matrix and reliable causal topology diagram by designing reasonable direction measure using information theory method[57]. In order to mine direct causality among the causal associations of many variables, references [58-59] proposed direct transfer entropy and its improved method. The Granger causality (GC) analysis method effectively excavates the causality between process variables from the perspective of the lag of the temporal digital characteristic quantity[60]. Reference [61] used kernel density estimation to calculate the probability density function in Bayesian network (BN) structure to realize fault tracing. Traditional BN has advantages in dealing with uncertainty. Based on dynamic BN (DBN) anomaly index, literature uses smooth reasoning to realize fault root diagnosis and propagation path identification[62].

Finally, considering that the process industrial process includes top-down vertical multi-level and horizontally interconnected coupled multi-processes, its internal mechanism knowledge and operation data have hierarchical cross-domain corresponding relationship, it is necessary to consider the internal state, external performance and historical experience of process operation. The mechanism knowledge reflecting the essence of the process, the experience knowledge reflecting the internal relationship between the production operation and the process, the process data and other dynamic information are integrated. Therefore, the method of fault root diagnosis and propagation path identification based onknowledge and data is a new direction at present. According to the time-varying characteristics of dynamic processes, reference [63] combined the Hidden Markov model (HMM) with BN to effectively quantify the causal relationship between variables. A method based on partial least squares and module contribution analysis provides a new idea for distributed root cause diagnosis and propagation path analysis[64].

In general, the research on fault root diagnosis and propagation path identification technology in process industry is still in the preliminary exploration stage. Most of the above researches are aimed at a specific working condition or a specific process, and lack of fault root diagnosis and propagation path identification methods from the perspectives of multi-level, multi-process and multi-working condition, and lack of hierarchical fault tracing strategies for hierarchy-process-working condition.

1.4 Review of operating performance assessment

Operating performance assessment refers to the judgment and evaluation of the quality of the process or equipment running state, the identification of the leading causes of non-optimal state, and the provision of control strategies to operators to keep the production process in the optimal state, providing a theoretical basis for the fine management of industrial processes, and also of great significance to improve the comprehensive economic benefits and achieve healthy and efficient production. Operating performance assessment can be divided into three types of methods: qualitative method, quantitative method, and the combination of qualitative and quantitative methods.

The qualitative method is used to deal with the condition evaluation problem based on semantic information. Early working condition evaluation studies mainly used qualitative methods, including expert system method[65], Bayesian network method[66], probabilistic rough set theory method[67], dynamic probability theory method[68], etc. The qualitative method has strong interpretation and simple modeling, but it has some disadvantages such as dependence on expert knowledge, low precision and poor real-time performance.

Quantitative method is used to deal with the problem of condition evaluation based on numerical information. Multivariate statistical methods such as PCA method[69] and multi-set PCA method[70] are widely used in working condition evaluation. For the process with single-peak and multi-peak data distribution, there are full latent structural projection method[71] and Gaussian process regression method[72], respectively. In recent years, artificial intelligence methods have gradually become popular methods[73]. The quantitative method has high precision, can establish the correlation between variables, and has good prediction performance, but it is poor in interpretation, difficult in modeling, and may cause overfitting phenomenon.

The manufacturing process has the characteristics of coexistence of qualitative and quantitative information, a single qualitative method will lose information due to discretization, reduce the evaluation accuracy, and quantitative technology is difficult to apply directly. The combination of qualitative and quantitative evaluation of operating conditions solves the problem well. Chang et al. divided the mixed model into molecular blocks and modeled it separately according to information types to evaluate the operating conditions of the whole process with the worst grade of the sub-block layer[74]. Combined with the random forest algorithm, Chang et al. proposed to use decision trees to process quantitative/qualitative information[75], which improved the accuracy and generalization ability of the model. Zou et al. used process knowledge and production experience and other knowledge auxiliary data to evaluate working conditions based on dynamic causal diagram[76].

Considering the large scale of complex industrial processes and the coupling of variables, the traditional method directly applied to the whole process operating condition assessment will have some problems such as poor accuracy. Zou et al. proposed a hierarchical and block structure to evaluate the operation conditions of the whole process, dividing the process industrial process into the whole process layer, functional zone layer and unit layer, and establishing the relationship among process variables, process indicators and comprehensive evaluation indicators[77]. Based on deep belief network and mixed sampling and boosting algorithm, Zhang et al. extracted each subblock feature input mixed sampling and lifting algorithm, established the working condition evaluation model, and calculated the difference coefficient for the non-optimal state to trace the non-optimal variables[78].

Most of the above researches focus on the problem of condition identification and evaluation through global/local model construction using statistical or machine learning methods. It rarely involves working condition pattern mining, and there are few reports on working condition pattern cognition using qualitative methods such as semantic description or combining qualitative and quantitative methods. The construction of operating condition assessment model based on the combination of qualitative knowledge and quantitative data under complex working conditions also lacks in-depth research.

1.5 Review of remaining useful life prediction

Generally, the existing RUL prediction methods can be classified into three categories: model-based, data-driven, and hybrid approaches[79]. The basic idea of the model-based method is to develop a mathematical model that can describe the physical characteristics and fault modes of the system to realize the prediction of the RUL. It usually achieves more accurate results but cannot be applied to systems that lack prior knowledge about the physical degradation. The widely used model-based approaches include the Paris-Erdogan model[80], the Kalman filter[81], and the particle filter[82]. Data-driven methods are able to construct a mapping relationship between input and output based on a large amount of historical data and then predict the RUL, which can avoid the disadvantages of model-based method. A large number of data-driven approaches, such as the autoregressive model[83], the proportional hazard model[84], the Wiener process[85], the support vector machine (SVM)[86], and the artificial neural networks[87], have been studied in the past few years and have made remarkable achievements. By combining several of the aforementioned methods, hybrid approaches are able to leverage advantages of different models while avoiding their disadvantages at the same time[88]. For instance, Wei et al. proposed a framework that integrates SVM and particle filter to estimate the RUL of batteries, which not only provides a possible fault time range, but also improved the accuracy of prediction results compared with conventional methods[89]. In reference [90], the particle filter algorithm is introduced to estimate the state of Wiener process to remove the influence of multisource variability and survival measurements. However, developing an effective and reliable hybrid approach still remains challenging, in particular the combination of model-based and data-driven methods.

References

[1] Zhang Y F, Qian C, Liu J X. Agent and cyber-physical system based self-organizing and self-adaptive intelligent shop floor [J]. IEEE Transactions on Industrial Informatics, 2017, 13 (2): 737-747.

[2] Graham W. GE advances analytical maintenance with digital twins. Aviation Week and Space Technology, 2015.

[3] Vachálek J, Bartalský L, Rovný O, et al. The digital twin of an industrial production line within the industry 4.0 concept. 21st International Conference on Process Control, 2017, 258-262.

[4] Schleich B, Anwer N, Mathieu L. Shaping the digital twin for design and production engineering [J]. CIRP Annals-Manufacturing Technology, 2017, 66 (1): 33-35.

[5] Greyce N, Charles S, Carlos E. Digital twin data modeling with automation ML and a communication methodology for data exchange [J]. IFAC PapersOnLine, 2016, 49 (30): 124-127.

[6] Glaessgen E, Stargel D. The digital twin paradigm for future NASA and US Air Force vehicles. 53rd Structures, Structural Dynamics and Materials Conference, 2012.

[7] Rios J, Hernandez J, Oliva M. Product avatar as digital counterpart of a physical individual product: literature review and implication in an aircraft [J]. ISPE CE, 2015: 657-666.

[8] Canedo A. Industrial IoT lifecycle via digital twin. 2016 International Conference on Hardware/Software Codesign and System Synthesis (CODES+ISSS), 2016.

[9] Kostenko D, Kudryashov N, Maystrishin M, et al. Digital twin applications: diagnostics, optimization and prediction. Annals of DAAAM & Proceedings, 2018.

[10] Cichon T, Roßmann J. Digital twin: assisting and supporting cooperation in human-robot teams. [C] //2018 15th International Conference on Control, Automation, Robotics and Vision (ICARCV) 2018: 486-491.

[11] Zheng Y, Yang S, Cheng H. An application framework of digital twin and its case study [J]. Journal of Ambient Intelligence and Humanized Computing, 2019, 10 (3): 1141-1153.

[12] Tao F, Zhang Z Y, Qi Q L, et al. Digital twin maturity model [J]. Computer Integrated Manufacturing Systems, 2022, 28 (5): 1267-1281.

[13] Li B C. Research on method of lean optimization design of 3D non-standard testing equipment based on digital twin model [J]. Shandong University, 2018.

[14] Zhuang C B, Liu J H, Xiong H, et al. Connotation, architecture and trends of product digital twin [J]. Computer Integrated Manufacturing Systems, 2017, 23 (4): 753-768.

[15] Chen Z, Ding X, Tang J J. Exploration of production control mode in aircraft assembly workshop based on Digital Twin [J]. Aeronautical Manufacturing Technology, 2018, 61 (12): 46-58.

[16] Qu G Q. Research on the models, paths, and breakthroughs to promote the development of intelligent manufacturing [J]. Policy Research & Exploration, 2017, 2: 34-37.

[17] Maybury M T. Global horizons final report: United States air force global science and technology vision [R]. US Air Force, 2013.

[18] Kraft E M. The air force digital thread/digital twin-life cycle integration and use of computational and experimental knowledge [C] //54th AIAA aerospace sciences meeting, 2016.

[19] Margaria T, Alexander S. The digital thread in industry 4.0. International Conference on Integrated Formal Methods [C] //Cham: Springer, 2019.

[20] Kraft E M. Approach to the development and application of a digital thread/digital twin authoritative truth source [C] //2018 Aviation Technology, Integration, and Operations Conferences, 2018.

[21] Daasea C, Haertela C, Nahhasa A, et al. Following the digital thread-a cloud-based observation [J]. Procedia Computer Science, 2023, 217: 1867-1876.

[22] Liu J F, Cao X W, Zhou H G, et al. A digital twin-driven approach towards traceability and dynamic control for processing quality [J]. Advanced Engineering Informatics, 2021, 50: 101395.

[23] Aheleroff S, Xu X, Zhong R Y, et al. Digital twin as a service (DTaaS) in Industry 4.0: an architecture reference model [J]. Advanced Engineering Informatics, 2021, 47: 101225.

[24] Julien N, Martin E. How to characterize a digital twin: a usage-driven classification [J]. IFAC-PapersOnLine, 2021, 54 (1): 894-899.

[25] Liu C, Jiang P Y, Jiang W L. Web-based digital twin modeling and remote control of cyber-

physical production systems [J]. Robotics and Computer-Integrated Manufacturing, 2020, 64: 101956.

[26] Leng J W, Wang D W, Shen W M, et al. Digital twin-based smart manufacturing system design in Industry 4.0: A review [J]. Journal of Manufacturing Systems, 2021, 60: 119-137.

[27] Kwon S, Monniera L V, Barbaua R, et al. Enriching standards-based digital thread by fusing as-designed and as-inspected data using knowledge graphs [J]. Advanced Engineering Informatics, 2020, 46: 101102.

[28] Xiang F, Huang Y Y, Zhang Z, et al. Digital twin driven smart design [J]. Academic Press, 2020, 165-184.

[29] Duan H B, Tian F. The development of standardized models of digital twin [J]. IFAC-PapersOnLine, 2020, 53 (5): 726-731.

[30] Zhou D H, Hu Y Y. Fault diagnosis techniques for dynamic systems [J]. Acta Automatica Sinica, 2009, 35 (6): 748-758.

[31] Ding S X. Model-based fault diagnosis techniques-design schemes, algorithms and tools [M]. 2nd Edition. London: Springer-Verlag, 2013.

[32] Zhong M Y, Ding S X, James L, et al. An LMI approach to design robust fault detection filter for uncertain LTI systems [J]. Automatica, 2003, 39 (3): 543-550.

[33] Zhong M Y, Ding S X, Ding E L. Optimal fault detection for linear discrete time-varying systems [J]. Automatica, 2010, 46 (8): 1395-1400.

[34] Wang J L, Yang G H, Liu J. An LMI approach to H∞ index and mixed H-/H∞ fault detection observer design [J]. Automatica, 2007, 43 (2): 1656-1665.

[35] Casavola A, Famularo D. Robust fault detection of uncertain linear systems via quasi-LMIs [J]. Automatica, 2008, 44 (1): 289-295.

[36] Zhang X, Polycarpou M M, Parisini T. Fault diagnosis of a class of nonlinear uncertain systems with Lipschitz nonlinearities using adaptive estimation [J]. Automatica, 2010, 46 (2): 290-299.

[37] Alwi H, Edwards C, Tan C P. Fault detection and fault-tolerant control using sliding modes [M]. London: Springer, 2011.

[38] Mondal S, Chakraborty G, Bhaattacharyya K. Robust unknown observer for nonlinear systems and its application to fault detection and isolation [J]. Journal of Dynamic Systems, Measurement and Control, 2008, 130 (4): 1-5.

[39] Li L L, Ding S X, Qiu J B, et al. Real-time fault detection approach for nonlinear systems and its asynchronous T-S fuzzy observer-based implementation [J]. IEEE Transactions on Cybernetics, 2017, 47 (2): 283-294.

[40] Gui W H, Yang C H, Chen X F, et al. Modeling and optimization Problems and Challenges Arising in Nonferrous Metallurgical Processes [J]. Acta Automatica Sinica, 2013, 39 (3): 197-207.

[41] Severson K, Chaiwatanodom P, Braatz R D. Perspectives on process monitoring of industrial

systems [J]. Annual Reviews in Control, 2016, 42: 190-200.

[42] Zhao C H, Yu W K, Gao F R. Data analytics and condition monitoring methods for nonstationary batch processes — current status and future [J]. Acta Automatica Sinica, 2020, 46 (10): 2072-2091.

[43] Peng K X, Ma L, Zhang K. Review of quality-related fault detection and diagnosis techniques for complex industrial processes [J]. Acta Automatica Sinica, 2017, 43 (3): 349-365.

[44] Jiang Y C, Yin S. Recent advances in key-performance-indicator oriented prognosis and diagnosis with a MATLAB toolbox: DB-KIT [J]. IEEE Transactions on Industrial Informatics, 2019, 15 (5): 2849-2858.

[45] Yin S, Gao H, Qiu J, et al. Fault detection for nonlinear process with deterministic disturbances: A just-in-time learning based data driven method [J]. IEEE Transactions on Cybernetics, 2017, 47 (11): 3649-3657.

[46] Zhou L, Li G, Song Z, et al. Autoregressive dynamic latent variable models for process monitoring [J]. IEEE Transactions on Control Systems Technology, 2017, 25 (1): 366-373.

[47] Zhao C, Wang W, Qin Y, et al. Comprehensive subspace decomposition with analysis of between-mode relative changes for multimode process monitoring [J]. Industrial & Engineering Chemistry Research, 2015, 54 (12): 3154-3166.

[48] Onel M, Kieslich C A, Guzman Y A, et al. Big data approach to batch process monitoring: Simultaneous fault detection and diagnosis using nonlinear support vector machine-based feature selection [J]. Computers & Chemical Engineering, 2018, 115: 46-63.

[49] Wen C L, Lv F Y, Bao Z J, et al. A review of data driven-based incipient fault diagnosis [J]. Acta Automatica Sinica, 2016, 42 (9): 1285-1299.

[50] Choi S W, Lee C, Lee J M, et al. Fault detection and identification of nonlinear processes based on kernel PCA [J]. Chemometrics and Intelligent Laboratory Systems, 2005, 75 (1): 55-67.

[51] Rashidi B, Singh D S, Zhao Q. Data-driven root-cause fault diagnosis for multivariate non-linear processes [J]. Control Engineering Practice, 2018, 70: 134-147.

[52] Zhu Q X, Gao H H, Xu Y. A survey on alarm management for industrial processes [J]. Acta Automatica Sinica, 2017, 43 (6): 955-968.

[53] Dong J, Zhang W, Peng K X, et al. A novel method of quality abnormality detection and fault quantitative assessment for industrial processes [J]. Acta Automatica Sinica, 2022, 48 (10): 2406-2415.

[54] Yang F, Xiao D Y. Review of SDG modeling and its application [J]. Control theory & Applications, 2005, 22 (5): 767-774.

[55] Ma L, Dong J, Peng K X, et al. Hierarchical monitoring and root-cause diagnosis framework for key performance indicator-related multiple faults in process industries [J]. IEEE Transactions on Industrial Informatics, 2019, 15 (4): 2091-2100.

[56] Yang B, Li H G, Wen B. A dynamic time delay analysis approach for correlated process

variables [J]. Chemical Engineering Research and Design, 2017, 122: 141-150.

[57] Naghoosi E, Huang B, Domlan E, et al. Information transfer methods in causality analysis of process variables with an industrial application [J]. Journal of Process Control, 2013, 23: 1296-1305.

[58] Duan P, Yang F, Chen T W, et al. Direct causality detection via the transfer entropy approach [J]. IEEE Transactions on Control Systems Technology, 2013, 21 (6): 2052-2066.

[59] Ma L, Dong J, Peng K X, et al. A novel data-based quality-related fault diagnosis scheme for fault detection and root cause diagnosis with application to hot strip mill process [J]. Control Engineering Practice, 2017, 67: 43-51.

[60] Yin J T, Xie Y F, Chen Z W, et al. Fault tracing method based on fault propagation and causality with its application to the traction drive control system [J]. Acta Automatica Sinica, 2020, 46 (1): 47-57.

[61] Chen X L, Wang J, Zhou J L. Probability density estimation and Bayesian causal analysis-based fault detection and root identification [J]. Industrial & Engineering Chemistry Research, 2019, 57 (43): 14656-14664.

[62] Wee Y Y, Cheah W P, Tan S C, et al. A method for root cause analysis with a Bayesian belief network and fuzzy cognitive map [J]. Expert Systems with Applications, 2015, 42 (1): 468-487.

[63] Don M G, Khan F. Dynamic process fault detection and diagnosis based on a combined approach of hidden Markov and Bayesian network model [J]. Chemical Engineering Science, 2019, 201: 82-96.

[64] Zhu Q X, Luo Y, He Y L. Novel distributed alarm visual analysis using multicorrelation block-based PLS and its application to online root cause analysis [J]. Industrial & Engineering Chemistry Research, 2019, 58 (45): 20655-20666.

[65] Tidriri K, Chatti N, Verron S, et al. Bridging data driven and model-based approaches for process fault diagnosis and health monitoring: A review of researches and future challenges [J]. Annual Reviews in Control, 2016, 42: 4263-4281.

[66] Hosack G R, Hayes K R, Dambacher J M. Assessing model structure uncertainty through an analysis of system feedback and Bayesian networks [J]. Ecological Applications, 2008, 18 (4): 1070-1082.

[67] Yao Y Y. Probabilistic rough set approximations [J]. International Journal of Approximate Reasoning, 2008, 49 (2): 255-271.

[68] Thunis P, Clappie A. Indicators to support the dynamic evaluation of air quality models [J]. Atmospheric Environment, 2014, 98: 402-409.

[69] Liu Y, Wang F L, Chang Y Q. Online fuzzy assessment of operating performance and cause identification of nonoptimal grades for industrial processes [J]. Industrial & Engineering Chemistry Research, 2013, 52 (50): 18022-18030.

[70] Liu Y, Wang F L, Chang Y Q. Operating optimality assessment based on optimality related

variations and nonoptimal cause identification for industrial processes [J]. Journal of Process Control, 2016, 39: 11-20.

[71] Fan H P, Wu M, Lai X Z, et al. A decentralized operating performance assessment for geological drilling process via multi-block total projection to latent structures and Bayesian inference [J]. Journal of Process Control, 2022, 117: 26-39.

[72] Du S, Wu M, Chen L F, et al. Operating performance improvement based on prediction and grade assessment for sintering process [J]. IEEE Transactions on Cybernetics, 2022, 52 (10): 10529-10541.

[73] Liu Y, Gong S Z, Wang F L, et al. Process operating performance assessment based on stacked supervised denoising auto-encoders [J]. Chinese Journal of Scientific Instrument, 2022, 43 (4): 271-281.

[74] Chang Y Q, Zou X Y, Wang F L, et al. Multi-mode plant-wide process operating performance assessment based on a novel two-level multi-block hybrid model [J]. Chemical Engineering Research & Design, 2018, 136: 721-733.

[75] Chang Y Q, Sun X T, Zhong L S, et al. Industrial Operation performance evaluation of industrial processes based on modified random forest [J]. Acta Automatica Sinica, 2021, 47 (9): 2214-2225.

[76] Zou X Y, Chang Y Q, Wang F L. Process operating performance optimality assessment with coexistence of quantitative and qualitative information [J]. The Canadian Journal of Chemical Engineering, 2018, 96 (1): 179-188.

[77] Zou X Y, Wang F L, Chang Y Q, et al. Plant-wide process operating performance assessment and non-optimal cause identification based on hierarchical multi-block structure [J]. Acta Automatica Sinica, 2019, 45 (2): 315-324.

[78] Zhang C F, Peng K X, Dong J, et al. Distributed DBN-HSBoost model for hot rolling process operating performance assessment with partial communication [C] //2021 33rd Chinese Control and Decision Conference (CCDC) . IEEE, 2021: 1284-1289.

[79] Meng H, Li Y F. A review on prognostics and health management (PHM) methods of lithium-ion batteries [J]. Renewable & Sustainable Energy Reviews, 2019, 116: 109405.

[80] Paris P, Erdogan F. A critical analysis of crack propagation laws [J]. Journal of Basic Engineering, 1963, 85 (4): 528-533.

[81] Plett G. Extended Kalman filtering for battery management systems of LiPB-based HEV battery packs: part 3. State and parameter estimation [J]. Journal of Power Sources, 2004, 134: 277-292.

[82] Liu Z, Sun G, Bu S, et al. Particle learning framework for estimating the remaining useful life of lithium-ion batteries [J]. IEEE Transactions on Instrumentation & Measurement, 2017, 66 (2): 280-293.

[83] Long B, Xian W, Jiang L, et al. An improved autoregressive model by particle swarm optimization for prognostics of lithium-ion batteries [J]. Microelectronics Reliability, 2013, 53

(6): 821-831.

[84] Hu J, Chen P. Predictive maintenance of systems subject to hard fault based on proportional hazards model [J]. Reliability Engineering System Safety, 2020, 196: 106707.

[85] Si X, Wang W, Hu C, et al. Estimating remaining useful life with three-source variability in degradation modeling [J]. IEEE Transactions on Reliability, 2014, 63 (1): 167-190.

[86] Khelif R, Chebel-Morello B, Malinowski S, et al. Direct remaining useful life estimation based on support vector regression [J]. IEEE Transactions on Industrial Electronics, 2017, 64 (3): 2276-2285.

[87] Zhao Z, Liang B, Wang X, et al. Remaining useful life prediction of aircraft engine based on degradation pattern learning [J]. Reliability Engineering System Safety, 2017, 164: 74-83.

[88] Liao L, Kottig F. Review of hybrid prognostics approaches for remaining useful life prediction of engineered systems, and an application to battery life prediction [J]. IEEE Transactions on Reliability, 2014, 63 (1): 191-207.

[89] Wei J, Dong G, Chen Z. Remaining useful life prediction and state of health diagnosis for Lithium-Ion batteries using particle filter and support vector regression [J]. IEEE Transactions on Industrial Electronics, 2018, 65 (7): 5634-5643.

[90] Zhang Y Y, Fang L Q, Qi Z Y, et al. A review of remaining useful life prediction approaches for mechanical equipment [J]. IEEE Sensors Journal, 2023, 23 (24): 29991-30006.

Chapter 2 HSMP Data Perception, Fusion, and Visualization

This chapter designs the research framework of HSMP production equipment data perception, fusion and visualization according to the knowledge requirements of intelligent factory production equipment data perception, fusion and visualization, and plans the corresponding technical route according to the research framework and focuses on the key technologies and related theories. This paper introduces the data fusion technology, the automatic data feature extraction technology based on deep neural network used in fusion analysis, and the visualization chart generation tool, intelligent data analysis BI tool and digital twin technology commonly used in data visualization, so as to prepare for the follow-up research.

2.1 Research framework and technical route

The research framework of HSMP data perception, fusion and visualization is shown in **Fig. 2.1**. This topic aims to study the data perception method of HSMP manufacturing field equipment and the data fusion and analysis technology of HSMP manufacturing field equipment under the background of intelligent manufacturing and industrial big data, taking HSMP equipment data as the research object and driven by equipment data. And research HSMP data visualization technology, HSMP data fusion, analysis results visualization, real-time monitoring and evaluation of HSMP equipment. The research objectives are to build a real-time operating status monitoring and evaluation platform for HSMP equipment, reduce the failure rate of equipment, extend the service life of equipment, do predictive maintenance of important equipment, and make independent production decisions in workshops, so as to provide guarantee for the realization of safe, stable, efficient and green production of HSMP.

According to the research framework in **Fig. 2.1**, the research technical route is designed, as shown in **Fig. 2.2**.

In the part of data perception, in order to solve the problem that the existing data collection scheme of data perception is not comprehensive enough to collect data and insufficient utilization of data, this paper designs a set of data intelligent perception

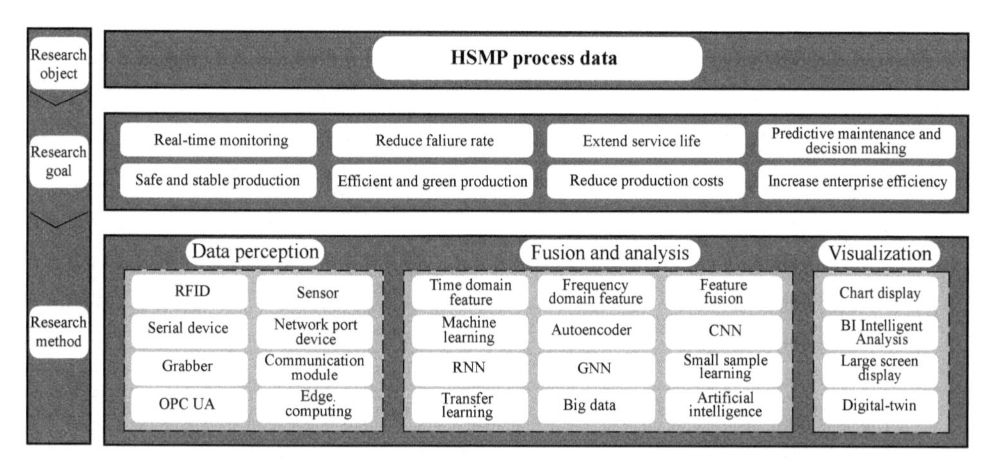

Fig. 2.1 Research framework for HSMP data perception, fusion, and visualization

method, integrating the existing data collection system of HSMP. It analyzes various industrial network protocols of HSMP data sources such as production equipment, sensors, control equipment, upper computer, and data collection system, and unified OPC UA data access interface to provide guarantee for real-time data transmission, secure access, storage and management. The status monitoring system of key equipment and key components is an important application of the comprehensive data collection method combined with the datafusion analysis results in real-time monitoring, predictive maintenance and independent decision-making of relevant equipment in hot rolling field.

In the part of data fusion and analysis, this paper designs a set of data intelligent fusion and analysis methods. Firstly, large data technology is used to process data real-time update, cleaning, fusion, correlation analysis, feature extraction, storage and management. The processed data is sent to the fusion and analysis model algorithm training or to the trained and deployed model for diagnosis, prediction, optimization or decision making.

In the two parts of visualization and application, research on HSMP data visualization related technologies, visualization of HSMP data fusion analysis results, real-time monitoring of equipment, predictive maintenance and autonomous decision-making, production optimization and marketing management and other important applications.

Next, this chapter will focus on key technologies such as data fusion, data feature extraction commonly used in deep neural networks, small sample learning, transfer learning, and data visualization from the research technology route, so as to prepare for subsequent research.

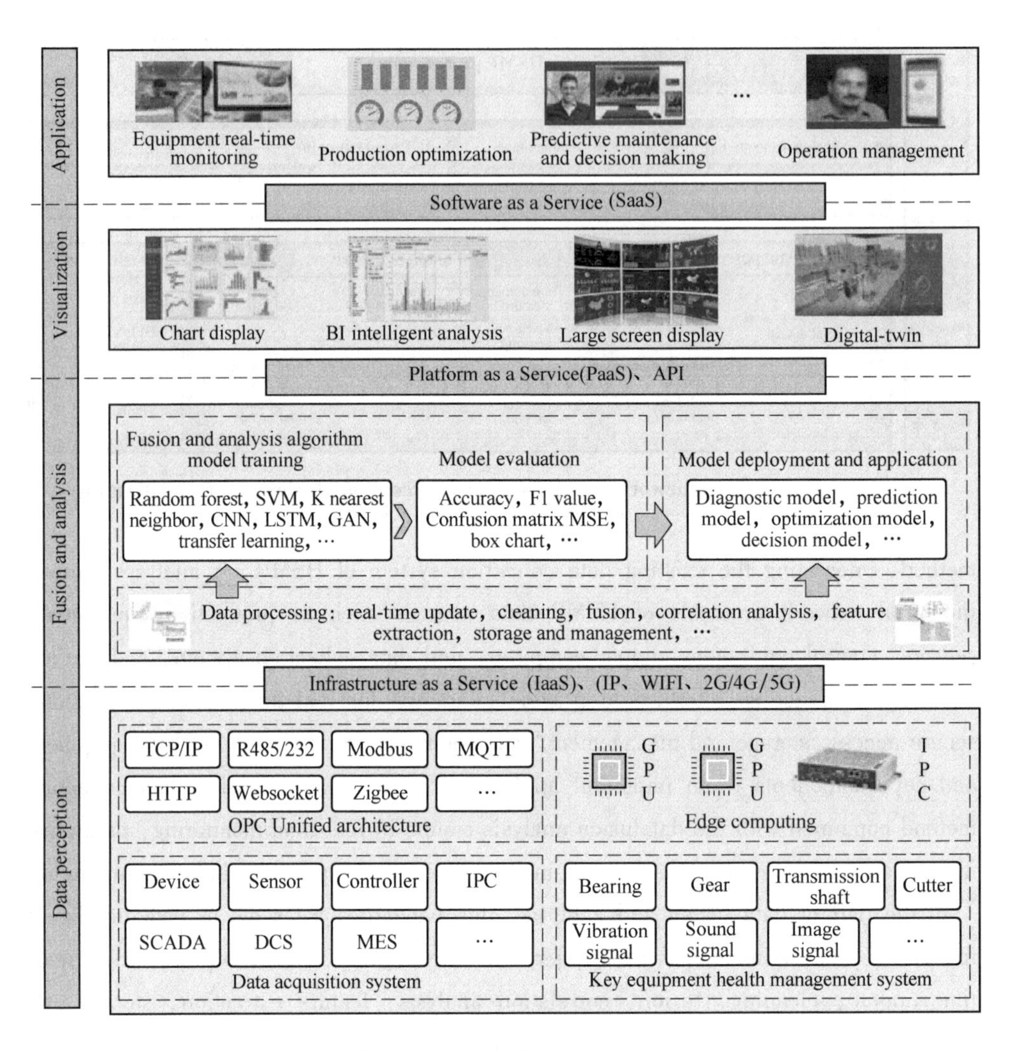

Fig. 2. 2 Technology roadmap

2. 2 HSMP data perception related theory

The current development of industrial data collection is becoming faster and wider, and the runway of the industrial internet is becoming wider. Industrial automation enterprises, network communication enterprises, and information technology enterprises have emerged one after another. At present, there are three main types of enterprises on the supply side of the industrial data collection industry:

(1) Industrial automation enterprises, starting from their core product capabilities, mainly provide access equipment for industrial data collection as the source of industrial

data collection, such as Siemens, Advantech, Honeywell, Security Control, etc.

(2) Industrial network service enterprises mainly provide industrial network protocol conversion, transmission, security and other supporting equipment and services for industrial data collection. Some enterprises are actively expanding and developing from their original advantages to the manufacturing industry, such as China Telecom, ZTE Communications, Huawei, etc.

(3) Industrial data collection solution enterprises mainly provide services such as industrial data collection solutions, system development, project implementation, and system integration, such as Beijing Automation Research Institute, Hollyssey, and Mingjiang Intelligent, etc.

2.2.1 HSMP data perception principles

To do a good job of data perception in the data workshop, two principles need to be clearly defined: first, purposefulness, and second, economy.

Purposefulness refers to the need to clarify whether the data to be collected has significant value before conducting data collection on the device. Although there are already many data collection systems in industrial sites, such as SCADA and HMI, which are widely used, how much data has been fully analyzed to explore its potential value? In most cases, after collecting data, one does not know how to utilize it. Most industrial sites in China are still in the stage of solving the visibility of production processes, and it is urgent to collect data first, at least to have real-time understanding of what is happening on site. In the context of intelligent manufacturing, it is important to have a long-term perspective before conducting data collection.

Economy refers to the return on investment. For data collection on industrial sites, many people's first idea is to install sensors. No matter what kind of project, the first thing we need to consider is the issue of return on investment, and the method of installing sensors may be the biggest investment. Therefore, in order to collect data on the equipment layer, we must combine our data collection objectives and make full use of the existing conditions of the equipment (such as the host computer system and communication protocol that the equipment already has) to design the data collection Technology roadmap in the most cost-effective way. Classification of industrial data collection devices: devices with upper computer systems (built-in monitoring systems), devices based on non-TCP/IP communication protocols, devices based on TCP/IP communication protocols, and devices without communication interfaces.

2.2.2 Classification of data collection methods

In HSMP, production information is distributed throughout various stages, and the collection of this information depends on the corresponding production needs. Specifically, the collection scope of HSMP management data mainly includes production task schedule, various equipment status data, processing personnel information, and process information of processing workpieces. Different data characteristics have certain differences, and different collection methods need to be adopted. As shown in **Fig. 2.3**, there are two main methods for HSMP data collection, including automated data collection and manual data collection.

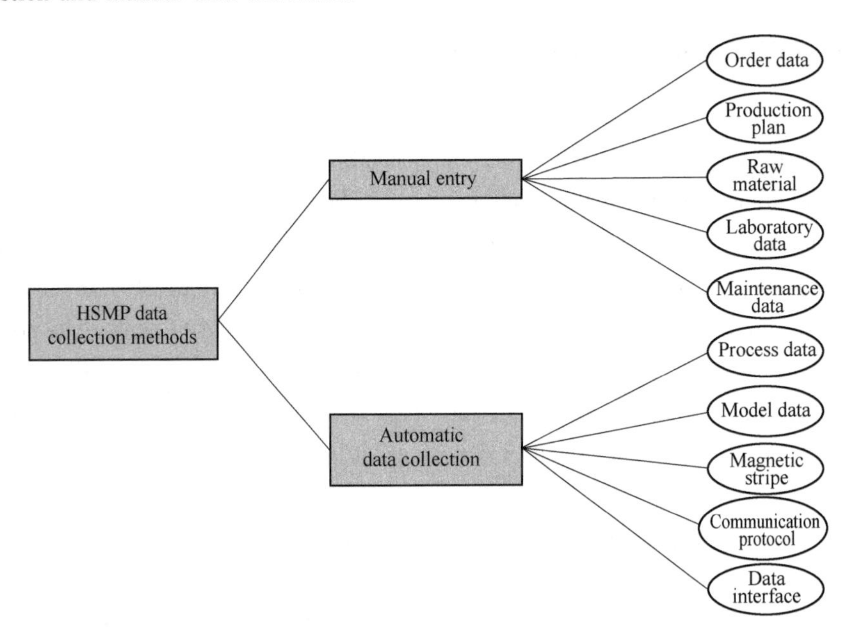

Fig. 2.3 Classification of data collection methods

Automatic data collection often relies on production equipment with good communication conditions and the application of various sensors. The HSMP equipment operation status, energy consumption and production shortage intelligent monitoring system belongs to the category of automatic data collection system. It is based on wireless sensor network, and uses environmental monitoring sensors, intelligent measurement and control devices, intelligent gateways, monitoring servers, MES Manufacturing execution system, etc. to realize wireless collection, transmission and early warning monitoring of the operation status, instruments, energy consumption and production shortage of equipment such as engines and frequency converters.

Manual data collection is mostly used as a supplement or alternative to automatic data collection, which is suitable for the field where it is difficult to realize automatic data collection and the economic cost is high. HSMP data collection methods and systems can be divided into the following levels:

(1) On site manufacturing and production equipment layer: mainly including various production elements on the production site, including workers and management personnel; equipment, tools, and workstation appliances; raw materials and auxiliary materials; processing and testing methods; environment, etc.

(2) Sensor and control equipment layer: Mainly includes sensors and control equipment required for automatically sensing various production factors on the production site, such as temperature and humidity sensors for collecting environmental information, network port devices, serial port devices, dedicated hardware acquisition modules for reading equipment status, PLC equipment, tools, fixtures, materials and other monitoring modules, image, sound, vibration and other detection and diagnostic equipment.

(3) SCADA/DCS layer: This layer is for each production element on the production site, including monitoring computer, remote terminal unit (RTU), Programmable logic controller (PLC), communication infrastructure, human-machine interface, etc.

(4) MES layer: This layer is responsible for the execution and management of the production process, and the management objects generally cover a series of production process data management modules from procurement to finished product delivery. Integrated and modular management provides a reliable, comprehensive, and safe manufacturing collaborative management platform for production enterprises.

(5) ERP layer: The ERP layer is used to manage enterprise resource information. It maximizes resource utilization by managing factories, warehouses, transportation, manufacturing, supply and sales, finance, equipment maintenance, process control/data collection interfaces, electronic communication, email, regulations and standards, projects, financial investments, market information, etc.

2.2.3 Intelligent manufacturing network architecture and OPC UA

We often discuss cloud platforms, big data analysis, artificial intelligence, and the Internet of Things, but data interconnection is the first obstacle in the implementation of topics or projects. At the same time, many people are not familiar with OPC UA as the basic standard and specification for data interconnection, and even many people who do so-called factory integration are not very clear. In China, in data collection, transmission, and production operation, it is necessary to collect machine status, production energy consumption, quality related, and production related parameters on site. However, if there is a lack of unified standards and information models, we will encounter a very big dilemma. The bright future of intelligent manufacturing requires the support of underlying technology, with standards and norms taking the lead. Otherwise, we will be further and further away from intelligent manufacturing.

The implementation of intelligent manufacturing is based on digital intelligent perception, which collects all production process data of the entire enterprise, including various production data, process parameters, and energy consumption from underlying equipment, sensors, to the site, providing a data analysis foundation for global production optimization. The scope of intelligent manufacturing is different from that of traditional fieldbus. The former only processes control level data, while the latter optimizes global production, routing, scheduling, and other issues. This process integrates heterogeneous networks and multi generation products, and OPC UA is a powerful tool to solve this problem.

Intelligent manufacturing is different from traditional vertical network architecture. The traditional pyramid structured data follows a bottom-up approach, and intelligent manufacturing will face a new situation where distributed computing units require different types of data, which can be freely combined end-to-end. As shown in **Fig. 2. 4**, starting from the control layer and passing through workshop management to multiple levels such as SCADA and MES/ERP, it belongs to the traditional automated data architecture. Therefore, the new automation technology based on CPS architecture requires a brand new architecture to support network interconnection. Traditional fieldbuses mainly solve interconnection and interoperability (communication protocols and specifications).

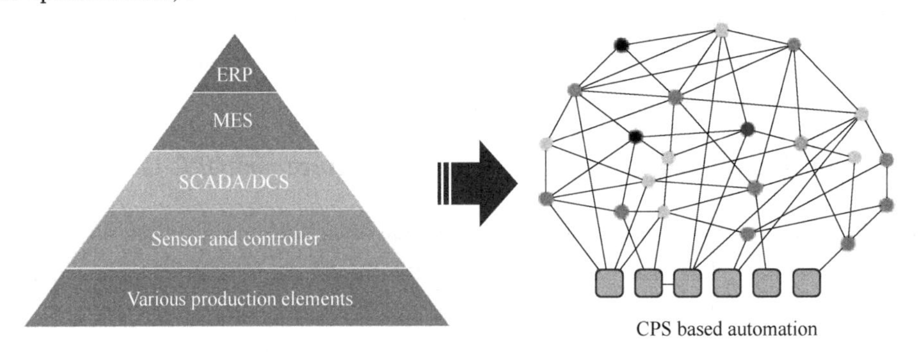

Fig. 2. 4 The network architecture of traditional manufacturing and intelligent manufacturing

Industrial communication is divided into several levels: interconnection-focusing on the connection of hardware interface, interworking-emphasizing the data format and specification of software level, Semantic interoperability-representing semantic definition and specification. From this perspective, fieldbus solves the connectivity problem in the industrial field, achieves interoperability and matching of various application programs, and the emergence of OPC UA enables semantic interoperability between different systems. Independence, security, international standards, modeling, and information modeling, and plug and play are all analyzed from a technical perspective to understand why data perception requires OPC UA.

The key reason for the above is:

(1) OPC UA has established an information model. No matter what industry, achieving interconnection and establishing coordination is based on establishing information models. Based on this, specific architecture related data can be collected, grouped, and analyzed.

(2) Security mechanism: OPC UA supports the X509 security information exchange standard, and implements permission management at different levels based on role and rule definitions.

(3) High independence: From the perspective of End User, user standards are different from the private communication standards and protocols of enterprises. As a public welfare organization, OPC UA leads the development of various technical organizations. As of now, some vertical industry technology organizations, such as MTConnect, OMAC/PackML, Euromap, POWERLINK, FDT/DTM, Profine and other fieldbus foundations, have joined the OPC UA Foundation.

Fig. 2.5 shows that the OPC UA fund has received strong support from various sectors worldwide. These organizations cover mainstream companies and organizations from various industries, including Huawei, OMAC, Euromap, Automation ML, ISA, FDT/DTM, MTConnect, BacNet, Microsoft, CISCO, PI, EPSG, ETG, SERCOSIII, etc. This international automation, IT, foundation organizations, and industry associations have all focused on OPC UA.

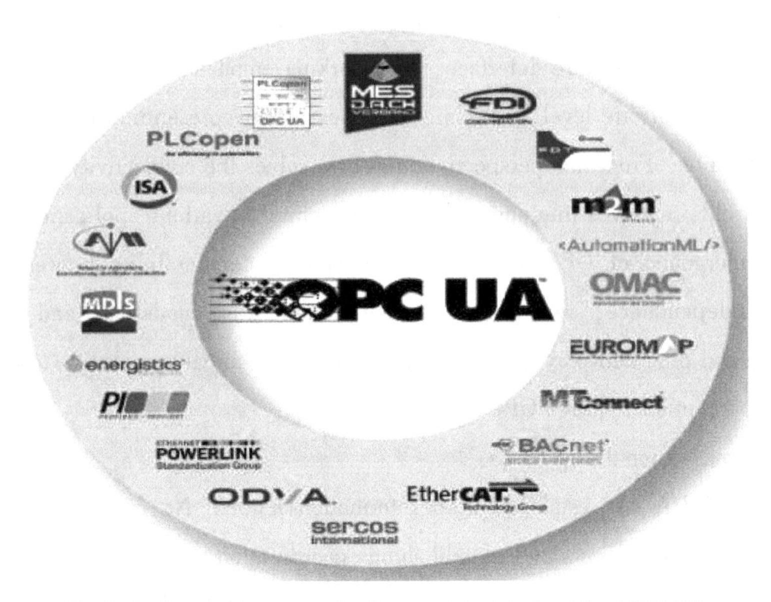

Fig. 2. 5 Supporting organizations and institutions for OPC UA

2. 3 HSMP data fusion methods

Data fusion technology refers to the information processing technology that utilizes electronic devices to automatically process, analyze, and synthesize several time-series observation information obtained under certain criteria to complete the required decision-making and evaluation tasks. There are many data collection systems available on industrial sites, but there are still issues where the data has not been fully analyzed and its potential value has been explored, and after collecting the data, one does not know how to use it. Therefore, in the context of intelligent manufacturing, it is necessary to analyze the types and characteristics of HSMP data fusion. Using data fusion technology, data mining techniques such as machine learning, deep learning, and other algorithms, typical HSMP data analysis scenarios such as anomaly detection, fault mode extraction, fault diagnosis, fault tracing, reliability analysis, etc. , need to provide analysis process templates from data processing to modeling, such as establishing an HSMP equipment health assessment model Production optimization and intelligent decision-making, etc.

2. 3. 1 Data fusion analysis framework

The system architecture design of HSMP data perception, fusion, and visualization research revolves around the "aggregation, communication, and utilization" of HSMP data, studying how to achieve data perception, fusion, analysis, sharing, and application of HSMP. **Fig. 2.6** is the data driven HSMP data fusion analysis framework. From bottom to top, it is the application architecture design from HSMP Big data perception, fusion, analysis to Big data. The system needs to solve the bottleneck problems encountered in the development of traditional business, such as large amount of data storage, fast acquisition speed and frequency, and complex structure, and adopt technologies such as time series data acquisition and governance technology related to Big data technology, massive data storage, etc. to solve the above problems.

The analysis and application of digital workshop data originated from the business needs of enterprises. There are multiple feasible solutions for the same business needs, and each solution has several possible implementation paths. For example, facing the business demand of reducing Product defect, it can be divided into two schemes: equipment fault diagnosis and process optimization. And equipment diagnosis can be further divided into clearer approaches based on different equipment and mechanisms, such as diagnosis for specific equipment faults. When encountering complex problems, these pathways may be further subdivided until they are clearly divided into several models. The first thing to understand is the input-output relationship, such as the relationship between specific parameters and device status, which is the embryonic form of knowledge. Then, it is necessary to find appropriate algorithms to extract and solidify this knowledge.

According to different business objectives, as shown in **Fig. 2.6**, data fusion analysis can be divided into the following four types:

(1) Statistical analysis: Descriptive analysis is used to answer "what happened" and reflect "what" knowledge. The weekly, monthly, and business intelligence (BI) analyses of industrial enterprises are typical descriptive analyses. Statistical analysis generally expresses various statistical features of data in a visual way that is easy for people to understand.

(2) Diagnostic analysis: used to answer "why did this happen" . The key to identifying the causes of problems and anomalies that occur during production, equipment operation, management, and other processes is to eliminate non-essential random correlations and various illusions in diagnostic analysis.

(3) Predictive analysis: used to return to "What is going to happen" . Predict possible future outcomes based on current visible factors for various problems in production and operation.

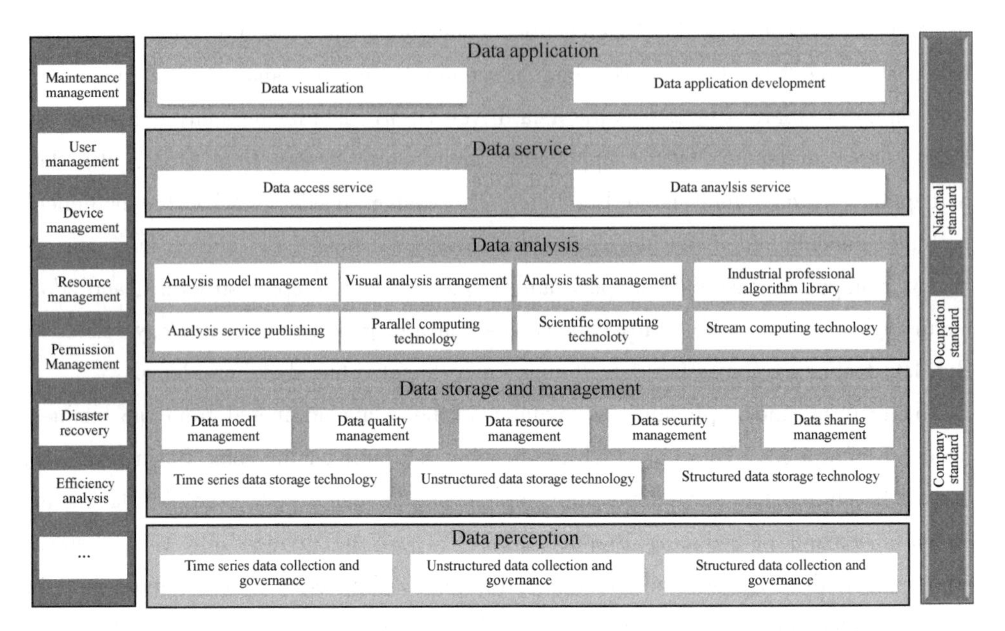

Fig. 2. 6　Data driven-based HSMP data fusion analysis framework

(4) Decision based analysis: used to answer questions about "what to do". Identify appropriate action plans for existing and upcoming problems, effectively solve existing problems or do better work.

Different business objectives require different conditions, requirements, and difficulties for data analysis. Generally speaking, the difficulty of the four problems described above is increasing: the goal of descriptive analysis is only to facilitate people's understanding; diagnostic analysis has clear goals and right or wrong; predictive analysis not only has clear goals and right or wrong, but also distinguishes between causality and correlation; decision based analysis often needs to be further combined with innovation in implementation methods and processes.

2. 3. 2　Traditional data fusion methods

The traditional data fusion technology has the following types according to the data abstraction level:

(1) Data layer-based fusion method: The fusion method based on the data layer is only applicable to the same type of data source, and it directly analyzes and processes the data obtained from the original sensor. The advantage of this method is that it retains the most original information and has good fusion performance. The disadvantage is that the real-time performance of the system is poor, which is due to the large amount of

data, resulting in a large amount of model analysis and calculation, belonging to the lowest level of fusion.

(2) Feature layer-based fusion method: The fusion method based on the feature layer extracts the features of the measured values of the sensors on the basis of the original data and obtains the corresponding feature vectors. Through comprehensive analysis and processing of the feature vectors, the main features of the information are retained, and a certain degree of information compression is achieved, ensuring real-time performance. It is the fusion of the middle level.

(3) Decision layer-based fusion method: The fusion method based on the decision-making layer preliminarily screens the feature vectors of each sensor based on specific decision-making problems, and then recombines and evaluates the primary results based on certain rules and credibility to obtain an optimal decision for specific decision-making objectives. This method has good fault tolerance and strong real-time performance and belongs to the highest level of fusion.

Data fusion can be divided into three structural methods based on structure: serial, parallel, and hybrid[1].

(1) Serial fusion method. As shown in **Fig. 2.7** (a), the current sensor needs to receive the fusion results of the previous level sensor, and each sensor not only receives data information, but also processes and fuses information.

(2) Parallel fusion method. As shown in **Fig. 2.7** (b), each sensor directly and independently transmits the received information to the fusion center, without affecting each other. The fusion center conducts comprehensive analysis according to appropriate rules to obtain the final result.

(3) Mixed mode method. As shown in **Fig. 2.7** (c), it combines the characteristics of serial and parallel. Each sensor also processes information on the basis of receiving it and sends the processed information to the fusion center for further fusion analysis to obtain the optimal result.

2.3.3 Deep autoencoder-based data feature extraction

In practical applications, the original time-domain signal is often affected by noise, making it difficult to extract effective features. In response to this issue, relevant researchers have proposed various solutions, commonly used methods include first denoising the original signal and then extracting features using deep learning models[2-4]. Traditional equipment fault diagnosis methods[2] mostly belong to "shallow learning methods" that require labeled samples, and their learning ability has certain

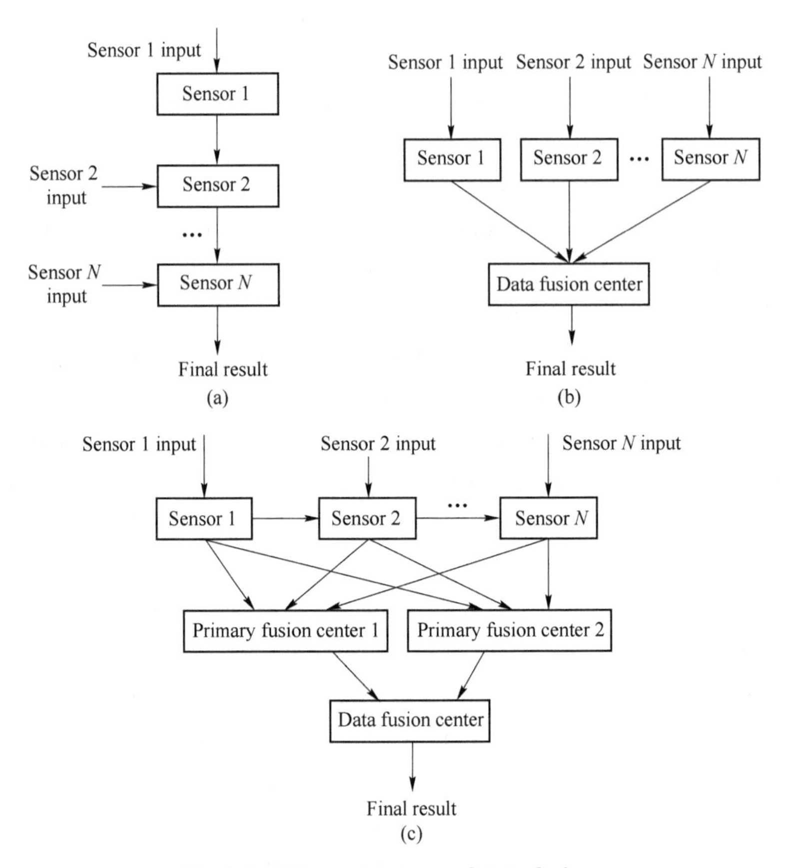

Fig. 2. 7　Three structures of data fusion

limitations, which cannot fully explore the deep features of the data. They lack representation ability in dealing with high-dimensional, high noise, and complex nonlinear problems. As an emerging method in the field of machine learning[5], deep learning has been widely used in image recognition, Natural language processing[6], speech recognition and other fields, which also provides new ideas for fault diagnosis. As a typical architecture in deep learning, stacked autoencoder (SA) has significant advantages in processing high-dimensional data and learning high-dimensional features of complex signals. In view of the problem of large noise in equipment fault data, Lu et al.[7] applied multiple noise reduction autoencoder stack to mechanical fault diagnosis and achieved good classification results.

Nowadays, deep autoencoder networks are also a widely used network reconstruction model in the field of deep learning. By studying signal features through the encoding and decoding process of signals, deep autoencoder can also be seen as a feature extraction tool that can handle complex nonlinear problems. It can directly obtain the most significant feature vectors from the original signal, effectively reducing feature information omissions caused by manual annotation. Therefore, it can learn more comprehensive multi-source signal features and achieve multi-source fusion fault diagnosis, this provides a new approach for fault diagnosis technology that integrates multi-sensor information[8]. The autoencoder network can learn the input data layer by layer through the constructed stacked autoencoder[9] (SAE) to achieve feature extraction and can also embed denoising autoencoder[10-12] (DAE) for signal data noise removal. Variational autocoding (VAE)[13] encodes and decodes signal data by comparing the mean and variance of the input signal, and its encoding process is called feature compression. In industrial processes, the operating conditions of mechanical equipment are usually non constant, and variational autoencoder is a generative network model that better characterizes the distribution characteristics of the original signal input by subjecting the hidden layer variables to a prior result of Gaussian distribution. This model has good generalization ability. Quantitative variational autoencoder (VQ-VAE)[14] is a recently proposed variational autoencoder improved model for low and high dimensions of input signals. The autoencoder network extracts features through different nonlinear structural combinations. The following will explain the feature extraction process of these four types of autoencoder in fault signals.

In supervised learning, data and tasks can be effectively classified, but it is difficult for us to reconstruct the original signal. People can effectively identify the true and false currency, but it is difficult to draw the currency, that is, there are some data to learn from in the classification task and can extract very rich features from it. Although we can distinguish the differences of this pile of banknotes, these features are not enough to reconstruct the original data, that is, for data sets and task classification, reasonable and sufficient features may not be able to complete image reconstruction, If the features of the input signal are encoded into high-dimensional space by constructing a multi-layer autoencoder method to obtain feature data, and then the corresponding decoder is used to decode the feature data, this stacked encoding method can reconstruct the data, as shown in **Fig. 2.8**.

The encoding and decoding of single layer autoencoder can be calculated as follows:

$$\begin{aligned} h &= f(W^{(1)}x + b^{(1)}) \\ y &= f(W^{(2)}x + b^{(2)}) \end{aligned} \tag{2.1}$$

The loss function of the model is calculated as:

$$J(W_1 b_1 b_2) = \sum_{i}^{n} \| X_i - \hat{X}_i \|^2 \tag{2.2}$$

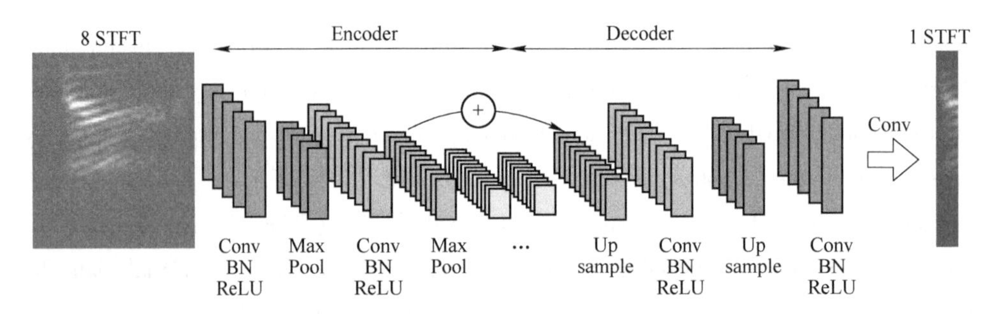

Fig. 2.8 Stacked autoencoder network structure

The initial parameters of the stacked sparse denoising autoencoder neural network are obtained through pre training with unlabeled data. The initial parameters are optimized using a BP neural network under the supervision of a small amount of labeled data, forming a deep denoising autoencoder neural network with feature extraction and pattern recognition functions.

The deep denoising autoencoder algorithm model can be used as a feature extractor, which interferes with model training by adding noise layers. In this way, the model can automatically extract abstract features of disturbance signals[15]. However, in actual training, single-layer autoencoder networks are often not sensitive to data noise, and shallow denoising autoencoder models are difficult to extract useful feature signals. Multiple denoising autoencoder models can be stacked together like stacked autoencoder to form stacked denoising autoencoder networks (SDAE) for network training. Add a classifier at the top level that can fine tune the entire network model from top to bottom, and use a layer by layer training method for feature extraction training, ultimately achieving the goal of denoising and feature recognition[16].

The noise layer of denoising autoencoder can be represented as:

$$\tilde{x} = x + \varepsilon, \ \varepsilon \sim N(0, \sigma^2 I) \tag{2.3}$$

The encoding and decoding of denoising autoencoder can be calculated as follows:

$$\begin{aligned} y &= f_\theta(x) = s(wx + b) \\ z &= g_{\theta'}(x) = s(w'y + b') \end{aligned} \tag{2.4}$$

The Loss function of the reconstructed signal is as follows:

$$\theta^*, \theta'^* = \arg\min_{\theta^*, \theta'^*} \frac{1}{n}\sum_{i=1}^{n} L(x^i, z^i) = \arg\min_{\theta^*, \theta'^*} \frac{1}{n}\sum_{i=1}^{n} L(x^i, g_{\theta'}(f_\theta(x^i))) \tag{2.5}$$

where, $\theta = \{w, b\}$, $\theta' = \{w', b'\}$.

Dropout can be introduced to significantly improve the classification accuracy of sparse denoising autoencoder networks. By improving the robustness and generalization of the model, the Dropout method is utilized to reduce overfitting in deep networks.

With deep autoencoder networks, the following problems have been solved:

(1) In the industrial field, there is a large amount of data, which includes both valuable information and many invalid information. Moreover, as the data dimension increases, effective information becomes very sparse. Deep autoencoder network is an algorithm that abstracts data at a high level through a multi-layer nonlinear structure, which can extract effective information from massive data. Through data-driven deep autoencoder networks, industrial data can be fully utilized.

(2) Mechanical systems in industrial systems are very complex, and manual feature construction requires a lot of domain knowledge, which increases the difficulty of feature construction. And deep autoencoder networks can learn to have strong automatic feature extraction capabilities, automatically extract features, perform unsupervised feature extraction, and do not require manual participation in feature construction. Through multi-level abstraction of the network, data feature mining is achieved, which has strong universality.

2.3.4 Convolutional neural network-based data feature extraction

The earliest convolutional neural network (CNN) LeNet-5 was proposed by French scientist Yann lecun in 1989. It is very good at processing data with similar grid structure. For example, time series data and image data. The "convolution" here is a special linear operation. Generally speaking, the convolution operation used in Convolutional neural network is not completely consistent with the definition in the engineering field and the pure Areas of mathematics. Convolutional layer and pooling layer are generally taken as several layers, and they are alternately set up, that is, one convolutional layer is connected to a pooling layer, and then another convolutional layer is connected to the pooling layer, and so on. The breakthrough of Convolutional neural network came from the AlexNet[17] proposed by Krizhevsky and others in the 2012 ImageNet Challenge. The great success of AlexNet set off a research upsurge of Convolutional neural network. Subsequently, Google Net[18], VGG[19], ResNet[20],

NAS Net[21], Dense Net[22] and others successively proposed. The most important structure of Convolutional neural network is convolution operation. The convolution here is slightly different from the convolution in image signal processing. The convolution layer composed of multiple convolution cores realizes the weight sharing on the same input feature map. Each convolution core detects the specific features on all positions of the input signal. The convolution operation implements the adaptive feature extraction process.

$$y^{l(l,\ J)} = K_i^{l} * X_i^{l(r)} = \sum_{j'=0}^{W-1} K_i^{l(j')} * X^{l(j+j')} \tag{2.6}$$

where, $K_i^{l(j')}$ represents the j'_{th} weight of the i_{th} convolutional kernel in the l_{th} layer; W represents the width of the convolutional kernel.

The principle of convolution calculation is shown in **Fig. 2. 9**.

The three important ideas for improving machine learning systems through convolutional operations are:

(1) sparse interactions;

(2) parameter sharing;

(3) equivariant representations.

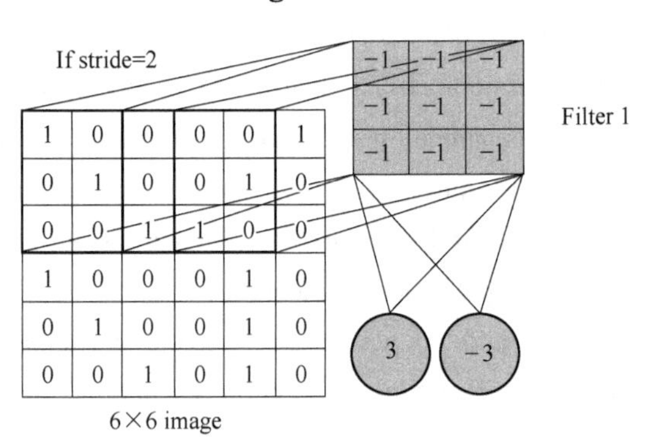

Fig. 2. 9 Calculation principle of convolutional neural network

Through sparse interactions (sparse connections), we can make the size of the kernel much smaller than the size of the input data, thereby reducing the storage requirements of the model and improving the statistical efficiency of the calculation. Parameter sharing ensures that we only need to learn one parameter set, rather than learning a separate parameter set for each position. The neural network layer has the property of being invariant to translation due to the special form of parameter sharing.

After convolution, there is a pooling layer, which mainly performs down sampling operations to reduce the parameters of the neural network. Common pooling methods include maximum pooling, average pooling, and weighted average pooling. Maximum pooling is the process of outputting the maximum value in the perception domain.

In the field of application, CNN has been widely used in the field of image processing. Krizhevsky et al. applied CNN for the first time in the LSVRC-12 competition and achieved astonishing classification results (this network structure is also known as AlexNet) . Subsequently, in the LSVRC-14 competition, Google's research team proposed Google LeNet with higher accuracy. In the LSVRC-15 competition, ResNet neural network proposed by He et al. surpassed humans for the first time in image classification tasks. In the field of audio retrieval, Abdel Hamid et al. combined hidden Markov models to establish CNN for speech recognition and conducted experiments on standard TIMIT speech databases. The experimental results showed that the error rate of this model was 10% lower than that of conventional neural network models with the same number of hidden layers and weights, indicating that the CNN model can improve the recognition accuracy of speech[23].

2.3.5 Recurrent neural network-based data feature extraction

The so-called recurrent neural network (RNN) is reflected in that thelower level input needs to be added with the current level output, that is, there is interaction between hidden layers. A multi-layer perceptron only characterizes the mapping relationship between input and output vectors, while RNN maps historical inputs to each output and can approach any measurable sequence to sequence mapping with any accuracy, as it has a sufficient number of hidden nodes. This corresponds to the global Approximation theory in MLPs. The most prominent feature of ring connections is their ability to retain the previous "memory" in the intermediate state of the network, which can affect the final output of the network. The forward propagation of MLP and RNN for a single hidden layer is basically the same, with the difference being that RNN calculates the activation value of the current point through a combination of the previous activation value of the hidden layer and the current input. Like the standard Backpropagation, the Chain rule is constantly used in the Backpropagation (BPTT) of Recurrent neural network. The difference lies in the fact that the activation value of the hidden layer applied by the recurrent network to the objective function not only transmits the activation value to the output layer, but also affects the hidden layer in the next time period. Its network structure is shown in **Fig. 2.10**.

As shown in the figure, the hierarchical structure of RNN is relatively simple compared to other neural networks, consisting of input layer, hidden layer, and output layer. In the hidden layer, there is an arrow indicating the cyclic update of data, which is the method of implementing time memory function. The hidden layers can be expanded as shown in **Fig. 2.11**.

RNN uses BPTT for backpropagation and adjusting parameters cannot effectively maintain long-term dependency relationships. Often, gradients disappear, making it

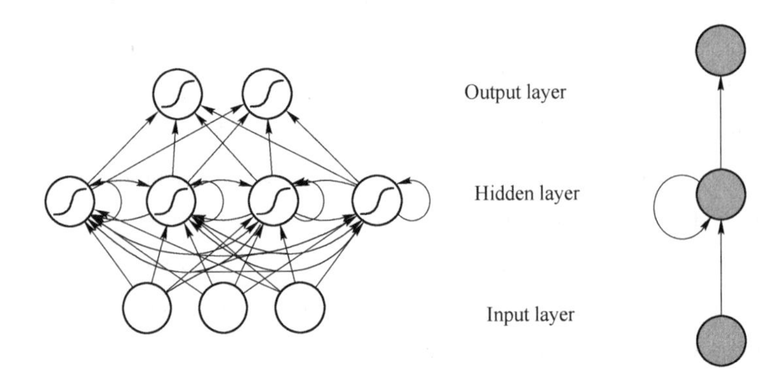

Fig. 2. 10 Structure diagram of recurrent neural network

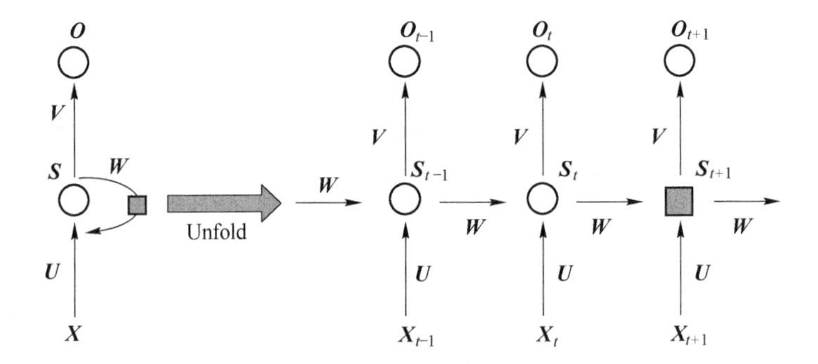

Fig. 2. 11 Hidden layer expansion diagram of recurrent neural network

difficult for the network to update parameters properly and effectively learn new samples. In response to the above issues, researchers have considered many ways, and RNN has gradually emerged with some variants to solve this problem.

In order to make full use of the temporal sequence information of sensor data, RNN is favored because of its ability to capture temporal sequence data relationships. However, traditional RNN has the problem of gradient vanishing and gradient explosion, which makes it extremely difficult to train the model. In order to solve these problems, long short-term memory neural network (LSTM) is proposed by Hochreiter and Schmidhuber[24], which is a gated RNN. LSTM introduces a more complex but effective hidden layer node activation method. LSTM introduces the concept of "gate" in artificial neurons. It can effectively solve the problem of insufficient depth in deep neural networks caused by gradient disappearance and gradient explosion. The algorithm structure diagram is shown in **Fig. 2. 12**.

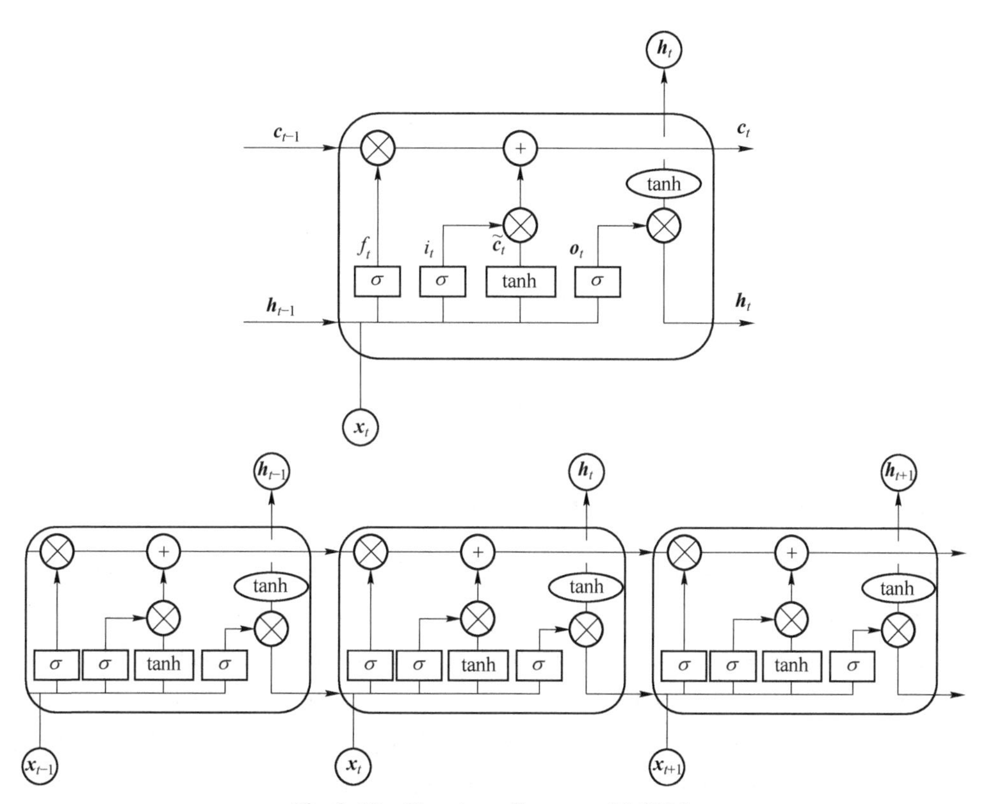

Fig. 2.12 Structure diagram of LSTM

To understand the key to LSTM networks, it is necessary to first understand the rectangular box shown in the figure below, which is called a memory block and mainly consists of three gates (input gate, output gate, forget gate) and a memory cell. The horizontal line above the rectangular box is called the cell state, which is like a pipeline that can control the transmission of information to the next moment. The cell state $\boldsymbol{c}_t$ and the hidden state $\boldsymbol{h}_t$ record the information in the time series. The updating of $\boldsymbol{c}_t$ decides what information will be added and how much information from previous states will be forgotten. The hidden state $\boldsymbol{h}_t$ decides the final output. The forward calculation formulas are shown as below:

$$\boldsymbol{f}_t = \sigma(\boldsymbol{W}_{\mathrm{f}} \cdot [\boldsymbol{h}_{t-1}, \boldsymbol{x}_t] + \boldsymbol{b}_{\mathrm{f}}) \tag{2.7}$$

$$\begin{aligned} \boldsymbol{i}_t &= \sigma(\boldsymbol{W}_i \cdot [\boldsymbol{h}_{t-1}, \boldsymbol{x}_t] + \boldsymbol{b}_i) \\ \tilde{\boldsymbol{c}}_t &= \sigma(\boldsymbol{W}_{\mathrm{c}} \cdot [\boldsymbol{h}_{t-1}, \boldsymbol{x}_t] + \boldsymbol{b}_{\mathrm{c}}) \end{aligned} \tag{2.8}$$

$$\boldsymbol{c}_t = \boldsymbol{f}_t^* \boldsymbol{c}_{t-1} + \boldsymbol{i}_t^* \tilde{\boldsymbol{c}}_t \tag{2.9}$$

$$o_t = \sigma(W_o \cdot [h_{t-1}, x_t] + b_o)$$
$$h_t = o_t^* \tanh(c_t) \tag{2.10}$$

where, W and b refer to the network weight matrix and bias vector, respectively; $\sigma()$ is sigmoid function, whose output is between 0 and 1; the output of function tanh( ) is between −1 and 1.

The forget gate f_t in equation (2.7) decides which information of cell state c_{t-1} will be forgot. The input gate it in Equation (2.8) decides how much candidate cell information $\tilde{c}_t$ in equation (2.9) is added to the cell state c_t. Then, the new cell state c_t is updated by equation (2.10). Finally, the output state h_t is calculated by the output gate o_t and the cell state c_t.

2.4 HSMP data visualization method

Data and information visualization is a scientific and technical method about how to use visual forms to represent data. The visual representation of this data is defined as information extracted in a certain abstract form, including different variables and attributes of each corresponding information unit. Mainly using image processing, graphics, computer vision, and user interfaces, through text expression, 3D modeling, and dynamic animation display, users can better understand the meaning of data through visualization. The research and application of Data and information visualization technology are mainly divided into the following categories.

2.4.1 Pure visual chart generation tool

(1) D3.js: It is a JavaScrip library. The project is open source. Its full name is Data Driven Documents. It is called an interactive and dynamic Data and information visualization library network. Various simple and easy to use functions are provided, which greatly reduces the difficulty of JS operating data. Especially at the Data and information visualization level, the complicated process from the original data to the visual presentation is simplified to several functions. Just input some simple data into the function, and the data can be converted into various colorful visual graphics.

(2) Echart: a pure Javascript Data and information visualization library, the project is open source, and Baidu's product is often used for software product development or Chart module of web pages. Visual charts can be highly customized on the Web side. There are many types of charts and dynamic visualization effects. All kinds of charts and forms are completely open source and free of charge. They can handle Big data and 3D

graphics as well.

(3) FineReport: It is actually a reporting software, an enterprise level application. Used for the development of business reports and data analysis reports in the system, it can also be integrated into application systems such as OA, ERP, CRM, etc. , to create data report modules, and can also be developed into financial analysis systems. It can create dashboards in various formats, even large visual screens. Known as: small screen for work, large screen for decision-making; Microsoft is used for office and Fansoft is used for business.

(4) AntV: It is a set of Data and information visualization syntax produced by Ant Financial (Alibaba series) . It seems to be the first visualization library in China that adopts the theory of The grammar of Graphics. Antv comes with a series of data processing APIs that can classify and analyze simple data and is used by many large companies as an underlying tool for their BI platforms.

(5) Highcharts: It is a visualization library developed abroad and requires payment for commercial use. Its advantage is that the document is detailed, the examples are also very detailed, and the JS scripts and CSS that are relied on in the document are very detailed. Learning and development are relatively timesaving and labor-saving, and the corresponding product has strong stability.

2.4.2 Business intelligence (BI) analysis tool

(1) Tableau: Well-known to data analysts, it has built-in commonly used analysis charts and some data analysis models, which can conduct rapid exploratory data analysis and make data analysis reports.

(2) FineBI: Built-in rich charts, no code calls, can be directly generated using dumb drag. It can quickly analyze business data, generate dashboard, and build a large visual screen. Different from Tableau, enterprises are more inclined to apply it because it can be combined with big data platforms, various multidimensional databases, and its focus is on the rapid analysis and visual presentation of business data. Therefore, it is widely used in enterprise-class BI applications, and it is free for personal use.

(3) PowerBI: The BI product launched by Microsoft after Excel can be used seamlessly with Excel to create personalized data Kanban.

2.4.3 Visualizing large-screen tool

(1) Ali DataV: Tmall Double 11 big screen is made with DataV, which is a drag-and-drop visualization tool of Ali Cloud, mainly used for big data visualization of

business data and geographic information integration, like some exhibition centers and enterprise control centers. No programming required, a simple drag-and-drop configuration can generate a large visual screen or dashboard.

(2) FineReport: This tool can also do visual reports, can also do large screen. Because the back end is usually connected to the business system data, it can be connected to the business data in real time and do some business data display of the enterprise. Such as exhibition center, BOSS cockpit, as well as urban traffic control center, trading hall and so on.

(3) Digital Hail: Focus on data image, 3D processing, data analysis and other related business, through the image visualization of data analysis, more used in smart city, industrial monitoring.

2.4.4 3D data visualization tool

Traditional monitoring system cannot well express the physical spatial logic relationship between assets and assets, equipment and equipment, and abstract data statistics are boring and difficult to understand, resulting in high understanding threshold, high learning cost, low efficiency in dealing with accidents and other problems. Therefore, through 3D visualization technology, all kinds of simulation, analysis, data accumulation, mining, and even artificial intelligence applications based on digital models, the most important point is thatit should be able to apply to real physical systems. The first step of intelligence in intelligent systems is perception and modeling, followed by the analysis and reasoning of data. Without the digital twin's accurate modeling of the real production system, the intelligent manufacturing system is like a house without a foundation. Web-based 3D data visualization technologies for digital twin application development are:

(1) Three. js: It is a 3D engine that can be run in the browser, with various objects such as cameras, shadows, materials, etc., you can create a variety of 3D scenes, and there are many interesting demonstrations on its home page.

(2) Blend4Web: Blend4Web is an open-source framework for creating and displaying interactive 3D computer graphics in web browsers. NASA developed an interactive web application called Experience Curiosity, which is based on the Blend4Web application, to celebrate the third anniversary of the Curiosity rover's landing on Mars, allowing it to operate the rover, control its camera and robotic arm and recreate some of the outstanding events of the Mars Science Laboratory mission.

(3) ThingJS: It is a 3D visualization PaaS development platform developed by

domestic enterprises and oriented to the Internet of Things. Based on WebGL, it is compatible with various browsers and mobile devices, and develops various 3D applications with zero threshold, high efficiency and low cost.

(4) Unity3D: Unity3D is a comprehensive game development tool developed by Unity Technologies that enables users to easily create interactive content types such as 3D video games, real-time 3D animation and architectural visualization. Support for multiple platforms can publish games to iPhone, Android, Windows, Mac, Wii, WebGL (requires HTML5) and other platforms, is a fully integrated professional game engine.

As for the requirements of 3D modeling technology in visualization, this system is currently considered to be mainly used for 3D visualization of equipment status in the digital workshop on the web client. For the requirementsof 3D modeling technology for HSMP data visualization, Three. js is considered to be a widely used and powerful JavaScript 3D library. From the creation of simple 3D animation to the creation of interactive 3D games, it can be realized, but considering the development and application costs and the economic benefits and workload of system development, no in-depth application research has been done on the requirements of 3D modeling technology for HSMP data visualization. If there is a more mature and economical 3D modeling visualization technology in the future, it can be used in system data visualization application.

2.5 Conclusions

This chapter introduces the relevant theories and technologies of device data perception, heterogeneous data fusion and data visualization in HSMP, designs the research framework of HSMP device data perception, fusion and visualization, proposes the corresponding technical route according to the research framework, and elaborates several key technologies in the technical route. It is a theoretical foundation for the subsequent research work of this book.

References

[1] Gravina R, Alinia P, Ghasemzadeh H, et al. Multi-sensor fusion in body sensor networks: State-of-the-art and research challenges [J]. Information Fusion, 2017, 35: 68-80.

[2] Meng Z, Yan X L, Wang Y C, et al. Rotating machinery fault diagnosis based on local mean decomposition and hidden Markov model [J]. China Mechanical Engineering, 2014, 25 (21): 2942-2946.

[3] Li Q W, Ma Y P, Zhou Y Q, et al. Saliency detection based on unsupervised SDAE network [J]. Acta Electronica Sinica, 2019, 47 (4): 871-879.

[4] Shi X, Zhu Y L, Ning X G, et al. Transformer fault diagnosis based on deep auto-encoder network [J]. Electric Power Automation Equipment, 2016, 36 (5): 122-126.

[5] Shi J R, Ma Y Y. Research progress and development of deep learning [J]. Computer Engineering and Applications, 2018, 54 (10): 1-10.

[6] Noda K, Yamaguchi Y, Nakadai K, et al. Audio-visual speech recognition using deep learning [J]. Applied Intelligence, 2015, 42 (4): 722-737.

[7] Lu C, Wang Z Y, Qin W L, et al. Fault diagnosis of rotary machinery components using a stacked denoising autoencoder-based health state identification [J]. Signal Processing, 2017, 130: 377-388.

[8] Qu J L, Du C F, Di Y Z, et al. Research and prospect of deep Auto-encoders [J]. Computer and Modernization, 2014, 8: 128-134.

[9] Jiang A G, Fu P L, Gu M, et al. Fault diagnosis of induction motors based on multimodal stacking autoencoder [J]. Journal of Electronic Measurement and Instrument, 2018, 8: 17-23.

[10] Chen H Y, Du J H, Zhang W N. Monitoring data repairing method based on deep denoising auto-encoder network [J]. Systems Engineering and Electronics, 2018, 40 (2): 435-440.

[11] Vincent P, Larochelle H, Bengio Y, et al. Extracting and composing robust features with denoising autoencoders [C] //Proceedings of the 25th International Conference on Machine Learning, 2008.

[12] Ochiai K, Takahashi S, Fukazawa Y. Arrhythmia detection from 2-lead ECG using convolutional denoising autoencoders [C] //KDD'18 Deep Learning Day, 2018.

[13] She B, Tian F Q, Liang W G. Fault diagnosis method based on deep convolutional variational autoencoders [J]. Chinese Journal of Scientific Instrument, 2018, 39 (10): 27-35.

[14] Tjandra A, Sisman B, Zhang M, et al. VQVAE unsupervised unit discovery and multi-scale Code2Spec inverter for zerospeech challenge 2019 [C] //Proceedings of the Annual Conference of the International Speech Communication Association, 2019.

[15] Chen J, Liu M, Xiong P, et al. ECG signal denoising based on convolutional auto-encoder neural network [J]. Computer Engineering and Applications, 2020, 56 (16): 148-155.

[16] Hong J Y, Wang H W, Che C C, et al. Gas path fault diagnosis for aero-engine based on improved denoising autoencoder [J]. Journal of vibration, Measurement & Diagnosis, 2019, 39 (3): 603-610.

[17] Krizhevsky A, Sutskever I, Hinton G E. Imagenet classification with deep convolutional neural networks [J]. Communications of the ACM, 2017, 60 (6): 84-90.

[18] Szegedy C, Liu W, Jia Y, et al. Going deeper with convolutions [C] //Proceedings of the IEEE conference on computer vision and pattern recognition, 2015.

[19] Simonyan K, Zisserman A. Very deep convolutional networks for large-scale image recognition [J]. arXiv preprint arXiv: 1409. 1556, 2014.

[20] He K, Zhang X, Ren S, et al. Deep residual learning for image recognition [C] //Proceedings of the IEEE conference on computer vision and pattern recognition, 2016.

[21] Zoph B, Vasudevan V, Shlens J, et al. Learning transferable architectures for scalable image recognition [C] //Proceedings of the IEEE conference on computer vision and pattern recognition, 2018.

[22] Huang G, Liu Z, Van DerMaaten L, et al. Densely connected convolutional networks [C] // Proceedings of the IEEE conference on computer vision and pattern recognition, 2017.

[23] Zhou F Y, Jin L P, Dong J. Review of convolutional neural network [J]. Chinese Journal of Computers [J]. 2017, 40 (6): 1229-1251.

[24] Hochreiter S, Urgen Schmidhuber J. Long short-term memory [J]. Neural Computation, 1997, 9 (8): 1735-1780.

Chapter 3 Design of Digital Thread for HSMP

HSMP has the characteristics of multi-stage coupling, heterogeneous data sources, and complex model mapping mechanisms, resulting in low data resource utilization, difficulty in digital representation of key production factors, difficulty in model interaction, and difficulty in knowledge modeling. How to construct a unified and standardized digital representation system for the lifecycle, establish a comprehensive and integrated common data source, reveal the correlation and coupling mechanism of all elements of the lifecycle information, and achieve cross manufacturing stage mapping services are key technical issues in digital thread design[1].

With the help of the hierarchical metadata structure model, research the hierarchical unified data description method of the manufacturing process, realize the form transformation and accessible configuration of data resources throughout the lifecycle, eliminate the information gap between each stage, study the model association, evolution feature extraction and Knowledge graph construction methods across stages, analyze the feature migration mechanism across manufacturing stages, study thespatio-temporal mapping association of each stage, and build the manufacturing process association network, Develop HSMP digital thread technology that can be traced across stages.

3.1 Research methods for constructing digital thread for HSMP

Aiming at the issues of data under circulation, model under fusion, and knowledge under association in various stages of HSMP, a consistent description model for multi-source heterogeneous data is proposed using the idea of standardized representation. Based on this, the correlation coupling relationship between data, models, and knowledge across manufacturing stages is studied, and the internal mapping mechanism of each manufacturing stage is revealed. Then, an HSMP association network is constructed to form a comprehensive analysis of all elements of HSMP, Construct the HSMP digital thread of model data knowledge collaboration, achieve the unity and coordination of manufacturing elements in the lifecycle, support data access, transformation, analysis, and traceability in the lifecycle, provide data, model, and knowledge support for theoretical research such as fault detection and diagnosis, operational state assessment, and residual life prediction in subsequent chapters, and establish an information environment that integrates and collaborates all elements throughout the lifecycle.

In response to the difficulty of effectively integrating and utilizing heterogeneous data from multiple sources in the lifecycle under non ideal conditions, data integration and augmentation are carried out through soft sensing and adversarial generation networks. Based on multi-level mapping relationships in the physical information space, combined with process mechanisms, a multi-level (multi-stage multi process multi system) quantitative, formal, and logical unified data description model is constructed. By analyzing the coupling between multi-dimensional and multi granularity models under dynamic temporal changes, combining time processing techniques with model topology diagrams, establish multi-stage model correlation relationships, analyze the mechanism of model linkage mapping under spatiotemporal patterns, establish collaborative matching relationships between models under spatiotemporal conditions, study cross domain transmission mechanisms and potential change trends, and describe the correlation and collaborative relationships between system models. Utilizing multi-stage spatiotemporal characteristic analysis technology to achieve hierarchical expression of multidimensional features, considering dynamic transmission and feedback mechanisms during the lifecycle process, researching data evolution feature extraction and cross domain linkage mapping technology, and designing full feature fusion methods using tensors, graph convolutional networks, and other techniques. Explore methods for extracting, transforming, and inferring implicit knowledge based on models and data, construct an HSMP association network, and achieve full element integration analysis and lifecycle traceability of HSMP through dynamic association mining. Establish a distributed global database based on the unified representation of multi-source heterogeneous data, study the collaborative scheduling method of cluster multiple computing nodes, and establish a digital thread oriented towards the lifecycle of HSMP, which is Shown in **Fig. 3.1**.

3.2 Unified description of multi-source heterogeneous data

Focusing on the multi-source heterogeneous characteristics of HSMP lifecycle data, this study focuses on:

(1) Comprehensive augmentation techniques for lifecycle data under non ideal

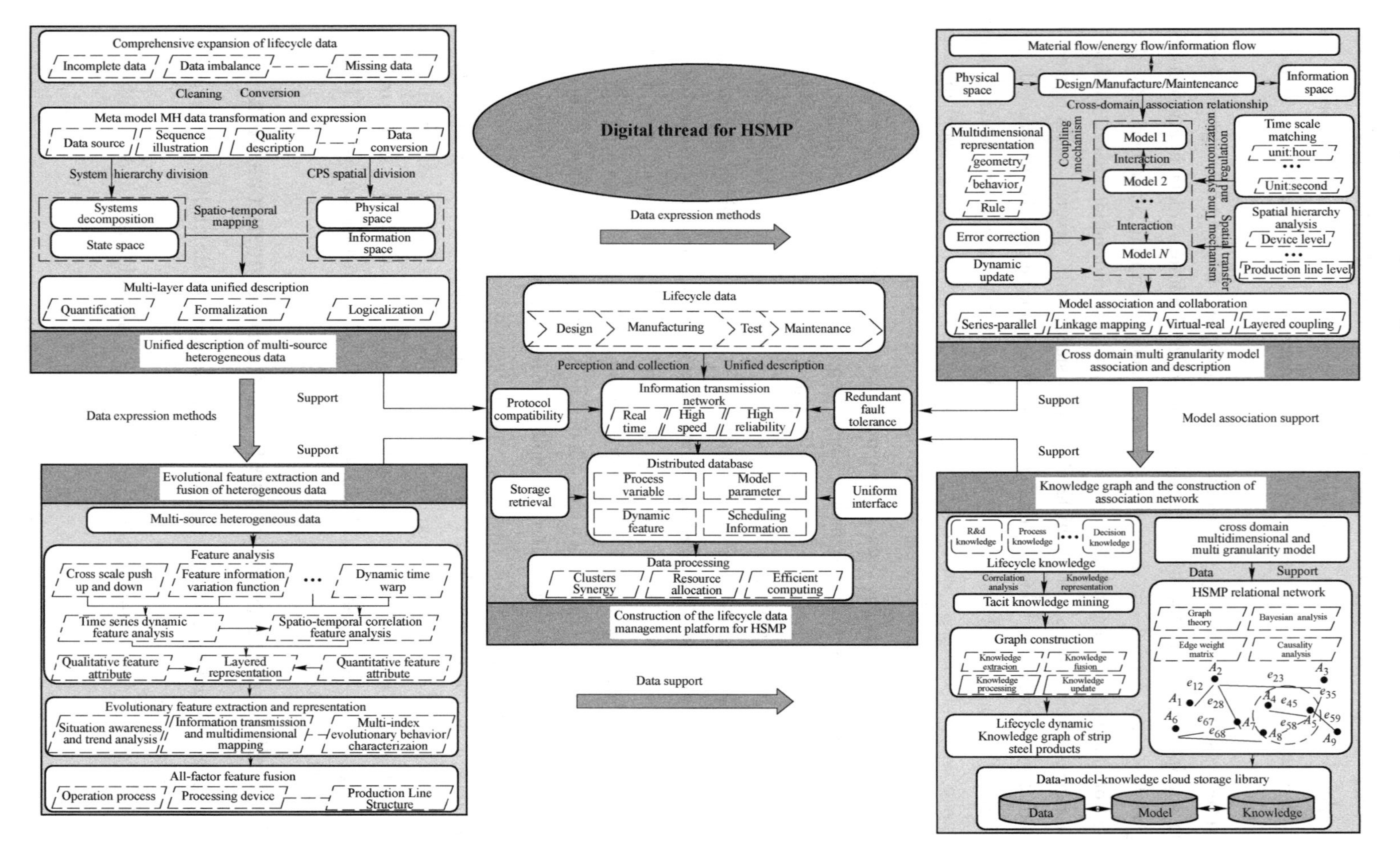

Fig.3.1 Schematic diagram of digital thread design for HSMP

conditions based on sufficient analysis of lifecycle data.

(2) Construct a metadata structure model for HSMP multi-source heterogeneous data, and study methods for transforming structured, semi structured, and unstructured multi-source heterogeneous data and unifying information description.

(3) Based on the multi-level spatiotemporal mapping relationship between physics and information space, a unified description method for multi-source heterogeneous data in HSMP with multi-layer quantification, formalization, and logicality is studied.

Aiming at the non-ideal characteristics of incomplete, imbalanced, locally missing, and performance mismatch in HSMP lifecycle data, guided by prior knowledge such as operational state information and model mechanism, this paper studies an idealized augmentation method based on technologies such as soft measurement, cycle synchronization, and semi supervised real-time learning, to achieve comprehensive augmentation of lifecycle data under non ideal conditions. Considering the characteristics of HSMP multi-source heterogeneous data, such as multi-scale, multi-level, and difficult to unify the form, we use technical metadata, process metadata, and business metadata to build a hierarchical metadata structure model. Through hypertext markup language, extensible markup language, resource description framework and other information description formats, we provide a unified description mechanism and operation model for data from different sources and structures, and then extract Plays a crucial role in the conversion process; On the basis of metadata model description, based on the multi-level information space mapping relationship from physical levels such as process and equipment to comprehensive production indicators, operational indicators, and control indicators, this paper studies the quantitative, formal, and logical unified description of HSMP multi-source heterogeneous data at multiple levels (multi-stage, multi-process, and multi-system), and combines process mechanism models and production operation methods to form a unified description method for HSMP multi-source heterogeneous data, Provide a foundation for feature extraction and fusion of subsequent multi-source heterogeneous data evolution. The schematic diagram of unified description of multi-source heterogeneous data is shown in **Fig. 3.2**.

3.2.1 Non-ideal data processing method

Non-ideal data augmentation method based on partial robust M-regression, multiple imputation strategy, and tensor N-tube rank decomposition is proposed to address the issues of incomplete, unbalanced, and local missing HSMP data, with the goal of low quality and optimal use of the data. This provides a data foundation for subsequent data-

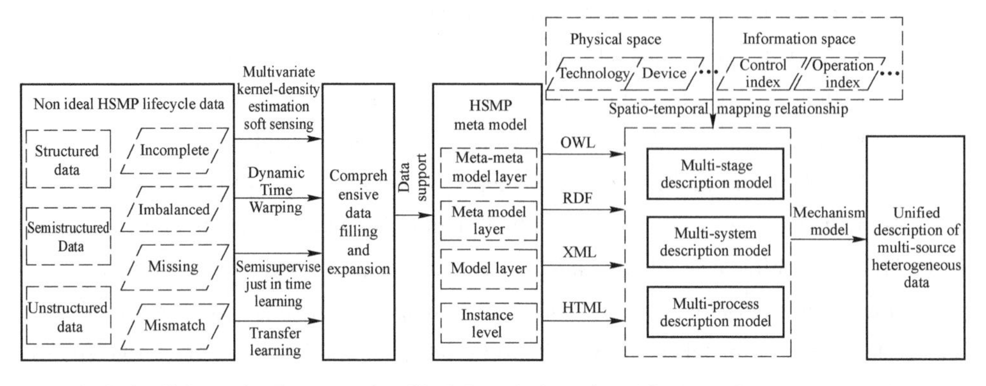

Fig. 3.2 Schematic diagram of unified description of multi-source heterogeneous data

driven modeling and association analysis methods.

3.2.1.1 Non-ideal data processing method Based on partial robust M-regression

Partial robust M-regression algorithm (PRM) is a robust mode of weighted iterative partial least squares (PLS) with a single output variable[2-3]. It eliminates the influence of outlier in the model by iteratively and adaptively allocating different sample weights. In PRM, outliers are divided into high residual samples and high leverage samples. Samples far from the center of the input sample are considered high leverage, i. e. abnormal input and normal output. Samples with significant absolute differences between predicted and observed values are considered to have high residuals, namely normal input and abnormal output. Different weighting methods are used for the two types of outliers mentioned above. The leverage weight w_i^{l} of the i_{th} sample can be defined as:

$$w_i^{\mathrm{l}} = g\left(\frac{\boldsymbol{t}_i - \mathrm{med}_{\mathrm{L}_1}(\boldsymbol{T})}{\mathrm{med}_i \parallel \boldsymbol{t}_i - \mathrm{med}_{\mathrm{L}_1}(\boldsymbol{T}) \parallel}, c\right), \quad g(z, c) = \frac{1}{\left(1 + \left|\frac{z}{c}\right|\right)^2} \tag{3.1}$$

where, $\boldsymbol{T}$ is the score matrix; $\parallel \cdot \parallel$ represents the Euclidean norm; c is the equilibrium constant.

The residual weight w_i^{r} of the i_{th} sample can be defined as:

$$w_i^{\mathrm{r}} = g\left(\frac{r_i}{\tilde{r}}, c\right) \tag{3.2}$$

where, r_i is the prediction error; $\tilde{r} = \mathrm{med}_i \mid r_i - \mathrm{med}_j(r_j) \mid$.

Considering both weights simultaneously, the weight of the i_{th} sample can be characterized as $w_i = \sqrt{w_i^{\mathrm{r}} w_{\mathrm{i}}^{\mathrm{l}}}$.

The original HSMP data should be normalized to eliminate the impact of different dimensions. Typically, the normalization process can be defined as:

$$\bar{u} = \frac{1}{n}\sum_{i=1}^{n} u_i, \qquad s^2 = \frac{1}{n-1}\sum_{i=1}^{n}(u_i - \bar{u}) \tag{3.3}$$

where, $\bar{u}$ and s^2 respectively represent the mean and variance values of the measurement data $\boldsymbol{U} = [u_1, u_2, \cdots, u_n]$. If the measured data contain Outlier, their mean and variance will change significantly and deviate from the true value.

To address the above issues, robust normalization is used to estimate the mean and variance. The average value in robust normalization is represented by the total square loss center and the robust center, as follows:

$$\bar{u} = \sum_{i=1}^{n} k_i u_i \quad k_i = \frac{1}{\sqrt{1+4u_i^2}} \Bigg/ \left(\sum_{j=1}^{n}\frac{1}{\sqrt{1+4u_j^2}}\right) \tag{3.4}$$

3.2.1.2 Non-ideal data processing method based on multiple interpolation

Industrial processes typically have hundreds of control loops, with different sensors collecting large amounts of process data. Sensor failure, data acquisition, transmission and storage all lead to outliers and missing values in industrial process data. Therefore, the statistical interpretation ability of subsequent monitoring models is largely dependent on incomplete data preprocessing. Based on the above PRM method, multiple imputation (MI) is introduced to interpolate the missing values[4].

The idea of MI comes from Bayesian inference, which makes up for the defect of single interpolation to a certain extent, while taking into account the uncertainty of missing data. Several possible interpolations can be generated by the MI algorithm, resulting in several complete data sets. Then, the same process is performed on each data set, and the estimator is obtained through comprehensive analysis to complete the statistical inference. Assume the dataset $\boldsymbol{X} = \begin{bmatrix}\boldsymbol{X}_{\text{obs}} \\ \boldsymbol{X}_{\text{mis}}\end{bmatrix} \in R^{n\times m}$ follows m-dimensional normal distribution, where $\boldsymbol{X}_{\text{obs}}$ is the observed part, $\boldsymbol{X}_{\text{mis}}$ is the missing part, and n is the number of samples. The detailed procedure of MI is as follows:

Step 1: Select vector θ^* randomly from the parameter vector θ to be evaluated.

Step 2: Obtain $\boldsymbol{X}^*_{i(\text{mis})}$ and form the conditional distribution $P(\boldsymbol{X}^*_{i(\text{mis})} \mid \boldsymbol{X}^*_{i(\text{obs})}, \theta^*)$.

Step 3: Let $\hat{\theta}_i$ and $\hat{U}_i$, $i = 1, 2, \cdots, I$ be I complete-data estimations and their associated variances for the parameter vector θ.

Step 4: According to the above results, the estimated value and total variance of MI parameter vector θ can be calculated as:

$$\bar{\theta} = \frac{1}{I}\sum_{i=1}^{I}\hat{\theta}_i \quad U_T = U_1 + \left(1 + \frac{1}{I}\right)U_2 \tag{3.5}$$

where, $U_1 = \sum_{i=1}^{I}\hat{U}_i/I$ is the within-imputation variance; $U_2 = \sum_{i=1}^{I}(\hat{\theta}_i - \bar{\theta})^2/(I-1)$ is the between-imputation variance.

3.2.1.3 Performance data restoration based on Tensor N-tubal rank decomposition

Aiming at the problems of incomplete, unbalanced, and local missing performance data in actual HSMP, a batch process performance data restoration method based on tensor sub rank decomposition is proposed with the goal of low quality and optimal use of data. HSMP is a typical batch production process, where data between adjacent rolling batches has internal correlations. A three-dimensional tensor model can maintain the correlation attributes of high-dimensional batch data. The production data of a single batch can form a two-dimensional matrix $\boldsymbol{X}(K \times J)$, where each row consists of all variable measurements at a fixed sampling time, and each column reflects the trajectory of a measurement variable at different sampling times within a batch operation. Assuming there is measurement data for I batch, I two-dimensional matrix $\boldsymbol{X}_i(K \times J)$ (i = 1, 2, ⋯, I) is obtained, which forms a three-dimensional tensor $\boldsymbol{X}(I \times J \times K)$. The three dimensions represent the batch (i = 1, 2, ⋯, I), variable (j = 1, 2, ⋯, J), and sampling time (k = 1, 2, ⋯, K), respectively. Usually, due to delayed testing and limited on-site measurement and control radian, there is an issue of incomplete data for the performance indicators in this data tensor. Based on the tensor structure of batch data, a corresponding tensor restoration model can be established by minimizing the tensor rank, thereby restoring potential target tensors. Considering that the tensor ' s N -tubal rank function is suitable for ($N \geqslant 3$) order form and can characterize the correlation of different intensities in each dimension of the tensor, minimizing the sum of tensor N -tubal rank elements is usually an NP difficult problem. Therefore, a convex relaxation form based on the weighted sum of the tensor nuclear norm (WSTNN) is defined, and a batch data restoration model is constructed in the case of incomplete performance data. The performance data restoration model objectives under the tensor architecture based on WSTNN can be expressed as:

$$\begin{aligned} &\min_{\boldsymbol{X}} \|\boldsymbol{X}\|_{\text{WSTNN}} \\ &\text{s.t.}\ \ P_{\Omega}(X - F) = 0 \end{aligned} \tag{3.6}$$

where, F is the missing data under some performance indicators of the observable non ideal data; $\boldsymbol{X}$ is the high-dimensional data to be solved; Ω is the location index of the missing performance elements in the target tensor; $P_{\Omega}(\boldsymbol{X})$ is the projection operator, which maps the elements in the corresponding position in the $\boldsymbol{X}$ to itself according to the index set Ω , while other location elements in the $\boldsymbol{X}$ are mapped to 0.

From the definition of the kernel norm weighted sum of the higher-order tensors, the above objectives can be reconstructed as:

$$\min_{\boldsymbol{X}} \sum_{\substack{1 \leqslant k_1 < k_2 \leqslant N \\ k_1,\ k_2 \in \mathbf{Z}}} \alpha_{k_1,\ k_2} \| X_{(k_1,\ k_2)} \|_{\text{TNN}} + \varphi_S(\boldsymbol{X}) \tag{3.7}$$

where, $\boldsymbol{\alpha} = (\alpha_{1,2},\ \cdots,\ \alpha_{1,N},\ \alpha_{2,3},\ \cdots,\ \alpha_{2,N},\ \cdots,\ \alpha_{N-1,N})$ is the weight vector, $\alpha_{k_1,k_2} \geqslant 0(1 \leqslant k_1 < k_2 \leqslant N,\ k_1,\ k_2 \in \mathbf{Z})$, $\sum_{\substack{1 \leqslant k_1 < k_2 \leqslant N \\ k_1,\ k_2 \in \mathbf{Z}}} \alpha_{k_1,k_2} = 1.$

The optimization problem of the above tensor is a convex optimization problem, which can be solved by introducing auxiliary variables using the Alternating direction method of multipliers (ADMM).

3.2.2 Meta model construction and MSH data description

Metadata, also known as relay data and intermediate data, is mainly used to describe data attribute information, representing storage locations, historical data, resource searches, file records, etc. Using metadata can identify resources, evaluate resources, track changes in resource usage, achieve simple and efficient management of large amounts of data, and effectively discover, search, and integrate information resources into organizations[5-7].

Based on mechanism analysis, system operation analysis, and expert knowledge, the rolling process data has the following characteristics:

(1) There are various types of data sources: hot rolling equipment data mainly consists of static data, dynamic data, and configuration data. Hot rolling equipment static data describes the static characteristics related to the hot rolling equipment, including equipment name, equipment model, equipment code, equipment manufacturer, equipment production date, installation location, geometric dimensions, etc. The dynamic data of hot rolling equipment includes the dynamic data of heating furnace equipment, rough rolling equipment, finishing mill equipment, and laminar cooling and coiling business equipment; Hot rolling equipment configuration data refers to the configuration items during the collection and access process of hot rolling equipment and equipment quantity.

(2) There are multiple methods for obtaining data: the hot rolling industrial process has different data acquisition channels, storage formats, and differences in metadata items, making it difficult to coordinate the application of data.

(3) There is no unified management format for data: the format of steel rolling process data is inconsistent, and there is no digital definition and unified description of the entire lifecycle products, processes, and resources, making it impossible to manage, query, and retrieve data in a unified way.

Based on the above data characteristics of the steel rolling process, the four-layer meta model of the steel rolling process is constructed. The four-layer meta model is the modeling language architecture designated by the OMG organization. It generally has four levels, namely, the meta model level, the meta model level, the model level, and the instance level. From top to bottom, it is from abstract to concrete, and from top to bottom. By recursively applying semantics to different levels, the definition of semantic structure is completed, providing an architectural foundation for the extension of the metamodel, and providing an architectural foundation for the integration of metamodel implementation with other standards based on the four-layer metamodel architecture.

Drawing inspiration from the knowledge description method of the unified resource description framework (RDF), which is a three tuple (subject, attribute, attribute value) knowledge description method, based on a four-layer meta model structure, a six-dimensional data model (time domain, spatial domain, entity object domain, relationship domain, attribute, attribute value) for steel rolling process data is proposed. Taking the knowledge description of the finish rolling mill process and its attribute outlet thickness as an example, the unified data model structure is described, the ultimate goal is to achieve the mutual conversion of structured and unstructured data and the unified description of information. On the basis of the Data element model of each system, process and stage, and considering the multi-level information space mapping relationship from the physical level of process and equipment to the comprehensive production index, operation index and control index, this paper studies the unified description method of multi-source heterogeneous data with multiple levels (multi-stage multi process multi system) of complex product manufacturing process.

The metadata of steel rolling process data is not just one type, but a metadata system. With the development of the steel rolling process data lifecycle, on the basis of sharing a core metadata, there are different extensions at different lifecycle stages. Metadata itself cannot represent substantive objects, but it can provide a unified description mechanism and operation model for data from different sources and structures, thereby playing a crucial role in the process of data extraction, transformation, and loading. The metadata structure model includes descriptions of data sources and descriptions, data owners and data sequences, data processing information, data quality, and data transformation methods. In response to the characteristics of multi-scale, multi-level, and difficult to unify forms of multi-source heterogeneous data, a metadata structure model is constructed using metadata. Based on the data management and service process of the steel rolling process, the metadata system of the steel rolling process is divided into technical metadata, management metadata, and business metadata according to application scenarios. Various metadata with different functions form an inherently interconnected metadata system[8-11].

Technical metadata refers to all data related to the information supply chain throughout the entire product lifecycle, mainly including data source metadata, storage metadata, calculation metadata, operation metadata, shared metadata, service metadata, process metadata, etc., such as the execution status, execution time, execution results, and metadata generated during the query and analysis process of data ETL tasks. Management metadata mainly refers to the description of the management attributes of data, including the processing, archiving, structure, technical processing, access control, management department, management responsible person, and related system information of digital objects. Clear management attributes are conducive to implementing data management responsibilities to departments and individuals and are the foundation of data security management. Business metadata refers to user-oriented metadata that describes technical metadata and related businesses, including data sources, types, connotations, relationships, constraints, and meanings, as well as related business rules, terminology, and attributes. It mainly includes model metadata, application metadata, analysis metadata, etc. Various metadata are interrelated and collaborative to achieve a unified description of data.

Based on the above characteristics and data relationships, it is proposed to first establish a steel rolling process meta database based on unified metadata standards, in order to achieve effective data management and solve the problem of unified data information and format in the steel rolling process. The meta database framework designed in this study is shown in **Fig. 3.3**.

The centralized management of steel rolling process data is the first step to achieve digital definition and unified description of products, processes, and resources throughout the entire lifecycle. In order to provide better data services to users, it is necessary to further research and establish a standard description and expression scheme

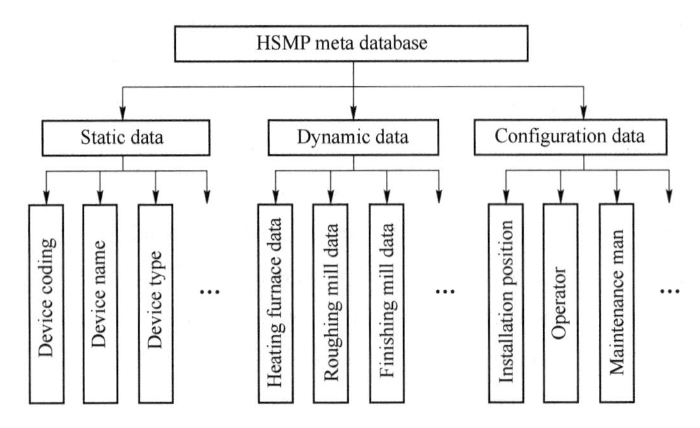

Fig. 3. 3 HSMP meta database

for steel rolling process data information, establish a comprehensive metadata model for steel rolling process, and achieve the circulation and sharing of data information among various systems. On the one hand, metadata can provide detailed and accurate descriptions of storage forms, data formats, access conditions and addresses, management information, and other related information; on the other hand, it can also provide a consistent data view for data management, facilitating the implementation of unified data management and data service operations. Construct a metadata structure model based on the characteristics of multi-level, multi-scale, and difficult to unify the form of steel rolling process data. Each model layer corresponds to different information description formats and instance characteristics.

3. 2. 3 KPI situation awareness and soft sensoring technology

3. 2. 3. 1 A soft sensoring model for mechanical properties based on Seq2Seq architecture

The working conditions of HSMP are complex, and there is a large lag phenomenon in the interaction between material flow, energy flow, and information flow. The collected data has strong coupling, dynamism, and multi-source characteristics, and the models interact with each other. Moreover, a large number of quality indicators can only be obtained through offline verification data. Mechanical performance, organizational performance indicators, and other indicators are usually obtained from laboratory analysis. Online time series prediction of the mechanical properties of strip steel has become an urgent issue that requires attention. A Seq2Seq time series soft sensor model based on two-stage attention mechanism, entity embedding layer, and convolutional bidirectional gated loop unit is proposed to predict the mechanical properties of strip steel in response to the above issues. The specific steps include:

First, in order to extract the correlation between variables, a Convolutional Bidirectional Gated Current (CBGRU) module is constructed. In order to consider the correlation of input variables, the CBGRU model constructs a CNN (Convolutional Neural Network) feature extraction layer to extract features between input variables and fully learn the correlation between features. The CNN feature extraction layer consists of two layers of convolution structure. Each layer includes a convolution layer, a ReLU Activation function layer and a pooling layer. Then, the BiGRU module is used to capture time series features. BiGRU extracts data features from the states of two GRUs that are unidirectional and opposite in direction[12-14].

Then, in order to adaptively extract relevant exogenous sequences and encoder hidden states, a two-stage attention mechanism is designed, which introduces attention mechanisms in the encoder and decoder sections respectively. In the encoder stage, the input attention mechanism combines the hidden layer state $\boldsymbol{h}_{t-1}$ of the previous stage of BiGRU and the input variable $\boldsymbol{x}^k$ of the current time to calculate the importance level $\boldsymbol{\alpha}_t^k$ of the k_{th} input feature at time t, as shown in the following equation:

$$\boldsymbol{e}_t'^k = \boldsymbol{v}_{\mathrm{e}}^T \tanh(\boldsymbol{W}_{\mathrm{e}} \boldsymbol{h}_{t-1} + \boldsymbol{U}_{\mathrm{e}} \boldsymbol{x}^k) \tag{3.8}$$

$$\alpha_t^k = \frac{\exp(\boldsymbol{e}_t'^k)}{\sum_{i=1}^{n} \exp(\boldsymbol{e}_t'^i)} \tag{3.9}$$

In the decoder stage, the decoder part calculates the context vector $\boldsymbol{\beta}_t$ through the temporal attention mechanism at time t, and then obtains the output $\boldsymbol{c}_t$ of the decoder's hidden layer, where $\boldsymbol{v}_t$, $\boldsymbol{W}_{\mathrm{d}}$ and $\boldsymbol{U}_{\mathrm{d}}$ are the parameters that need to be learned.

$$\boldsymbol{l}_t^z = \boldsymbol{v}_{\mathrm{d}}^T \tanh(\boldsymbol{W}_{\mathrm{d}} \boldsymbol{h}_t + \boldsymbol{U}_{\mathrm{d}} \boldsymbol{h}_z) \tag{3.10}$$

$$\boldsymbol{\beta}_t = \frac{\exp(\boldsymbol{l}_t^z)}{\sum_{i=1}^{n} \exp(\boldsymbol{l}_t^i)} \tag{3.11}$$

The calculation process of the final output $\hat{\boldsymbol{y}}_T$ of the model is calculated according to the following equation. $\boldsymbol{v}_y$, $\boldsymbol{W}_y$, $\boldsymbol{b}_{\mathrm{W}}$, and $\boldsymbol{b}_{\mathrm{v}}$ are parameters to be learned. $\boldsymbol{b}_{\mathrm{W}}$ and $\boldsymbol{b}_{\mathrm{v}}$ are bias terms.

$$\tilde{\boldsymbol{y}}_t = \tilde{\boldsymbol{w}}_t [\boldsymbol{\beta}_t; \boldsymbol{y}_t] + \tilde{\boldsymbol{b}} \tag{3.12}$$

$$\hat{\boldsymbol{y}}_T = \boldsymbol{v}_y^T (\boldsymbol{W}_y [\boldsymbol{h}'_{T-1}; \boldsymbol{\beta}_{T-1}] + \boldsymbol{b}_{\mathrm{W}}) + \boldsymbol{b}_{\mathrm{v}} \tag{3.13}$$

Finally, a DA-CBGRU-Seq2Seq model is constructed based on the CBGRU module and two-stage attention mechanism. In the encoder section, the entity embedding layer and merging layer are used to fuse multiple external factors and auxiliary variable sets, and the overall input is used as the CNN feature extraction layer. The BiGRU layer extracts time series correlation from the output variables of CNN, and then calculates the attention weight of the k_{th} input feature at time t based on the hidden layer state $\boldsymbol{h}_{t-1}$ at the previous time and the input variable $\boldsymbol{x}^k$ at the current time, training the variable $\tilde{\boldsymbol{x}}_t$ of the input decoder's attention layer.

The decoder stage of the model uses the GRU network layer to decode the output information of the encoder. In order to prevent the prediction performance from deteriorating when the input sequence is too long, the model introduces a temporal attention mechanism to select the corresponding hidden layer state. Obtain the importance of the hidden state of the i_{th} encoder for the final prediction $\boldsymbol{\beta}_t$ through the Softmax layer, and then sum the weights to obtain the text vector $\boldsymbol{c}_t$. Combining the text vector with the target sequence, the GRU is used as the Activation function, and the predicted value is calculated. The calculation process is similar to that of the input attention mechanism. The network structure diagram is shown in **Fig. 3. 4**.

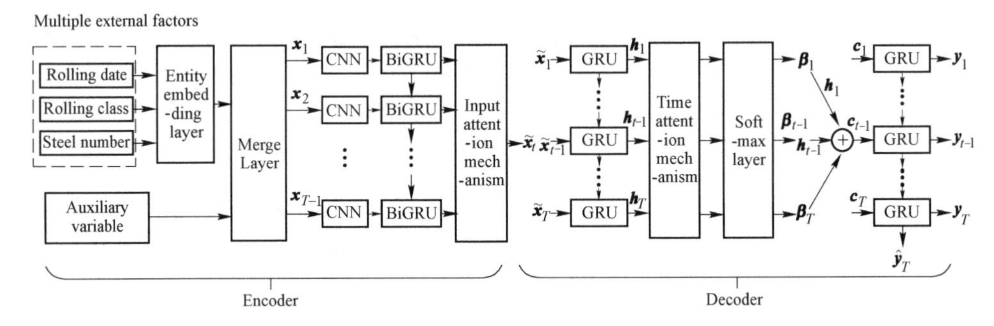

Fig. 3. 4 The network structure of DA-CBGRU-Seq2Seq

3. 2. 3. 2 A shape performance perception model based on Transformer

Thickness and flatness are important quality indicators for measuring strip steel products. However, the working environment of HSMP is harsh, with high temperature, high humidity, and high pressure. And with the continuous expansion of production capacity and the continuous improvement of product quality requirements, the requirements for the automation level of HSMP are also increasing. Therefore, precise control of thickness and shape during the rolling process has become an urgent issue to be solved in HSMP. At present, the control of thickness and flatness is achieved through the mechanism model of a secondary system or by operators, which is not precise enough. There are thickness gauges and flatness gauges at the rolling exit to measure the thickness and shape of the strip steel, and a large number of process data and quality data have not been fully used in the control of HSMP. The use of artificial intelligence methods to improve the accuracy of prediction and control has become a hot research topic in HSMP. By accurately predicting the thickness and flatness of the rolling exit, it is beneficial to optimize the control of HSMP.

Transformer network has been widely used in Natural language processing, behavior recognition, stock trend prediction and residual service life prediction due to its fast training speed, strong learning ability, and the ability to learn the dependency between dynamic continuous non Gaussian. Based on data related to rolling quality, establish a joint prediction model for rolling exit thickness and flatness using a Multi output Transformer with embedded sliding window (SW-MTrans) embedded sliding window. Firstly, in order to address the impact of time delay between rolling process variables on quality index prediction, a sliding window is embedded in the input layer of the model. Then, the Transformer network is improved to achieve accurate prediction of thickness and flatness simultaneously. Through experimental verification, SW-MTrans can accurately predict the thickness and flatness of strip steel simultaneously.

The structure of the prediction model is shown in **Fig. 3.5**. Firstly, the processed data of each variable is embedded into the input layer of the model with a window length of m, which contains delay information. Then, the input data is positional encoded. After multi-layer encoding and decoding, the final thickness and flatness prediction values are obtained through linear transformation and Softmax function[15-20].

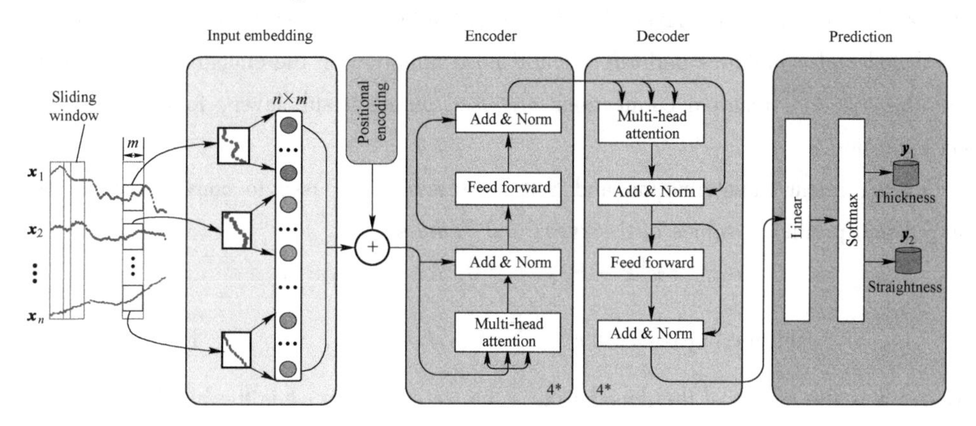

Fig. 3.5 The network structure of SW-MTrans

In this subsection, 46 relevant variables affecting thickness and flatness were selected, and the input matrix of the model embedded in the sliding window is as follows:

$$\boldsymbol{X}_i(t) = [\boldsymbol{x}_i(t-m), \cdots, \boldsymbol{x}_i(t-1), \boldsymbol{x}_i(t)] \quad (3.14)$$

$$\boldsymbol{X}(t) = [\boldsymbol{X}_1(t), \boldsymbol{X}_2(t), \cdots, \boldsymbol{X}_{46}(t)] \quad (3.15)$$

where, $\boldsymbol{x}$ represents the model input variable; $\boldsymbol{X}$ represents the model input matrix.

The encoder and decoder parts are both composed of Multi Head Attention, Feed Forward Networks, and residual summation and normalization. Multi head attention is composed of multiple attention layers running in parallel, which enables the model to jointly focus on information from different representation subspaces at different locations. The calculation formula is as follows.

$$\text{head}_i = \text{Attention}(Q\boldsymbol{W}_i^Q, K\boldsymbol{W}_i^K, V\boldsymbol{W}_i^V) \quad (3.16)$$

$$\text{MultiHead}(Q, K, V) = \text{Conct}(\text{head}_1, \cdots, \text{head}_h)W^O \quad (3.17)$$

where, $\boldsymbol{W}_i^Q \in R^{d_{\text{model}} \times d_q}$, $\boldsymbol{W}_i^K \in R^{d_{\text{model}} \times d_k}$, $\boldsymbol{W}_i^V \in R^{d_{\text{model}} \times d_v}$, $\boldsymbol{W}^O \in R^{hd_v \times d_{\text{model}}}$.

Each layer of encoder and decoder contains a fully connected feedforward network, which includes two linear transformations and a ReLU Activation function in the middle, as shown:

$$\text{FFN}(\boldsymbol{x}) = \max(0, \boldsymbol{x}\boldsymbol{W}_1 + \boldsymbol{b}_1)\boldsymbol{W}_2 + \boldsymbol{b}_2 \quad (3.18)$$

The encoder is composed of a stack of four identical layers, each consisting of a multi head self-attention mechanism and two sub layers of a simple, fully connected feedforward network by position. A residual connection is used between the multi head self-attention mechanism and the feedforward network, followed by layer normalization. The output of each sublayer is $\text{LayerNorm}(\boldsymbol{x} + \text{Sublayer}(\boldsymbol{x}))$, where $\text{Sublayer}(\boldsymbol{x})$ is a function implemented by the sublayer itself. The decoder is also composed of four stacked layers of the same layer. In addition to the two sub layers in each encoder layer, the decoder also includes performing multi head attention on the encoder output. Then, like the encoder, residual connections are used for each sub layer, followed by layer normalization.

Finally, linear transformation and Softmax function are used to convert the decoder output into predicted values of thickness and flatness.

$$z_k = \boldsymbol{w}_k^{\mathrm{T}}\boldsymbol{x} + \boldsymbol{b}_k, \quad k = 0, 1, \cdots, K-1 \quad (3.19)$$

$$\text{softmax}(z_k) = a_k = e^{z_k} \Big/ \sum_{i=1}^{K} e^{z_i}, \quad k = 0, 1, \cdots, K-1 \quad (3.20)$$

where, $\boldsymbol{x}$ is the output of the decoder; $\boldsymbol{w}$ is the weight matrix; $\boldsymbol{b}$ is the bias; e is the base number.

3.3 Cross domain multi granularity model association and description

Focusing on the characteristics of HSMP, such as complex spatiotemporal coupling, hierarchical collaborative correlation, and inter model interaction and linkage, this study focuses on:

(1) Consistency representation based on multidimensional and multi granularity models, studying the correlation between multi-stage models under dynamic temporal changes.

(2) Considering the linkage mapping and hierarchical differences of spatial models, studying the evolution and evolution of models under spatiotemporal linkage and spatial collaborative matching techniques.

(3) Based on the premise of the interaction between material flow energy flow information flow and HSMP, study the cross-domain transmission mechanism and potential change trends between models, and establish direct or indirect correlation and collaborative relationships between system models.

Based on the analysis of the frequency distribution, behavior patterns, rule information, and other typical characteristics of multidimensional and multi granularity models under dynamic temporal changes, such as slow drift, asynchronous operation, or local strong time-varying, this study investigates time processing methods such as key time node localization, time synchronization and regulation, and interactive time window selection to obtain the correlation between the multi-stage model and the time series, Further study the topological structure between models and establish the correlation between multi-stage models under dynamic temporal changes; Analyze the characteristics of spatiotemporal linkage mapping between models, use nodes and directed edges to jointly describe the evolutionary behavior between models, and combine HSMP mechanism knowledge to study the evolutionary reasoning method based on fuzzy cognitive maps to obtain the evolutionary laws of models under spatiotemporallinkage. Based on this, consider the physical spatial location of each model, utilize process parameter constraints between models, and use clustering, robust learning, and other methods, Implement spatial collaborative matching of models; According to the exchange, transmission and processing of material flow, energy flow and information flow in the HSMP, analyze the difference information and state change of the model, use approximate entropy to describe the potential change trend of the model, combine the system level information, reveal the trans regional transmission mechanism between the models through the in-depth causality diagram and Glenmorangie distillery causality analysis, and use information entropy and association rules to analyze and establish the direct or indirect association and synergy between the system models, Provide model support for the construction of HSMP correlation networks. The schematic diagram of cross domain multi granularity model association and description is shown in **Fig. 3. 6**.

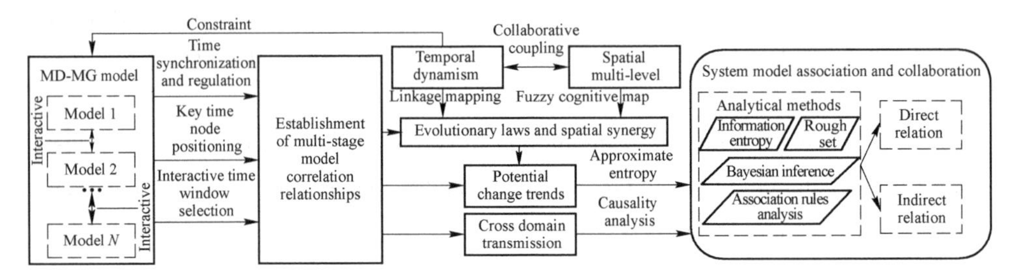

Fig. 3. 6 Schematic diagram of cross domain multi granularity model association and description

3. 3. 1 Multi process quality correlation model

In view of the characteristics of different but interrelated and coupled processes of HSMP, taking the typical steel manufacturing process of Angang Steel Iron and Steel Co. , Ltd. as the actual production background, a multi process association model considering quality heredity was constructed. The model fully excavates the quality association information and realizes the hierarchical fusion and extraction of quality related features by using variable screening method and deep learning algorithm.

The functions of each process in HSMP are different but interrelated and coupled, and the process quality problems that occur in the upstream process will be inherited along the cascade interconnection system to the downstream process. Taking the actual industrial process as an example, the rolling process is a typical continuous and serial process, and quality defects are a multi process coupling and accumulation process. On the one hand, the coupling of processes and the large scale of production lines limit the mining of key features, making it necessary to decompose multiple processes into sub blocks. On the other hand, a series of chain reactions occur during the dissemination of quality information, which makes it difficult to characterize the propagation mechanism of quality flow between various sub blocks. At the same time, the redundant correlation generated by the variable information within the concatenated sub blocks during the sequential transmission process also poses great challenges for characterizing genetic effects[21].

A quality genetic effect characterization model based on cascade interconnected systems was constructed to address the above issues. The model uses prior knowledge to divide large-scale industrial processes containing different processes into multiple sub blocks according to the product processing order to facilitate the mining of local quality information. Secondly, Mutual information is used to evaluate the correlation between variables and build new quality related and quality independent data sets. Finally, a parallel deep variational information bottleneck method Variational autoencoder maximum correlation minimum redundancy (DVIB-VAE-mRMR) local monitoring model is constructed to characterize the quality genetic effect of cascade interconnected systems.

Firstly, combining production data and empirical knowledge, comprehensively analyze the relationship between terminal product quality and process data, and carry out the decomposition work of complex multi process processes. During process decomposition, the rolling process is divided into upstream, midstream, and downstream according to the product processing sequence and spatial topology structure. For example, the finishing rolling process has 7 operating units $A \to B \to C \to D \to E \to F \to G$, which are divided into: $A \to B \to C$ (upstream process), $D \to E$ (midstream process), and $F \to G$ (downstream process). On this basis, the coupling relationship between multiple processes is reflected through the sequential transmission of quality information.

Secondly, evaluate the correlation between upstream, midstream, and downstream process variables and quality indicators. In the finishing rolling process, the Mutual information matrix is used to express the correlation between process variables and quality indicators. One process (upstream or midstream or downstream process) is described as follows:

$$\mathrm{MI}_b = \begin{bmatrix} \mathrm{MI}_b(\boldsymbol{x}_{b1}, \boldsymbol{y}_1) & \mathrm{MI}_b(\boldsymbol{x}_{b1}, \boldsymbol{y}_2) & \cdots & \mathrm{MI}_b(\boldsymbol{x}_{b1}, \boldsymbol{y}_r) \\ \mathrm{MI}_b(\boldsymbol{x}_{b2}, \boldsymbol{y}_1) & \mathrm{MI}_b(\boldsymbol{x}_{b2}, \boldsymbol{y}_2) & \cdots & \mathrm{MI}_b(\boldsymbol{x}_{b2}, \boldsymbol{y}_r) \\ \vdots & \cdots & \ddots & \vdots \\ \mathrm{MI}_b(\boldsymbol{x}_{bM_b}, \boldsymbol{y}_1) & \mathrm{MI}_b(x_{\boldsymbol{bM}_b}, \boldsymbol{y}_2) & \cdots & \mathrm{MI}_b(\boldsymbol{x}_{bM_b}, \boldsymbol{y}_r) \end{bmatrix} \tag{3.21}$$

where, $\boldsymbol{x}_{bj}$ represents any process variable and quality variable $\boldsymbol{y}_i$ in a certain process.

The impact of different process variables on product quality varies. Thresholds α_b and $\beta_b (0 < \alpha_b < \beta_b < 1)$ are set to evaluate the correlation between process variables and quality indicators, which are:

Level 1: If $\beta_b < \mathrm{MI}_b(\boldsymbol{x}_{bj}, \boldsymbol{y}_i) < 1$, then $\boldsymbol{x}_{bj}$ is highly correlated with the quality.

Level 2: If $\alpha_b < \mathrm{MI}_b(\boldsymbol{x}_{bj}, \boldsymbol{y}_i) < \beta_b$, then $\boldsymbol{x}_{bj}$ is moderately correlated with the quality.

Level 3: If $0 < \mathrm{MI}_b(\boldsymbol{x}_{bj}, \boldsymbol{y}_i) < \alpha_b$, then $\boldsymbol{x}_{bj}$ is weakly correlated with the quality.

Merge variables that meet Level 1 and 2 to construct a quality related process variable dataset, and merge process variables that meet Level 2 and 3 to construct a quality independent process variable dataset.

Finally, construct a quality genetic effect model for cascade interconnected systems. When modeling, mining the deep-seated features of process variables from hidden, fragmented, and low-quality data and considering the issue of quality inheritance. In addition, different process variables in the system have different effects on product quality, so it is necessary to consider the correlation between process variables and quality variables. In order to meet the above requirements, when building the local model, the parallel deep variable information bottleneck (DVIB)-Variational autoencoder (VAE) deep learning network is used to extract quality related Latent and observable variables and quality independent features for different processes. At the same time, the quality related Latent and observable variables extracted from the previous process is transferred to the next process as the output sequence sequence. In this process, in order to eliminate the adverse effects of redundant information, the maximum correlation and minimum redundancy (mRMR) algorithm is used to screen the quality related Latent and observable variables. The quality related Latent and observable variables screened from the previous process is used as a constraint condition in the next process to characterize the quality information transmission process throughout all sub blocks[22].

3.3.2 Multi-scale data association model

The HSMP horizontal full process production line and vertical integrated automation system interact and collaborate, and information resources exhibit complex spatiotemporal and multi-scale characteristics under the interweaving of material flow, energy flow, and information flow. The heating furnace, rough rolling, finishing rolling, laminar cooling, and coiling are distributed in series on the production line according to certain process specifications. Many heavy equipment is coupled and connected with complex control circuits, and the comprehensive automation system layers such as equipment layer (L0), real-time control layer (L1), and process control layer (L2) are interconnected through information system collaboration. The numerous data resources in the hot rolling production process exhibit typical spatiotemporal and multi-scale characteristics under the interweaving of material flow, energy flow, and information flow; At the same time, multi-dimensional attributes such as process data, quality data, and process data exacerbate the difficulty of feature association analysis and hierarchical feature expression.

A multi-scale data association analysis and feature expression model is proposed to address the above issues. The reference variables for the first level data (process data dimension) in this model include: total rolling force of the frame, rolling force difference of the frame, loop tension, loop position control, loop height, roll gap of the frame, power, rolling speed, and roll shifting force (F1 frame has no bending roll force). The reference variables for the secondary data (process parameter dimension) include: mill load distribution, inlet thickness setting, rolling force setting, roll gap reference value, speed reference value, forward slip value, stand distance, and X-ray thickness gauge distance. On the basis of mechanism models such as rolling force model, thickness model, and rolling speed model, the production process of hot rolling and finishing mill of strip steel relies on feedback adjustment of automatic thickness control system, and various equipment, links, and control circuits work together to roll the intermediate slab into strip steel products that meet the preset specifications. Among them, the process data dimension information directly reflects the highly dynamic behavior of real-time operation, while the process parameter dimension information plays a key regulatory role on the underlying equipment and control loop.

Firstly, based on process mechanisms and knowledge, the entire manufacturing process is decomposed into physical spaces to form multiple production processes or operational units constrained by quality flow. Among them, each process or operating unit contains specific production equipment and relatively independent control systems. Collect operation signals, control signals, physical measurement signals, and other data reflecting the operational status of each process or unit under normal working conditions from industrial sites as reference variables for process variables.

Secondly, based on the integrated management and control of enterprises and the actual information interconnection architecture, the manufacturing system is decomposed into information spaces to form multiple system levels under workflow constraints. Among them, each level cooperates with adjacent levels, conveying job information and production performance to the upper level, and conveying production plans and operation instructions to the lower level. Export model setting data and process parameter information for specific product specifications/batches/operating conditions from the production information management system database.

On this basis, spatiotemporal collaborative matching is performed on process data and process parameters, and the missing spatiotemporal data is filled using fully connected network tensor decomposition.

Subsequently, consider the highly dynamic information contained in the process data, guided by upper-level information and constrained by closed-loop control loops. The process parameters contain relative static information that plays an important role in regulating the real-time production operation status. Using the dynamic internal principal component analysis method, taking into account the temporal autocorrelation properties of process variables and the local dynamic time-varying effects caused by process inertia, the spatiotemporal strong dynamic behavior during the process is extracted to form small-scale feature expression. In parallel, the slow feature analysis method is used to extract the relatively static behavior of the process with strong spatiotemporal Strongly correlated material and form large-scale feature expression, taking into account the gradual change trend of process information and set values and eliminating the adverse effects of local noise and interference as far as possible.

Based on multi-scale feature representation, a double-layer hidden Markov model is used for feature association and fusion, and the mechanism of information association and matching between levels is excavated. The first layer hidden Markov model corresponds to the dynamic operation status of the underlying device and real-time control layer, where the observation sequence is the extracted small-scale temporal features. The second layer of Hidden Markov model corresponds to the control state of the process control layer, where the observation sequence is the extracted large-scale feature. The double-layer hidden Markov model uses a two-layer topological structure to describe the process operation status at different scales (large scale to small scale). Based on multi-scale spatiotemporal features, train and estimate the parameters of a two-layer hidden Markov model under normal operating conditions. Among them, the small-scale feature set is the input of the first layer hidden Markov model, and the large-scale feature set is the input of the second layer hidden Markov model. Process production is usually a continuous process with feedback regulation, and the real-time operation status is closely related to the previous state under the constraints of set values and process settings. Two different scale features reflect the spatiotemporal behavior of processes at different levels, and a two-layer hidden Markov model is used to establish linkage mapping relationships at different scales. The double-layer hidden Markov model describes the correlation mechanism between the real-time control layer and the process parameter layer by mining the potential dynamic behavior under the constraint of process parameter dimensions.

3. 3. 3 Layered fuzzy signed directed graph model

In the field of artificial intelligence, signed directed graph (SDG) is known as a deep knowledge model. Its model construction is relatively simple and intuitive, and it can express complex causal relationships and contain large-scale potential information. It has achieved good applications in many fields. The SDG model can be obtained by expressing process variables as graphical nodes and causal relationships as directed branches.

Fig. 3. 7 shows a typical SDG model, which typically consists of the following three parts:

(1) Nodes: Each node in SDG represents a specific physical variable, and the states of each variable are represented by node symbols { "+" "–" "0" } . Indicates whether the node is outside or below the normal operating range;

(2) Edge: an edge that connects different nodes and can represent the direction of transmission and influence between nodes;

(3) Internode influence relationship: The relationship between nodes is divided into positive and negative influences, represented by "+" and "–" respectively.

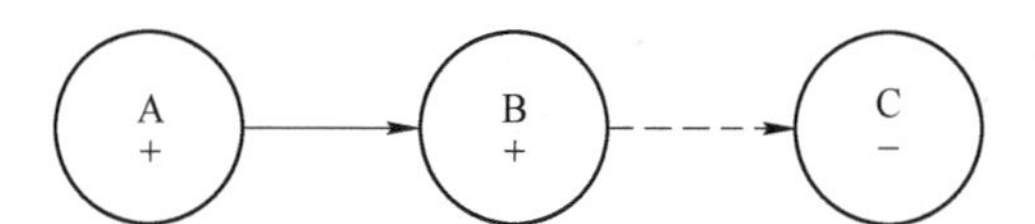

Fig. 3. 7 Schematic diagram of nodes and directed edges in SDG model

The SDG in **Fig. 3. 7** consists of three nodes: A, B, and C. A and B with a sign of "+" indicate that the sampling value of their corresponding variable is above the upper threshold, while C with a sign of "–" indicates that the sampling value of its corresponding variable is below the set lower threshold. The solid line connecting arrows between A and B indicates that A node has a positive impact on B node, that is, an increase in A's state value will also increase B's state value, and a decrease in A's state value will also decrease B's state value. The connecting arrow between B and C is a dashed line indicating that B has a negative impact on C, meaning that an increase in the state value of B will cause a decrease in the state value of C, while a decrease in the state value of B will cause an increase in the state value of C.

In traditional SDG models, the quantitative information of the model is insufficient.

Variable nodes only have three states: { "+" "−" "0" }, which contain less information. Therefore, the concept of dual thresholds for alarm and early warning is introduced to quantify the node state as {−1, −s, 0, +s, +1}, −s and +s the calculation formula for is:

$$-s = \frac{x - \varepsilon_{l1}}{\varepsilon_{l1} - \varepsilon_{l2}}, \quad +s = \frac{x - \varepsilon_{h1}}{\varepsilon_{h2} - \varepsilon_{h1}} \tag{3.22}$$

where, $[\varepsilon_{l1}, \varepsilon_{h1}]$ and $[\varepsilon_{l2}, \varepsilon_{h2}]$ are the first and second threshold values, respectively.

Quantifying node states can reflect the of deviation of nodes, which has better risk assessment and compensates for the shortcomings of traditional SDG quantitative information. The higher the absolute value of s, the higher the degree of anomaly in nodes, which can determine which nodes need to be monitored in order to prioritize elimination.

The interaction between nodes contains nonlinear conditions, which cannot be described by clear values. The results of simple five level node state diagnosis are not perfect, so the fuzzy set theory is introduced. By introducing a Fuzzy set with a fuzzy language variable {NB, NS, PS, PB}, the new node state can be expressed as $V = (\nu_i, \mu, \psi(\nu_i))$, where μ is the membership $\{\mu_{NB}, \mu_{NS}, \mu_{PS}, \mu_{PB}\}$ corresponding to the node deviation and the language variable. The membership function used is a normal type, and the following membership function expressions are obtained based on the 3 σ principle:

$$\mu_{NB}(x) = \begin{cases} 1 & x < \varepsilon_{l2} \\ e^{-\frac{(x-\varepsilon_{l2})^2}{2\sigma_{NB}^2}} & \varepsilon_{l1} < x < \varepsilon_{l2},\ \sigma_{NB} = \dfrac{\varepsilon_{l2} - \varepsilon_{l1}}{3} \\ 0 & x > \varepsilon_{l1} \end{cases} \tag{3.23}$$

$$\mu_{NS}(x) = \begin{cases} 0 & x < \varepsilon_{l2},\ x > x_0 \\ e^{-\frac{(x-\varepsilon_{l1})^2}{2\sigma_{NS1}^2}} & \varepsilon_{l2} < x < \varepsilon_{l1},\ \sigma_{NS1} = \dfrac{\varepsilon_{l1} - \varepsilon_{l2}}{3} \\ e^{-\frac{(x-\varepsilon_{l1})^2}{2\sigma_{NS2}^2}} & \varepsilon_{l1} < x < x_0,\ \sigma_{NS2} = \dfrac{x_0 - \varepsilon_{l1}}{3} \end{cases} \tag{3.24}$$

$$\mu_{PS}(x) = \begin{cases} 0 & x < x_0,\ x > \varepsilon_{h2} \\ e^{-\frac{(x-\varepsilon_{h1})^2}{2\sigma_{PS1}^2}} & x_0 \leqslant x < \varepsilon_{h1},\ \sigma_{PS1} = \dfrac{\varepsilon_{h1} - x_0}{3} \\ e^{-\frac{(x-\varepsilon_{h1})^2}{2\sigma_{PS2}^2}} & \varepsilon_{h1} \leqslant x \leqslant \varepsilon_{h2},\ \sigma_{PS2} = \dfrac{\varepsilon_{h2} - \varepsilon_{h1}}{3} \end{cases} \tag{3.25}$$

$$\mu_{PB}(x) = \begin{cases} 0 & x < \varepsilon_{h1} \\ e^{-\frac{(x-\varepsilon_{h2})^2}{2\sigma_{PB}^2}} & \varepsilon_{h1} \leqslant x < \varepsilon_{h2},\ \sigma_{PB} = \dfrac{\varepsilon_{h2} - \varepsilon_{h1}}{3} \\ 1 & x > \varepsilon_{h2} \end{cases} \tag{3.26}$$

The corresponding distribution function model is shown in **Fig. 3.8**.

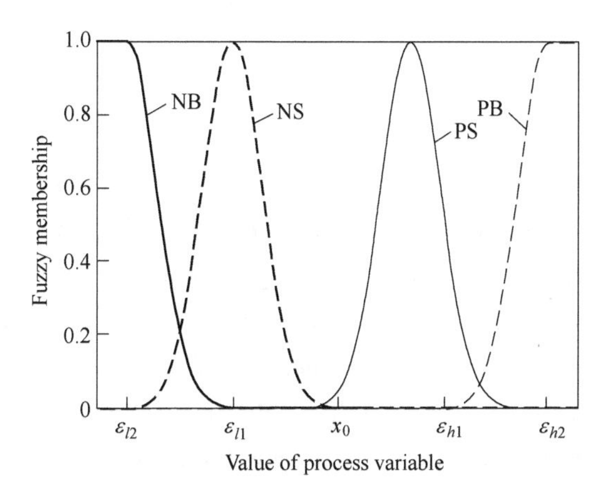

Fig. 3.8 Fuzzy membership function corresponding to fuzzy sets

3.3.4 Case study on the correlation of HSMP model

The finishing mill is the core equipment of HSMP, and the quality control function of finished strip steel is mainly concentrated in the finishing mill area. The finishing mill unit is generally composed of 7 continuous stands, and each stand is not independent, but connected together using various control methods. Each rack is equipped with a hydraulic system to provide the required bending and rolling force to control the roll gap, ensuring that the strip steel is rolled through the rack to obtain the corresponding outlet thickness of the strip steel. When constructing the SDG model, the main considerations are the thickness, convexity, and temperature of the strip steel. Next, we will introduce the process setting model one by one[23].

3.3.4.1 Rolling force setting

Rolling force (RF) is related to factors such as steel grade, strip temperature, width, surface condition, reduction, and deformation speed. However, due to the complex production conditions of on-site rolling, it is very difficult to fully consider each link. At present, the rolling force model adopts the following basic form:

$$p = Bl_c Q_p K_F K_T \tag{3.27}$$

where, p is the rolling force; l_c is the horizontal projection length of the contact arc between the flattened roller and the rolled piece; Q_p is the influence coefficient of the stress state caused by the friction force on the contact arc; K_F is the metal deformation resistance; K_T is the influence coefficient of the pre and post tensile stress on the rolling force; B is the width of the strip steel.

3.3.4.2 Bounce equation

For finishing mill processes with a reduction of only a few millimeters, it is necessary to consider the impact of bounce and accurately calculate the bounce in order to obtain products that meet the thickness tolerance requirements. All that can be adjusted for thickness during the operation of a plate and strip mill is the empty gap between the rolls, and the biggest challenge in the operation of a plate and strip mill is how to estimate the rolling force and how to achieve the required plate and strip thickness by adjusting the gap between the rolls. The relationship between bounce phenomenon is as follows:

$$h = S_p = S' + \frac{p}{C'} \tag{3.28}$$

where, h is the outlet thickness of the rolled piece; S_p is the loaded roll gap of the rolling mill; S' is the unloaded roll gap; p is the rolling force; C' is the total stiffness of the rolling mill base.

The rolling mill stiffness refers to the amount of rolling force required to increase the gap between the rolling rolls by 1 mm under the action of rolling force. In practical operation, in order to eliminate the influence of the non-linear relationship between the elastic deformation of the small rolling force machine seat and the rolling force, it is necessary to use the method of manually pressing the zero position. Now, after pressing the roller to a certain pre pressing force p_0, the indication of the roller gap gauge is reset to zero. Taking into account the effects of temperature rise and wear of the rollers on roll gap drift, changes in oil film thickness of the roller bearings, and changes in plate width on the stiffness of the rolling mill, the spring equation can be further expressed as:

$$h = S_0 + \frac{p - p_0}{C_p} + O + G \tag{3.29}$$

where, S_0 is the roller gap indicator for manual zero position; C_p is the stiffness coefficient of the rolling mill; O is the oil film thickness of the roller bearing; G is the zero position of the roller gap.

3.3.4.3 Roll gap shape equation

The cross-sectional shape (including convexity) of the rolled strip steel is the shape of the loaded roll gap, and the control of the cross-sectional shape of the strip steel is actually the control of the shape of the loaded roll gap. There are many factors that affect the shape of the loaded roller gap, mainly including:

(1) The rolling force that causes bending deformation (including some shear deformation) of the roller system;

(2) The bending force that causes the bending deformation of the roller system;

(3) The original roll shape of the roll system;

(4) Changes in hot and worn roll profiles during rolling;

(5) Continuous Variable Convexity (CVC), Linear Variable Convexity (LVC), and other controllable roll profile technologies that change the shape of the roll gap.

Therefore, the shape of the loaded roller gap can be described by the following equation:

$$CR = \frac{p}{K_p} + \frac{F}{K_F} + E_C\omega_C + E_\Sigma(\omega_H + \omega_W + \omega_0) + E_0\Delta \tag{3.30}$$

3.3.4.4 Temperature control of plate and strip final rolling

During the hot strip rolling process, the temperature of the strip is one of the important factors that directly affect the dimensional accuracy, mechanical properties, and reasonable load distribution of the rolling mill. If the final rolling temperature is too low, it will affect the plasticity of the metal and reduce the processing performance of the strip steel; Excessive final rolling temperature can reduce the mechanical properties of the steel and may also cause the formation of oxide scales on the surface of the strip, affecting the surface quality of the finished plate and strip. Therefore, determining the temperature changes of the strip steel at different positions can effectively improve the quality level of the final product.

The outlet temperature of each stand during the hot rolling process is generally represented by the following function:

$$T = f(T_0,\ H_0,\ h,\ v,\ q_w) \tag{3.31}$$

where, T_0 is the temperature of the strip steel at the entrance of the rolling mill; H_0 is the thickness of the billet at the entrance of the rolling mill; v is the rolling speed; q_w is the cooling water flow rate.

Due to the different thermophysical properties of different steel grades, the coefficients in the function will depend on the steel grade. Models generally tend to adopt simplified theoretical formulas or statistical empirical formulas. This section establishes an SDG model for the four process models mentioned above for the rolling mill frame, which is shown in **Fig. 3. 9**.

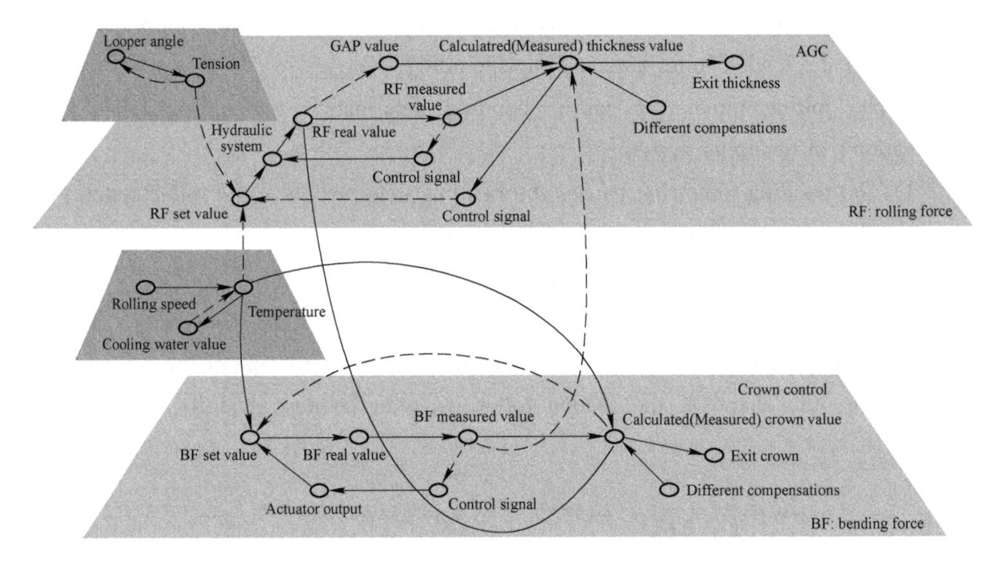

Fig. 3. 9 SDG model of finishing mill stand

By using the layering method to layer the model, the four-layer SDG diagnostic model shown in **Fig. 3. 10** was obtained. The first layer of nodes includes rolling speed, cooling water valve opening, tension, loop angle, rolling force compensation, and bending roll force compensation. The second layer of nodes includes temperature. The third layer of nodes includes rolling force setting value, hydraulic system output, hydraulic control signal, rolling force measurement value, rolling force actual value, and bending roll force measurement value. The fourth layer of nodes includes bending roll force actual value, bending roll force setting value, roll gap value, and actuator control signal Output of actuator, thickness calculation value, convexity calculation value, rolling force control signal, output convexity, output thickness[24-25].

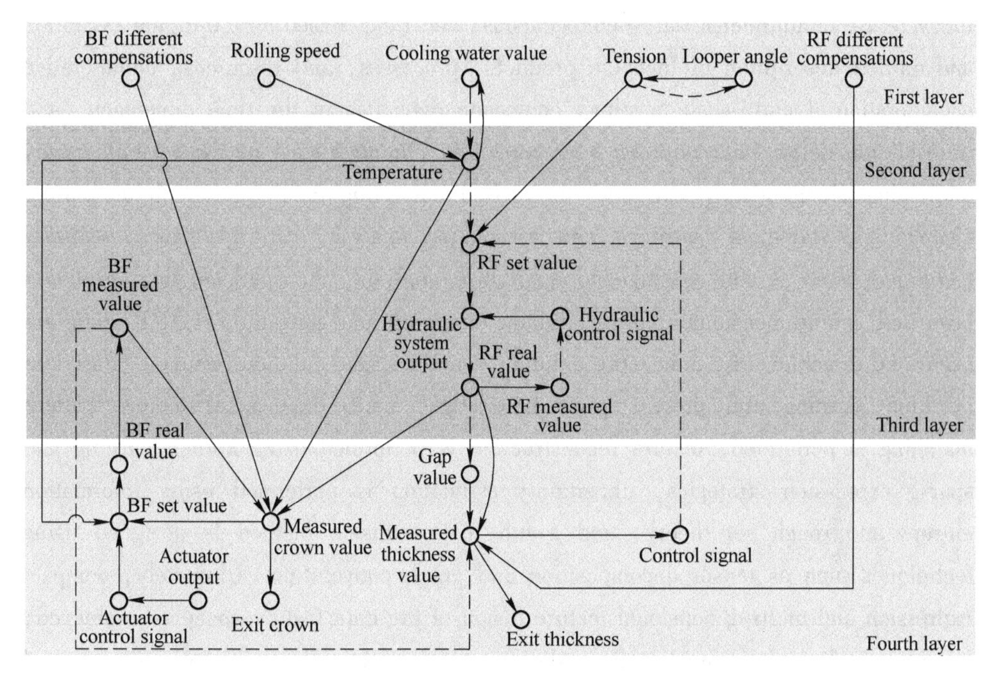

Fig. 3.10 Layered SDG model of finishing mill stand

3.4 Evolutional feature extraction and fusion of heterogeneous data

Focusing on the characteristics of multi-stage, multimodal, and complex spatiotemporal dynamic evolution behavior of HSMP lifecycle data, this study focuses on:

(1) researching multi-stage spatiotemporal feature extraction methods based on the unified representation of heterogeneous data from multiple sources such as research and development design, production manufacturing, and operation maintenance.

(2) Consider the perception feedback and interaction linkage of lifecycle processes, study the data driven HSMP spatiotemporal trend analysis technology, and then achieve dynamic characterization of linkage evolution behavior.

(3) Based on multi-dimensional information such as geometry, physics, behavior, and rules, research the mechanism of information transmission and feature association mapping, and construct a full feature fusion method.

HSMP usually covers the stages of R&D, design, production and manufacturing, operation and maintenance, and involves Multiphysics simulation, multiple information systems, and complex human-computer interaction. The information resources in the lifecycle are multidimensional, heterogeneous, and rich. Based on the digital definition and unified description of lifecycle products, processes, and resources, collaborative consideration of multi-scale nesting, composite dynamics in the time dimension, and transmission delay and coupling interconnection characteristics in the spatial range, multi-dimensional feature extraction is carried out using techniques such as pattern aggregation, statistical learning, and association analysis. On this basis, utilizing techniques such as soft sensing and qualitative analysis, the spatiotemporal evolution laws and multidimensional feature linkage mapping mechanism of HSMP data are excavated to achieve dynamic representation of multi-stage evolution behavior. Based on machine learning and process prior knowledge, multi-dimensional feature pattern matching is performed, feature reconstruction is performed using kernel learning and sparse expression strategies, uncertainty estimation is performed using information entropy and rough set theory, and a full feature fusion method is designed using techniques such as tensor decomposition and graph convolution. Ultimately, compact expression and multi-dimensional feature fusion of the data feature space are achieved, providing information support for subsequent research. The schematic diagram of Evolutional feature extraction and fusion of heterogeneous data is shown in **Fig. 3. 11**.

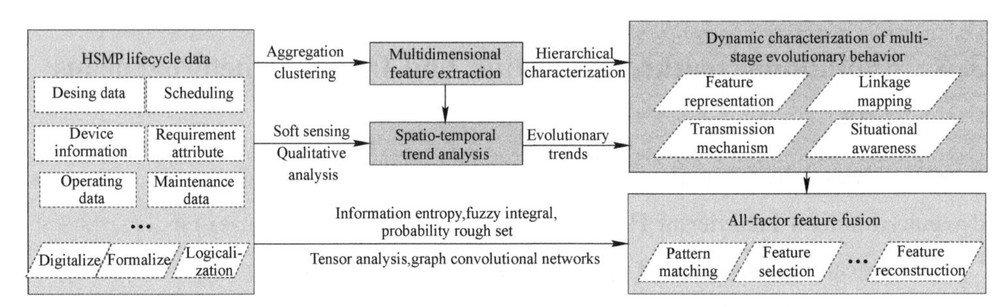

Fig. 3. 11 Schematic diagram of Evolutional feature extraction and fusion of heterogeneous data

3. 4. 1 Spatiotemporal feature extraction and fusion expression

HSMP typically consists of numerous processes, complex process flow, long production lines, and numerous local operating units. Within each process, there is a relatively independent process model to develop process procedures, process parameters, and set values for each control loop. Mechanism/statistical models and single process relatively independent control are used to ensure the economy of the production process and the stability of product quality. From the perspective of the entire production process, under the constraints of material flow, energy flow, and information flow, multiple local units distributed in different physical spaces interact, couple, and collaborate with each other. The spatiotemporal characteristics are not only reflected in the coupling of numerous process variables within local units, but also in the spatial correlation between multiple units. In addition, multi specification product varieties, frequent production conditions, and even conflicting and balanced control objectives make it difficult to ensure quality stability in actual production processes. There are many factors that affect product quality and performance, and in most cases, it is difficult to perceive them in real-time. Local performance anomalies may evolve and propagate throughout the entire process, and performance related factors exhibit concealment and dynamism under spatiotemporal constraints. In addition, performance related dynamic behaviors in the spatiotemporal domain also interact and influence each other. Therefore, it is necessary to study the extraction and fusion methods of multi process spatiotemporal features under performance supervision, which can help reveal potential performance fluctuations throughout the entire process.

A multi process data association model based on parallel spatiotemporal feature analysis is proposed to address the above issues. Time domain features focus on the dynamic autocorrelation characteristics of fixed sampling points; The spatial domain features focus on the cross correlation changes of sampling points at different positions. Under the parallel framework, time-domain feature extraction module, spatial feature extraction module, and spatiotemporal feature fusion module were designed respectively. The core is to solve the genetic effect characterization of quality performance between processes and the problem of mining multi process association mechanisms, aiming to analyze the association patterns of manufacturing process data from the perspective of multiple processes.

The time-domain feature extraction module is cascaded by multiple attention mechanism long short term memory (LSTM) networks. The process data and quality data of each process are respectively regarded as the input and output of the network. The local quality prediction of each process in the training process is used as the constraint of the network module of the next process, and is input into the model together with the training data for training. Due to the recursive mechanism and temporal learning properties of LSTM, the attention mechanism can make the training process focus more on time-domain features that have a greater impact on quality information. Therefore, the features extracted through this network structure can capture more time-domain dynamic features under quality performance supervision.

Spatial information is reflected in the interdependence between process variables within local operating units. Spatial feature extraction is dedicated to mining the implicit spatial correlations among different process variables and capturing process fluctuations under closed-loop control. Firstly, the correlation coefficients of all variables of the sliding window are obtained through Mutual information (MI), and position coding is performed to form the spatial information representation based on MI matrix. Due to the sharp increase in data dimensions after unfolding by position, there is information redundancy. Therefore, KPCA is used to capture principal components and achieve spatial feature extraction.

Assuming the entire process consists of K subprocesses, the temporal dynamic features extracted from the k_{th} unit are denoted as $\boldsymbol{X}_k^{\text{tem}}$, the spatial features are denoted as $\boldsymbol{X}_k^{\text{spa}}$, and the spatiotemporal support domain of the $\boldsymbol{k}_{\text{th}}$ unit under quality performance supervision is constructed as $\boldsymbol{X}_{[k]}^{\text{tem-spa}} = [\boldsymbol{X}_k^{\text{tem}};\ \boldsymbol{X}_k^{\text{spa}}] \in R^{N\times(d+r)}$. Considering all operating units, the third-order multi process spatiotemporal tensor is constructed as $\underline{\boldsymbol{X}}^{\text{tem-spa}} = [\boldsymbol{X}_{[1]}^{\text{tem-spa}},\ \boldsymbol{X}_{[2]}^{\text{tem-spa}},\ \cdots,\ \boldsymbol{X}_{[K]}^{\text{tem-spa}}]$. We plan to use the Canonical Polyadic (CP) decomposition of higher-order tensors to explore the coupling mechanism of multi process association under quality performance constraints, fuse the spatiotemporal features of multiple processes, and decompose the original tensor into the additive form of β rank 1 sub tensors. The objective function is as follows:

$$\min_{\boldsymbol{t}_{A,i},\ \boldsymbol{p}_{A,i},\ \boldsymbol{q}_{A,i}} \left\| \underline{\boldsymbol{X}}^{\text{tem-spa}} - \sum_{i=1}^{\beta} \boldsymbol{t}_{A,i} \circ \boldsymbol{p}_{A,i} \circ \boldsymbol{q}_{A,i} \right\|_F^2 \tag{3.32}$$

where, $\|\cdot\|_F$ is a Frobenius norm operator.

The above objective function contains three parameters to be solved, which are transformed into the following three sub objectives using the idea of alternating least squares:

$$\begin{cases} \min\limits_{\boldsymbol{T}_A} \left\| \boldsymbol{X}_{(1)}^{\text{tem-spa}} - \boldsymbol{T}_A (\boldsymbol{Q}_A \odot \boldsymbol{P}_A)^T \right\|_F^2 \\ \min\limits_{\boldsymbol{P}_A} \left\| \boldsymbol{X}_{(2)}^{\text{tem-spa}} - \boldsymbol{P}_A (\boldsymbol{Q}_A \odot \boldsymbol{T}_A)^T \right\|_F^2 \\ \min\limits_{\boldsymbol{Q}_A} \left\| \boldsymbol{X}_{(3)}^{\text{tem-spa}} - \boldsymbol{Q}_A (\boldsymbol{P}_A \odot \boldsymbol{T}_A)^T \right\|_F^2 \end{cases} \tag{3.33}$$

where, $\odot$ is Khatri Rao multiplier.

The above triple sub optimization problem can be solved by alternating least squares algorithm, and a multi process Strongly correlated material subspace and independent subspace of each process can be obtained.

In summary, through parallel spatiotemporal feature extraction and decomposition based on third-order multi-process spatiotemporal tensors, the inter process association mechanism under quality performance supervision has been excavated, achieving refined feature expression of multi-subspace collaboration.

3.4.2 Feature level fusion method for non-stationary data

Due to the dynamic operation of manufacturing processes, their process data exhibits typical statistical non-stationary characteristics. For non-stationary features, a feature level fusion method for non-stationary data based on slow feature analysis and cointegration analysis is proposed, providing a feature basis for data association pattern analysis.

For the non-stationary process, the following factors should be considered:

(1) Due to the production mode of variable batches and strong dynamics, some variables of industrial process data show typical non-stationary characteristics, while other variables may remain stable;

(2) Although operating conditions may vary over time, stationary variables may follow some equilibrium relationship beyond the current time.

Based on consideration factor (1), the variable is first divided into stationary and non-stationary variables by enhancing the augmented Dickey Fuller (ADF) test. Based on factor (2), a cointegration analysis (CA) is performed on non-stationary variables to obtain the stationary residual sequence of the variables, which is then fused with the stationary variables to form a new feature and fused dataset. The specific technical steps are shown in **Fig. 3.12**.

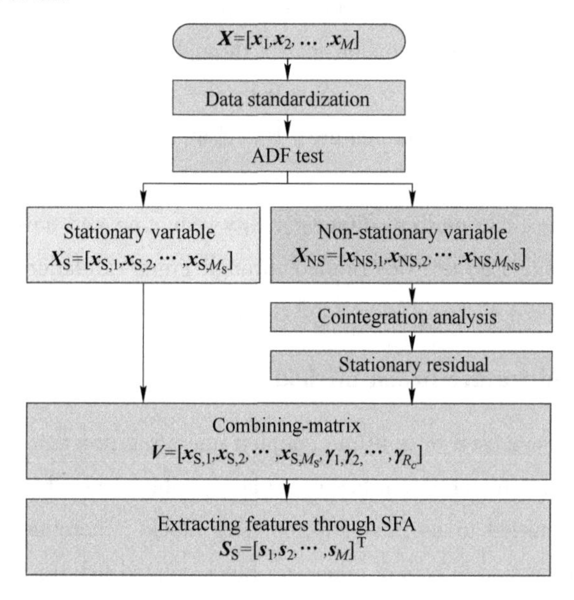

Fig. 3.12 The flowchart of SFA-CA feature fusion

(1) Standardization of datasets $\boldsymbol{X} = [\boldsymbol{x}_1, \boldsymbol{x}_2, \cdots, \boldsymbol{x}_M] \in R^{N\times M}$:

$$\boldsymbol{x}^* = \frac{\boldsymbol{x} - \bar{\boldsymbol{x}}}{\sigma} \tag{3.34}$$

where, $\boldsymbol{x}$ is the original training set data; $\bar{\boldsymbol{x}}$ is the mean of the training set; σ is the standard deviation of the training set.

(2) Variable stationarity test. Based on the above consideration factor (2), the variables are divided into non-stationary process variables $\boldsymbol{X}_{\mathrm{NS}} = [\boldsymbol{x}_{\mathrm{NS},1}, \boldsymbol{x}_{\mathrm{NS},2}, \cdots, \boldsymbol{x}_{\mathrm{NS},M_{\mathrm{NS}}}] \in R^{N\times M_{\mathrm{NS}}}$ and stationary process variables $\boldsymbol{X}_{\mathrm{S}} = [\boldsymbol{x}_{\mathrm{S},1}, \boldsymbol{x}_{\mathrm{S},2}, \cdots, \boldsymbol{x}_{\mathrm{S},M_{\mathrm{S}}}] \in R^{N\times M_{\mathrm{S}}}$ through ADF test, where $M = M_{\mathrm{NS}} + M_{\mathrm{S}}$.

(3) Cointegration analysis. Based on the above considerations factor (1), the cointegration analysis algorithm is applied to the identified non-stationary variable $\boldsymbol{X}_{\mathrm{NS}}$ to obtain the cointegration model:

$$\boldsymbol{\gamma}_t = \boldsymbol{B}^{\mathrm{T}} \boldsymbol{X}_{\mathrm{NS}}(t), \ t = 1, \cdots, N \tag{3.35}$$

where, $\boldsymbol{\gamma}_t \in R^{R_c}$ is a stationary residual sequence; $\boldsymbol{X}_{\mathrm{NS}}(t)$ is a non-stationary variable sample at time t ; $\boldsymbol{B} = (\beta_1, \cdots, \beta_r) \in R^{N_{\mathrm{NS}}\times R_c}$ is a cointegration matrix; R_c is the number of cointegration vectors, which can be determined through Johansen's test.

By fusing the stationary residual sequence with stationary variables, a data feature fusion matrix $\boldsymbol{V} = [\boldsymbol{x}_{\mathrm{S},1}, \boldsymbol{x}_{\mathrm{S},2}, \cdots, \boldsymbol{x}_{\mathrm{S},M_{\mathrm{S}}}, \boldsymbol{\gamma}_1, \boldsymbol{\gamma}_2, \cdots, \boldsymbol{\gamma}_{R_c}] \in R^{M\times(M_{\mathrm{S}}+R_c)}$ is obtained. The new matrix contains information about both stationary and non-stationary variables, achieving feature level information fusion.

(4) Extract the data feature fusion matrix through slow Feature analysis (SFA):

$$\boldsymbol{S} = \boldsymbol{W}\boldsymbol{V} \tag{3.36}$$

where, $\boldsymbol{S} \in R^{M\times(M_s+R_c)}$ is a slow feature; $\boldsymbol{W} = [\boldsymbol{w}_1, \boldsymbol{w}_2, \cdots, \boldsymbol{w}_{N_s+R_c}]^{\mathrm{T}}$ is a feature projection matrix.

SFA can simultaneously perform dimensionality reduction and noise reduction. Here, the reserved dimensions R_s are determined through cross validation, i. e. the reserved slow features are $\boldsymbol{S}_s = (\boldsymbol{s}_1, \boldsymbol{s}_2, \cdots, \boldsymbol{s}_N)^{\mathrm{T}} \in R^{N\times R_s}$

3.4.3 Distributed feature fusion method

HSMP is typically composed of multiple coupled operating programs, and heritability is an inherent characteristic of the process industry, where the characteristics of a sub block may be transmitted to its next adjacent sub block. Therefore, it is necessary to incorporate the connection effects of other sub block related features into the modeling of the stator block. Each sub block can extract internal private features and external connected features, and pass them on to the feature extraction process of the next sub block. Internal private features only contain fault information within sub blocks, while external connection features contain information about the impact of sub blocks on other sub blocks. Local Tangent space alignment (LTSA) algorithm is used to reduce data dimensions and extract internal features. Canonical correlation analysis (CCA) algorithm is used to extract external features. Finally, the features of multiple sub blocks are fused to obtain the internal and external features of the whole process. The main steps include:

Step 1: Considering the operational characteristics, fault mechanisms, and variable meanings of the process, the entire process is divided into several sub blocks with physical significance.

Step 2: Reduce the dimensionality of each sub block variable and extract internal features. Assuming that the process industry can be divided into B typical operating units. For the b_{th} sub block, its internal variable group is $\boldsymbol{x}_b$, and the corresponding data is $\boldsymbol{X}_b \in R^{l_b \times N}$, $b = 1, 2, \cdots, B$, where l_b is the number of variables in the b_{th} sub block, and N is the number of samples. In order to find the maximum information contained in the data, LTSA is used to reduce the dimensionality of $\boldsymbol{x}_b$ and extract the internal features of each sub block. The data after dimension reduction is $\boldsymbol{t}_x$, the projection matrix $\boldsymbol{P}_{X_{\mathrm{pc}}} \in R^{l_b \times k_{x,b}}$ and the diagonal matrix $\boldsymbol{\Lambda}_{X_{\mathrm{pc}}}$ containing the eigenvector $\lambda_1, \cdots, \lambda_{k_{x,b}}$ can be obtained. The projection will be carried out from the l_b dimensional space to determine the number of principal elements $k_{x,b}$, which is the dimensionality of the reduced low dimensional space. The internal features after dimensionality reduction contain information within the sub blocks and do not transmit or affect other sub blocks.

Step 3: The dimension of external variable is reduced. All the variables of other sub blocks except sub block b are taken as the External variable of sub block b, recorded as $\boldsymbol{y}_b$, the corresponding sample data is $\boldsymbol{Y}_b \in R^{l_{\bar{b}} \times N}$, and $l_{\bar{b}}$ is the number of external variables of sub block b. The external variable group $\boldsymbol{y}_b$ contains information related to $\boldsymbol{x}_b$, and the variables in $\boldsymbol{y}_b$ are also related. Therefore, by analogy with $\boldsymbol{x}_b$, $\boldsymbol{y}_b$ can be processed in the same way to obtain a low dimensional matrix $\boldsymbol{t}_y$. $k_{y,b}$ is the dimension projected by $\boldsymbol{y}_b$. Reduce the dimensionality of the sample data corresponding to $\boldsymbol{x}_b$ and $\boldsymbol{y}_b$ respectively to obtain the corresponding score matrices as $\boldsymbol{T}_x \in R^{k_{x,b} \times N}$ and $\boldsymbol{T}_y \in R^{k_{y,b} \times N}$.

Step 4: External feature extraction. The principal component space of the internal variable and the principal component space of the external variable of sub block b are obtained, which respectively contain the maximum information of the internal and external variable. In order to extract the most relevant information between the external variable and the internal variable, the Canonical correlation analysis is carried out on $\boldsymbol{T}_{\mathrm{x}}$ and $\boldsymbol{T}_{\mathrm{y}}$, and the most relevant Canonical correlation feature is determined according to the contribution rate of the Canonical correlation coefficient, so as to construct the external correlation feature.

Step 5: External feature transmission. In order to transfer the external features of sub block b to the model construction of the next adjacent sub block, the extracted external feature $\boldsymbol{f}_b$ is input as the external variable of sub block $b+1$, and the feature extraction process of sub block is carried out, that is, the external variable of sub block $b+1$ is the combination of the original external variable $\boldsymbol{y}_{b+1}$ and the external related feature $\boldsymbol{f}_b$ of the previous sub block b, that is, $\boldsymbol{y}'_{b+1} = [\boldsymbol{y}_{b+1}, \boldsymbol{f}_b]$. So, when the sub block $b+1$ is modeled, the external variable replaces the original $\boldsymbol{y}_{b+1}$, that is, the inputs are $\boldsymbol{x}_{b+1}$ and $\boldsymbol{y}'_{b+1}$ respectively. After the same processing, the relevant features of sub block $b+1$ are extracted and transferred to adjacent sub blocks, and so on, forming a feature transfer chain of sub blocks.

Step 6: Fusion of internal and external features. Combine the internal and external features of each sub block to obtain the internal and external features of the entire process.

3.5 Knowledge graph and the construction of association network

In response to the difficulties in transforming implicit knowledge and tracing cross stage information in the evolution and update of the life cycle, the focus is on researching: (1) based on the spatiotemporal evolution characteristics of HSMP hierarchical cross domain element interconnection, research the construction method of HSMP causal correlation network; (2) considering the real-time updating and evolution of multi-stage and multi-source knowledge, research the transformation method of tacit knowledge, and build a dynamic life cycle Knowledge graph; (3) based on the dynamic evolution of the life cycle Knowledge graph and the HSMP causal correlation network, the organizational management form and storage update strategy of cloud information are studied to achieve cross stage traceability of information[26].

Each stage of the lifecycle has its own unique data flow, models, and related knowledge, which creates obstacles to information sharing between each stage. Upstream data cannot support downstream engineering activities, and downstream data is difficult to flow back to the upstream for guidance and optimization, resulting in low operational efficiency, high maintenance costs, and difficult interaction and comprehensive state tracing between digital and physical spaces. Based on HSMP multi-source heterogeneous data and cross domain multidimensional and multi granularity models, methods such as graph theory, causal analysis, and neural networks are used to clarify the nonlinear causal relationships between models and data at different levels. Combined with the deep causal graph modeling method based on probability prediction models, an HSMP causal relationship network is constructed to mine implicit knowledge in the data and model; In view of the multi-stage dynamic evolution characteristics of knowledge in the life cycle of R&D design, production manufacturing, overhaul and maintenance, the domain ontology based tacit knowledge transformation is adopted, combined with Natural language processing, machine learning and other methods, to realize the construction of the time series dynamic life cycle Knowledge graph. Based on the dynamic and time-varying Knowledge graph and HSMP associated network, a real-time updated and adjusted life cycle data model knowledge Cloud storage library is built to support the access, transformation, analysis and traceability of information from various systems in the HSMP life cycle. The schematic diagram of knowledge graph and the construction of association network is shown in **Fig. 3.13**.

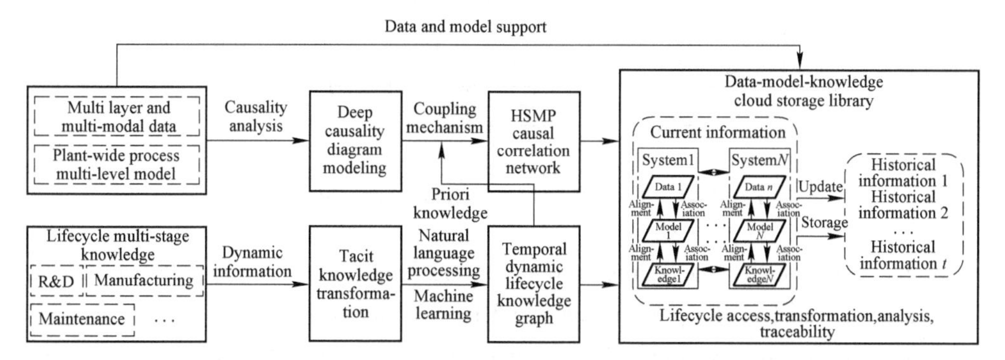

Fig. 3.13 Schematic diagram of knowledge graph and the construction of association network

3.5.1 Knowledge graph-based tacit knowledge mining and transformation

The knowledge graph is a structured representation composed of entities, relationships and their semantic descriptions. It is usually represented by triple groups, in which entities represent concepts or objects in reality, and relationships represent semantic descriptions between entities. The construction of knowledge graph is the key to the structural representation of domain knowledge. The main sources of knowledge are structured data, semi-structured data, and unstructured data.

Structured data and semi-structured data, such as production logs and fault logs, need to be transformed into structured knowledge through Named-entity recognition and relationship extraction. Pipeline knowledge extraction has the problem of error transmission. Joint extraction solves the problem of error transmission by constructing two subtasks of named-entity recognition and relationship extraction into multi task learning. The Bert-BiLSTM-CRF joint extraction model based on parameter sharing method is shown in **Fig. 3. 14**.

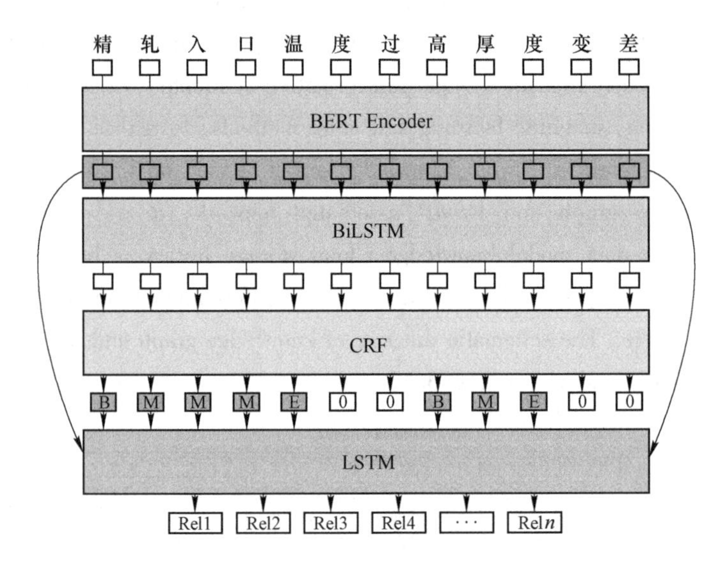

Fig. 3. 14 Bert-BiLSTM-CRF joint extraction model

In the named-entity recognition sub task, the label of the current data is mainly related to the label of the previous data and the characteristics of the current data. In order to improve the reliability and accuracy of entity sequence labeling, the linear chain conditional Random field is used to solve the best sequence of named entities. Assuming that the input data is $\boldsymbol{x}$, the entity sequence is labeled as $\boldsymbol{y}$, the linear chain conditional Random field is represented as $p(\boldsymbol{y}|\boldsymbol{x})$, and $f_k(\boldsymbol{y}_{i-1}, \boldsymbol{y}_i, \boldsymbol{x}, i)$ is used to represent the characteristic function:

$$f_k(\boldsymbol{y}, \boldsymbol{x}) = \sum_{i=1}^{n} f_k(\boldsymbol{y}_{i-1}, \boldsymbol{y}_i, \boldsymbol{x}, i) \tag{3.37}$$

The weight coefficients corresponding to $f_k(\boldsymbol{y}_{i-1}, \boldsymbol{y}_i, \boldsymbol{x}, i)$ are as follows:

$$\omega_k = \begin{cases} \lambda_k & k = 1, 2, \cdots, K_1 \\ \mu_l & k = K_1 + l, l = 1, 2, \cdots, K_2 \end{cases} \tag{3.38}$$

So the linear chain condition random field can be simplified as:

$$p(\boldsymbol{y} \mid \boldsymbol{x}) = \frac{1}{Z(\boldsymbol{x})} \exp \sum_{k=1}^{K} \omega_k f_k(\boldsymbol{y}, \boldsymbol{x}) \tag{3.39}$$

where, $Z(\boldsymbol{x})$ is the normalization factor:

$$Z(\boldsymbol{x}) = \sum_{y} \exp \sum_{k=1}^{K} \omega_k f_k(\boldsymbol{y}, \boldsymbol{x}) \tag{3.40}$$

Finally, $\hat{\boldsymbol{y}}$ is taken as the last output series label, and its maximum probability, i.e. the global optimal sequence, is solved through the Dynamic programming algorithm:

$$\hat{\boldsymbol{y}} = \arg \max_{y} p(\boldsymbol{y} \mid \boldsymbol{x}, \boldsymbol{\omega}) \tag{3.41}$$

Structured data, such as process data, features, and states stored in MySQL, can generally be used as the knowledge structure required in the knowledge graph after simple transformation, and often be added to the information of corresponding nodes and edges as attributes and attribute values.

3.5.2 Construction of causal correlation network for HSMP

The finishing mill control system mainly includes thickness control subsystem, shape control subsystem, speed control subsystem, and temperature control subsystem. Based on the above mechanism analysis and association pattern mining, a causal relationship network between variables in each subsystem was established using the idea of joint driving of data and mechanism, providing a model foundation for the subsequent analysis of the entire process operation mechanism and the revelation of multidimensional and multi granularity model mapping associations[27].

3.5.2.1 Thickness control system

There are many factors that affect the thickness change of sheet metal. To design a well-designed thickness control system, it is necessary to first understand the main factors that affect the thickness change, and then take corresponding measures. The factors that affect the thickness of sheet metal mainly come from the rolling mill and the rolled piece.

A The factors from the rolling mill

(1) The variation of rolling mill constant. The rolling mill constant is mainly determined by the manufacturer and varies with the production of the rolling mill constant.

Nowadays, hot continuous rolling units usually have a relatively large rolling mill constant, which to some extent reduces the amount of bounce changes during the rolling process, which is beneficial for improving the accuracy of thickness control. (2) Roll thermal expansion and wear. During the rolling process, the thermal expansion of the work roll can reach 200. The thermal expansion is a function of the roll material and rolling time. Even when rolling the same strip, the thermal expansion is different. The thermal expansion during threading is relatively small, reaching its maximum during stable rolling; Wear and tear are also related to the material of the roller, and also to the length of the strip steel. The longer the strip steel, the more severe the wear, and the better the material of the roller, the less wear. (3) The influence of roll eccentricity. Roll eccentricity, especially the eccentricity of the support roller, is one of the main factors affecting thickness accuracy. It has the opposite effect to automatic gauge control (AGC), as roll eccentricity can easily mislead AGC operations, reduce thickness accuracy, and even cause accidents. (4) The influence of rolling speed. Rolling speed has a certain impact on rolling temperature, deformation resistance, and oil film. If the accuracy of rolling speed setting is low, it is easy to cause significant fluctuations in the final rolling temperature, and the temperature fluctuations of each frame are also large, leading to a decrease in thickness control accuracy. (5) Changes in the oil film of the support roller. The change in oil film thickness during production can reach 200-300. The oil film thickness is a function of rolling speed and rolling force. Generally speaking, the oil film thickness increases with the increase of rolling speed, but decreases with the increase of rolling force. In general, hot rolling mills are equipped with oil film compensation circuits.

B The factors from rolling

(1) Changes in incoming material thickness. According to rough calculations, when the thickness of the slab after rough rolling is about 30, the actual deviation in thickness after rough rolling can reach 500-1000. Nowadays, some manufacturers have installed thickness measurement systems behind rough rolling mills, which have to some extent solved the thickness fluctuations of rough rolling slabs changes in material properties. (2) During the heating process, there is a certain temperature difference (caused by watermarks, etc.), and uneven cooling during the rolling process can lead to different temperatures in different zones of the same steel strip. This will cause differences in the characteristics of the strip steel, specifically manifested as fluctuations in the plasticity coefficient changes in tension. (3) Tension is a relatively active factor that can cause changes in rolling force, causing fluctuations in strip thickness. Some manufacturers

also compensate for this change by compensating for tension and have achieved certain results.

C Causal correlation network

The factors listed above affect the accuracy of thickness control. In order to improve the accuracy of thickness control, it is necessary to correctly understand the impact of these factors on the thickness control system. Based on the above mechanism knowledge, expert experience, and data feature analysis results, corresponding variables are found in the actual control system, and a causal correlation network is constructed using the idea of jointly driving mechanism and data, as shown in **Fig. 3.15**.

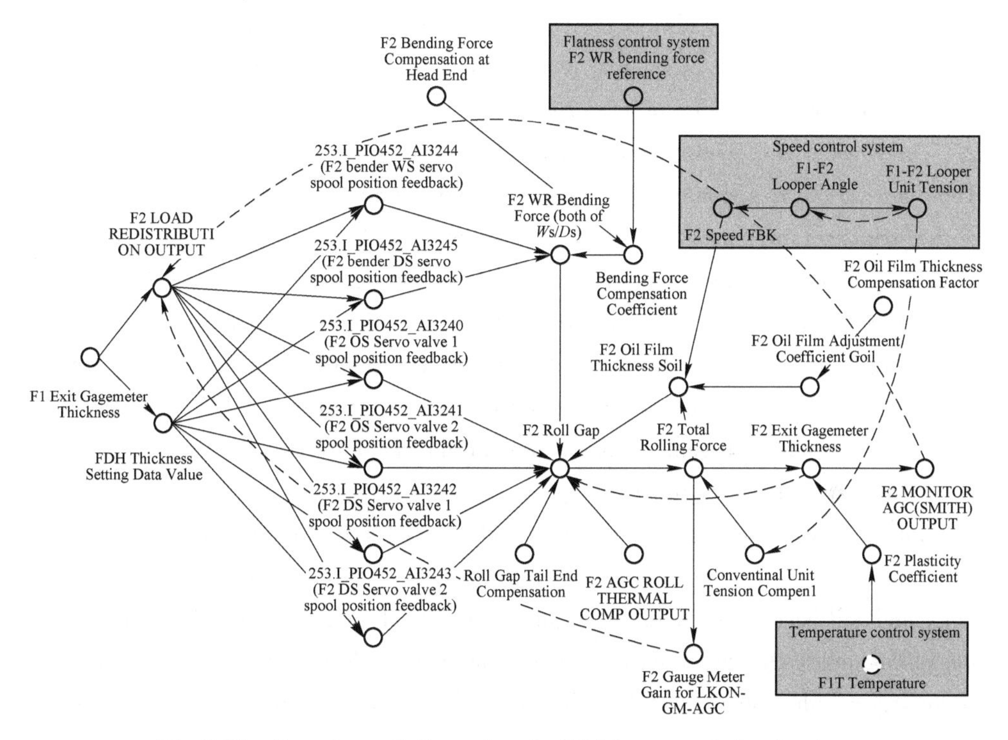

Fig. 3.15 Causal correlation network of thickness control subsystem

3.5.2.2 Shape control system

Plate thickness and shape are two important indicators for the accuracy of hot-rolled strip products. With the widespread application of AGC technology, the thickness control of strip steel has achieved high accuracy, while the issue of plate shape is receiving increasing attention. Especially for finished thin plates such as automotive steel plates, tinned steel plates, and silicon steel plates, the requirements for plate shape are very high. If the steel plate has poor shape, defects such as waves, warping, and edge thinning will seriously affect the quality and service life of the product.

In order to achieve high-precision control of plate shape, the study of plate shape control models is particularly important. In plate shape control, the cross-sectional shape and strip flatness are the two most important indicators, and these two indicators affect each other. When changing the cross-sectional shape of the plate and strip, it often causes flatness defects such as warping and waviness; Moreover, the cross-sectional shape is influenced by various factors, such as the thermal expansion and wear of the rollers, rolling force, etc., making it difficult to achieve precise control. Therefore, the control of hot rolled strip shape has always been a challenge, and research on shape can provide technical support for the production of high value-added, high-quality, ultra-thin strip steel and the implementation of advanced rolling technology.

A The concept and control characteristics of plate shape

There are various descriptions of plate shape, and the cross-sectional shape and flatness of the strip steel are currently the two most important indicators for describing the shape of the strip steel plate. The cross-sectional shape reflects the external geometric characteristics of the strip steel along the width direction, mainly described by convexity; Flatness reflects the external geometric characteristics of the strip steel along the length direction. These two indexes are not isolated from each other, but affect and transform each other. Together, they determine the shape quality of strip steel.

B Plate shape control equipment

The rolling mill model is a key link in plate and strip rolling production, which determines the first factor and foundation of the shape control performance of the rolling mill, and will have a long-term impact on the quality of the shape control performance of the rolling mill. On the basis of conventional four high rolling mills, some modern technologies with more powerful flatness control capabilities have been developed one after another. The commonly used flatness control devices currently include two types: bending roller devices and changing roller shape.

C Crown calculation of roll system

The convexity of the roll system includes the convexity between the work roll systems and the convexity from the work roll to the support roll system, which directly determines the outlet shape of the strip steel. When studying the crown ofthe roll system, there are two situations: unloaded and loaded. Empty load is defined as the absence of steel strip in the roll gap, and its comprehensive roll shape is influenced by factors such as the initial roll shape, thermal crown, wear crown, roll grinding crown, and roll shifting position related to grinding crown; In addition to the above factors, the loaded roll shape also needs to consider the impact of rolling force and bending force on the roll.

D Causal Association Network

Based on mechanism knowledge and empirical knowledge, two important indicators of plate shape, convexity and straightness, are identified. They involve numerous feedback and related variables in the measurement process. For example, rolling speed, rolling temperature, bending force, and shifting force can all simultaneously affect crown and straightness. And these two important indicators will affect each other, ultimately affecting the flatness indicators of steel rolling. Based on the above series of mechanism experience knowledge, draw the control loop of the flatness control system, find the corresponding variables in Iba, and use the idea of mechanism and data driven jointly to construct a local causal network for a single rack. Through the analysis of system operation mechanism, construct the final causal correlation network, as shown in **Fig. 3.16**.

3.5.2.3 Temperature control system

Temperature is also an important process parameter in HSMP, which directly affects the rolling force. Therefore, in order to ensure the thickness and shape of the plate, it is necessary to accurately predict the rolling temperature of each frame of the finishing mill. In the setting calculation, the inlet temperature of each finishing mill frame should be pre estimated as soon as the steel billet reaches the final rough rolling frame. Considering that different specifications of steel require different final rolling temperatures, all factors that affect the final rolling temperature (such as descaling, inter frame water spraying, threading speed, acceleration, and maximum speed) require corresponding values to be set accordingly, which is called temperature setting.

A Thermal radiation model

In the setting model studied, the thermal radiation structural unit establishes the temperature transfer model through the thickness (cross section) of the strip steel, and calculates the heat transferred from the strip steel surface to the surrounding based on the Steven Boltzmann radiation equation, using the Finite difference method in the model calculation. For steel plates whose thickness cannot be ignored, the radiation surface area should also include its side edges. The model can simulate convective heat transfer to the rolling mill, as well as other heat transfer using conventional heat transfer coefficients. The model can also simulate the temperature distribution through the thickness of strip steel.

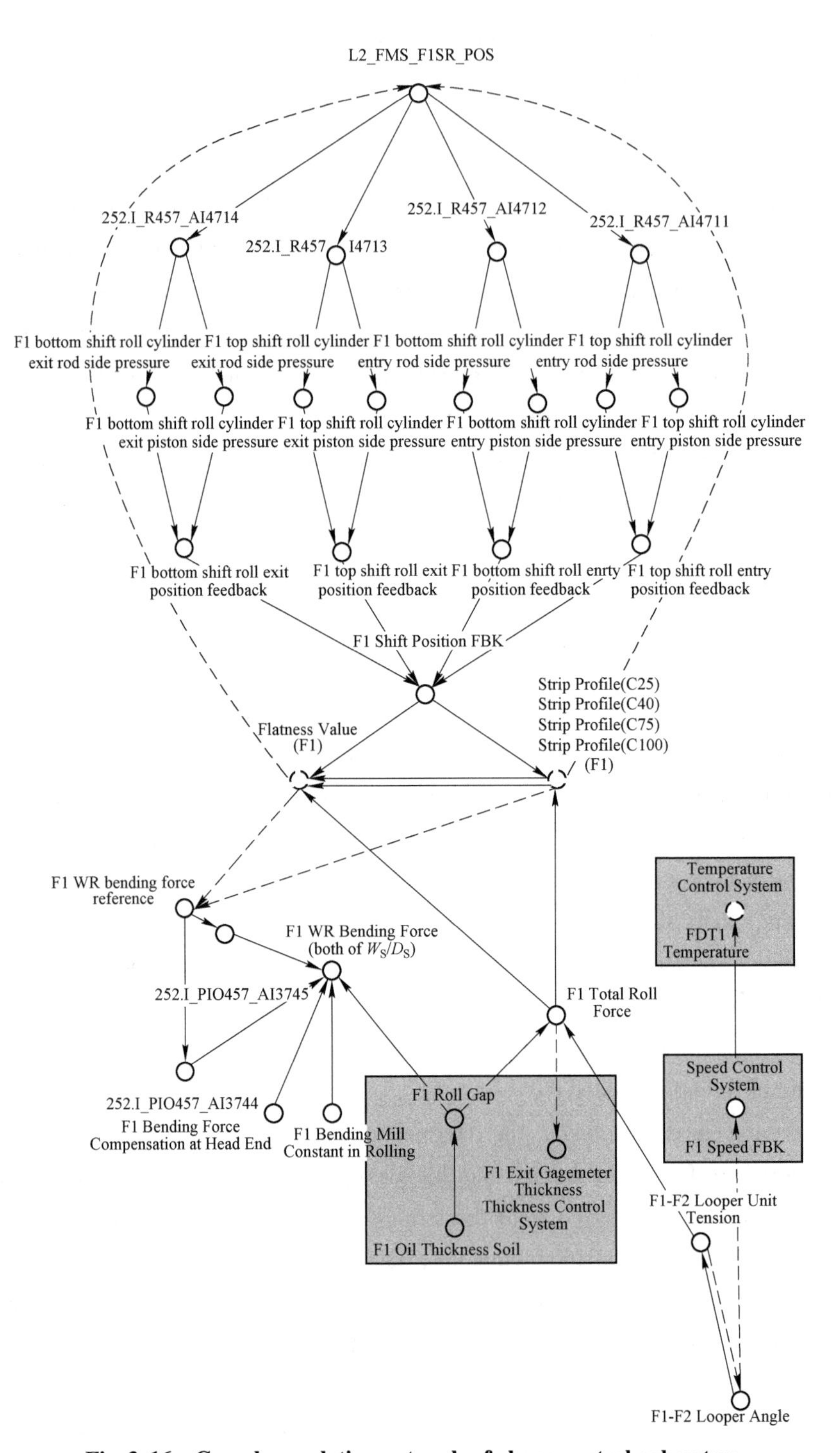

Fig. 3. 16 Causal correlation network of shape control subsystem

B Spray model

The precision rolling setting uses a spray model to calculate the temperature drop after spraying during the movement of the strip steel. The heat removal rate is directly proportional to the thermal conductivity coefficient, the temperature difference between the rolled piece and the spray temperature, and the area where the rolled piece is exposed to air. This model is based on the Newton convection equation. The "jet efficiency", as a function of flow, is used to calculate the convection coefficient, and the Finite difference method is used in the calculation.

C Conduction, deformation, and friction temperature models

When the roller bites into the rolled piece, it is necessary to calculate the temperature change of the rolled piece as the roller bites into the internal temperature of the rolled piece. The heat transfer of the roll biting into the rolled piece consists of three parts: heat conduction, deformation heat, and friction heat. Heat conduction is caused by the temperature difference between the rolled piece and the working roll; Deformation heat is the energy demand that causes deformation of the rolled piece; Friction heat is generated between the rolled piece and the working roll as it moves. This calculation process can simulate the temperature distribution through the thickness of rolled piece. The Finite difference method is used in the calculation

D Causal correlation network

In summary, the means of regulating temperature are mainly divided into speed and cooling water between racks. When adjusting the speed, when the finish rolling head reaches the finish rolling outlet temperature, feedback calculation begins and the acceleration is adjusted based on the temperature difference. The controller uses temperature deviation, linear speed, sampling speed, and reference acceleration as input signals to control the main transmission device to change the acceleration of the rolled piece. Adaptive control is reflected in its use of real-time information to calibrate the model, making it closer to the actual state and improving the accuracy of model prediction and control.

When using cooling water regulation, it is also divided into feedforward and feedback control. Feedforward control starts when the strip steel enters the phosphorus removal position during precision rolling. By predicting the deviation of the finish rolling inlet temperature and the current velocity time data through the finish rolling inlet temperature and intermediate roller conveyor model, the water spray mode is pre adjusted before the strip enters the rolling process to correct the deviation caused by the inlet temperature at the finish rolling stand position. During feedback control, the deviation between the outlet temperature and the secondary setting generates a control quantity, which acts on the coolingwater valve to control the amount of water sprayed, thereby changing the temperature. The causal correlation network of the rack is constructed through a mechanism driven and data driven approach, as shown in **Fig. 3. 17**.

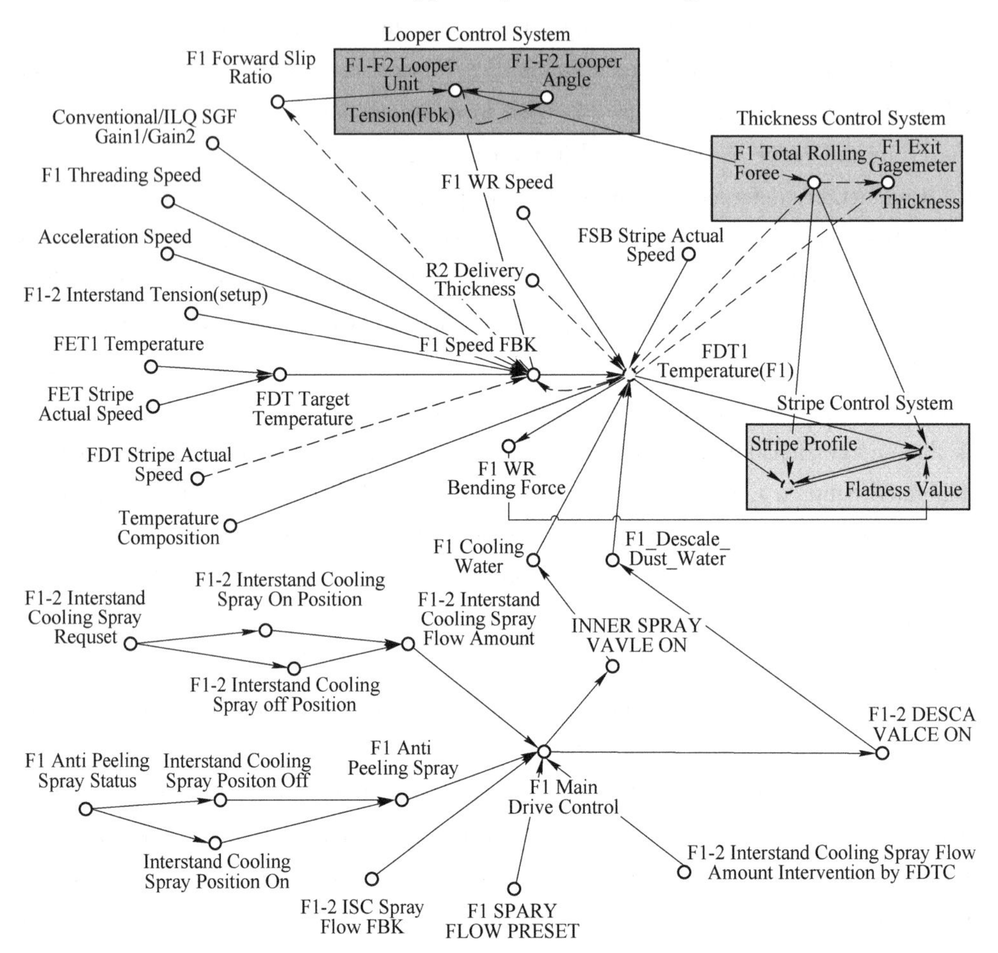

Fig. 3. 17 Causal correlation network of temperature control subsystem

3. 5. 2. 4 Speed control system

The modern HSMP finishing mill has two basic characteristics: constant loop quantity and constant low tension rolling; During the rolling process, due to the dynamic biting speed drop of the main transmission system, there may be various external disturbances in the stable state during rolling. It is impossible to maintain a constant speed ratio relationship between each frame. Therefore, a loop is set up to detect changes in the number of loops of the rolled piece between frames due to speed deviation, Able to absorb changes in the quantity of these loops and maintain a constant small tension, ensuring smooth production. The looper device not only controls the looper quantity, but also controls a constant small tension.

A Tension model

Under normal circumstances, the tension received during the continuous rolling process is actually very small, just to maintain normal threading and prevent deviation, the tension is about 10% less than the high-temperature deformation strength.

Tension affects thickness accuracy to a certain extent. There is a certain interference between tension control and thickness control, especially during the threading process, excessive tension fluctuation can cause vibration of the rolling mill, leading to AGC not being put into normal use. Moreover, during the steady-state period of hot rolling, AGC adjustment of thickness changes will lead to changes in tension, and on the contrary, changes in tension also cause changes in thickness to some extent. The lateral tension and loop height control of the loop system ensure the goal of constant tension and equal flow rate per second.

B Automatic control of loop height

a Calculation of looper quantity

In the hot continuous rolling unit, the amount of steel loop formed between two stands is an undetectable quantity, which can only be indirectly obtained through the angle of the loop. So, the control accuracy of the loop control system largely depends on selecting a high-precision loop model.

Due to the dynamic speed drop of the motor during the continuous rolling process, a fixed loop quantity is inevitably generated, where the loop quantity is defined as $l = AB + BC - AC$:

$$l = AB + BC - L = \frac{L1 + R\cos\theta}{\cos\alpha} + \frac{L1 + R\cos\theta}{\cos\alpha} \tag{3.42}$$

$$\cos\alpha = \frac{L1 + R\cos\theta}{\sqrt{(L1 + R\cos\theta)^2 + (R\sin\theta - L3 + r)^2}} \tag{3.43}$$

$$\cos\beta = \frac{L - L1 - R\cos\theta}{\sqrt{(L - L1 - R\cos\theta)^2 + (R\sin\theta - L3 + r)^2}} \tag{3.44}$$

$$\alpha = \mathrm{tg}^{-1} \frac{R\sin\theta - L3 + r}{L1 + R\cos\theta} \tag{3.45}$$

$$\beta = \mathrm{tg}^{-1} \frac{R\sin\theta - L3 + r}{L - L1 - R\cos\theta} \tag{3.46}$$

$$l = \sqrt{(L1 + R\cos\theta)^2 + (R\sin\theta - L3 + r)^2} + \sqrt{(L - L1 - R\cos\theta)^2 + (R\sin\theta - L3 + r)^2} - L \tag{3.47}$$

where, R is the arm length of the loop roller; L is the distance between frames; r is the radius of the loop roller; θ is the swing angle of the loop roller; α and β are the included angle between the strip steel and the rolling line; $L3$ is the distance between the loop torque support point and the rolling line.

b Loop height control principle

In the loop height control system, the loop support is used as a loop quantity detection device to achieve loop quantity feedback. The difference between the feedback value and the set value is input to the loop height adjuster, which is calculated according to the control algorithm to generate the adjustment quantity output to the main transmission control system of the upstream frame of the loop. By adjusting the speed of the upstream frame, the strip steel loop quantity approaches the set value. It mainly consists of four parts: benchmark setting Ⅰ, height detection Ⅱ, control link Ⅲ, and control object Ⅳ. When the loop height control is in a closed loop state, the tension control system is in a current single closed loop state, with fast response and no delay in angle detection. The inertia of the loop tension control system can also be ignored.

C Causal correlation network

First, according to the mechanism knowledge and experience knowledge, for example, the important indicator of looper system control is the constant micro tension of strip steel between different racks, and the looper angle control is to adjust the servo valve opening of looper system according to the speed cascade relationship. On this basis, using the idea of mechanism and data combined drive, the causal correlation network of the rack is constructed, as shown in **Fig. 3. 18**.

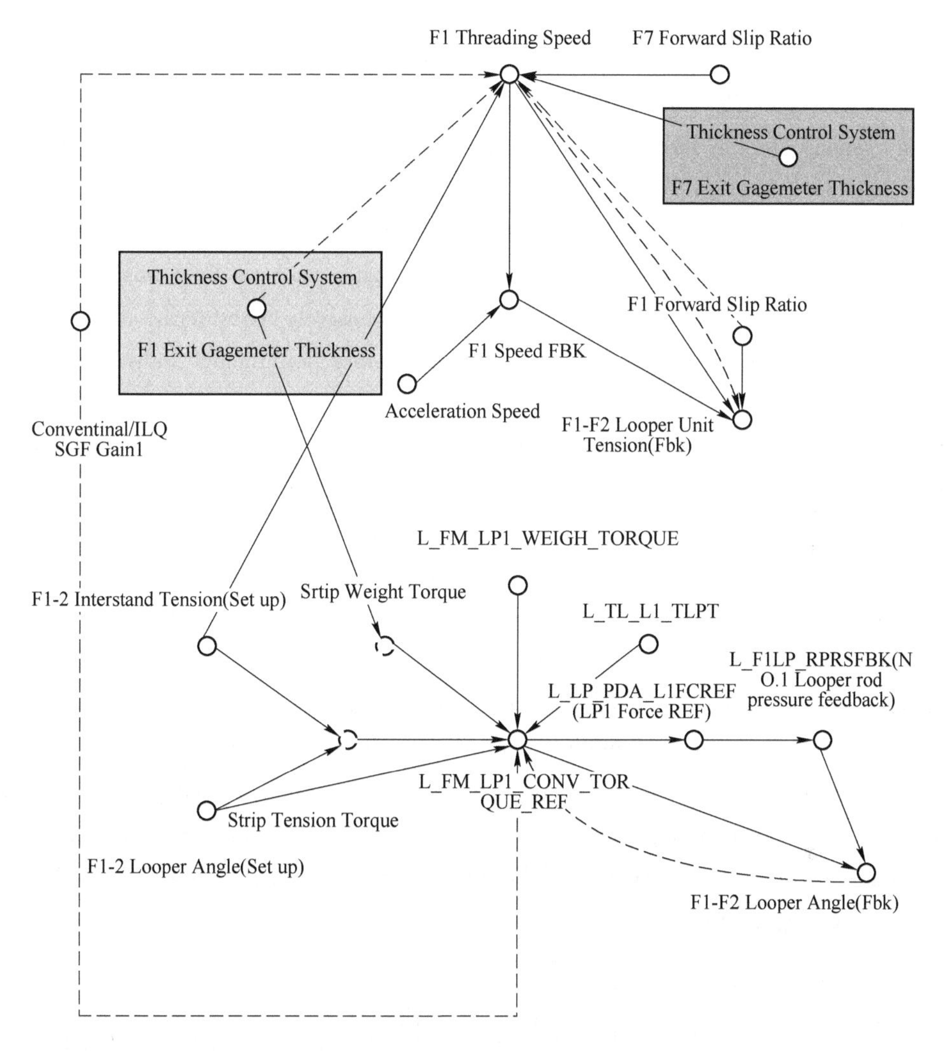

Fig. 3.18 Causal correlation network of loop control subsystem

3.6 Construction of the lifecycle data management platform for HSMP

Focusing on the characteristics of multiple sources, large scale, and high real-time requirements of HSMP lifecycle data, the research focuses on: (1) the construction of a real-time, high-speed, and highly reliable transmission network for multi-source concurrent data, supporting the parsing, conversion, and compatibility of various communication protocols. (2) Establish a Distributed database to achieve global consistency and interface unification of cross stage data, and support efficient data storage, retrieval and parallel computing. (3) Efficient collaborative computing and dynamic resource allocation in clusters under large-scale data processing tasks.

In response to the large volume, multiple sources, high real-time and reliability requirements of HSMP lifecycle data, data management platforms have been established for satellite manufacturing and steelmaking continuous casting continuous rolling processes. Combining transmission technologies such as field bus, Industrial Ethernet, wireless network, etc., and comprehensively considering transmission efficiency, energy balance, reliability, delay and other performance indicators, build a "cloud edge end" real-time data transmission network, use redundant fault tolerance technology to ensure strong real-time, high robustness and high security of multi-source concurrent data transmission, and use OPC UA architecture to achieve cross system Cross level interconnectivity and compatibility and conversion of multiple communication protocols; Based on the Hadoop architecture, the HDFS multi replica distributed storage architecture is used to build a Distributed database, giving consideration to efficient massive data throughput and low latency random access capabilities, to achieve data organization, storage, analysis, calculation and query; Fully utilize the resources of multiple computing nodes in the cluster, improve platform computing efficiency based on Spark distributed computing technology, and utilize Yarn distributed scheduling technology to ensure balanced cluster computing load. The schematic diagram of construction of the lifecycle data management platform for HSMP is shown in **Fig. 3. 19**.

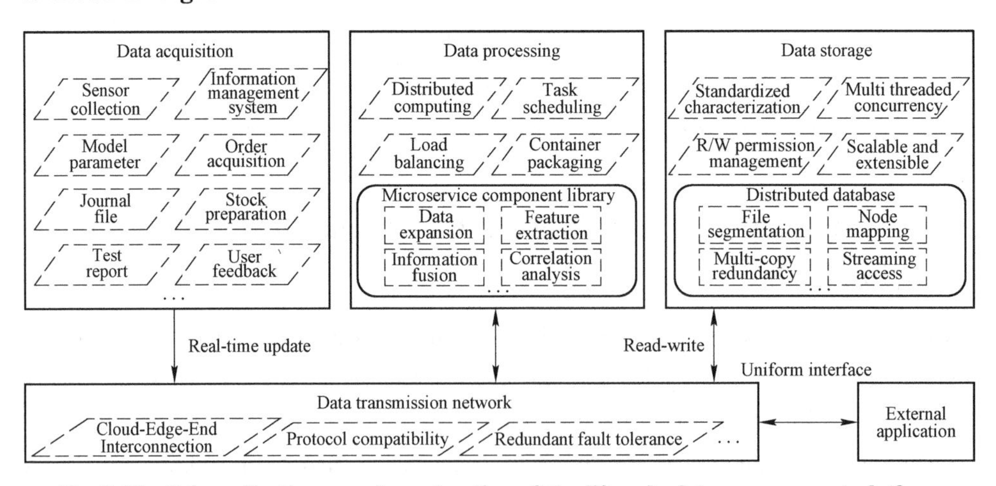

Fig. 3. 19 Schematic diagram of construction of the lifecycle data management platform

3.6.1 Overall architecture of digital thread platform

The overall architecture of the digital thread platform is shown in **Fig. 3.20**. The digital thread is composed of a high-performance server, with pSpace real-time database as the core. Through its embedded interface tools such as ODBCRouter and SQLRouter, various databases such as the temporal database InfluxDB, relational database MySQL, and graphics database Neo4j are interconnected for read and write access on the end, edge, and cloud sides. Assists multiple PC end workstations to deploy the functional modules involved in this topic, such as multi-source heterogeneous data conversion, non ideal data expansion, spatio-temporal feature extraction, etc. After effective integration and utilization of various mechanism models, process knowledge, and multi-source heterogeneous data throughout the entire lifecycle, they are stored in corresponding databases. For example, the Knowledge graph and associated network of the manufacturing process are stored in the Graph database Neo4j to realize the association and traceability between the elements of the whole life cycle. The topology diagram and setting information of the manufacturing process design mechanism model are stored in the relational database MySQL, laying the foundation for the study of correlation and collaborative analysis between multi-stage, multi-process, and multi-system models[28].

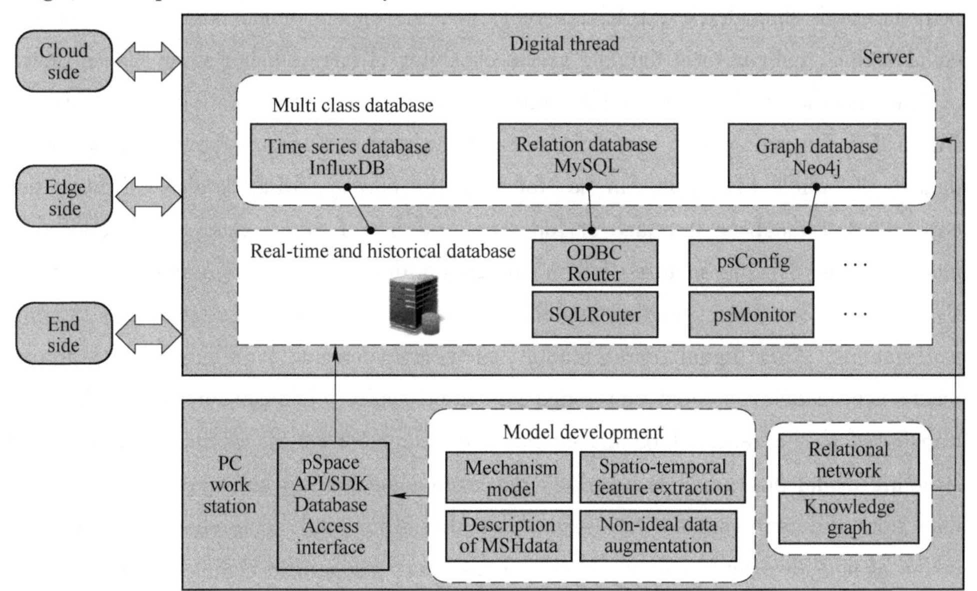

Fig. 3.20 Basic architecture of digital thread platform

3.6.2 Specific deployment plan of digital thread platform

The first step of a digital thread platform should be information acquisition, utilizing existing information systems and data collection equipment to obtain data from various stages, systems, and processes of the entire lifecycle of complex products, providing data support for subsequent collaborative scheduling and optimization. Improving the construction of data transmission networks requires the analysis, conversion, and compatibility of various communication protocols. In the communication mode of equipment end, it is proposed to use OPC UA, Modbus TCP and other communication protocols under the standard Industrial Ethernet and field bus specifications in combination with the existing communication standards in the actual production process. In addition, different industrial hardware devices should also consider their own communication protocols, including PLC, board module, intelligent instrument, etc. The overall communication deployment is shown in **Fig. 3.21**. It is proposed to deploy an integrated embedded integrated Industrial internet of things gateway at the edge end. It is proposed to use high-performance embedded hardware gateways to assist in providing different access schemes for data acquisition, pre-processing, protocol conversion, multiplexing and forwarding of different Industrial internet of things devices by means of industrial software gateway communication, etc., so as to achieve edge data communication. Cloud integration can connect various IoT gateways or terminal devices in both directions through OPC UA, Ethernet, etc. It supports cloud edge integration and collaboration, and can meet the data access capability of large number of concurrent links.

The specific deployment architecture of the digital thread platform is shown in **Fig. 3.22**. The mechanism model, process knowledge and multi-source heterogeneous data in the whole life cycle can be stored in the corresponding database through the corresponding database interface or Industrial internet of things gateway. In addition, based on the existing mature Python Integrated development environment (such as PyCharm, etc.), corresponding model development is carried out in multiple PC end workstations. After digital representation, element fusion and knowledge association of mechanism models, process knowledge and multi-source heterogeneous data, they are transferred to the required database. Some development interfaces includ a metadata structure model that provides a unified description mechanism and operation model for data from different sources and structures; the filling and comprehensive expansion model of non-ideal multi-source heterogeneous data; mechanism model linkage mapping and spatio-temporal collaboration Relational model; the spatiotemporal feature extraction model of manufacturing process data, as well as the Knowledge graph and associated network of manufacturing process are constructed.

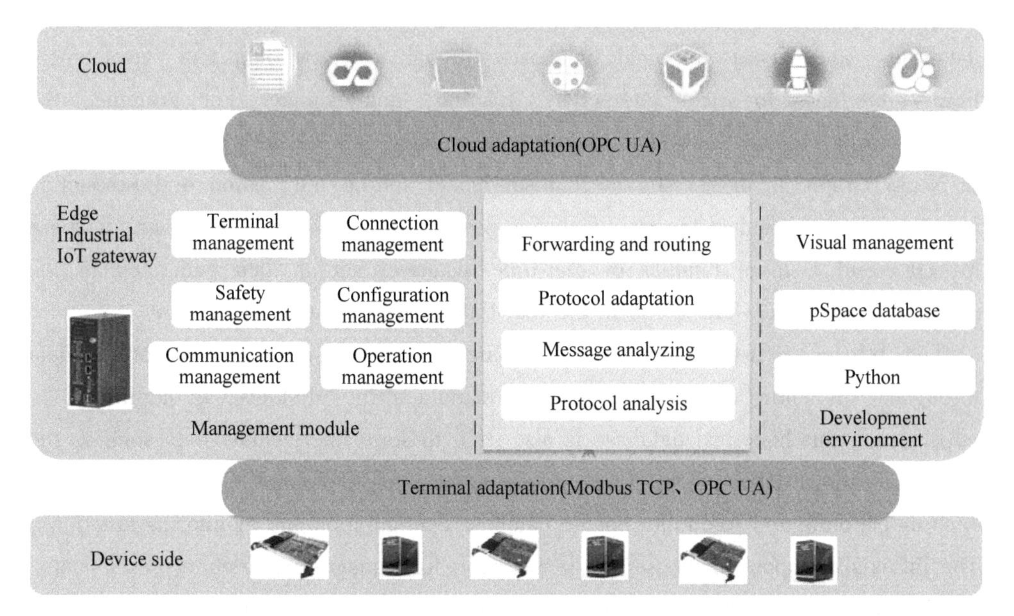

Fig. 3.21 Overall communication deployment diagram

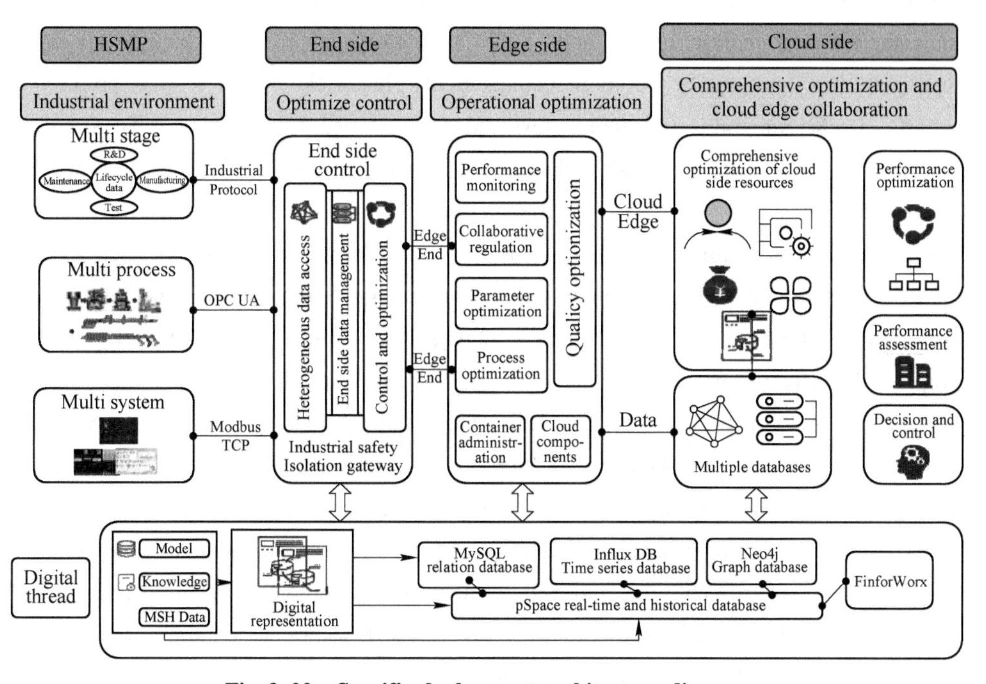

Fig. 3.22 Specific deployment architecture diagram

The existing various development models can be extended and connected to pSpace real-time historical databases, InfluxDB temporal databases, MySQL relational databases, and Neo4j graphical databases through corresponding tool development library interfaces for data interaction, reading, and storage. For example, the manufacturing process Knowledge graph and associated network are uploaded and stored in Neo4j Graph database, the mechanism model setting information and production scheduling information in the manufacturing process are stored in the relational database MySQL, and a large number of real-time/historical sensor data generated in the manufacturing process are stored in the real-time historical database pSpace.

The pSpace real-time historical database stores real-time and historical data throughout the entire lifecycle of complex manufacturing products. In addition, the pSpace real-time historical database is also used to store intermediate data such as the results of various development model processing and model parameters. Its built-in tools such as psConfig and psMonitor enable database configuration and monitoring functions. The InfluxDB temporal database is a key support for managing production process data. It automatically collects and stores data, knowledge, and models processed from multiple stages such as research and development design, production manufacturing, and operation maintenance. Combined with visualization tools, it can provide multiple management and monitoring methods such as factory models, production operation management, equipment operation management, historical review, and production reports. It also integrates and interacts with MySQL relational database and Neo4j Graph database to form a comprehensive database. In the interactive integration of MySQL relational database and pSpace real-time historical database, data is saved in different tables, and these tables are placed in different databases for quick access and management through SQL structured query statements. Neo4j is an embedded, high-performance, lightweight Graph database management system used to store Knowledge graph and associated networks of manufacturing processes. It supports various database connections and can be configured to query various databases. It supports querying the whole process production data and master data through data filters, covering model/semantic/spatio-temporal metadata, matching the process data spatio-temporal with Knowledge graph, and integrating Neo4j visual interface. The dump tools embedded in pSpace, such as ODBC Router and SQL Router, can complete data dump and unified docking with the InfluxDB temporal database, MySQL relational database, and Neo4j graphics database. In addition, the entire lifecycle data management platform of the manufacturing process adopts the FinforWorx integrated control platform, which is based on online low code development as an integrated control platform. It is based on the fusion of multiple types of databases and supports comprehensive data management. It enables simultaneous query of different types of databases such as InfluxDB chronological database, MySQL relational database, and Neo4j graphic database, integrating platform basic data, business data, process data Unified management and integration of statistical data and alarm data, achieving the sharing and exchange of data resources, can provide a collaborative information environment for further digital thread platform construction in the future.

3.7 Conclusions

This chapter has driven important technical progress in HSMP non ideal data expansion and feature fusion technology, unified description of multi-source heterogeneous data and multi-stage information digital representation, association mode analysis driven by data and mechanism collaboration, tacit knowledge mining and transformation, etc. It has preliminarily broken through the difficulties in model data network structure and knowledge modeling, completed the design of digital thread platform architecture, and achieved comprehensive integration of data sources, this provides theoretical and technical support for the construction of digital thread and the integration and deployment of related algorithms for the deep integration of model data knowledge in the future.

References

[1] Chu F, Zhao X, Dai W, et al. Data-driven robust evaluation method for optimal operating status and its application [J]. Acta Automatica Sinica, 2020, 46 (3): 439-450.

[2] Serneels S, Croux C, Filzmoser P, et al. Partial robust M-regression [J]. Chemometrics & Intelligent Laboratory Systems, 2000, 79: 55-64.

[3] Xie X, Sun W, Cheung K C. An advanced PLS approach for key performance indicator-related prediction and diagnosis in case of outliers [J]. IEEE Transactions on Industrial Electronics, 2015, 63 (4): 2587-2594.

[4] Murray J S. Multiple imputation: A review of practical and theoretical findings [J]. Statistical Science, 2018, 33 (2): 142-159.

[5] Yang C, Zhou K, Liu J. SuperGraph: Spatial-temporal graph-based feature extraction for rotating machinery diagnosis [J]. IEEE Transactions on Industrial Electronics, 2021, 69 (4): 4167-4176.

[6] Zhao C, Huang B. A full-condition monitoring method for nonstationary dynamic chemical processes with cointegration and slow feature analysis [J]. AIChE Journal, 2018, 64 (5):

1662-1681.

[7] Zhao C. Perspectives on nonstationary process monitoring in the era of industrial artificial intelligence [J]. Journal of Process Control, 2022, 116: 255-272.

[8] Zhao C, Wang W, Tian C, et al. Fine-scale modeling and monitoring of wide-range nonstationary batch processes with dynamic analytics [J]. IEEE Transactions on Industrial Electronics, 2021, 68 (9): 8808-8818.

[9] Song B, Shi H, Tan S, et al. Multisubspace orthogonal canonical correlation analysis for quality-related plant-wide process monitoring [J]. IEEE Transactions on Industrial Informatics, 2021, 17 (9): 6368-6378.

[10] Fan W, Zhu Q, Ren S, et al. Robust probabilistic predictable feature analysis and its application for dynamic process monitoring [J]. Journal of Process Control, 2022, 112: 21-35.

[11] Yao L, Shao W, Ge Z. Hierarchical quality monitoring for largescale industrial plants with big process data [J]. IEEE Transactions on Neural Networks and Learning Systems, 2021, 32 (8): 3330-3341.

[12] Marino R, Wisultschew C, Otero A, et al. A machine- learning-based distributed system for fault diagnosis with scalable detection quality in industrial IoT [J]. IEEE Internet Things Journal, 2021, 8 (6): 4339-4352.

[13] Zheng Y, Huang T, Zhao X, et al. Tensor N-tubal rank and its convex relaxation for low-rank tensor recovery [J]. Information Sciences, 2020, 532: 170-189.

[14] Yao L, Ge Z. Scalable semi-supervised GMM for big data quality prediction in multimode processes [J]. IEEE Transactions on Industrial Electronics, 2019, 66 (5): 3681-3692.

[15] Zheng J, Song Z. Semi-supervised learning for probabilistic partial least squares regression model and soft sensor application [J]. Journal of Process Control, 2018, 64: 123-131.

[16] Dong J, Zhang W, Peng K, et al. A novel method of quality abnormality detection and fault quantitative assessment for industrial processes [J]. Acta Automatica Sinica, 2022, 48 (10): 1-10.

[17] Wu P, Lou S, Zhang X, et al. Quality-relevant process monitoring based on dynamic locally linear embedding concurrent canonical correlation analysis [J]. Industrial & Engineering Chemistry Research, 2020, 59: 21439-21457.

[18] Bao Y, Wang B, Guo P, et al. Chemical process fault diagnosis based on a combined deep learning method [J]. The Canadian Journal of Chemical Engineering, 2022, 100 (1): 54-66.

[19] Zhao R, Yan R, Chen Z, et al. Deep learning and its applications to machine health monitoring [J]. Mechanical Systems and Signal Processing, 2019, 115: 213-237.

[20] Dong J, Sun R, Peng K, et al. Industrial process quality monitoring method and application joint-driven by automatic encoder and canonical correlation analysis method [J]. Control Theory & Applications, 2019, 36 (9): 1493-1500.

[21] Zhu Y, Geng L. Research on SDG fault diagnosis of ocean shipping boiler system based on fuzzy granular computing under data fusion [J]. Polish Maritime Research, 2018, 25 (s2): 92-97.

[22] Berger U, Schwichtenberg H, Seisenberger M. The warshall algorithm and dickson's lemma:

Two examples of realistic program extraction [J]. Journal of Automated Reasoning, 2001, 26 (2): 205-221.

[23] Li P, Cheng K, Jiang P, et al. Investigation on industrial dataspace for advanced machining workshops: Enabling machining operations control with domain knowledge and application case studies [J]. Journal of Intelligent Manufacturing, 2022, 33 (1): 103-119.

[24] Lin X, Sun R, Wang Y. Improved key performance indicator-partial least squares method for nonlinear process fault detection based on just-in-time learning [J]. Journal of the Franklin Institute, 2023, 360 (1): 1-17.

[25] Zhang J, Zhou D, Chen M. Adaptive cointegration analysis and modified RPCA with continual learning ability for monitoring multimode nonstationary processes [J]. IEEE Transactions on Cybernetics, 2022, doi: 10.1109/TCYB.2021.3140065.

[26] Zheng Y, Huang T, Zhao X, et al. Tensor N-tubal rank and its convex relaxation for low-rank tensor recovery [J]. Information Science, 2020, 532: 170-189.

[27] Geng Z, Chen Z, Meng Q, et al. Novel transformer based on gated convolutional neural network for dynamic soft sensor modeling of industrial processes [J]. IEEE Transactions on Industrial Informatics, 2022, 18 (3): 1521-1532.

[28] Yuan X, Li L, Shardt Y, et al. Deep learning with spatiotemporal attention-based LSTM for industrial soft sensor model development [J]. IEEE Transactions on Industrial Electronics, 2021, 68 (5): 4404-4414.

Chapter 4 Quality-Related Distributed Fault Detection Based on QMRSFA

With the deep integration of informatization and industrialization as well as the increasing global competition, process industries such as semiconductor, and steelmaking are developing towards large-scale, complexity, and integration. In these process industries, multiple sub-processes are connected in series, and the output of the previous sub-process is used as the input of the latter one. They are distributed in different positions, connected and transmitted in the form of quality flow and information flow. Taking the HSMP as an example, the thickness, width, flatness, and other indicators, which characterize the strip quality, run through every sub-process in the form of intermediate products[1-3]. Under the constraint of quality and information flow, the variables within sub-processes are dynamically time-varying, and the quality inheritance among sub-processes is sequentially connected[4-6]. The above characteristics make it possible that the fault of any sub-process may affect the whole sequential process. Therefore, advanced and reasonable fault detection methods become particularly important in large-scale sequential processes, which can distinctly determine operating conditions and provide reference information for subsequent diagnosis.

In recent decades, data-driven fault detection methods have been widely studied due to the quick development of computer and automation technologies. As one of the most representative methods, the key idea of multivariate statistical methods is to initially extract features, project high-dimensional data into low-dimensional space, and then establish the monitoring statistics in the low-dimensional feature space[7-10]. Among them, principal component regression (PCR), partial least squares (PLS), canonical correlation analysis (CCA), and their extensions are considered as the most popular methods for quality-related fault detection[11-12]. In the practical process, dynamic characteristics are inherent and inevitable problems due to the existence of feedback systems, random noise, and environmental fluctuations. However, those traditional fault detection methods are purely algebraic structures based on steady-state data, which are not suitable for dynamic processes. Several fault detection and diagnosis methods are already extended to dynamic versions, such as dynamic principal component analysis

(DPCA), dynamic PLS (DPLS), and dynamic independent component analysis (DICA)[13]. Dynamic information can be extracted by augmenting each input vector with lagged samples in those methods. Meanwhile, dynamic inner PCA (DiPCA), dynamic inner PLS (DiPLS) and dynamic inner CCA (DiCCA) have been proposed, which explore the process dynamic properties through the internal covariance between latent variables and predicted values[14-16]. The shortcomings of the original methods still exist, such as the quality-related subspace still contains many parts unrelated to quality in the DiPLS method. In addition, the state space model based methods are also commonly used in dynamic process modeling, and canonical variate analysis (CVA) is a typical representation. CVA has been widely used in dynamic fault detection by maximizing the correlation between past and future variables, mining the autocorrelation and dynamic cross-correlation between variables and realizing the description of dynamic characteristics[17-18].

Recently, slow feature analysis (SFA) method has been studied and applied to dynamic fault detection[19-23]. As an unsupervised dimension reduction method, the goal of SFA is to capture the slow changing features carrying key information. And two pairs of statistics are respectively constructed to monitor static and dynamic information, which can clearly reflect the dynamics of the process. However, SFA may be not suitable for quality-related fault detection. To overcome this problem, a supervised approach to extract slow features (SFs) is proposed[24], which combines the temporal slowness element of SFA and the output correlation element of PLS. In addition, quality-related slow feature regression is also proposed in reference [25], which maximizes the correlation between latent and quality variables while minimizing the slowness of latent variables. Several iterative calculations will also bring higher computational complexity in above methods. Besides, process and dynamic behavior are fully considered in a performance-relevant SFA, which is devised and decomposed by CCA[18]. SFA has been expanded into different quality-related forms and achieved good results. Nevertheless, SFA is to minimize the overall time variation of features, and the ability to retain local process information is insufficient[26].

The process industries are usually characterized by large-scale, multiple operating sub-processes and complex correlations. The above-centralized fault detection methods may suppress or overwhelm local information, and it is difficult to ensure monitoring performance. Associated with these trends, some multi-block and decentralized fault detection methods have been introduced[27]. Based on the idea of "divide and rule", most existing methods consist of three steps: process decomposition, local monitoring, and integrated decision[28-29]. In the process decomposition, process variables are divided into several blocks or sub-processes according to certain rules that mainly include process knowledge-based and data-based methods. Appropriate local monitoring methods are established according to local process characteristics, where the classical fault detection methods are generally adopted. Fusion decision is mainly to establish a global-oriented decision to accomplish a comprehensive evaluation of the state of the whole process. As a popular probabilistic inference method, Bayesian fusion has been adopted by many distributed fault detection methods due to its good interpretability. Multiple local detection results are transformed into fault probability indices, and the fault statistics are effectively reflected in the final results[29-30].

However, the information interaction between sub-processes is generally ignored. It is a common strategy of decentralized fault detection to synthesize information from each local monitoring model in the fusion stage. Actually, each sub-process is not physically independent[31].

Recently, distributed CCA and distributed PLS[32] algorithms have been proposed, which fully considered the communication between sub-processes. The neighbor information is merged into the local sub-process, which reduced the uncertainty of the local sub-process and a more accurate distributed detection can be realized. It is worth mentioning that most distributed fault detection methods considering the information interaction between sub-processes are still in the exploration stage and more in line with the actual needs of process industries. The quality related fault detection method proposed in this chapter belongs to one special case of the distributed methods, the information interaction among sub-processes is considered to be unidirectional, and the main difference can be explained by **Fig. 4. 1**[33-34]. It is because that the main feature of the large-scale sequential process is that multiple sub-processes are connected in tandem. And the quality inheritance among sub-processes is also in sequence, that is, the output of the former sub-processes serves as the input of the latter sub-processes.

Motivated by the above observations, and considering the dynamic coupling variables within sub-processes, quality inheritance among sub-processes, and uncertain outliers, a new quality-related distributed fault detection method is proposed in this chapter. The dynamic characteristics of local proximity in sub-processes and unidirectional information interaction among sub-processes are considered simultaneously. And the robust preprocessing methods combining robust normalization with partial robust M-regression (PRM) are used to eliminate outliers in data. Specifically, the main contributions are as follows:

(1) A new quality-related SFA method is proposed to analyze the local dynamic behavior in each sub-process, in which quality-related and quality-unrelated statistics are also constructed.

(2) The expression of sequential connection relation between sub-processes is given, which takes the quality-related SFs extracted from the previous sub-process as part constraint conditions of the current sub-process.

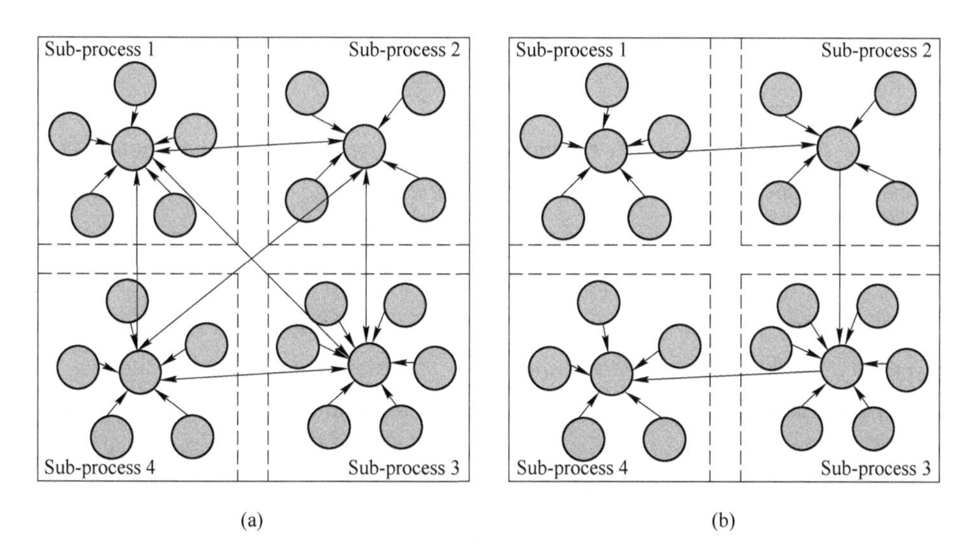

Fig. 4.1 The main difference between commonly distributed (a) and distributed with sequential connection (b)

4.1 Review of PRM and SFA

4.1.1 Partial robust M-regression

PRM is a robust mode of weighted iterative PLS with a single output variable, which is calculated by iteratively and adaptively assigning different weights of samples to eliminate the effect of outliers in the model[35-37]. In PRMR, outliers are divided into high residual and high leverage samples. Samples with large absolute differences between the predicted and observed values are considered as high residual, i.e., normal input and abnormal output. Samples that are far from the center of the input samples are considered as high leverage, i.e., abnormal input and normal output. Different weighting methods are used to weight the above two types of outliers. The leverage weight w_i^l of the i_{th} sample can be defined as:

$$w_i^{\mathrm{l}} = g\left(\frac{\boldsymbol{t}_i - \mathrm{mec}_{\mathrm{L}_1}(\boldsymbol{T})}{\underset{i}{\mathrm{med}} \parallel \boldsymbol{t}_i - \mathrm{med}_{\mathrm{L}_1}(\boldsymbol{T}) \parallel}, c\right), \quad g(z, c) = \frac{1}{\left(1 + \left|\frac{z}{c}\right|\right)^2} \tag{4.1}$$

where, $\boldsymbol{T}$ is the score matrix; $\boldsymbol{t}_i$ is the i_{th} of $\boldsymbol{T}$; the $\mathrm{mec}_{\mathrm{L}_1}$ is the median value of $\boldsymbol{T}$; $\parallel \cdot \parallel$ is the Euclidean metric; c is a tuning constant[36].

The residual weight w_i^{r} of the i_{th} sample can be given as:

$$w_i^{\mathrm{r}} = g\left(\frac{r_i}{\tilde{r}}, c\right) \tag{4.2}$$

where, r_i is the prediction error; $\tilde{r} = \mathrm{med}_i \left| r_i - \mathrm{med}_j(r_j) \right|$ is the median value of r_i .

Considering both weights together, the weight h_i of the i_{th} sample can be determined by:

$$h_i = \sqrt{w_i^{\mathrm{r}} w_i^{\mathrm{l}}} \tag{4.3}$$

The sampling weights are updated iteratively until the algorithm satisfies the convergence condition. Considering the effect of different magnitudes, the raw data should be normalized before using PRMR. When there are no outliers, the mean and variance can be used to determine the central value and dispersion of the data. However, when outliers are present, this may be inaccurate in monitoring models. Fortunately, robust normalization may provide a feasible method to solve this problem.

4.1.2 Slow feature analysis

SFA algorithm is a rising unsupervised method that extracts slowly varying latent variables from time-series data[38]. Its aim is to find a feature function $f(\cdot)$, so that the output signals vary as slowly as possible but carry important information. Mathematically, for a given temporal input signal $\boldsymbol{x}(t) = [x_1(t), x_2(t), \cdots, x_m(t)]^{\mathrm{T}}$, the SFA optimization problem can be formalized as

$$\min \Delta(s_j) := \langle \dot{s}_j^2 \rangle_t \tag{4.4}$$

under the constraints

$$\begin{gathered} \langle s_j \rangle_t = 0 \quad (\text{zeromean}) \\ \langle s_j^2 \rangle_t = 1 \quad (\text{unitvariance}) \\ \forall i \neq j: \ \langle s_i s_j \rangle_t = 0 \quad (\text{decorrelationandorder}) \end{gathered} \tag{4.5}$$

where, $\langle \cdot \rangle_t$ denotes time averaging and $\dot{s}_j$ is the first-order derivative of s with respect to time, expressed as $\dot{s}_j(t) = s_j(t) - s_j(t-1)$. The first two constraints guarantee the generality of minimization problem and avoid trivial solution $s_j(t) \equiv \mathrm{const}$. The third constraint explains that each SF carries different information instead of simply copying other SFs. The objective function in equation (4.3) can also be rewritten as

$$\arg\min_{S}\sum_{i=1}^{n-1}\|\dot{s}_j\|_2^2,\quad \text{s.t.}\quad \boldsymbol{S}\boldsymbol{S}^{\mathrm{T}}=\boldsymbol{I} \tag{4.6}$$

Generally, $f(\cdot)$ can be selected as a linear transformation function. In the linear mapping, each SF can be defined as

$$\boldsymbol{S}=\boldsymbol{W}^{\mathrm{T}}\boldsymbol{X} \tag{4.7}$$

where, $\boldsymbol{W}=[\boldsymbol{w}_1,\boldsymbol{w}_2,\cdots,\boldsymbol{w}_m]^{\mathrm{T}}$ is the mapping coefficient matrix; $\boldsymbol{S}$ denotes all extracted SFs.

Then, we have

$$\sum_{i=1}^{n-1}\|\dot{s}_j\|_2^2=\mathrm{trace}(\dot{\boldsymbol{S}}\dot{\boldsymbol{S}}^{\mathrm{T}})=\mathrm{trace}(\boldsymbol{W}^{\mathrm{T}}(\dot{\boldsymbol{X}}\dot{\boldsymbol{X}}^{\mathrm{T}})\boldsymbol{W}) \tag{4.8}$$

where, trace($\mathring{a}$) represents the trace of $\mathring{a}$.

Furthermore, the optimization problem in equation (4.6) can be further expressed by

$$\begin{aligned}&\arg\min_{W}\mathrm{trace}(\boldsymbol{W}^{\mathrm{T}}(\dot{\boldsymbol{X}}\dot{\boldsymbol{X}}^{\mathrm{T}})\boldsymbol{W}),\\&\text{s.t.}\quad \boldsymbol{S}\boldsymbol{S}^{\mathrm{T}}=\boldsymbol{W}^{\mathrm{T}}(\dot{\boldsymbol{X}}\dot{\boldsymbol{X}}^{\mathrm{T}})\boldsymbol{W}=\boldsymbol{W}^{\mathrm{T}}\boldsymbol{W}=\boldsymbol{I}\end{aligned} \tag{4.9}$$

The above optimization problem can be transformed into two singular value decompositions (SVDs). The mapping coefficient matrix $\boldsymbol{W}$ can be calculated by

$$(\dot{\boldsymbol{X}}\dot{\boldsymbol{X}})^{\mathrm{T}}\boldsymbol{W}=\boldsymbol{\Omega}\boldsymbol{W}^{\mathrm{T}} \tag{4.10}$$

where, $\boldsymbol{\Omega}$ is the diagonal matrix of eigenvalues.

Finally, the SFs can be obtained by equation (4.7). In the process industries, the slowest features usually tend to capture the changing trend of the process, and the fastest features can be regarded as environmental noise. According to the different slowness, the obtained SFs can be further divided into two parts

$$\boldsymbol{S}=[\boldsymbol{S}_{\mathrm{d}},\boldsymbol{S}_{\mathrm{e}}] \tag{4.11}$$

where, $\boldsymbol{S}_{\mathrm{d}}$ represents dominant slowest SFs with reduced dimension; $\boldsymbol{S}_{\mathrm{e}}$ denotes residual fastest SFs. The number of $\boldsymbol{S}_{\mathrm{d}}$ can be selected by a reconstruction criterion. However, SFA is an unsupervised method that only considers process variables, which cannot be directly used for quality-related fault detection. In addition, SFA only extracts global SFs, and its ability to retain local process information and handle dynamics is limited.

4.2 Distributed detection method for quality-related faults

In this section, the proposed quality-related distributed fault detection method is described in detail.

4.2.1 Robust preprocessing and data decomposition

As the actual process industry is generally influenced by the complexity and diversity of the production environment, data collection errors in measuring equipment, data transmission, and management will all lead to the introduction of outliers. The presence of outliers has a serious influence on modeling, which may affect the construction of statistics and lead to inaccurate monitoring results. Thus, in this chapter, the combination of robust normalization and PRM is used as the robust data preprocessing.

The raw data should be normalized to eliminate the influence of different dimensions. Generally, the normalization can be defined as

$$\bar{u} = \frac{1}{n}\sum_{i=1}^{n} u_i, \qquad s^2 = \frac{1}{n-1}\sum_{i=1}^{n} (u_i - \bar{u}) \tag{4.12}$$

where, $\bar{u}$ and s^2 denote the mean and variance value of the measured data $U = [u_1, u_2, \cdots, u_n]$, respectively.

However, if the measured data contains some outliers, the mean and variance may change and deviate from the true value significantly. To solve this problem, robust normalization is used to estimate the mean and variance. The mean value in robust normalization is expressed by the total square loss center and the robust center $\bar{e}$ that can be expressed as

$$\bar{e} = \sum_{i=1}^{n} k_i u_i \quad k_i = \frac{1}{\sqrt{1+4u_i^2}} \Bigg/ \left(\sum_{j=1}^{n} \frac{1}{\sqrt{1+4u_j^2}}\right) \tag{4.13}$$

Similarly, the variance value Q_n can be calculated as

$$Q_n = \eta \{ | u_i - u_j | ; i < j \}_{(k)} \tag{4.14}$$

where, $k = h(h-1)/2$, $h = [n/2] + 1$, $[\cdot]$ is an integer function; η is the correction factor. Different $| u_i - u_j |$ can be calculated under $i < j$. The data are arranged in ascending order, and Q_n can be obtained by multiplying the sorted k_{th} with η. Up to now, the mean $\bar{e}$ and variance Q_n are replaced by the traditional normalization.

Assume that the process and quality variables after robust normalization are $\boldsymbol{X} \in R^{n\times m}$ and $\boldsymbol{Y} \in R^{n\times l}$, respectively. Then, the PRM algorithm is used to further process $\boldsymbol{X}$ and $\boldsymbol{Y}$ to eliminate outliers. The new $\boldsymbol{X}' = \boldsymbol{H}_i\boldsymbol{X}$ and $\boldsymbol{Y}' = \boldsymbol{H}_i\boldsymbol{Y}$ can be obtained, where $\boldsymbol{H}_i$ is a diagonal weight matrix determined by equation (4.3). In order to simplify the scale and make models have strong physical significance, a large-scale sequential process can be divided into different sub-processes by mechanism knowledge. The decomposition strategy based on mechanism knowledge generally includes the physical structure and production technology of the equipment,} which can have lower computational complexity and better monitoring performance. Therefore, process variables $\boldsymbol{X}'$ can be divided into B sub-processes: $\boldsymbol{X}_1$, $\boldsymbol{X}_2$, $\cdots$, $\boldsymbol{X}_B$, where the b_{th} sub-process $\boldsymbol{X}_b$ contains m_b variables, $b = 1, 2, \cdots, B$.

4.2.2 Quality-related SFA based local modeling

The traditional SFA method is based on minimizing the overall time variation of the features. In this way, the ability to retain local process information is inadequate, and extracted SFs cannot be fully characterized the process data. In the presence of faults, the global or local relationships of normal data may change. Therefore, to explain the nature of process dynamics comprehensively, the modified regularized SFA (MRSFA) method is proposed, which takes the time variation and local neighbor information of the raw data into account[39].

By introducing modified regularized constraints in the optimization problem, the extracted SFs can preserve local neighbor relationships of raw data while exploring the time-series relationship of variables. The objective function of MRSFA can be defined as

$$\arg\min_{W} \sum_{j=1}^{n-1} \| \dot{s}_j \|_2^2 + \lambda \sum_{i=1}^{n-1} \sum_{j=1}^{n-1} C_{ij} \| \dot{s}_i - \dot{s}_j \|_2^2 \quad \text{s.t.} \quad \boldsymbol{W}\boldsymbol{W}^{\mathrm{T}} = \boldsymbol{I} \tag{4.15}$$

where, λ is the parameter used to balance the weight between the SF and the modified regularization.

The first term in equation (4.15) is similar to SFA, which is used to ensure the slow change of features. The second term is modified regularized, which is used to constrain the local neighborhood structure of raw data. C_{ij} is an element of the similarity matrix $\boldsymbol{C}$, which can be determined by the initial state of raw data. Specifically, if $\boldsymbol{x}_i$ is the k nearest neighbors of $\boldsymbol{x}_j$, or $\boldsymbol{x}_j$ is the k nearest neighbors of $\boldsymbol{x}_i$, we have

$$C_{ij} = \exp\left(- \frac{\| \boldsymbol{x}_i - \boldsymbol{x}_j \|^2}{\gamma} \right) \tag{4.16}$$

where, γ is a weight parameter for adjusting the similarity, and $C_{ij} = 0$ otherwise.

Through above derivations, equation (4.15) can be further formulated as

$$\begin{aligned}&\arg\min_{W}\sum_{j=1}^{n-1}\|\dot{s}_j\|_2^2+\lambda\sum_{i=1}^{n-1}\sum_{j=1}^{n-1}C_{ij}\|\dot{s}_i-\dot{s}_j\|_2^2\\&=\mathrm{trace}(\dot{S}\dot{S}^{\mathrm{T}})+\lambda\,\mathrm{trace}(\dot{S}(D-C)\dot{S}^{\mathrm{T}})\\&=\mathrm{trace}(\dot{S}(I+\lambda(D-C))\dot{S}^{\mathrm{T}})\\&=\mathrm{trace}(\dot{S}L\dot{S}^{\mathrm{T}})\end{aligned} \tag{4.17}$$

where, $\boldsymbol{D}$ is a diagonal matrix with $D_{ii}=\sum_{j}C_{ij}$, $\boldsymbol{L}=\boldsymbol{I}+\lambda(\boldsymbol{D}-\boldsymbol{C})$. Then, a new object function is rewritten as

$$\begin{aligned}&\arg\min_{W}\mathrm{trace}(\dot{S}L\dot{S}^{\mathrm{T}}),\\&\mathrm{s.t.}\quad WW^{\mathrm{T}}=I,\ \dot{S}D\dot{S}^{\mathrm{T}}=I\end{aligned} \tag{4.18}$$

Similar to SFA, we can get the mapping $\boldsymbol{W}$ by solving generalized feature decomposition

$$(\dot{X}L\dot{X}^{\mathrm{T}})W=\Omega(\dot{X}D\dot{X}^{\mathrm{T}})W^{\mathrm{T}} \tag{4.19}$$

The traditional SFA and MRSFA model are unsupervised methods and only consider process variables, which cannot be directly used for quality-related fault detection. In addition, due to the existence of feedback system, random noise, and environmental fluctuation, the variables show dynamic behavior, and the current values of variables may be dependent on the past values. It is insufficient to analyze the dynamic characteristics of process data only considering the first-order time difference. Quality variables usually have static characteristics, while process variables have strong dynamics. Therefore, a novel dynamic quality-related MRSFA (QMRSFA) is proposed in this chapter. The time-delay values of process variables are introduced to construct an augmented matrix, and then the MRSFA algorithm and latent variable projection are reasonably combined to produce quality-related SFs (QSFs) and quality-unrelated SFs (QUSFs).

First, the dynamic model can be constructed in each sub-process, and its augmented input matrix can be expressed as

$$X_{Ab}=\begin{bmatrix}x_{b(1+d)}^{\mathrm{T}} & x_{b(d)}^{\mathrm{T}} & \cdots & x_{b(1)}^{\mathrm{T}}\\ x_{b(2+d)}^{\mathrm{T}} & x_{b(d+1)}^{\mathrm{T}} & \cdots & x_{b(2)}^{\mathrm{T}}\\ \vdots & \vdots & \cdots & \vdots\\ x_{b(n)}^{\mathrm{T}} & x_{b(n-1)}^{\mathrm{T}} & \cdots & x_{b(n-d)}^{\mathrm{T}}\end{bmatrix}\in R^{(n-d)\times m_b(d+1)} \tag{4.20}$$

where, d is the lag time. To ensure that the quality variables $\boldsymbol{Y}'$ and $\boldsymbol{X}_{Ab}$ have the same number of rows, $\boldsymbol{Y}'$ can be corrected as

$$\boldsymbol{Y}_A = \begin{bmatrix} \boldsymbol{y}_{1+d}^{\mathrm{T}} \\ \boldsymbol{y}_{2+d}^{\mathrm{T}} \\ \vdots \\ \boldsymbol{y}_n^{\mathrm{T}} \end{bmatrix} \in R^{(n-d)\times l} \tag{4.21}$$

Then, the MRSFA model is established for $\boldsymbol{X}_{Ab}$, and all the SFs with local structure can be described as

$$\boldsymbol{S}_b = [\boldsymbol{S}_{b,k}, \ \boldsymbol{S}_{b,e}] = \boldsymbol{X}_{Ab} \boldsymbol{W}_b \tag{4.22}$$

where, $\boldsymbol{S}_{b,k} \in R^{n\times k_b}$ are the k_b dominant slowest SFs.

Because of the transmission and feedback control among materials and information in process industries, the product quality should change slowly under normal conditions. Therefore, it is reasonable to use SFs with relatively slow change to establish regression model. In the case of complete decomposition, the regression relation between $\boldsymbol{S}_{b,k}$ and quality variables $\boldsymbol{Y}_A$ can be expressed as

$$\boldsymbol{M}_b = (\boldsymbol{S}_{b,k}^{\mathrm{T}} \boldsymbol{S}_{b,k})^{\dagger} \boldsymbol{S}_{b,k}^{\mathrm{T}} \boldsymbol{Y}_A \tag{4.23}$$

where, $(*)^{\dagger}$ is the pseudo-inverse.

The SVD is performed on $\boldsymbol{M}_b \boldsymbol{M}_b^{\mathrm{T}}$ that can be given by

$$\boldsymbol{M}_b \boldsymbol{M}_b^{\mathrm{T}} = \begin{bmatrix} \hat{\boldsymbol{P}}_{M_b} & \hat{\boldsymbol{P}}_{M_b} \end{bmatrix} \begin{bmatrix} \Lambda_{M_b} & 0 \\ 0 & 0 \end{bmatrix} \begin{bmatrix} \hat{\boldsymbol{P}}_{M_b} \\ \hat{\boldsymbol{P}}_{M_b} \end{bmatrix} \tag{4.24}$$

where, $\hat{\boldsymbol{P}}_{M_b} \in R^{k_b\times l}$, $\hat{\boldsymbol{P}}_{M_b} \in R^{k_b\times(k_b-l)}$, and $\Lambda_{M_b} \in R^{l\times l}$. The orthogonal projection matrices Π_{M_b} and $\Pi_{M_b}^{\perp}$ can be constructed by

$$\begin{aligned} \Pi_{M_b} &= \hat{\boldsymbol{P}}_{M_b} \hat{\boldsymbol{P}}_{M_b}^{\mathrm{T}} \\ \Pi_{M_b}^{\perp} &= \hat{\boldsymbol{P}}_{M_b} \hat{\boldsymbol{P}}_{M_b}^{\mathrm{T}} \end{aligned} \tag{4.25}$$

Then, $\boldsymbol{S}_{b,k}$ can be projected into

$$\begin{aligned} \boldsymbol{S}_{b,y} &= \boldsymbol{S}_{b,k} \boldsymbol{\Pi}_{M_b} \in \operatorname{span}\{M_b\} \\ \boldsymbol{S}_{b,\tilde{y}} &= \boldsymbol{S}_{b,k} \boldsymbol{\Pi}_{M_b}^{\perp} \in \operatorname{span}\{M_b\}^{\perp} \end{aligned} \tag{4.26}$$

where, $\boldsymbol{S}_{b,y}$ and $\boldsymbol{S}_{b,\tilde{y}}$ are mutually orthogonal for the QSFs and QUSFs, respectively.

Finally, it is necessary to divide monitoring statistics into two orthogonal subspaces to realize quality-related and quality-unrelated fault detection. The statistical indicators for quality-related faults based on QMRSFA model can be calculated by

$$T_{b,y}^2 = \boldsymbol{S}_{b,y}^{\mathrm{T}} \boldsymbol{S}_{b,y}, \quad S_{b,y}^2 = \dot{\boldsymbol{S}}_{b,y}^{\mathrm{T}} \boldsymbol{\Omega}_{b,y}^{-1} \dot{\boldsymbol{S}}_{b,y} \tag{4.27}$$

and statistical indicators for quality-unrelated faults can be computed by

$$T_{b,\tilde{y}}^2 = \boldsymbol{S}_{b,\tilde{y}}^{\mathrm{T}} \boldsymbol{S}_{b,\tilde{y}}, \quad S_{b,\tilde{y}}^2 = \dot{\boldsymbol{S}}_{b,\tilde{y}}^{\mathrm{T}} \boldsymbol{\Omega}_{b,\tilde{y}}^{-1} \dot{\boldsymbol{S}}_{b,\tilde{y}} \tag{4.28}$$

In addition, the faster SFs in equation (4.11) are usually related to noise. Once $\boldsymbol{S}_{b,e}$ is abnormal, it is also considered as a quality-unrelated fault. Then, the corresponding statistics can be defined as

$$T_{b,e}^2 = \boldsymbol{S}_{b,e}^{\mathrm{T}} \boldsymbol{S}_{b,e}, \quad S_{b,e}^2 = \dot{\boldsymbol{S}}_{b,e}^{\mathrm{T}} \boldsymbol{\Omega}_{b,e}^{-1} \dot{\boldsymbol{S}}_{b,e} \tag{4.29}$$

$T_{b,y}^2$, $T_{b,\tilde{y}}^2$, and $T_{b,e}^2$ can be used as static statistics to detect the deviation of operating conditions from the steady distribution, $S_{b,y}^2$, $S_{b,\tilde{y}}^2$, and $S_{b,e}^2$ can be used as dynamic statistics to detect dynamic anomalies from the dynamic distribution. Combined with quality indicators, fine quality-related fault detection can be obtained under different process conditions. And, the control limit of above statistics can be obtained by kernel density estimation (KDE) with a certain confidence limit.

4.2.3 Sequential connection relation construction

When the local model is established, the interconnection of adjacent sub-process is considered. The QSFs extracted from the previous sub-process are taken as the part constraint conditions of QMRSFA model in the current sub-process. As the transmission information between sub-process, QSFs are more in line with the actual performance of industrial systems, and the accuracy and credibility of modeling can be improved. Therefore, the basic QMRSFA model has been modified, where $\boldsymbol{S}_{b,k}$ in equation (4.23) can be supplemented. The new $\boldsymbol{S}'_{b,k}$ with interconnection can be determined by the joint action of the $\boldsymbol{S}_{b,k}$ in sub-process and $\boldsymbol{S}_{b-1,y}$ in the previous sub-process, which can be expressed as

$$\begin{aligned} \boldsymbol{S}'_{1,k} &= \boldsymbol{S}_{1,k} \\ \boldsymbol{S}'_{2,k} &= [\boldsymbol{S}_{2,k}, \ \boldsymbol{S}_{1,y}] \\ &\vdots \\ \boldsymbol{S}'_{b,k} &= [\boldsymbol{S}_{b,k}, \ \boldsymbol{S}_{b-1,y}] \end{aligned} \tag{4.30}$$

The least square regression matrix is also expressed in a new way: $\boldsymbol{M}_b^{\mathrm{T}} = (\boldsymbol{S}_{b,k}'^{\mathrm{T}} \boldsymbol{S}'_{b,k})^{-1} \boldsymbol{S}_{b,k}'^{\mathrm{T}} \boldsymbol{Y}_A$. After that, other steps of the QMRSFA model are the same as before. According to the construction strategy of sequential connection, the QMRSFA model in each sub-process is connected in series. Finally, the distributed QMRSFA with sequential connection is formed through QSFs.

4.2.4 Bayesian fusion based global monitoring

In order to obtain the overall comprehensive monitoring results, Bayesian fusion strategy is introduced between monitoring statistics and fault occurrence probabilities, which can be expressed as

$$P_{\phi_i}(F|\boldsymbol{x}_b) = \frac{P_{\phi_i}(\boldsymbol{x}_b|F)P_{\phi_i}(F)}{P_{\phi_i}(\boldsymbol{x}_b)}$$
$$P_{\phi_i}(\boldsymbol{x}_b) = P_{\phi_i}(\boldsymbol{x}_b|F)P_{\phi_i}(F) + P_{\phi_i}(\boldsymbol{x}_b|N)P_{\phi_i}(N) \tag{4.31}$$

where, ϕ_i represents any monitoring statistic in this paper; N and F denote normal and abnormal conditions; $P_{\phi_i}(\boldsymbol{x}_b|F)$ and $P_{\phi_i}(\boldsymbol{x}_b|N)$ are the relevant prior probabilities, respectively. When $\boldsymbol{P}_{\phi_i}(N)$ is chosen as the confidence level α, then the $P_{\phi_i}(F)$ can be simply calculated as $1-\alpha$. Moreover, the conditional probabilities $P_{\phi_i}(\boldsymbol{x}_b|F)$ and $P_{\phi_i}(\boldsymbol{x}_b|N)$ can be given as

$$P_{\phi_i}(\boldsymbol{x}_b|F) = e^{-\phi_{i,\mathrm{th}}/\phi_{i,\mathrm{new}}} \tag{4.32}$$

$$P_{\phi_i}(\boldsymbol{x}_b|N) = e^{-\phi_{i,\mathrm{new}}/\phi_{i,\mathrm{th}}} \tag{4.33}$$

where, $\phi_{i,\mathrm{new}}$ and $\phi_{i,\mathrm{th}}$ are statistic and corresponding control limit of ϕ_i.

Then, the local online monitoring indices (LoMI) can be defined as

$$\mathrm{LoMI}_{b,\mathrm{re}} = \sum_{\phi_i \in \Phi_1} \frac{P_{\phi_i}(\boldsymbol{x}_b|F)P_{\phi_i}(F|\boldsymbol{x}_b)}{\sum_{\phi_i \in \Phi_1} P_{\phi_i}(\boldsymbol{x}_b|F)}, \quad \Phi_1 = \{T^2_{b,y}, S^2_{b,y}\}$$
$$\mathrm{LoMI}_{b,\mathrm{un}} = \sum_{\phi_i \in \Phi_2} \frac{P_{\phi_i}(\boldsymbol{x}_b|F)P_{\phi_i}(F|\boldsymbol{x}_b)}{\sum_{\phi_i \in \Phi_2} P_{\phi_i}(\boldsymbol{x}_b|F)}, \quad \Phi_2 = \{T^2_{b,\tilde{y}}, S^2_{b,\tilde{y}}, T^2_{b,e}, S^2_{b,e}\} \tag{4.34}$$

where, $\mathrm{LoMI}_{b,\mathrm{re}}$ and $\mathrm{LoMI}_{b,\mathrm{un}}$ are quality-related and quality-unrelated comprehensive monitoring statistics in the b_{th} sub-process. Similarly, all local statistics can be obtained, and the control limit ℓ_{L} of LoMI is determined by the confidence level.

Then, the global online monitoring indices (GoMI) can be expressed by

$$\mathrm{GoMI}_{\mathrm{re}} = \sum_{\phi_i \in \Phi_3} \frac{P_{\phi_i}(\boldsymbol{x}_b|F)P_{\phi_i}(F|\boldsymbol{x}_b)}{\sum_{\phi_i \in \Phi_3} P_{\phi_i}(\boldsymbol{x}_b|F)}, \quad \Phi_3 = \{\mathrm{LoMI}_{1,\mathrm{re}}, \mathrm{LoMI}_{2,\mathrm{re}}, \cdots, \mathrm{LoMI}_{B,\mathrm{re}}\}$$
$$\mathrm{GoMI}_{\mathrm{un}} = \sum_{\phi_i \in \Phi_4} \frac{P_{\phi_i}(\boldsymbol{x}_b|F)P_{\phi_i}(F|\boldsymbol{x}_b)}{\sum_{\phi_i \in \Phi_4} P_{\phi_i}(\boldsymbol{x}_b|F)}, \quad \Phi_4 = \{\mathrm{LoMI}_{1,\mathrm{un}}, \mathrm{LoMI}_{2,\mathrm{un}}, \cdots, \mathrm{LoMI}_{B,\mathrm{un}}\} \tag{4.35}$$

where, the control limit ℓ_G of GoMI is similar as LoMI.

As a summary, the overall flowchart of the proposed distributed detection method for quality-related faults is shown in **Fig. 4. 2**, and the procedures are as follows.

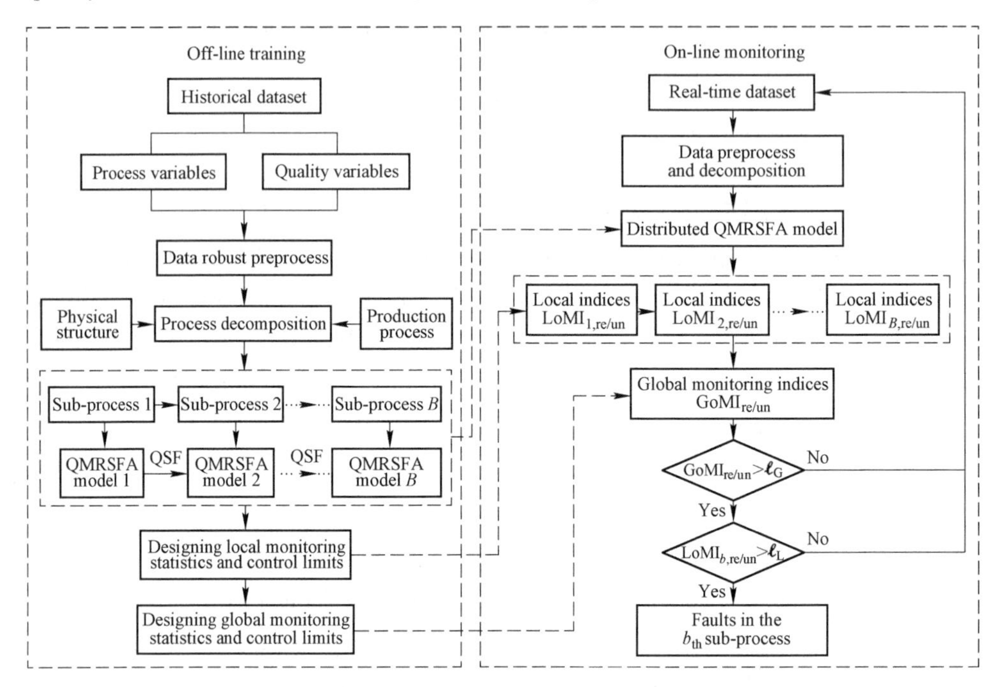

Fig. 4. 2 Flowchart of QMRSFA-based monitoring method

4. 2. 4. 1 Offline training

Step 1: Historical training dataset $\boldsymbol{X}'$ and $\boldsymbol{Y}'$ under normal conditions are collected and robust preprocessed.

Step 2: Different sub-processes are divided by mechanism knowledge.

Step 3: The QMRSFA model is built in each sub-process, and sequential connection relation is also given.

Step 4: The local and global two-level indexes are established by equation (4. 34) and equation (4. 35), respectively.

4. 2. 4. 2 Online monitoring

Step 5: A new online sample $\boldsymbol{x}_{\text{new}}$ is collected and preprocessed.

Step 6: $\boldsymbol{x}_{\text{new}}$ is divided into B sub-processes.

Step 7: The distributed QMRSFA is constructed in $\boldsymbol{x}_{b,\text{new}}$.

Step 8: Two-level monitoring statistics are calculated, and the final detection results are given.

4.3 Case study

4.3.1 Variable description and sub-process division

In order to verify the effectiveness of the proposed method, normal and fault data were collected and recorded from the commissioning and production of the hot strip rolling line in a steel company. The used process and quality variables are listed in **Table 4.1**, and the sampling period is 0.01s. When considering the division of finishing mill, the variables in each sub-process have been shown in **Table 4.2**. In the simulation study, 3000 normal samples are used for the training of the proposed method. Two frequently occurring quality-related faults are used to verify the detection performance, and 5% of them are randomly selected to introduce disturbance. In addition, the confidence level is set to 95%, $c = 4$, $\eta = 2.2219$, the lag time $d = 2$, and the number of k_b is 19,14,15.

Table 4.1 Process and quality variables in finishing mill

Variable	Type	Description	Unit
$x_1 - x_7$	Process	Average roll gap at the i_{th} stand, $i=1, \cdots, 7$	mm
$x_8 - x_{14}$	Process	Total force at the i_{th} stand, $i=1, \cdots, 7$	MN
$x_{15} - x_{20}$	Process	Work roll bending force at the i_{th} stand, $i=2, \cdots, 7$	MN
y	Quality	Finishing mill exit strip thickness	mm

Table 4.2 Sub-process division results

Sub-process No.	1	2	3
Variable No.	x_1, x_2, x_3, x_8, x_9, x_{10}, x_{15}, x_{16}	x_4, x_5, x_{11}, x_{12}, x_{17}, x_{18}	x_6, x_7, x_{13}, x_{14}, x_{19}, x_{20}

4.3.2 Results and analysis

To illustrate the detection performance of different models for quality-related faults, global-local preserving projections (GLPP), dynamic PLS (DPLS), multi-block global and local partial least-squares projection dynamic PLS (MB-GLPLS), multi-block QSFA (MB-QSFA), and distributed QMRSFA without robust preprocessing (NR-QMRSFA) are selected for comparison with the proposed method (D-QMRSFA). It should be noted that in MB-GLPLS and MB-QSFA, the basic GLPLS or QSFA is built in each sub-process, and then the quality-related and quality-unrelated statistics of each sub-process are synthesized to obtain the global monitoring results.

4.3.2.1 Distributed detection results for Fault 1

Fault 1 is the fault of the hydraulic screw-down roll gap control system of the 4_{th} stand, which occurs in the second sub-process. When this fault occurs, it will cause the set value of the roll gap to be abnormal, and the rolling force will also be affected. Then, this fault will also affect the roll gap, the sampling value of the rolling force in back stands, and the final outlet thickness. Fault 1 is considered to be a quality-related fault, which occurs at the 2001_{th} sample and continues until the end of sampling.

Fig. 4.3 shows the global detection results for Fault 1. It can be seen that DPLS, NR-QMRSFA, and D-QMRSFA methods can detect the fault in real-time at the 2001_{st} sampling point, and clearly show the abnormal fluctuation of the process. However, DPLS and MB-GLPLS are prone to fault alarm at the beginning of sampling and the fault stage. In addition, there are some false positives and false negatives in GLPP and MB-

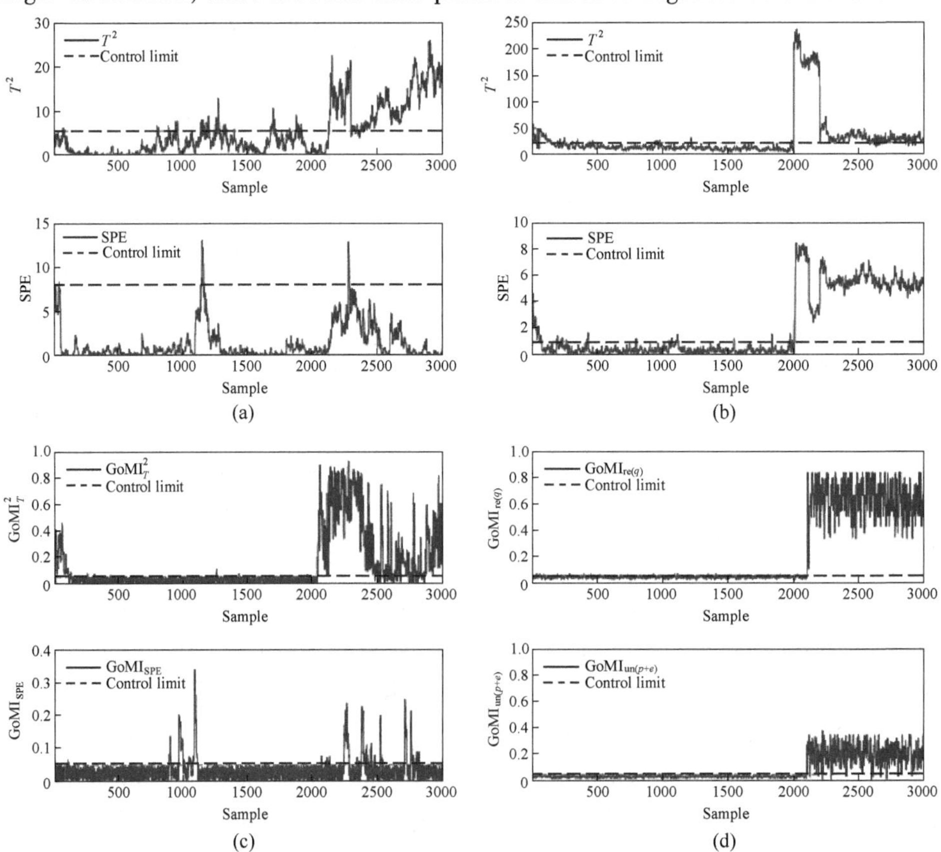

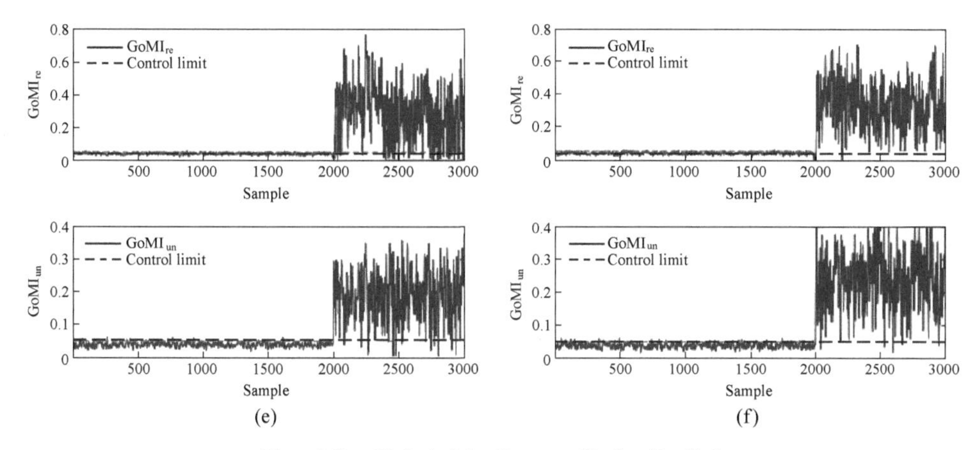

Fig. 4.3 Global detection results for Fault 1

(a) GLPP; (b) DPLS; (c) MB-GLPLS; (d) MB-QSFA; (e) NR-QMRSFA; (f) D-QMRSFA

GLPLS, because two methods ignore the dynamic characteristics. Compared with DQMRSFA, MB-QSFA has delay in the detection of Fault 1, which reduces fault detection rate (FDR). It is because the features of nearest neighbor relationship and the information interaction of related subsystems are not considered in these methods. Due to the introduction of disturbance, NR-QMRSFA has a higher fault alarm rate (FAR) and some omissions. In comparison, the proposed method can detect faults in time with higher FDR and lower FAR, the specific comparisons are shown in **Table 4.3**.

Table 4.3 Quality-related detection performance comparison results

Fault No.	GLPP (T^2)		DPLS		MB-GLPLS		MB-QSFA		NL-QMRSFA		D-QMRSFA	
	FAR	FDR	FAR	FDR	FAR	FDR	FAR	FDR	FAR	FDR	FAR	FDR
Fault 1	9.00	82.50	9.65	92.60	7.60	88.10	8.49	91.48	8.64	94.59	**4.35**	**98.40**
Fault 2	8.60	84.93	12.53	91.67	12.00	85.80	9.40	95.27	7.80	95.87	**5.67**	**97.40**

Fig. 4.4 shows the local detection results of the second sub-process of Fault 1. It can be seen that the proposed D-QMRSFA method can detect the fault occurrence in time at the 2001_{st} sampling point. The fault is quality-related and the static statistics are abnormal, which is consistent with the deviation of the actual roll gap setting value of the 4_{th} stand. Quality inheritance information can provide more useful monitoring basis. Detailed and accurate detection results can provide more useful operation information for field operators. The other three methods have also got the corresponding fault information, but in terms of detection performance, MB-QSFA has a certain detection delay, while NR-QMRSFA has a lower detection rate. Poor fault detection performance

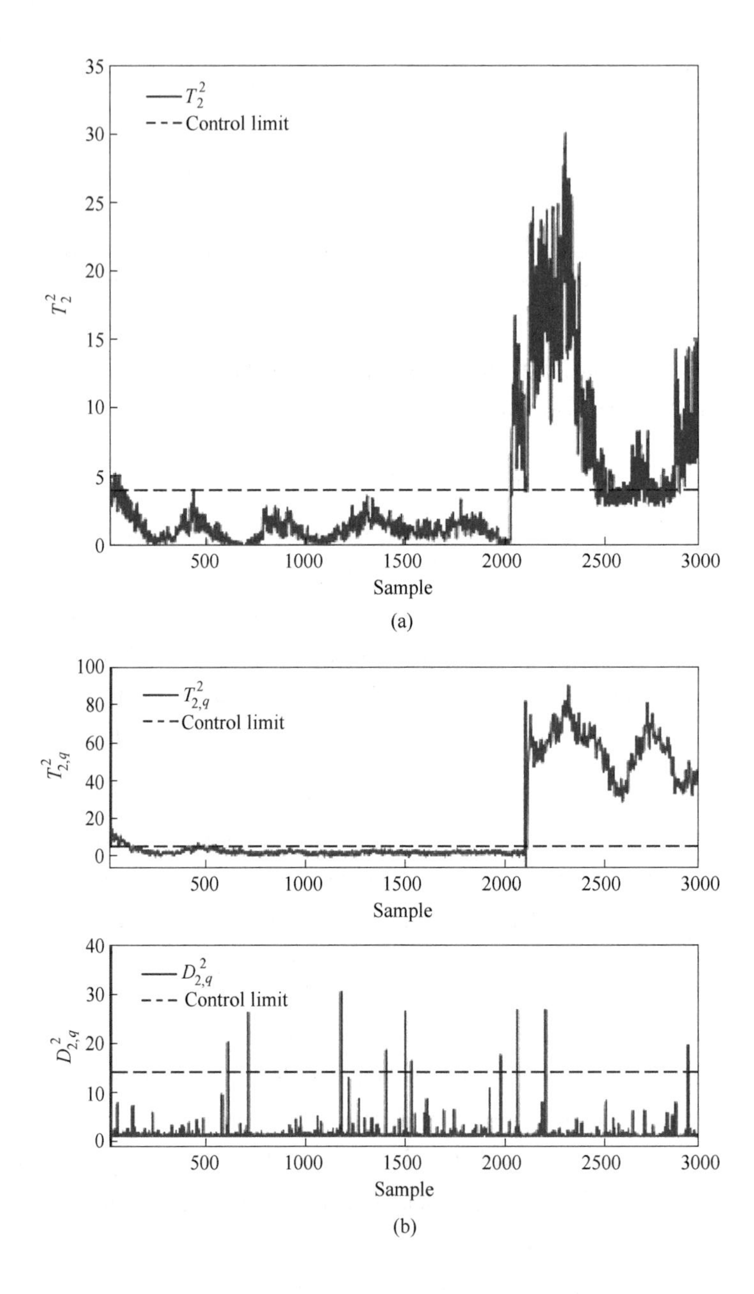

T_2^2
Control limit
35
30
25
20
15
10
5
0
500
1000
1500
2000
2500
3000
Sample
$T_{2,q}^2$
Control limit
100
80
60
40
20
0
$D_{2,q}^2$
Control limit
40
30
20
10

(a)

(b)

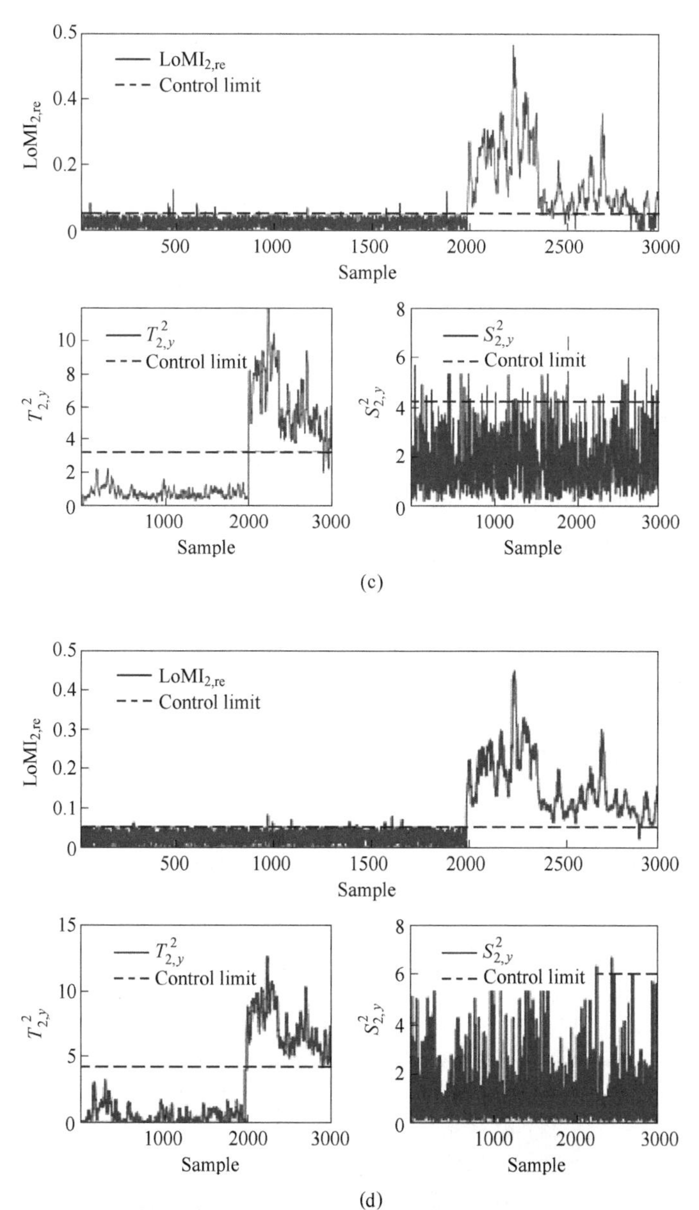

Fig. 4.4 Local detection results for Fault 1

(a) MB-GLPLS; (b) MB-QSFA; (c) NR-QMRSFA; (d) D-QMRSFA

is shown in MB-GLPLS, especially for a period of time after the fault occurs, which is that dynamics is not considered.

4.3.2.2 Distributed detection results for Fault 2

Fault 2 is the actuator fault of cooling water control valve between the second stand and the third stand, which occurs in the first sub-process. This will affect the rolling force and roll gap of the subsequent stands, thus leading to the positive deviation of the thickness of the outlet strip steel and affecting the final product quality. Fault 2 is also a quality-related fault, which occurs at the 1001_{st} sample and ends in the 2500_{th} sample.

The global detection results of six algorithms are shown in **Fig. 4.5**. In comparison, it can be seen that the method based on DPLS cannot reflect the abnormal fluctuation of dynamic process, and there is a high FAR. GLPP and MB-GLPLS have undesirable detection performance and the lowest FAR. MB-QSFA still has detection delay and a lower FDR. The proposed method shows higher FDR and lower FAR, considering the nearest neighbor relationship of dynamic features and information interaction between sub-processes. The comparison of the detection results are shown in **Table 4.3**. Aiming at the quality-related faults in HSMP, the model established by D-QMRSFA method has better detection performance.

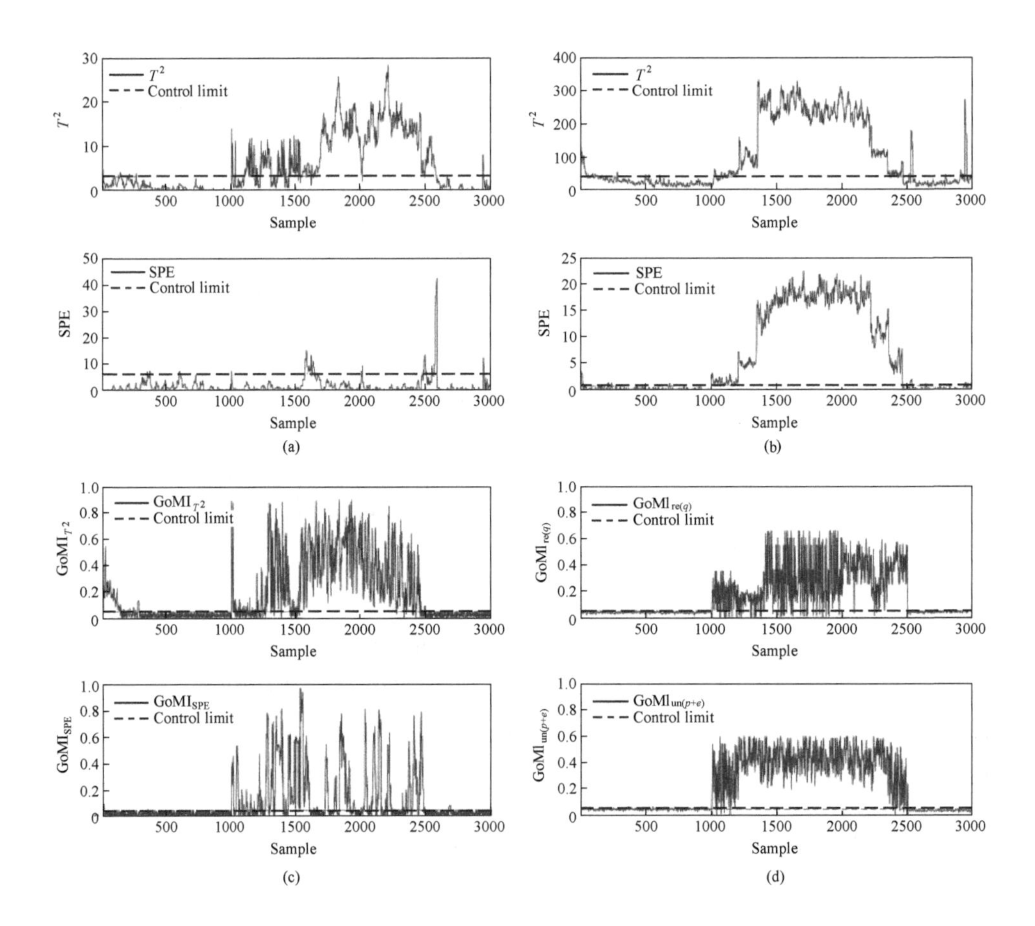

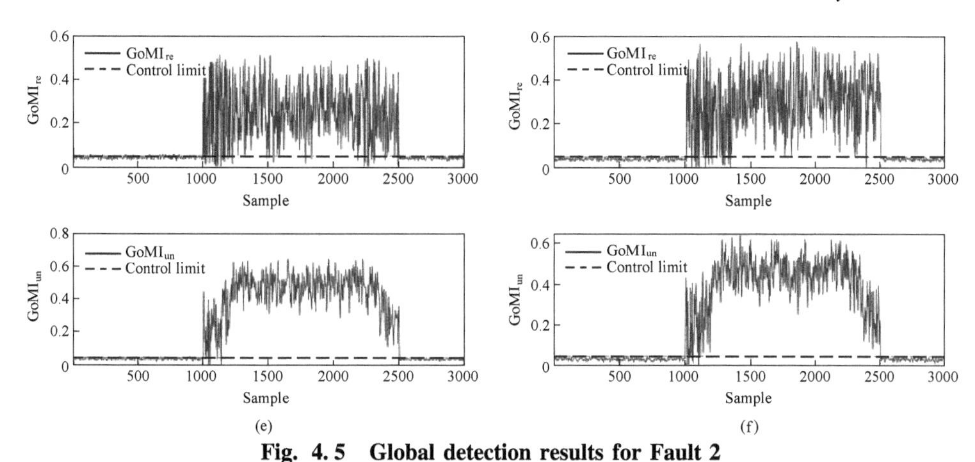

Fig. 4.5 Global detection results for Fault 2

(a) GLPP; (b) DPLS; (c) MB-GLPLS; (d) MB-QSFA; (e) NR-QMRSFA; (f) D-QMRSFA

Fig. 4.6 shows the local detection results of the third sub-process of Fault 2. The quality-related fault information is transmitted in the subsequent sub-process in D-QMRSFA, and finally accumulated in the 3_{rd} sub-process, which can provide more useful monitoring basis for detection. It can be seen that the local synthesis and quality-related dynamic statistics of D-QMRSFA method have been detected. The fault is considered to be a quality-related and real dynamic abnormality, which is consistent with the fault of the cold actuator between the second and the third stands. Compared with other competitive methods, D-QMRSFA method not only has better robustness to disturbances, but also can provide more detailed and accurate information about process state.

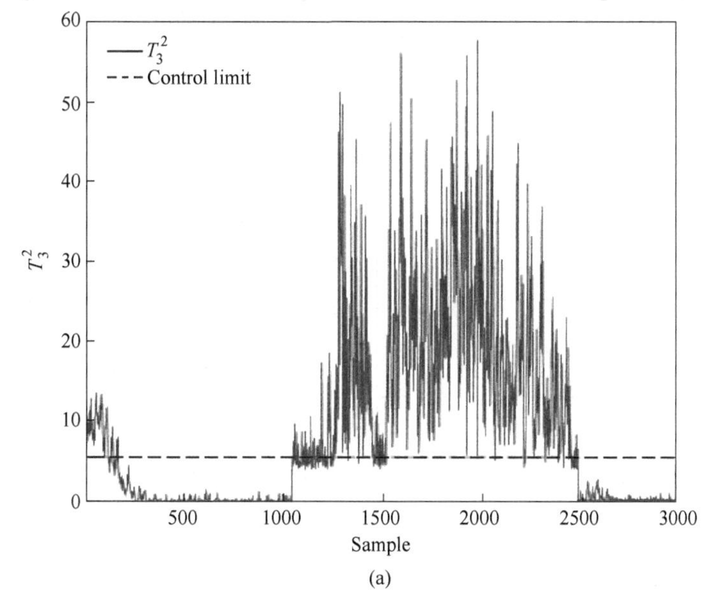

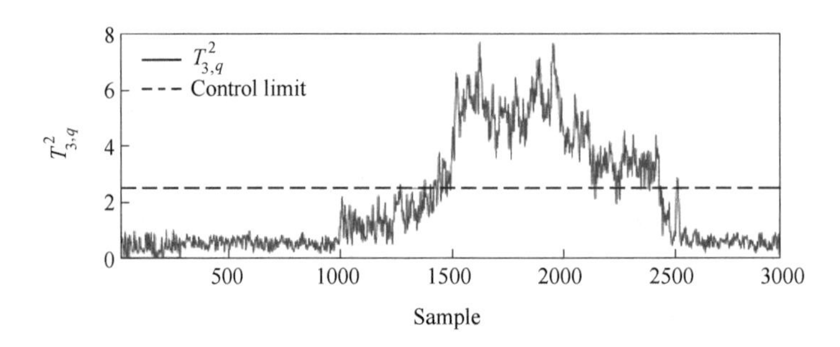

$T^2_{3,q}$
Control limit
Sample

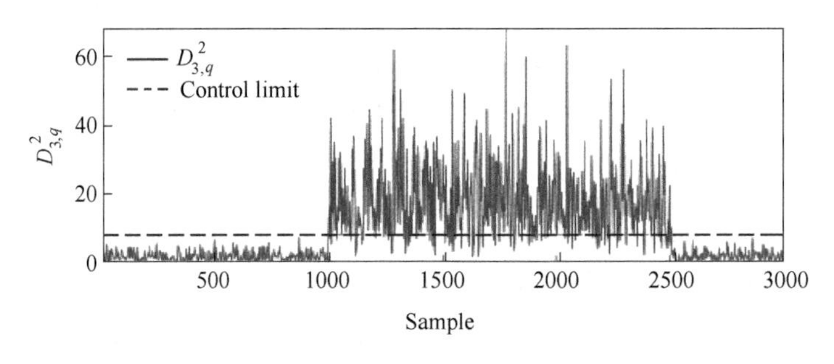

$D^2_{3,q}$
Control limit
Sample

(b)

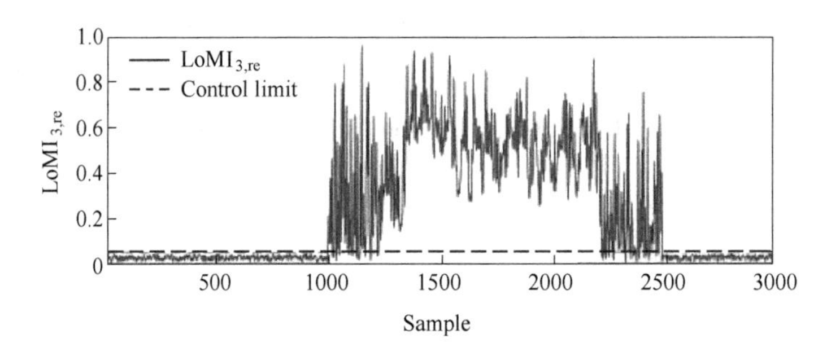

$LoMI_{3,re}$
Control limit
Sample

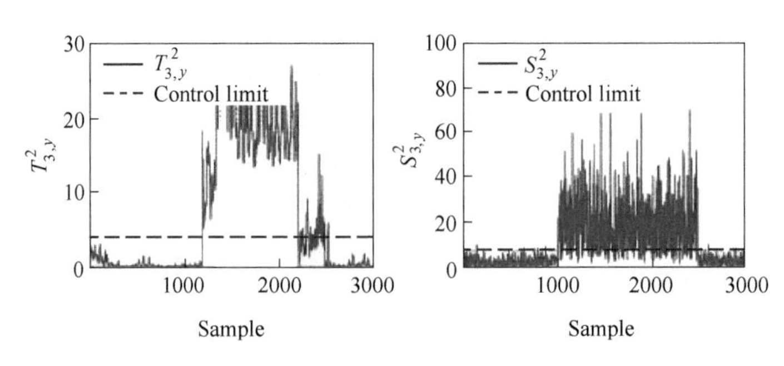

$T^2_{3,y}$
Control limit
Sample
$S^2_{3,y}$
Control limit
Sample

(c)

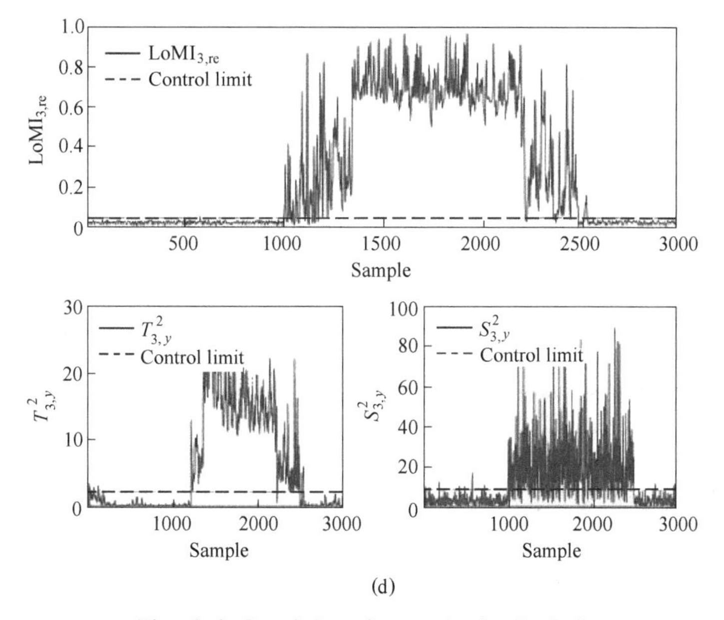

(d)

Fig. 4. 6 Local detection results for Fault 2

(a) MB-GLPLS; (b) MB-QSFA; (c) NR-QMRSFA; (d) D-QMRSFA

4. 4 Conclusions

In this work, a novel quality-related distributed fault detection method is developed to solve the problem of dynamic coupling and quality inheritance in large-scale sequential processes. A new QMRSFA model is built for analyzing the local dynamic behavior in each sub-process. Meanwhile, an expression of sequentially interconnected relation between sub-processes is given and unified monitoring indexes are established based on Bayesian fusion. The effectiveness of the proposed method is verified by a large-scale HSMP, where better detection performance is achieved compared with other competitive methods. And more timely and refined information can provide for field engineers. The proposed method only considers the dynamic relationship between variables and is suitable for detection of single quality-related faults.

References

[1] Cai P P, Deng X G. Incipient fault detection for nonlinear processes based on dynamic multi-block probability related kernel principal component analysis [J]. ISA Transactions, 2020, 105: 210-220.

[2] Cao Y, Yuan X F, Wang Y L, et al. Hierarchical hybrid distributed PCA for plant-wide

monitoring of chemical processes [J]. Control Engineering Practice, 2021, 111: 104784.

[3] Chen J H, Liu K C. On-line batch process monitoring using dynamic PCA and dynamic PLS models [J]. Chemical Engineering Science, 2002, 57: 63-75.

[4] Chen Z W, Cao Y, Ding S X, et al. A distributed canonical correlation analysis-based fault detection method for plant-wide process monitoring [J]. IEEE Transactions on Industrial Informatics, 2019, 15: 2710-2720.

[5] Peng K X, Zhang K, Li G, et al. Contribution rate plot for nonlinear quality-related fault diagnosis with application to the hot strip mill process [J]. Control Engineering Practice, 2013, 21: 360-369.

[6] Li Z M, Bao S Y, Peng X, et al. Fault detection and diagnosis in multivariate systems using multiple correlation regression [J]. Control Engineering Practice, 2021, 116: 104916.

[7] Qin S J. Survey on data-driven industrial process monitoring and diagnosis [J]. Annual Reviews in Control, 2012, 36: 220-234.

[8] Yin S, Ding S X, Xie X C, et al. A review on basic data-driven approaches for industrial process monitoring [J]. IEEE Transactions on Industrial Electronics, 2014, 61: 6418-6428.

[9] Ge Z Q. Review on data-driven modeling and monitoring for plant-wide industrial processes [J]. Chemometrics and Intelligent Laboratory Systems, 2017, 171: 16-25.

[10] Zheng J L, Zhao C H, Gao F R. Retrospective comparison of several typical linear dynamic latent variable models for industrial process monitoring [J]. Computers and Chemical Engineering, 2022, 147: 107587.

[11] Peng K X, Ma L, Zhang K. Review of quality-related fault detection and diagnosis techniques for complex industrial processes [J]. Acta Automatica Sinica, 2017, 43: 349-365.

[12] Sun R R, Wang Y Q. C-IPLS-IKPLS for modeling and detecting nonlinear multimode processes [J]. Industrial & Engineering Chemistry Research, 2021, 60: 1684-1698.

[13] Lee J M, Yoo C K, Lee I B. Statistical monitoring of dynamic processes based on dynamic independent component analysis [J]. Chemical Engineering Science, 2004, 59: 2995-3006.

[14] Dong Y N, Qin S J. A novel dynamic PCA algorithm for dynamic data modeling and process monitoring [J]. Journal of Process Control, 2018, 67: 1-11.

[15] Dong Y N, Qin S J. Dynamic latent variable analytics for process operations and control [J]. Computers and Chemical Engineering, 2018, 114: 69-80.

[16] Liu Q, Fang D, Dong Y N, et al. Dynamic modeling and reconstruction based fault detection and location of train bearings [J]. Acta Automatica Sinica, 2019, 45: 2233-2241.

[17] Deng X G, Liu X Y, Cao Y P, et al. Incipient fault detection for dynamic chemical processes based on enhanced CVDA integrated with probability information and fault-sensitive features [J]. Journal of Process Control, 2022, 114: 29-41.

[18] Zheng J L, Zhao C H. Online monitoring of performance variations and process dynamic anomalies with performance-relevant full decomposition of slow feature analysis [J]. Journal of Process Control, 2019, 80: 89-102.

[19] Shang C, Huang B, Yang F, et al. Slow feature analysis for monitoring and diagnosis of control performance [J]. Journal of Process Control, 2016, 39: 21-34.

[20] Dong J, Wang Y Q, Peng K X. A novel fault detection method based on the extraction of slow features for dynamic nonstationary processes [J]. IEEE Transactions on Instrumentation and Measurement, 2022, 71: 3500611.

[21] Puli V K, Raveendran R, Huang B. Complex probabilistic slow feature extraction with applications in process data analytics [J]. Computers and Chemical Engineering, 2021, 154: 107456.

[22] Fan L, Kodamana H, Huang B. Semi-supervised dynamic latent variable modeling: I/O probabilistic slow feature analysis approach [J]. AIChE Journal, 2019, 65: 964-979.

[23] Zhang H Y, Deng X G, Zhang Y C, et al. Dynamic nonlinear batch process fault detection and identification based on two-directional dynamic kernelslow feature analysis [J]. The Canadian Journal of Chemical Engineering, 2021, 99: 306-333.

[24] Chiplunkar R, Huang B. Output relevant slow feature extraction using partial least squares [J]. Chemometrics and Intelligent Laboratory Systems, 2019, 191: 148-157.

[25] Qin Y, Zhao C H. Comprehensive process decomposition for closed-loop process monitoring with quality-relevant slow feature analysis [J]. Journal of Process Control, 2019, 77: 141-154.

[26] Zhong W M, Jiang C, Peng X, et al. Online quality prediction of industrial terephthalic acid hydropurification process using modified regularized slow-feature analysis [J]. Industrial & Engineering Chemistry Research, 2018, 57: 9604-9614.

[27] Zhang Y W, Zhou H, Qin S J, et al. Decentralized fault diagnosis of large-scale processes using multiblock kernel partial least squares [J]. IEEE Transactions on Industrial Informatics, 2010, 6: 3-10.

[28] Ge Z Q, Song Z H. Distributed PCA model for plant-wide process monitoring [J]. Industrial and Engineering Chemistry Research, 2013, 52: 1947-1957.

[29] Jiang Q C, Yan X F, Huang B. Review and perspectives of data-driven distributed monitoring for industrial plant-wide processes [J]. Industrial & Engineering Chemistry Research, 2019, 58: 12899-12912.

[30] Ge Z Q, Zhang M G, Song Z H. Nonlinear process monitoring based on linear subspace and Bayesian inference [J]. Journal of Process Control, 2010, 20: 676-688.

[31] Peng X, Ding S X, Du W L, et al. Distributed process monitoring based on canonical correlation analysis with partly-connected topology [J]. Control Engineering Practice, 2020, 101: 104500.

[32] Chen X L, Wang J, Ding S X. Complex system monitoring based on distributed least squares method [J]. IEEE Transactions on Automation Science and Engineering, 2021, 18: 1892-1900.

[33] Ma L, Dong J, Hu C J, et al. A novel decentralized detection framework for quality-related faults in manufacturing industrial processes [J]. Neurocomputing, 2021, 428: 30-41.

[34] Li L L, Ding S X, Peng X. Distributed data-driven optimal fault detection for large-scale systems [J]. Journal of Process Control, 2020, 39: 94-103.

[35] Xie X C, Sun W, Cheung K C. An advanced PLS approach for key performance indicator-related prediction and diagnosis in case of outliers [J]. IEEE Transactions on Industrial Electronics, 2016, 63: 2587-2594.

[36] Chu F, Zhao X, Dai W, et al. Data-driven robust evaluation method for optimal operating status and its application [J]. Acta Automatica Sinica, 2020, 46: 439-450.

[37] Serneels S, Croux C, Filzmoser P, et al. Partial robust M-regression [J]. Chemometrics and Intelligent Laboratory Systems, 2005, 79: 55-64.

[38] Wiskott L, Sejnowski T J. Slow feature analysis: Unsupervised learning of invariance [J]. Neural Computation, 2002, 14: 715-770.

[39] Xu X, Ding J L. Decentralized dynamic process monitoring based on manifold regularized slow feature analysis [J]. Journal of Process Control, 2021, 98: 79-91.

Chapter 5 Full Condition Process Monitoring for HSMP

Process industries are always characterized by large-scale and complex correlations. Nowadays, feedback control can hardly maintain industrial safety and product quality[1-2]. Process monitoring has become an important complement of industrial process control. In past decades, with the continuously development of computer hardware and sensor technology, data-driven methods have been widely studied in process industries[3-4]. However, for nonlinear process industries, traditional monitoring methods, such as principal component analysis (PCA)[5] and partial least squares (PLS)[6] have poor monitoring performances due to their linear assumptions[7].

To handle the problem posed by process nonlinearity, the kernel-based methods are widely used, whose core idea is to transform the nonlinear space into a linear projection space by implementing a kernel mapping. Peng et al. developed a modified kernel PLS (KPLS)-based monitoring model for HSMP[8]. Zhang took advantages of three kernel methods to enhance the nonlinear detection ability[9]. Since last decade, manifold learning methods have draw much attention and been proven to be powerful in nonlinear process monitoring. Zhou et al. proposed a nonlinear monitoring model with PLS and locally linear embedding (LLE)[10]. Compared with kernel-based methods, manifold learning methods can effectively discover the intrinsic structure hidden in nonlinear data and don't need a prior kernel function for low dimension mapping. However, LLE merely preserves the local features of nonlinear samples. Fortunately, *t*-SNE can not only preserve local geometry, but also reveal global structure information[11]. Thus, *t*-SNE will be utilized to extract nonlinear principal components (NPCs) from the hot rolling data in this chapter.

Moreover, nonlinear processes always have dynamic characteristic. To address dynamics in the process, dynamic models have been derived from traditional multivariate statistical ones. Chen developed dynamic PCA (DPCA) and dynamic PLS (DPLS) models for online batch process monitoring[12]. To overcome Gaussian distribution assumption of DPCA, dynamic ICA (DICA) was proposed by Lee et al. and showed more powerful monitoring performance[13]. Canonical variate analysis (CVA) is another efficient tool for fault detection in dynamic process[14]. The latent states of CVA can reflect process dynamics. However, above methods cannot effectively capture temporal behaviors of the process. Besides, many nonstationary variables exist in process industries and their statistical properties are time variant. Traditional multivariate statistical techniques can not sufficiently extract latent information of nonstationary variables. Recently, slow feature analysis (SFA) and cointegration analysis (CA) have become useful techniques for fault detection in dynamic process[15-16]. Shang et al. proposed a monitoring model based on SFA, which could detect process dynamics anomalies[17]. CA can be used to analyze long-run dynamic relations among nonstationary variables. SFA can not only detect static variations, but also find temporal dynamic changes from time distribution. The slowest features obtained by SFA reveal static variations of process and the fastest ones are viewed as temporal dynamic variations. Because of complementarities between CA and SFA, Zhao developed a full-condition monitoring method based on CA and SFA for dynamic chemical processes[18]. However, SFA is a linear algorithm and has a poor ability to deal with nonlinear data. CA also may not function well for nonlinear process industries.

In addition, manufacturing production systems are not always under loaded condition. The process may be in idle condition for a long period of time. For example, strip steels are produced coil by coil with each coil representing a batch as shown in **Fig. 5. 1**. There is a long idle period between coils. Meanwhile, the rolling speed can reach up to 20 m/s and the producing time of each coil is usually a few minutes. It can be concluded from **Fig. 5. 1** that the duration of idle condition is close to that of loaded condition. Previous works only pay attention to fault detection in loaded condition and ignore monitoring idle condition of the HSMP. However, strip production usually begins from idle condition. If an incipient fault occurs at this moment, fault severity will be aggravated along the fault propagation path. The equipment can hardly produce qualified strip steels. More seriously, it may result in equipment damage, environmental pollution, property losses and even major safety accidents. Thus, monitoring idle condition is of great significance, although it is not as complex as the loaded one.

In this chapter, a full condition monitoring model is developed for nonlinear dynamic process industries. Not only loaded condition of the process is fully analyzed, but also idle condition is effectively monitored. In addition, for different products, establishing corresponding models can also get good monitoring results. The contributions of this work are as follows:

(1) By robust preprocessing, a novel dissimilarity index (DI) is proposed which can effectively identify the process condition;

(2) A nonlinear cointegration analysis (NCA) model based on t-SNE and CA is developed which can extract and analyze the long-run dynamic relations of nonstationary variables;

(3) A nonlinear slow feature analysis (NSFA) model based on t-SNE and SFA is developed which can extract and analyze the temporal dynamic and static variations of stationary variables as well as the steady residual after NCA modeling.

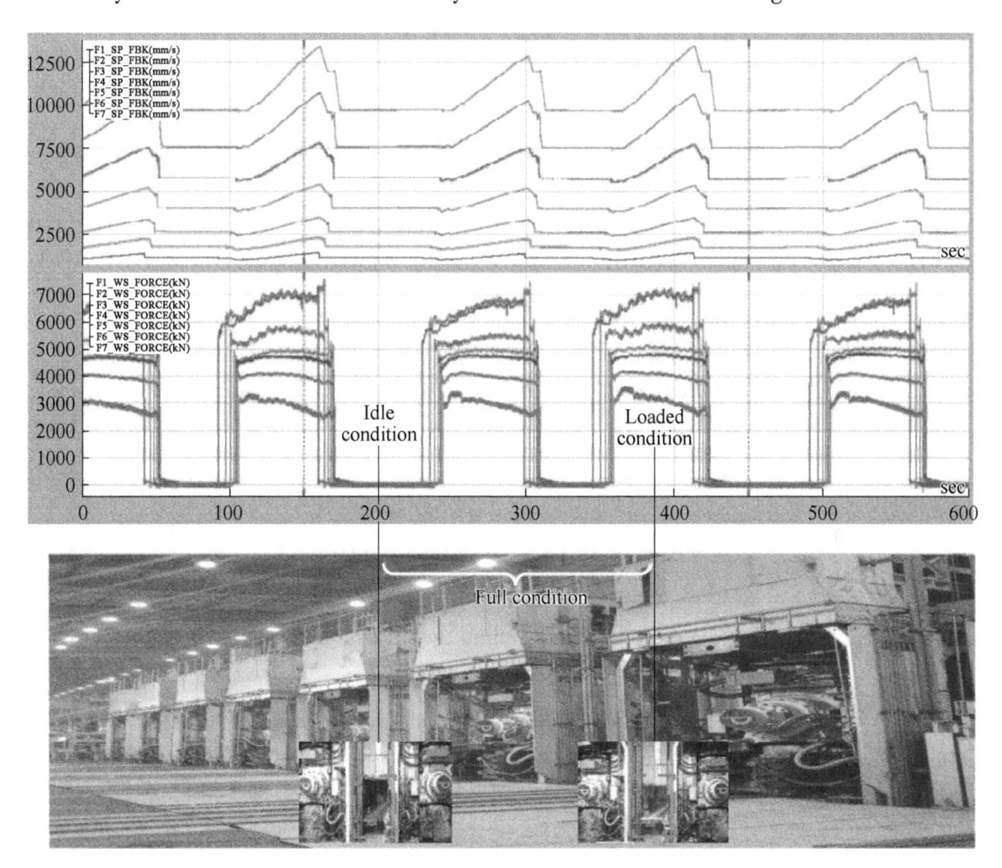

Fig. 5.1 A real datalog of HSMP

5.1 Review of *t*-SNE, CA, and SFA

5.1.1 Basic theory of *t*-SNE

Maaten and Hinton proposed a manifold learning algorithm for nonlinear dimension reduction, which is called t-SNE. t-distribution is used instead of Gauss distribution in projection space. Given a set of data $X \in R^{m\times n}$, the cost function of t-SNE can be obtained as:

$$C = \sum_i KL(P_i \parallel Q_i) = \sum_i \sum_j p_{ij} \lg \frac{p_{ij}}{q_{ij}} \tag{5.1}$$

where, p_{ij} and q_{ij} are the joint probability of two samples in original data space ($X(i)$ and $X(j)$) and mapped samples in projection space ($Y(i)$ and $Y(j)$), respectively.

p_{ij} and q_{ij} are calculated as:

$$p_{ij} = \frac{\exp(-\parallel X(i) - X(j) \parallel^2 / 2\sigma_i^2)}{2n\sum_{k\neq i}\exp(-\parallel X(i) - X(k) \parallel^2 / 2\sigma_i^2)} + \frac{\exp(-\parallel X(j) - X(i) \parallel^2 / 2\sigma_j^2)}{2n\sum_{k\neq j}\exp(-\parallel X(j) - X(k) \parallel^2 / 2\sigma_j^2)} \tag{5.2}$$

$$q_{ij} = \frac{(1 + \parallel Y(i) - Y(j) \parallel^2)^{-1}}{\sum_{k\neq l}(1 + \parallel Y(k) - Y(l) \parallel^2)^{-1}} \tag{5.3}$$

where, σ_i is variance of the Gaussian that is centered over each original data point $X(i)$. A t-distribution is used in q_{ij} to solve the crowding problem.

5.1.2 Basic theory of CA

Engel and Granger first developed CA in the 1980s[19]. The latent relations among nonstationary variables can be effectively extracted by CA. Assuming $X = [x_1, x_2, \cdots, x_m]^T \in R^{m\times n}$ is a nonstationary dataset. The cointegration relationship of nonstationary variables can be represented as follows:

$$\xi = \beta_1 x_1 + \beta_2 x_2 + \cdots + \beta_m x_m = \beta^T - X \tag{5.4}$$

where, β is a cointegration vector; ξ is the equilibrium error.

When equation (5.4) contains more than two variables, there may be multiple cointegration relationships. If Engle-Granger (E-G) two-step method is used in this situation, the co-integration vectors cannot be found. Johansen et al. proposed a vector auto-regression (VAR) test method to solve this problem[20]. The VAR (p) model of $X(t)$ is as follows:

$$\begin{aligned} X(t) &= \sum_{k=1}^{p} \Psi_k X(t-k) + \varphi(t) \\ \Delta X(t) &= \Psi X(t-1) + \sum_{k=1}^{p-1} \Phi_k \Delta X(t-k) + \varphi(t) \end{aligned} \tag{5.5}$$

where, t denotes sampling time; p denotes order, $\Psi_k(m \times m)$ are coefficient matrices; $\varphi(t)(m \times 1)$ is white noise; $\Psi = \sum_{k=1}^{p} \Psi_k - I$, $\Phi_k = \sum_{j=k+1}^{p} \Psi_j$, $k = 1, 2, \cdots, p-1$.

Then $\boldsymbol{\Psi}$ can be factorized into two matrices, that is, $\boldsymbol{\Psi}=\boldsymbol{\alpha}\boldsymbol{\beta}^{\mathrm{T}}$, where $\boldsymbol{\alpha}(m \times r)$ denots a weight matrix, $\boldsymbol{\beta}(m \times r)$ denotes a cointegration matrix, r is the number of cointegration vectors. $\boldsymbol{\beta}$ is calculated by maximum likelihood estimation.

5.1.3 Basic theory of SFA

In 2002, Wiskott et al. first developed SFA model[21]. It can extracts slow features from quickly varying time series. Assuming $\boldsymbol{X}(t)=[x_1(t), x_2(t), \cdots, x_m(t)]^{\mathrm{T}}$ to be an m-dimensional input data, SFA aims to find a propoer function $\boldsymbol{g}(\boldsymbol{X}(t))=[g_1(\boldsymbol{X}(t)), \cdots, g_m(\boldsymbol{X}(t))]^{\mathrm{T}}$ which can produce an output $\boldsymbol{S}(t)=[s_1(t), s_2(t), \cdots, s_m(t)]^{\mathrm{T}}$ with $\{s_i(t)=g_i(\boldsymbol{X}(t))\}_{i=1}^{m}$. The optimization function can be defined as:

$$\min\Delta(s_i)=\min\langle \dot{s}_i^2\rangle \tag{5.6}$$

under the constraints

$$\begin{aligned} &\langle s_i\rangle_t=0 && (\text{zero mean})\\ &\langle s_i^2\rangle_t=1 && (\text{unit variance})\\ &\forall i\neq j:\ \langle s_i s_j\rangle=0 && (\text{decorrelation}) \end{aligned} \tag{5.7}$$

where, $\dot{s}$ is the derivative of s and the angle brackets denote temporal averaging. When $g_i(\cdot)$ is a linear function, $s_i(t)$ is written as $s_i(t)=g_i(\boldsymbol{X}(t))=\boldsymbol{w}_i^{\mathrm{T}}\boldsymbol{X}(t)$. Thus, the objective function can be rewritten as follows:

$$\begin{aligned} &\min\langle \dot{s}_i(t)^2\rangle=\min \boldsymbol{w}_i^{\mathrm{T}}\langle \dot{\boldsymbol{X}}(t)\dot{\boldsymbol{X}}(t)^{\mathrm{T}}\rangle \boldsymbol{w}_i=\min \boldsymbol{w}_i^{\mathrm{T}}\boldsymbol{A}\boldsymbol{w}_i\\ &\text{s.t.}\quad \langle s_i^2\rangle_t=\boldsymbol{w}_i^{\mathrm{T}}\langle \boldsymbol{X}(t)\boldsymbol{X}(t)^{\mathrm{T}}\rangle \boldsymbol{w}_i=\boldsymbol{w}_i^{\mathrm{T}}\boldsymbol{B}\boldsymbol{w}_i=1 \end{aligned} \tag{5.8}$$

where, $\{\boldsymbol{w}_i\}_{i=1}^{m}$ are coefficient vectors; $\boldsymbol{A}$ is the covariance matrix of $\dot{\boldsymbol{X}}(t)$; $\boldsymbol{B}$ is the covariance matrix of $\boldsymbol{X}(t)$.

5.2 Nonlinear full condition process monitoring

In this section, the nonlinear full condition process monitoring for HSMP will be presented. First, a condition identification based on DI is discussed to determine the current process condition. If the process is under idle condition, support vector data description (SVDD) is used for process monitoring. Otherwise, NCA is proposed to extract the long-run dynamic relations in loaded condition. Afterward, NSFA is introduced for further exploring the temporal dynamic and static variations. From the above three interrelated models in the whole process, four statistics can be obtained for the nonlinear dynamic process monitoring, which is illustrated in **Fig. 5.2**.

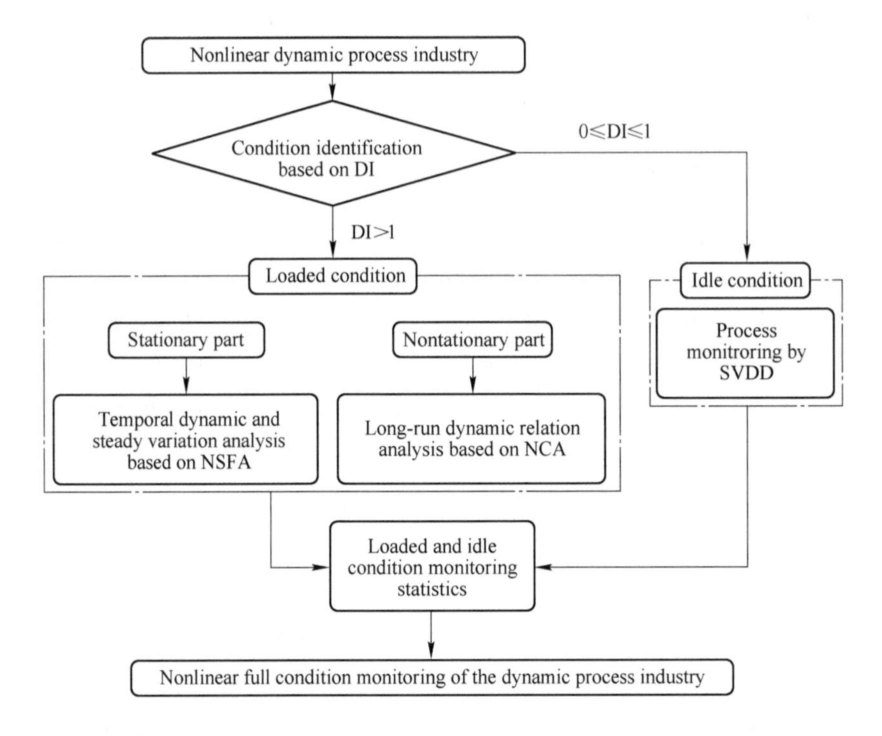

Fig. 5. 2 Illustration of the nonlinear full condition process monitoring method

5. 2. 1 DI based condition identification

The raw data of hot rolling are usually non-stationary and contain outliers, which may affect the condition identification and the subsequent process monitoring. In order to improve data stationarity and eliminate outliers, raw data should be first preprocessed with the following robust method[22]:

$$\boldsymbol{r}' = \frac{\boldsymbol{r} - \text{median}(\boldsymbol{r})}{\text{MAD}(\boldsymbol{r})} \tag{5.9}$$

where, $\boldsymbol{r}$ denotes a vector of raw data, median $(\boldsymbol{r})$ is the median of $\boldsymbol{r}$, and $\text{MAD}(\boldsymbol{r}) = \text{median}(|\boldsymbol{r} - \text{median}(\boldsymbol{r})|)$ is the median absolute deviation. Since median and MAD statistics are both insensitive to outliers, the MAD based scaling can be regarded as the robust normalization. The MAD method is simple and very effective in most occasions.

Assume that $\boldsymbol{X} = [\boldsymbol{x}_1, \boldsymbol{x}_2, \cdots, \boldsymbol{x}_m]^{\mathrm{T}} \in R^{m\times n}$ are preprocessed by MAD method. $\boldsymbol{X}$ consists of two subset: the idle condition data set $\boldsymbol{X}_I \in R^{m\times n_I}$ (reference data set) and the loaded condition data set $\boldsymbol{X}_W \in R^{m\times n_W}$, where $n_I + n_W = n$. A dissimilarity index is proposed for condition identification:

$$\mathrm{DI} = \sqrt{\frac{1}{m}\sum_{i=1}^{m}\frac{(n_I - 1)\left[2x_i(t) - Q_1(x_i) - Q_3(x_i)\right]^2}{4\sum_{j=1}^{n_I}\left[x_i(j) - \mathrm{median}(x_i)\right]^2}} \tag{5.10}$$

where, $x_i(t)$ is the ith variable at sampling time t ; $\mathrm{median}(x_i)$, $Q_1(x_i)$ and $Q_3(x_i)$ are the median, bottom and top quartile of the ith variable in $\boldsymbol{X}_I$, respectively.

When the computed value lies between 0 and 1, it indicates that the relative distribution of $\boldsymbol{X}(t)$ is the same as the average of $\boldsymbol{X}_I$. Whereas a value is much greater than 1, it indicates that there is no overlap between them. In this chapter, a value of 1 is set as the identification threshold. If DI is beyond the threshold, the HSMP is in loaded condition and NCA is applied to it for further analysis in the following subsection. Otherwise, the HSMP works under idle condition and fault detection is conducted with SVDD.

5.2.2 SVDD-based idle condition process monitoring

SVDD tries to seek an optimal hypersphere with radius R and center a, which contains normal samples in $\boldsymbol{X}_I$ as much as possible. In order to establish a model in idle condition, the data are mapped to the feature space by a transformation $\boldsymbol{\Phi}$: $\boldsymbol{X}_I \rightarrow \boldsymbol{F}$. The optimization function of SVDD is as follows[23] :

$$\begin{gathered}\min\left(R^2 + C\sum_{t=1}^{n_I}\xi_t\right)\\ \text{s.t.}\ \ \|\boldsymbol{\Phi}(\boldsymbol{X}_I(t)) - a\|^2 \leqslant R^2 + \xi_t,\quad \xi_t \geqslant 0,\quad t = 1,\ \cdots,\ n_I\end{gathered} \tag{5.11}$$

where, parameter C makes a trade-off between the volume of hypersphere and errors; ξ_t is the slack variable of $\boldsymbol{\Phi}(\boldsymbol{X}_I(t))$.

The corresponding dual problem of equation (5.11) can be represented as:

$$\begin{gathered}\max L = \sum_{t=1}^{n_I}\alpha_t K(t,\ t) - \sum_{t=1}^{n_I}\sum_{v=1}^{n_I}\alpha_t\alpha_v K(t,\ v)\\ \text{s.t.}\ \sum_{t=1}^{n_I}\alpha_t = 1,\ 0 \leqslant \alpha_t \leqslant C,\ t = 1,\ \cdots,\ n_I\end{gathered} \tag{5.12}$$

where, α_t is the Lagrange multiplier of $\boldsymbol{X}_I(t)$; $K(t,\ v) = \boldsymbol{\Phi}(\boldsymbol{X}_I(t))^{\mathrm{T}}\boldsymbol{\Phi}(\boldsymbol{X}_I(v))$ is a kernel function.

Support vectors (SVs) are samples with $0 \leqslant \alpha_t \leqslant C$. R can be calculated as:

$$R = \sqrt{1 - 2\sum_{t}^{n_I} \alpha_t K(t, \ \mathrm{SV}) + \sum_{t=1}^{n_I} \sum_{v=1}^{n_I} \alpha_t \alpha_v K(t, \ v)} \tag{5.13}$$

where, $K(t, \ \mathrm{SV}) = \boldsymbol{\Phi}(X_I(t))^{\mathrm{T}} \boldsymbol{\Phi}(X_I(\mathrm{SV}))$, $X_I(\mathrm{SV})$ is one of SVs.

When a test sample $\boldsymbol{X}(\mathrm{new})$ is obtained, its distance to a is computed as:

$$D = \sqrt{1 - 2\sum_{t}^{n_I} \alpha_t K(t, \ \mathrm{new}) + \sum_{t=1}^{n_I} \sum_{v=1}^{n_I} \alpha_t \alpha_v K(t, \ v)} \tag{5.14}$$

A fault occurs when D is larger than R , otherwise the process is under normal condition.

5.2.3 Long-run dynamic relation analysis based on NCA

When the HSMP is in loaded condition, the relations among process variables become more complex due to nonlinearity and dynamic of the process. Despite that the values of process variables may vary with the loaded condition, the process variables may have some certain dynamic relations from a long-run perspective. Such long-run dynamic relations of nonstationary variables can be captured by CA. Because stationary and nonstationary variables are mixed in process data $\boldsymbol{X}_W$, augmented Dickey-Fuller (ADF) test is employed to distinguish stationary variables $\boldsymbol{X}_{W_S} \in R^{m_S \times n_W}$ and nonstationary ones $\boldsymbol{X}_{W_N} \in R^{m_N \times n_W}$, where $m_S + m_N = m$.

In order to address the nonlinear behavior in loaded condition, t-SNE is used on the nonstationary variables $\boldsymbol{X}_{W_N}$ to extract NPCs $\boldsymbol{Y}_{W_N} \in R^{m_n \times n_W}$, where $m_n < m_N$. By minimizing equation (5.1), $\boldsymbol{Y}_{W_N}$ can be obtained as follows:

$$\frac{\partial C}{\partial \boldsymbol{Y}_{W_N}(i)} = 4\sum_{j} \frac{(p_{ij} - q_{ij})(\boldsymbol{Y}_{W_N}(i) - \boldsymbol{Y}_{W_N}(j))}{1 + \| \boldsymbol{Y}_{W_N}(i) - \boldsymbol{Y}_{W_N}(j) \|^2} \tag{5.15}$$

Then, CA is performed on $\boldsymbol{Y}_{W_N}$ to obtain long-run dynamic relations as follows:

$$\boldsymbol{E} = \boldsymbol{\beta}_W^{\mathrm{T}} \boldsymbol{Y}_{W_N} \tag{5.16}$$

where, $\boldsymbol{E} \in R^{r_W \times n_W}$ is the equilibrium errors reflecting the long-run dynamic relations; $\boldsymbol{\beta}_W \in R^{m_n \times r_W}$ is the cointegration matrix; r_W is the number of cointegration vectors for $\boldsymbol{Y}_{W_N}$.

Then squared prediction error is utilized to monitor the long-run dynamic relations of nonstationary variables, which is defined as follows:

$$L^2 = \boldsymbol{E}(t)^{\mathrm{T}}\boldsymbol{E}(t) \tag{5.17}$$

where, $\boldsymbol{E}(t)$ denotes the value of equilibrium errors at time t.

And the threshold of L^2 can be obtained as:

$$L_\alpha^2 = g\chi_{h,\alpha}^2, \quad g = \frac{s}{2\mu}, \ h = 2\frac{2\mu^2}{s^2} \tag{5.18}$$

where, $g\chi_{h,\alpha}^2$ denotes a chi-squared distribution with h degrees of freedom at signification level α; g is a weighted parameter; μ and s^2 are the sample mean and variance of L^2. If $L^2 > L_\alpha^2$, it reveals that the long-run dynamic relations among the nonstationary variables are out of equilibrium. Mode transitions or incipient faults may have occurred in loaded condition.

However, L^2 statistic only describe the long-run dynamic variations. After NCA modeling, stationary information in the remaining part of $\boldsymbol{Y}_{W_N}$ can be extracted by difference operation. The steady residual $\widetilde{\boldsymbol{Y}}_{W_N} \in R^{m_n \times n_W}$ can be obtained as follows:

$$\begin{aligned}\widetilde{\boldsymbol{Y}}_{W_N} &= (\boldsymbol{I} - \boldsymbol{\beta}_W\boldsymbol{\beta}_W^T)\Delta\boldsymbol{Y}_{W_N} \\ &= (\boldsymbol{I} - \boldsymbol{\beta}_W\boldsymbol{\beta}_W^T)\begin{bmatrix}\Delta\boldsymbol{Y}_{W_N}(1) \\ \vdots \\ \Delta\boldsymbol{Y}_{W_N}(t) \\ \vdots \\ \Delta\boldsymbol{Y}_{W_N}(n_W)\end{bmatrix}^{\mathrm{T}}\end{aligned} \tag{5.19}$$

where, $\Delta\boldsymbol{Y}_{W_N}(t) = \boldsymbol{Y}_{W_N}(t) - \boldsymbol{Y}_{W_N}(t-1)$ denotes the difference between two neighboring samples at time t.

5.2.4 Temporal dynamic and static variation analysis based on NSFA

The steady residual $\widetilde{\boldsymbol{Y}}_{W_N}$ may still contain some valuable information, which can be combined with stationary variables for further analysis. First, t-SNE is used on $\boldsymbol{X}_{W_S}$ to extract NPCs, which is fundamentally the same as above subsection. Assuming that $\boldsymbol{Y}_{W_S} \in R^{m_s \times n_W}$ is the NPC matrix, where $m_s < m_S$. Then a new stationary combination matrix $\boldsymbol{Y}_S = \begin{bmatrix}\widetilde{\boldsymbol{Y}}_{W_N} \\ \boldsymbol{Y}_{W_S}\end{bmatrix} \in R^{m_C \times n_W}$ is constructed with $\widetilde{\boldsymbol{Y}}_{W_N}$ and $\boldsymbol{Y}_{W_S}$, where $m_C = m_n + m_s$. It may contain both temporal dynamic and static variation information, which can be extracted by SFA. Wiskott proposed an effective algorithm for finding the coefficient matrix $\boldsymbol{W} =$

$[\boldsymbol{w}_1, \boldsymbol{w}_2, \cdots, \boldsymbol{w}_{m_C}]^{\mathrm{T}}$. Similar to independent component analysis (ICA), SFA requires the following singular value decomposition for $\langle \boldsymbol{Y}_S \boldsymbol{Y}_S^{\mathrm{T}} \rangle_t$ to obtain a whitening transformation:

$$\begin{aligned} \langle \boldsymbol{Y}_S \boldsymbol{Y}_S^{\mathrm{T}} \rangle_t &= \boldsymbol{U} \boldsymbol{\Lambda} \boldsymbol{U}^{\mathrm{T}} \\ \boldsymbol{Z} &= \boldsymbol{\Lambda}^{-1/2} \boldsymbol{U}^{\mathrm{T}} \boldsymbol{Y}_S = \boldsymbol{Q} \boldsymbol{Y}_S \end{aligned} \tag{5.20}$$

where, $\boldsymbol{Q} = \boldsymbol{\Lambda}^{-1/2} \boldsymbol{U}^{\mathrm{T}}$ is a whitening matrix.

Then another singular value decomposition is applied to $\langle \dot{\boldsymbol{Z}} \dot{\boldsymbol{Z}}^{\mathrm{T}} \rangle_t$ and coefficient matrix is obtained as follows:

$$\begin{aligned} \langle \dot{\boldsymbol{Z}} \dot{\boldsymbol{Z}}^{\mathrm{T}} \rangle_t &= \boldsymbol{P}^{\mathrm{T}} \boldsymbol{\Omega} \boldsymbol{P} \\ \boldsymbol{W} &= \boldsymbol{P} \boldsymbol{\Lambda}^{-1/2} \boldsymbol{U}^{\mathrm{T}} \end{aligned} \tag{5.21}$$

Finally, the NSFA model can be expressed as:

$$\begin{aligned} \boldsymbol{S} &= \boldsymbol{W} \boldsymbol{Y}_S = \boldsymbol{P} \boldsymbol{\Lambda}^{-1/2} \boldsymbol{U}^{\mathrm{T}} \boldsymbol{Y}_S \\ \dot{\boldsymbol{S}} &= \boldsymbol{W} \dot{\boldsymbol{Y}}_S = \boldsymbol{P} \boldsymbol{\Lambda}^{-1/2} \boldsymbol{U}^{\mathrm{T}} \dot{\boldsymbol{Y}}_S \end{aligned} \tag{5.22}$$

where, $\boldsymbol{S}$ and $\dot{\boldsymbol{S}}$ are slow features and temporal features of $\boldsymbol{Y}_S$.

According to the criterion defined in reference [17], $\boldsymbol{S}$ and $\dot{\boldsymbol{S}}$ can be further divided into 2 subsets as follows:

$$\boldsymbol{S} = \begin{bmatrix} \boldsymbol{S}_S \\ \boldsymbol{S}_F \end{bmatrix} \quad \dot{\boldsymbol{S}} = \begin{bmatrix} \dot{\boldsymbol{S}}_S \\ \dot{\boldsymbol{S}}_F \end{bmatrix} \tag{5.23}$$

where, $\boldsymbol{S}_S \in R^{J \times n_W}$ denotes the J slowest features; $\boldsymbol{S}_F \in R^{(m_C - J) \times n_W}$ denotes the fastest ones; $\dot{\boldsymbol{S}}_S$ denotes the J slowest temporal features; $\dot{\boldsymbol{S}}_F$ denotes the fastest ones.

If NSFA is applied to HSMP, $\boldsymbol{S}_S$ tends to reflect the general static tendency, while $\dot{\boldsymbol{S}}_F$ is regarded as the short-run dynamic variations. In order to analyze temporal dynamic and static variation of the HSMP, two monitoring statics based on $\boldsymbol{S}_S$ and $\dot{\boldsymbol{S}}_F$ are defined as below:

$$
\begin{aligned}
T_S^2 &= \boldsymbol{S}_S^{\mathrm{T}}(t)\, \boldsymbol{S}_S(t) \\
T_F^2 &= \dot{\boldsymbol{S}}_F^{\mathrm{T}}(t)\boldsymbol{\Omega}_F^{-1}\, \dot{\boldsymbol{S}}_F(t)
\end{aligned}
\tag{5.24}
$$

where, $\boldsymbol{S}_S(t)$ and $\dot{\boldsymbol{S}}_F(t)$ denote the slow and temporal features at the sampling time t; $\boldsymbol{\Omega}_F = \langle \dot{\boldsymbol{S}}_F \dot{\boldsymbol{S}}_F^{\mathrm{T}} \rangle_t$ is the empirical covariance matrix of $\dot{\boldsymbol{S}}_F$.

Assuming that $\boldsymbol{S}_S$ and $\dot{\boldsymbol{S}}_F$ follow multivariate Gaussian distributions, with $(1-\alpha)$ confidence level, the threshold of T_S^2 follows a chi-squared distribution with J degrees of freedom, the threshold of T_F^2 follows a F-distribution with $(m_C - J)$ and $(n_W - m_C + J - 1)$ degrees of freedom:

$$
\begin{aligned}
T_{S,\alpha}^2 &= \chi_{J,\alpha}^2 \\
T_{F,\alpha}^2 &= \frac{(m_C - J)(n_W^2 - 2n_W)}{(n_W - 1)(n_W - m_C + J - 1)} F_{S-J,n_W-m_C+J-1,\alpha}
\end{aligned}
\tag{5.25}
$$

For a new sample $\boldsymbol{X}(\text{new})$, if $T_S^2 > T_{S,\alpha}^2$, then the static equilibrium is broken, and the process is deviated from the current process condition (i.e., step changes or drifts). If $T_F^2 > T_{F,\alpha}^2$, then a random variation may affect the process dynamics and hence control performance changes.

However, it should be noted that the t-SNE algorithm is suitable for offline analysis, but the intrinsic structure is not suitable for online monitoring of new samples. Based on the pseudoinverse and ridge regression learning algorithms[24], the transformation matrix from input variables to the NPCs can be obtained as:

$$
\begin{aligned}
&\underset{\boldsymbol{Q}_c}{\operatorname{argmin}} \; \| \boldsymbol{X}_{W_c}\boldsymbol{Q}_c - \boldsymbol{Y}_{W_c} \|_2^2 + \lambda_c \, \| \boldsymbol{Q}_c \|_2^2 \\
&\boldsymbol{Q}_c = (\lambda_c \boldsymbol{I}_c + \boldsymbol{X}_{W_c}\boldsymbol{X}_{W_c})^{-1} \boldsymbol{X}_{W_c}^{\mathrm{T}} \boldsymbol{Y}_{W_c}
\end{aligned}
\tag{5.26}
$$

where, $c = N$ or S, $\boldsymbol{Q}_N$ and $\boldsymbol{Q}_S$ are the transformation matrix of nonstationary variables and stationary variables respectively.

When new samples are coming, the NPCs can satisfy the needs of real-time computation. In addition, the number of NPCs is an important parameter. Too many or too few features will affect the performance of offline modeling. In this work, the mean squared error (MSE) between process data and reconstructed data is used to analyze how many NPCs should be retained for the following modelling. The flowchart of off-line training and on-line monitoring is shown in **Fig. 5.3**.

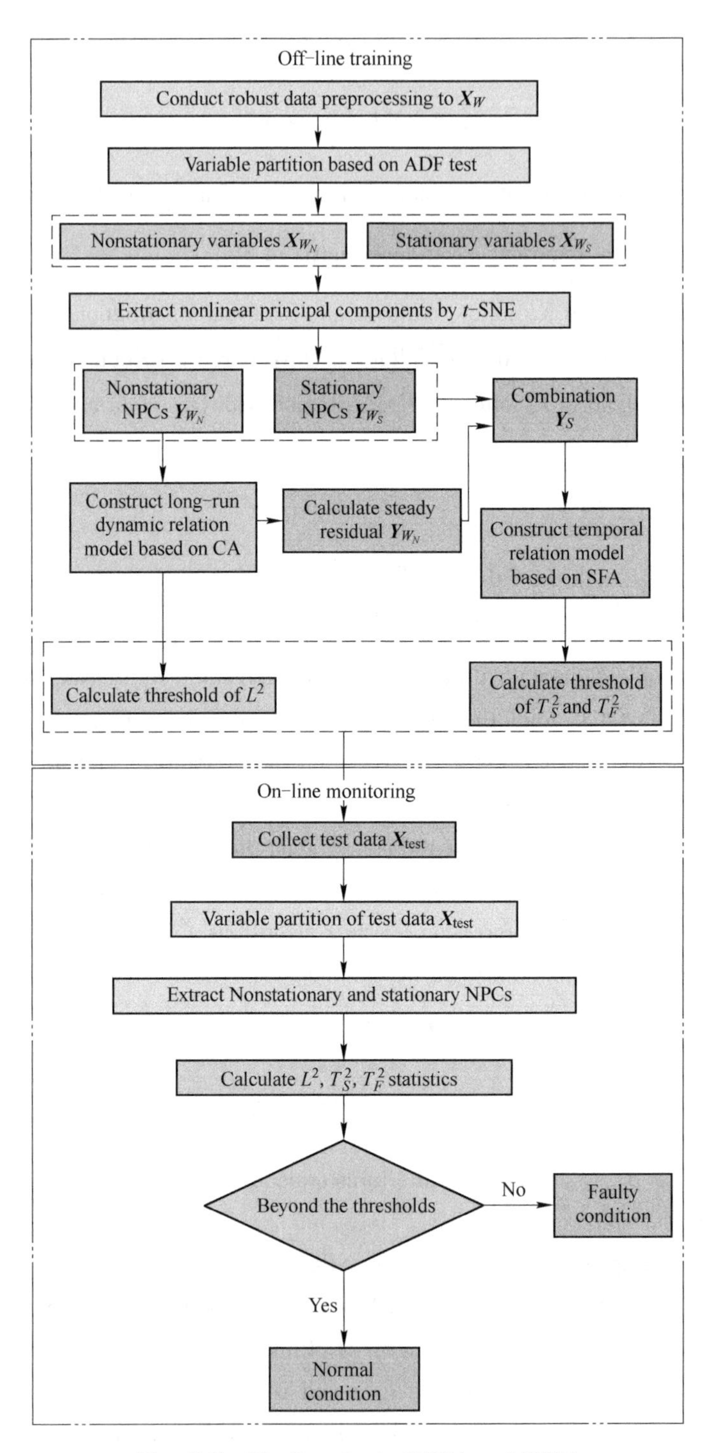

Fig. 5.3 The flowchart of NCA and NSFA

5.3 Case study

In this chapter, two products with thickness 2.70 mm and 3.95 mm are used for experimental study. For the specific description of HSMP, readers can get more details in previous work[25]. For simplicity, stationary variables and nonstationary variables with integrated of order one are considered in this work. The ADF test is applied to variables in the finishing mill area (FMA). **Table 5.1** shows relevant information of the variables, including 14 nonstationary variables and 18 stationary variables. The relationship between MSE and the number of the NPCs is shown in **Fig. 5.4**. Therefore, the NPCs of nonstationary variables (m_n) and stationary variables (m_s) are selected as 8 and 11, respectively.

Table 5.1 Descriptions of variables in the FMA

No.	Description	Unit
1	Strip flatness of the finishing mill exit	I
2	Strip width of the finishing mill exit	mm
3	Strip crown of the finishing mill exit	mm
4	Strip temperature	℃
5	Strip thickness	mm
6-12	Rolling speed of the 1_{st}-7_{th} stand	mm/s
13-19	Gap of the 1_{st}-7_{th} stand	mm
20-26	Rolling force of the 1_{st}-7_{th} stand	MN
27-32	Bending force of the 2_{nd}-7_{th} stand	MN

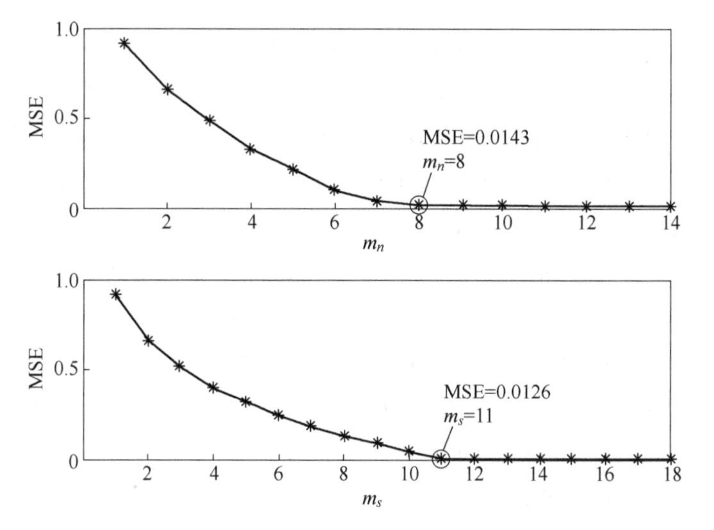

Fig. 5.4 The relationship between MSE and the number of the NPCs

In this chapter, the proposed method is applied to FMA for analyzing dynamic characteristic of the nonlinear process. KPCA, CVA, dynamic kernel principal component analysis (DKPCA), and kernel canonical variate analysis (KCVA) are used for comparison in this work. Four groups of normal and faulty datasets are collected with sampling interval of 10 ms. 1000 training samples in idle condition are used for condition identification and SVDD modeling, 3000 training samples in loaded condition are used for NCA and NSFA modeling, 4000 testing samples (1000 samples in idle condition and 3000 samples in loaded condition) are used for online test. Three typical faults under loaded condition are chosen in our research, which are shown in **Table 5.2**.

Table 5.2 Quality-related faults of the FMA

Fault No.	Description	Occurrence
1 (3.95 mm)	Malfunction of gap control loop in the 4_{th} stand	2001_{st}
2 (2.70 mm)	Sensor fault of bending force in the 5_{th} stand	1001_{st}
3 (3.95 mm)	Fault of cooling valve between the 2_{nd} and 3_{rd} stands	1001_{st}

In order to carry out full condition monitoring of the entire process, idle and loaded conditions are identified by the dissimilarity index. **Fig. 5.5** shows the identification results of a test dataset. From the 1001_{st} sample, slabs enter into loaded condition form idle condition and the DI statistic surpass the threshold which agrees well with the actual situation. Compared with the loaded condition, idle condition is relatively stable, and the values of process variables change little. **Fig. 5.6** shows that SVDD can monitor idle condition well and the false alarm rate (FAR) is 0.2%. Because of the process complexity and changeability under loaded condition, it is more important to monitor such condition. Therefore, all selected faults occur in loaded condition. The fault detection rates (FDRs) and FARs of different methods are calculated and tabulated in **Table 5.3** and **Table 5.4**, respectively. The highest FDR and the lowest FAR of each fault are signed in bold type. Obviously, the proposed method is more efficient than the other four methods when detecting Fault 2 and Fault 3. For instance, FDR of the proposed method for Fault 3 is the highest, and moreover, it has the lowest false alarms than the other four methods.

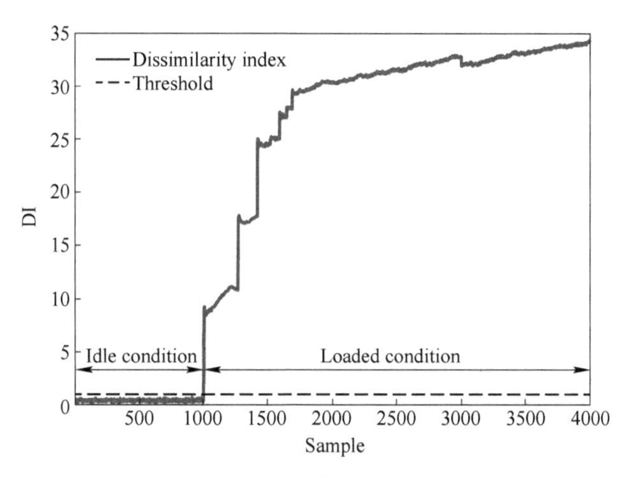

Fig. 5.5 Condition identification

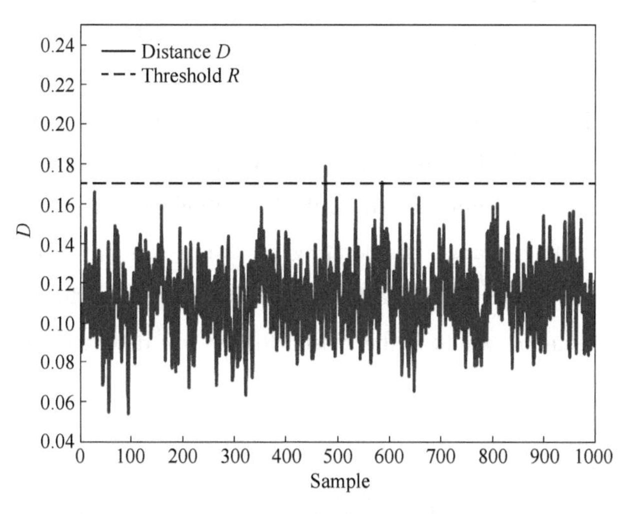

Fig. 5.6 Process monitoring in idle condition

Table 5.3 The FDRs of the 3 faults in loaded condition

No.	KPCA		CVA		DKPCA		KCVA		Propsed		
	T^2	SPE	T^2	SPE	T^2	SPE	T^2	SPE	L^2	T_S^2	T_F^2
1	**1**	**1**	**1**	**1**	**1**	**1**	**1**	**1**	**1**	**1**	0.169
2	0.782	**1**	0.909	**1**	0.944	**1**	0.978	**1**	**1**	**1**	0.195
3	0.689	0.998	0.945	**1**	0.962	0.995	0.992	**1**	**1**	**1**	0.132

Table 5.4　The FARs of the 3 faults in loaded condition

No.	KPCA		CVA		DKPCA		KCVA		Proposed		
	T^2	SPE	T^2	SPE	T^2	SPE	T^2	SPE	L^2	T_S^2	T_F^2
1	0.083	0.020	0.073	0.010	0.066	0.016	0.049	0.007	**0**	0.021	0.011
2	0.051	0.035	0.045	0.037	0.032	0.029	0.025	0.027	0.025	**0.012**	0.023
3	0.261	0.252	0.212	0.226	0.194	0.198	0.156	0.173	0.127	0.181	**0.048**

(1) **Fault case 1**: Once this fault occurs, the setting value of rolling gap is seriously affected, and the strip quality can also be immediately deteriorated. With the feedback control system, rolling forces and bending ones in the 5_{th} stands are changed in response to such anomaly. The monitoring results are shown in **Fig. 5.7**(a)-(f). The fault detection performances of the five methods are almost the same, but the thresholds of SPE in KPCA, CVA, DPKCA, and KCVA are not explicit, which must be magnified to find false alarm samples. Moreover, KPCA and CVA have more false alarms at the beginning stage, which is caused by the initial setup of the system. As nonlinear dynamic methods, DPKCA and KCVA have better fault detection capability than KPCA and CVA, but their T^2 statistics still have some false alarm samples. In comparison, the proposed method shows the best monitoring performance, where the L^2 and T_S^2 statistics can completely detect this fault. When Fault 1 is introduced at the 2001_{st} sample, the strong temporal dynamic change of operating condition can be well captured by T_F^2. After the process fully enters into the faulty condition, it is worth noting that T_F^2 goes back below the threshold, which is the reason for the low detection rate of T_F^2.

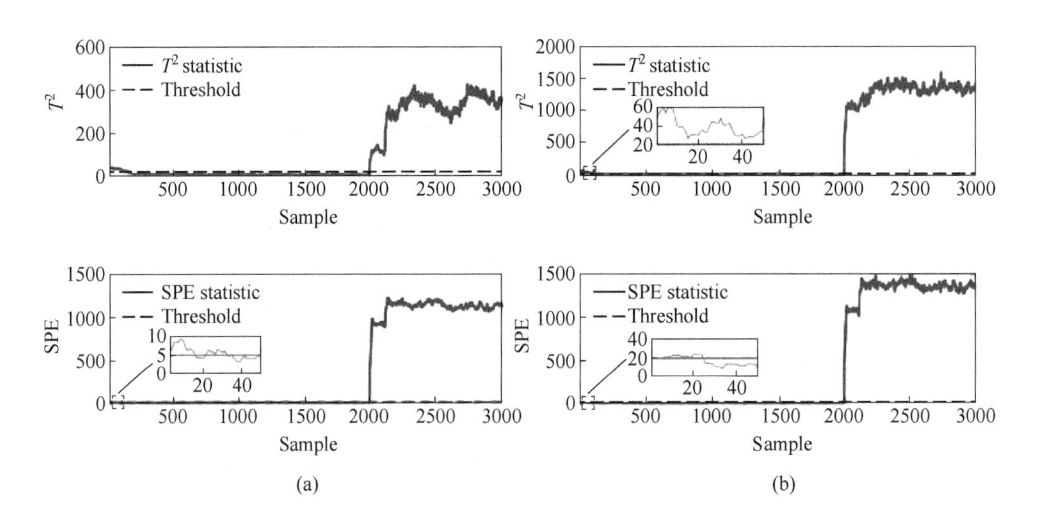

(a)　(b)

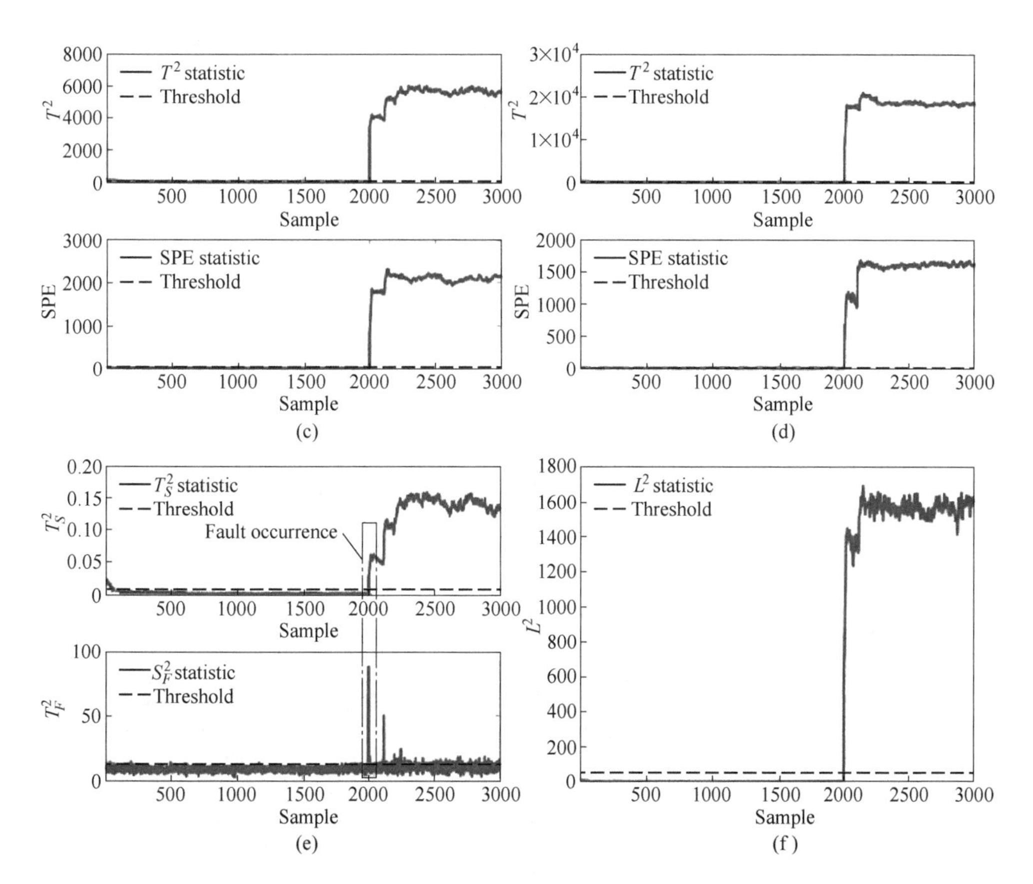

Fig. 5.7 Monitoring results in loaded condition with Fault 1

(a) KPCA; (b) CVA; (c) DKPCA; (d) KCVA; (e)-(f) NCA-NSFA

(2) **Fault case 2**: It is a sensor anomaly located in the 5_{th} stand which involves a step change in the measurement of bending force. When Fault 2 occurs, there is a sudden increase in the bending force of the 5_{th} stand. Owing to the feedback control, it subsequently drops the bending force of the 6_{th} stand as well as the 7_{th} stand. Thus, the crown and flatness of the strips will be influenced. KPCA-based method cannot effectively detect this fault as shown in **Fig. 5.8** (a). When the process is under fault-free condition, there are many false alarms. What's more, plenty of faulty samples can't be detected by T^2 statistic in the fault duration. By contrast, CVA, DKPCA, and KCVA show much better performance with a higher detection rate in **Fig. 5.8**(b), whilst the FAR of CVA is nearly the same as that of KPCA. In addtion, DKPCA and KCVA have fewer false alarm samples than KPCA and CVA, which are shown in **Fig. 5.8**(c)-(d). Among the five methods, the proposed method can better detect the occurrence of this

fault than the other four methods. As shown in **Fig. 5.8**(e)-(f), the L^2 and T_S^2 of the proposed method have the highest FDR and lowest FAR. Meanwhile, the temporal dynamic statistic return to normal when the fault is manually eliminated.

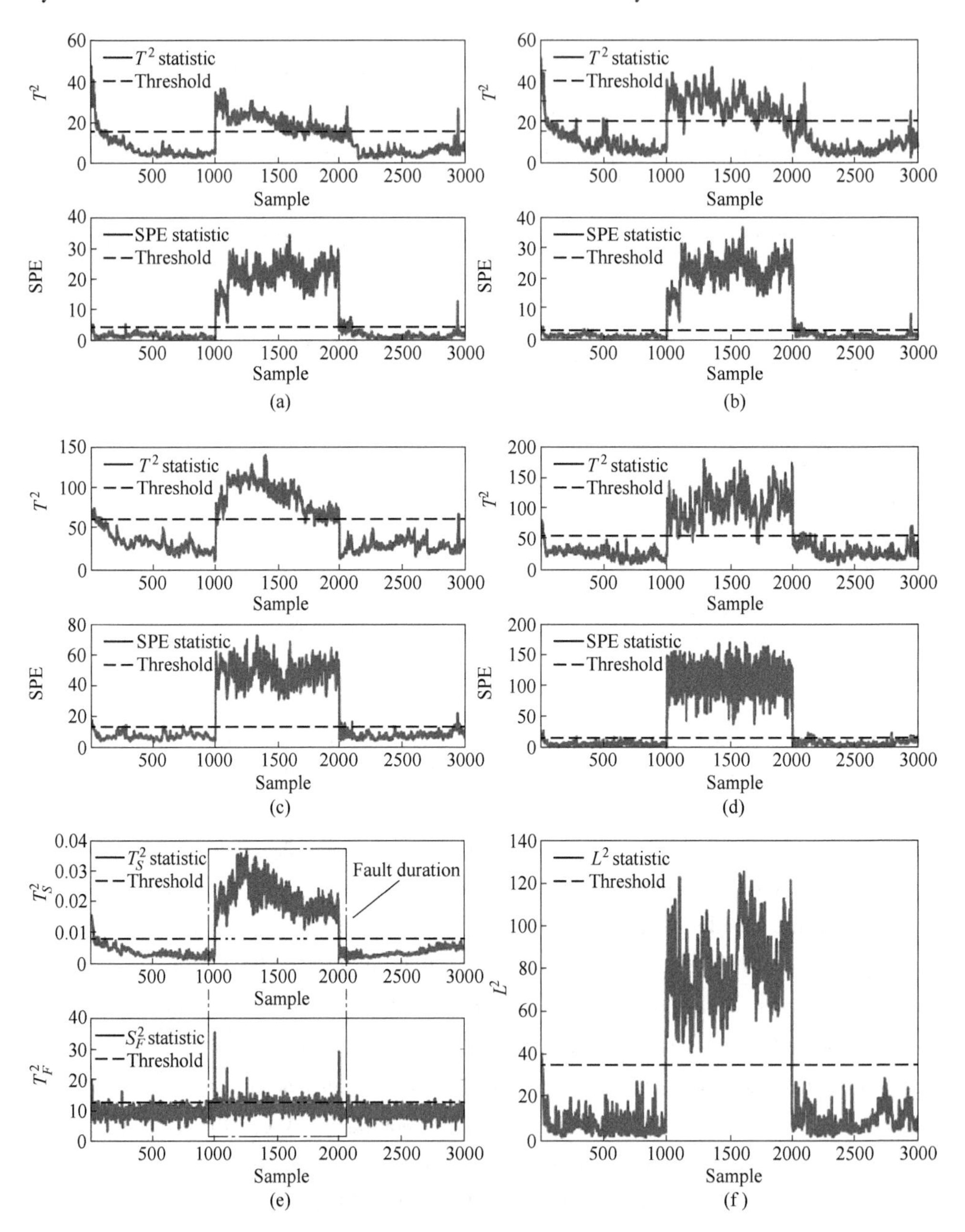

Fig. 5.8 Monitoring results in loaded condition with Fault 2

(a) KPCA; (b) CVA; (c) DKPCA; (d) KCVA; (e)-(f) NCA-NSFA

(3) **Fault case 3**: It is a cooling valve malfunctions. The strip temperature of the 3_{rd} stand is abnormal, which causes roll gap changes at the 4_{th} stands. Finally, the strip thickness is unqualified. KPCA cannot successfully detect such fault. Lots of missing alarms exist in **Fig. 5.9**(a) . A large number of false alarms in T^2 are flooded at the beginning stage. Besides, after the removal of the fault, T^2 statistic has still many false

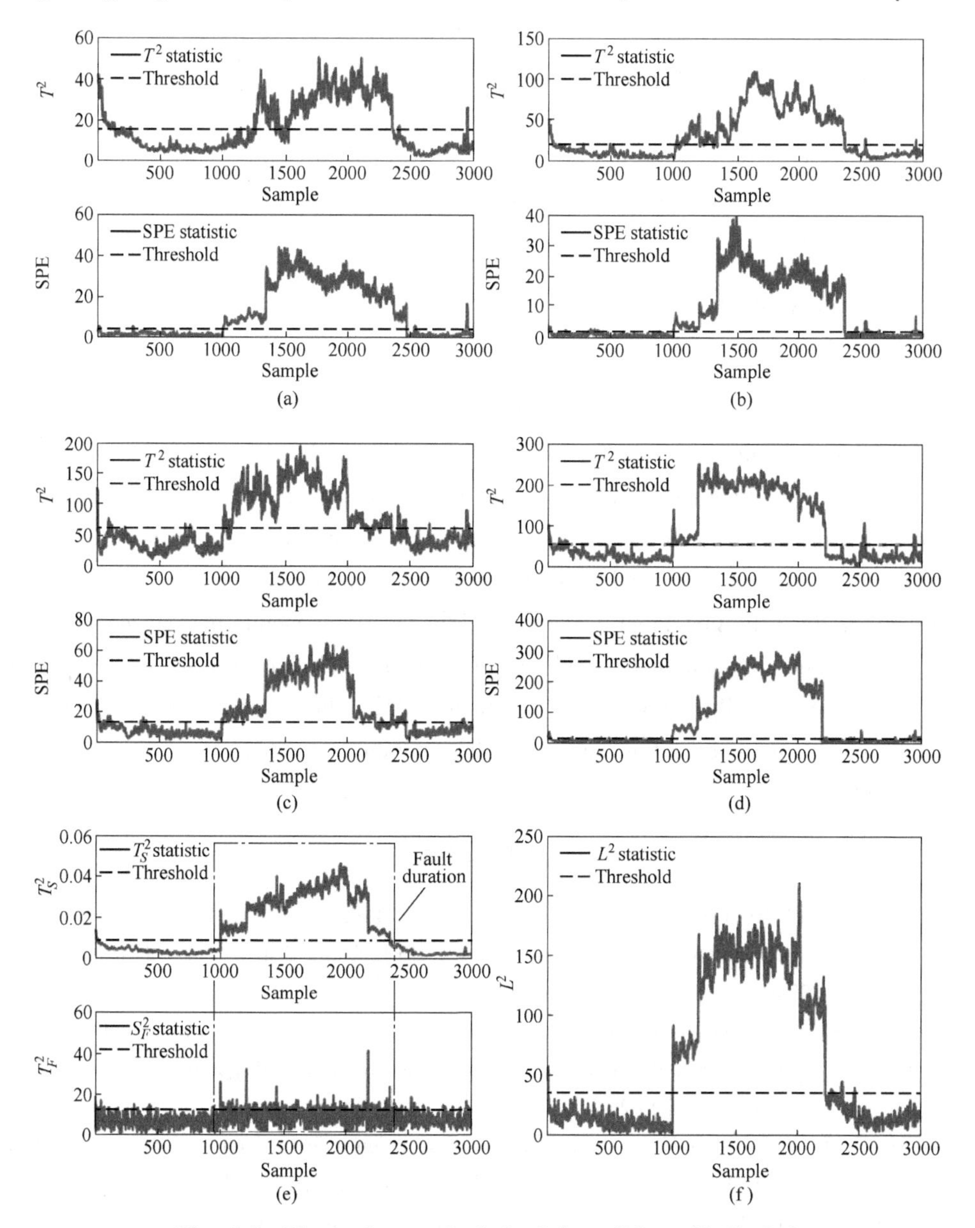

Fig. 5.9 Monitoring results in loaded condition with Fault 3

(a) KPCA; (b) CVA; (c) DKPCA; (d) KCVA; (e)-(f) NCA-NSFA

alarms. This circumstance can be regarded that the downstream stands need some time to recover from the anomaly and the measurements of subsequent stands remain abnormal during a short period of time. The monitoring results of CVA is presented in **Fig. 5.9** (b) . Similar to KPCA, there are plenty of false alarms, even though the detection rate is much higher in T^2. **Fig. 5.9** (c)-(d) shows the monitoring results of DKPCA and KCVA. Because of their ability to deal with dynamic nonlinearity issue, DKPCA and KCVA have higher FDRs and lower FARs than KPCA and CVA. Evidently, the proposed method performs the best among the five methods, which has the highest FDR and the lowest FAR in L^2 and T_S^2. Meanwhile, it offers temporal dynamic changes represented by the large value of T_F^2. From above analyses, it can be concluded that the proposed monitoring model has better fault detection capability for HSMP.

5.4 Conclusions

In this chapter, a nonlinear full condition monitoring model is proposed for dynamic HSMP. Different from existing process monitoring methods, the objective of this method is the whole rolling process, including both idle condition and loaded one. In particular, it can simultaneously extract and analyze the long-run dynamic equilibrium relations, temporal dynamic and static variations in loaded condition. The well-known *t*-SNE, CA and SFA are jointly incorporated in the proposed method. It has three subsection: the condition identification and idle condition monitoring subsection, the long-run dynamic relation analysis subsection, the temporal dynamic and static variation analysis subsection. By applying to a practical HSMP, the proposed method functions well for condition identification and has better monitoring performance compared with other methods.

References

[1] Li L L, Chadli M, Ding S X, et al. Diagnostic observer design for T-S fuzzy systems: Application to real-time weighted fault detection approach [J]. IEEE Transactions on Fuzzy Systems, 2018, 26 (2): 805-816.

[2] Ding S X. Model-based fault diagnosis techniques [M]. Berlin: Springer- Verlag, 2008.

[3] Qin S J. Survey on data-driven industrial process monitoring and diagnosis [J]. Annu Rev Control, 2012, 36 (2): 220-234.

[4] Yin S, Li X W, Gao H J, et al. Data-based techniques focused on modern industry: An overview [J]. IEEE Transactions on Industrial Electronics, 2015, 62 (1): 657-667.

[5] Macgregor J F, Kourti T. Statistical process control of multivariate processes [J]. Control Engineering Practice, 1995, 3 (3): 403-414.

[6] Li G, Qin S J, Zhou D H. Geometric properties of partial least squares for process monitoring [J]. Automatica, 2010, 46 (1): 204-210.

[7] Lee J M, Yoo C K, Choi S W, et al. Nonlinear process monitoring using kernel principal component analysis [J]. Chemical Engineering Science, 2004, 59 (1): 223-234.

[8] Peng K X, Zhang K, Li G, et al. Contribution rate plot for nonlinear quality-related fault diagnosis with application to the hot strip mill process [J]. Control Engineering Practice 2013, 21 (4): 360-369.

[9] Zhang Y W. Enhanced statistical analysis of nonlinear processes using KPCA, KICA and SVM [J]. Chemical Engineering Science, 2003, 64 (5): 801-811.

[10] Zhou J L, Ren Y W, Wang J. Quality-relevant fault monitoring based on locally linear embedding orthogonal projection to latent structure [J]. Industrial & Engineering Chemistry Research, 2019, 58 (3): 1262-1272.

[11] Maaten L V D, Hinton G E. Visualizing data using *t*-SNE [J]. Journal of Machine Learning Research, 2008, 9: 2579-2605.

[12] Chen J H, Liu K C. On-line batch process monitoring using dynamic PCA and dynamic PLS models [J]. Chemical Engineering Science, 2002, 57 (1): 63-75.

[13] Lee J M, Yoo C K, Lee I B. Statistical monitoring of dynamic processes based on dynamic independent component analysis [J]. Chemical Engineering Science, 2004, 59 (14): 2995-3006.

[14] Russell E L, Chiang L H, Braatz R D. Fault detection in industrial processes using canonical variate analysis and dynamic principal component analysis [J]. Chemometrics & Intelligent Laboratory Systems, 2000, 51 (1): 81-93.

[15] Zhang S M, Zhao C H. Slow feature analysis-based batch process monitoring with comprehensive interpretation of operation condition deviation and dynamic anomaly [J]. IEEE Transactions on Industrial Electronics, 2019, 66 (5): 3773-3783.

[16] Chen Q, Kruger U, Leung A Y T. Cointegration testing method for monitoring nonstationary processes [J]. Industrial & Engineering Chemistry Research, 2009, 48 (7): 3533-3543.

[17] Shang C, Yang F, Gao X Q, et al. Concurrent monitoring of operating condition deviations and process dynamics anomalies with slow feature analysis [J]. AIChE Journal, 2015, 61 (11): 3666-3682.

[18] Zhao C H, Huang B. A full-condition monitoring method for nonstationary dynamic chemical processes with cointegration and slow feature analysis [J]. AIChE Journal, 2018, 64 (5): 1662-1681.

[19] Engle R F, Granger C W J. Cointegration and error correction: Representation, estimation and testing [J]. Econometrica 1987, 55: 251-276.

[20] Johansen S, Juselius K. Maximum likelihood estimation and inference on cointegration with applications to the demand for money [J]. Oxford Bulletin of Economics & Statistics, 1990, 52 (2): 169-210.

[21] Wiskott L, Sejnowski T J. Slow feature analysis: Unsupervised learning of invariances [J]. Neural Computation, 2002, 14 (4): 715-770.

[22] Zhu J L, Ge Z Q, Song Z H, et al. Review and big data perspectives on robust data mining approaches for industrial process modeling with outliers and missing data [J]. Annual Review of Control, 2018, 46: 107-133.

[23] Tax D M J, Duin R P W. Support vector data description [J]. Machine Learning, 2004, 54 (1): 45-66.

[24] Hoerl A E, Kennard R W. Ridge regression: Biased estimation for nonorthogonal problems [J]. Technometrics, 2000, 42 (1): 80-86.

[25] Ma L, Dong J, Peng K X, et al. Hierarchical monitoring and root-cause diagnosis framework for key performance indicator-related multiple faults in process industries [J]. IEEE Transactions on Industrial Informatics, 2019, 15 (4): 2091-2100.

Chapter 6 Exergy-Related Process Monitoring with Spatial Information

As an energy and resource intensive industry, the iron and steel industry have high carbon emissions. China' s iron and steel industry accounts for 60% of the carbon emissions of the global iron and steel industry. Taking the hot strip mill process (HSMP) as an example, the slab temperature in the furnace is 1250 ℃, the final rolling temperature is about 850 ℃, and the coiling temperature is generally 630 ℃. There is a lot of energy dissipation in the HSMP. Therefore, accelerating the low-carbon transformation of China' s iron and steel can promote the early realization of industry carbon peaking and carbon neutrality. Modern HSMP has the characteristics of complex mechanism, many production equipment, coupling of various subsystems, and changeable working conditions. As the end process of steel manufacturing, the manufacturing capacity of HSMP largely determines the quality of strip products. Once a fault occurs in the process, it will spread and evolve among different equipment along the energy flow, which will seriously affect the normal production, and even cause casualties[1-3]. However, with the rapid development of sensor technology and data storage technology, the scale of data collection is exploding in various complex forms. Traditional process monitoring methods based on first-principle models can no longer meet practical production demands[4-6]. By contrast, multivariate statistical analysis (MSA) methods have received increasing research attention since 1990s, including PCA, PLS, and ICA, which lays the foundation for data driven-based process monitoring and promotes its rapid development[7-9]. MSA methods do not need to acquire knowledge of the mechanism of processes, but only rely on process data modeling[10]. They can effectively extract key information from process data, eliminate redundancy, and significantly reduce data dimension, thereby directly displaying process condition in two-dimensional statistical monitoring charts[11].

In recent years, machine learning methods have achieved success in a large number of applications[12]. From the perspective of classification, process monitoring is one-class classification problem, which can be solved by classification methods in machine learning[13]. As a well-known machine learning-based classifier, support vector domain description (SVDD) was proposed by Tax and Duin in 2004[14]. The basic idea of SVDD is to build a hypersphere to contain positive samples as much as possible and minimize the volume of the hypersphere. When samples are inside the hypersphere, the process is in a normal condition. When samples are outside the hypersphere, the faults may occur in the process. Wang et al. proposed a SVDD monitoring model based on kernel similarity for the irregularity of hypersphere in high-dimensional space[15]. In order to monitor the batch process of time-varying dynamic characteristics, Lv et al. proposed an improved SVDD algorithm combined with just-in-time learning strategy[16]. Zhang et al. developed a two-step SVDD model to deal with the dynamic, nonlinear, and non-Gaussian characteristics of industrial process data simultaneously[17]. Liu et al. proposed a semi-supervised SVDD method to overcome the limitations of sample labeling in rolling bearing fault detection[18]. To deal with the degradation of monitoring performance caused by outliers, Yuan et al. presented a pruned SVDD model to improve the robustness of fault detection system[19].

Although current research has shown the feasibility of SVDD in process monitoring, there are still some problems worthy of in-depth study. On the one hand, the existing SVDD-based methods rarely consider the concept of energy. However, in real industrial processes, the material flow will run dynamically and orderly along the specific process network according to the set program under the driving and action of energy flow. Marais et al. proposed an energy-based process monitoring method and verified that the energy-based method can effectively reduce the modeling workload and increase the computational efficiency[20]. According to the thermodynamic irreversibility of the system, Durand et al. applied the fault detection method based on entropy and enthalpy to the petrochemical process[21]. Marais et al. extended the energy-based fault detection method using exergy information and outlined the advantages of fault detection method based on exergy information extraction in petrochemical industry[22]. As the concept combining the first and second laws of thermodynamics, exergy can be used to better quantify the direction of low efficiency and distinguish energy quality. Based on exergy information and SVDD, Zhou et al. proposed an incipient fault detection method for the industry distillation column[23]. On the other hand, traditional SVDD-based algorithms are usually aimed at vector data and seldom consider the spatial information of different equipment in the process. When spatial information is introduced, the initial process vector data will be expanded to higher-order tensor data. Take the HSMP as an example, we assume that J variables are measured at K time instances. Then, the I spatial positions of process variables can be organized as a third-order tensor $\boldsymbol{X} \in$

$R^{I\times J\times K}$. Traditional SVDD-based methods can not directly deal with such tensor data, they usually unfold the third-order tensor to a matrix in six possible ways. However, in practical problems, many data need to be better represented in the form of tensor. Tensor structure has good expressive and computational properties. Based on SVDD, Deng et al. proposed support tensor data description (STDD) to deal with tensor data[24].

In order to solve above problems and promote energy conservation and emission reduction, a novel exergy-related process monitoring method based on improved support tensor data description (ISTDD) is proposed for HSMP with analytics of energy flow and spatial information. The main objectives of this chapter are summarized as follows:

(1) Exergy-related process variables are selected by the minimal redundancy maximal relevance (mRMR), and the third-order tensor is constructed with the spatial information of different equipment in the process.

(2) Based on the exergy-related tensor and historical fault tensor, the ISTDD and its robust version are developed for fault detection. After the fault is detected, the fault type can be identified by the multiple fault identification model.

6.1 Review of CP decomposition, STDD, and exergy calculation

6.1.1 CANDECOMP/PARAFAC (CP) decomposition

CP decomposition is to decompose a higher-order tensor into the sum of several rank one tensors[25]. For a third-order tensor $\boldsymbol{X} \in R^{I\times J\times K}$, the CP decomposition of $\boldsymbol{X}$ can be expressed as:

$$\boldsymbol{X} \approx \sum_{r=1}^{\beta} \boldsymbol{X}_r = \sum_{r=1}^{\beta} \boldsymbol{a}_r \circ \boldsymbol{b}_r \circ \boldsymbol{c}_r = [\boldsymbol{A},\ \boldsymbol{B},\ \boldsymbol{C}] \tag{6.1}$$

where, $\circ$ represents cross product; $\boldsymbol{a}_r \in R^I$, $\boldsymbol{b}_r \in R^J$, and $\boldsymbol{c}_r \in R^K$ are corresponding factor vector; β represents the feature dimension.

If $\boldsymbol{A}$, $\boldsymbol{B}$, and $\boldsymbol{C}$ are regularized, there will be a weight vector $\boldsymbol{\lambda}$ after decomposition, and equation (6.1) can be transformed into:

$$\boldsymbol{X} \approx \sum_{r=1}^{\beta} \boldsymbol{X}_r = \sum_{r=1}^{\beta} \boldsymbol{\lambda}_r \cdot \boldsymbol{a}_r \circ \boldsymbol{b}_r \circ \boldsymbol{c}_r = [\boldsymbol{\lambda},\ \boldsymbol{A},\ \boldsymbol{B},\ \boldsymbol{C}] \tag{6.2}$$

Then CP decomposition is shown in **Fig. 6.1**.

6.1.2 Principle of STDD

STDD is a one-class classification method that extends SVDD. STDD adopts tensors as

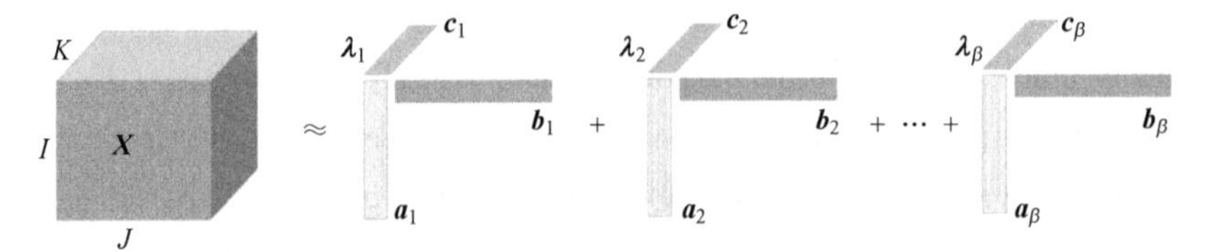

Fig. 6. 1 CP decomposition of a third-order tensor

input and works in tensor space directly. Like SVDD, the kernel trick can be used in STDD. Let $\{\boldsymbol{X}_i\}$, $i=1, 2, \cdots, M$ be a set of training tensors and ϕ be a mapping of transforming input space $\boldsymbol{X} \in R^{I\times J\times K}$ to Hilbert space $\phi(\boldsymbol{X}) \in R^{I'\times J'\times H'}$. The optimization problem of STDD is defined as:

$$
\begin{aligned}
&\min\left(R^2 + C\sum_i \xi_i\right) \\
&\text{s. t. } \ \|\phi(\boldsymbol{X}_i) - A\|_F^2 \leqslant R^2 + \xi_i, \ \xi_i \geqslant 0
\end{aligned}
\tag{6.3}
$$

where, R is the radius; A is the center of the hypersphere; penalty parameter C controls the trade-off between the volume of the hypersphere and the errors; ξ_i is the slack variable.

The Lagrange's dual problem of equation (6. 3) can be written as:

$$
\begin{aligned}
&\max\left(\sum_i \alpha_i\langle\phi(\boldsymbol{X}_i), \phi(\boldsymbol{X}_i)\rangle - \sum_{i,j}\alpha_i\alpha_j\langle\phi(\boldsymbol{X}_i), \phi(\boldsymbol{X}_j)\rangle\right) \\
&\text{s. t. } \sum_i \alpha_i = 1, \ 0 \leqslant \alpha_i \leqslant C
\end{aligned}
\tag{6.4}
$$

Let the CP decompositions of $\boldsymbol{X}_i$ and $\boldsymbol{X}_j$ be $\boldsymbol{X}_i = \sum_{r=1}^{\beta} \boldsymbol{a}_{ir} \circ \boldsymbol{b}_{ir} \circ \boldsymbol{c}_{ir}$ and $\boldsymbol{X}_j = \sum_{r=1}^{\beta} \boldsymbol{a}_{jr} \circ \boldsymbol{b}_{jr} \circ \boldsymbol{c}_{jr}$. Hence, the tensor product kernels can be derived as:

$$
\begin{aligned}
K(\boldsymbol{X}_i, \boldsymbol{X}_j) &= K\left(\sum_{r=1}^{\beta} \boldsymbol{a}_{ir} \circ \boldsymbol{b}_{ir} \circ \boldsymbol{c}_{ir}, \sum_{r=1}^{\beta} \boldsymbol{a}_{jr} \circ \boldsymbol{b}_{jr} \circ \boldsymbol{c}_{jr}\right) \\
&= \sum_{r=1}^{\beta} K(\boldsymbol{a}_{ir} \circ \boldsymbol{b}_{ir} \circ \boldsymbol{c}_{ir}, \boldsymbol{a}_{jr} \circ \boldsymbol{b}_{jr} \circ \boldsymbol{c}_{jr})
\end{aligned}
\tag{6.5}
$$

In this chapter, the radial basis function (RBF) kernel is used as the kernel function, equation (6. 5) can be calculated as:

$$
\begin{aligned}
K(\boldsymbol{X}_i, \boldsymbol{X}_j) &= \sum_{r=1}^{\beta} K(\boldsymbol{a}_{ir} \circ \boldsymbol{b}_{ir} \circ \boldsymbol{c}_{ir}, \boldsymbol{a}_{jr} \circ \boldsymbol{b}_{jr} \circ \boldsymbol{c}_{jr}) \\
&\sum_{r=1}^{\beta} \exp\left(-\frac{\|\boldsymbol{a}_{ir} - \boldsymbol{a}_{jr}\|^2 + \|\boldsymbol{b}_{ir} - \boldsymbol{b}_{jr}\|^2 + \|\boldsymbol{c}_{ir} - \boldsymbol{c}_{jr}\|^2}{2\sigma^2}\right)
\end{aligned}
\tag{6.6}
$$

For a test tensor $\boldsymbol{X}_{\text{new}}$, a discriminant function can be expressed as follows:

$$
\begin{aligned}
f(\boldsymbol{X}_{\text{new}}) &= \| \phi(\boldsymbol{X}_{\text{new}}) - A \|_F^2 - R^2 \\
&= K(\boldsymbol{X}_{\text{new}}, \boldsymbol{X}_{\text{new}}) - 2\sum_i \alpha_i K(\boldsymbol{X}_i, \boldsymbol{X}_{\text{new}}) + \\
&\quad \sum_{i,j} \alpha_i \alpha_j K(\boldsymbol{X}_i, \boldsymbol{X}_j) - R^2 \\
&= \text{sign}(E) \\
&= \begin{cases} 1 & E \leqslant 0 \\ 0 & E > 0 \end{cases}
\end{aligned} \tag{6.7}
$$

When discriminant function is equal to 1, which indicates $\boldsymbol{X}_{\text{new}}$ belongs to the target class.

6.1.3 Exergy calculation

The first law of thermodynamics explains the rule of energy conservation, and the second law of thermodynamics points out the direction of energy conversion. In the past, the evaluation of process energy consumption can only reflect the gain and loss of energy in quantity according to the first law of thermodynamics and cannot reflect the differences of various forms of energy in quality. In 1956, Rant first coined the term exergy to solve the above problems. As the combination of the first and second laws of thermodynamics, exergy reflects the unity of quality and quantity in energy. For the HSMP, heat exergy (HE) and enthalpy exergy (EE) are generally considered:

$$
\text{HE} = \int \left(1 - \frac{T_0}{T}\right) \partial Q \tag{6.8}
$$

$$
\text{EE} = (h - h_0) - (s - s_0) T_0 + \frac{1}{2} c^2 + gz \tag{6.9}
$$

where, T_0 is the ambient temperature; T is the system temperature; Q is the exchange of heat between the system and outside; h and s are the enthalpy and entropy of material, respectively; c is the flow velocity of the material; z is the height of the system.

Set the process input exergy is $E_{x,\text{in}}$ and the process output exergy is $E_{x,\text{out}}$, the difference between them is the internal exergy consumption $E_{xl,\text{in}}$. According to the practical analysis needs, $E_{xl,\text{out}}$ can be further divided into the external exergy consumption $E_{xl,\text{out}}$ and effective output exergy $E_{x,\text{ef}}$, and the exergy balance model is shown in **Fig. 6.2**.

According to **Fig. 6.2**, the exergy balance equation can be established as follows:

$$
E_{x,\text{in}} = E_{xl,\text{in}} + E_{x,\text{out}} = E_{xl,\text{in}} + E_{xl,\text{out}} + E_{x,\text{ef}} \tag{6.10}
$$

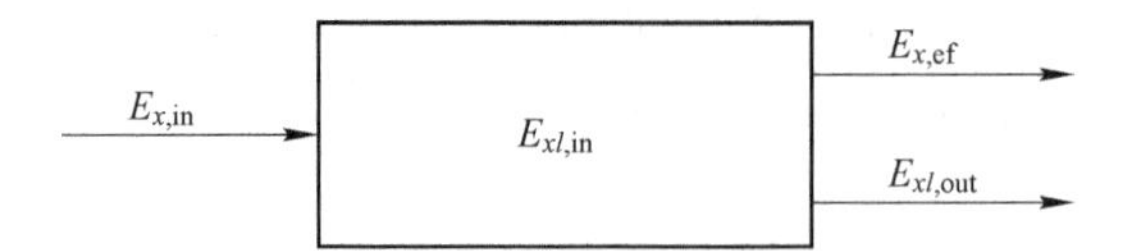

Fig. 6. 2 Commen exergy analysis model

The ratio of effective output exergy to the input exergy indicates the effective utilization of input exergy by the process, which is called exergy efficiency η_{ex} :

$$\eta_{ex} = \frac{E_{x,\text{ef}}}{E_{x,\text{in}}} \tag{6.11}$$

As an important index to measure the effective energy utilization of the process, exergy efficiency η_{ex} can represent the main characteristics of process energy information and can be used to reveal the performance changes of the process, so as to reflect the fault change information. Therefore, using exergy efficiency η_{ex} as energy feature can not only reduce the dimension of input space, but also improve the effect of fault feature extraction.

6. 2 Exergy-related process monitoring based on ISTDD

6. 2. 1 Exergy analysis model for HSMP

The energy consumption of HSMP in iron and steel production is not very large, but this can not hide its important position in the plant-wide production chain. As the end of iron and steel production chain, its role is equivalent to the top of biological chain. The incoming material of HSMP is the slab obtained by multiple processes and consuming a lot of energy. The metal yield of the final process directly affects the energy and material consumption level of the plant-wide production chain. As the unity of quality and quantity reflecting energy, exergy contains the performance change information of the process, which can explain the fault change information better than the general attribute variables. Therefore, it is necessary to consider the energy flow information in developing the process monitoring model. The energy of incoming and outgoing materials in the HSMP is shown in **Fig. 6. 3**, where $E_{x,\text{sup}}$ is the exergy value of the continuous casting slab, $E_{x,\text{e}}$ is the exergy value of electricity consumed by the process, $E_{x,\text{mp}}$ is the exergy value of the strip steel, $E_{x,\text{bp}}$ is the exergy value of the scrap steel, iron oxide scale and recovered steam, and $E_{xl,\text{out}}$ is the exergy value of waste water discharged from the process.

In modern HSMP, hundreds of process variables are measured in different production

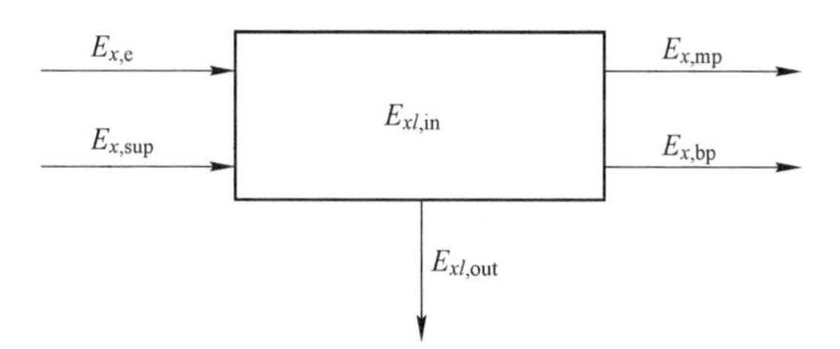

Fig. 6.3 Exergy analysis model for HSMP

equipment to reflect the operating performance of HSMP. For traditional methods, latent features are extracted from all process variables to develop monitoring models. Without considering the spatial information of different production equipment, the variable correlations of each equipmentcan not be effectively used. It is necessary to add one dimension to describe the spatial information of the process variables, for example, in the form of one-hot encoding, as shown in **Fig. 6.4**. Assume that J variables are measured at K time instances from I different production equipment. Then, the process data can be organized as a third-order tensor $\boldsymbol{X} \in R^{I \times J \times K}$. Therefore, considering the spatial information of process variables, the tensor-based method should be used for feature extraction.

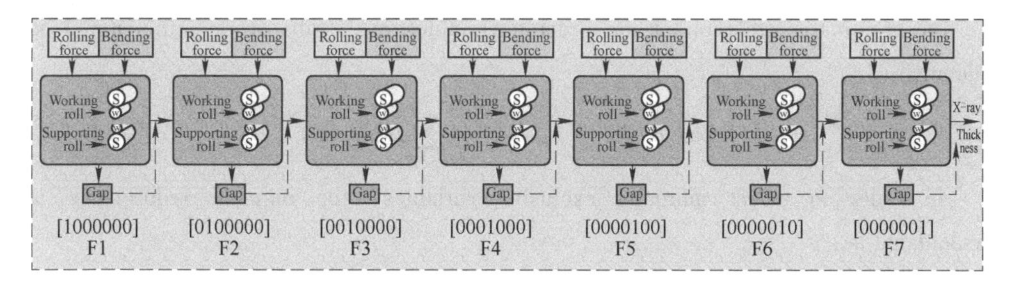

Fig. 6.4 Spatial information with the form of one-hot encoding for HSMP

Exergy efficiency is not only the specific form of the exergy information in HSMP, but also the evaluation index of the overall energy consumption. However, the calculation of exergy efficiency is complex, which is usually difficult to obtain online. Meanwhile, there are measurement errors in the practical process. In this work, the minimal redundancy maximal relevance algorithm is used to select the exergy-related variables. On this basis, a third-order tensor is constructed according to the location information of different production equipment, and then the fault detection model is developed by using the support tensor data description, so as to realize process monitoring of HSMP integrating energy flow and spatial information.

6.2.2 Exergy-related variable selection

Assume that the historical process data $\boldsymbol{X}$ contains J variables and the exergy efficiency $\boldsymbol{Y}$ is obtained by off-line analysis. Both $\boldsymbol{X}$ and $\boldsymbol{Y}$ have been normalized. Mutual information (MI) is a basic evaluation index in information theory, which quantifies the interdependence between two variables in the point of view of entropy[26]. Considering two process variables $\boldsymbol{x}_i$ and $\boldsymbol{x}_j$, the MI is defined as:

$$
\begin{aligned}
I(\boldsymbol{x}_i, \boldsymbol{x}_j) &= H(\boldsymbol{x}_i) + H(\boldsymbol{x}_j) - H(\boldsymbol{x}_i, \boldsymbol{x}_j) \\
&= \sum_{\boldsymbol{x}_i} p(\boldsymbol{x}_i)\lg\frac{1}{p(\boldsymbol{x}_i)} + \sum_{\boldsymbol{x}_j} p(\boldsymbol{x}_j)\lg\frac{1}{p(\boldsymbol{x}_j)} - \sum_{\boldsymbol{x}_i}\sum_{\boldsymbol{x}_j} p(\boldsymbol{x}_i, \boldsymbol{x}_j)\lg\frac{1}{p(\boldsymbol{x}_i, \boldsymbol{x}_j)}
\end{aligned} \tag{6.12}
$$

where, $p(\boldsymbol{x}_i)$ and $p(\boldsymbol{x}_j)$ denote the marginal probability density function (MPDF) of $\boldsymbol{x}_i$ and $\boldsymbol{x}_j$, and $p(\boldsymbol{x}_i, \boldsymbol{x}_j)$ denotes the joint probability density function (JPDF). However, for the process variable set $\boldsymbol{X}$, the maximal interdependency on the target exergy efficiency $\boldsymbol{Y}$ is hard to obtain because of the complex computations of JPDF. By reducing redundancy and maintaining relevance between different variables, the minimal redundancy maximal relevance (mRMR) algorithm can deal with this knotty issue[27].

The maximal correlation of the process variables in $\boldsymbol{X}$ with exergy efficiency $\boldsymbol{Y}$ is denoted as:

$$
\max D(\boldsymbol{X}, \boldsymbol{Y}),\ D = \frac{1}{J}\sum_{\boldsymbol{x}_i \in \boldsymbol{X}} I(\boldsymbol{x}_i, \boldsymbol{Y}) \tag{6.13}
$$

In order to select mutually exclusive variables, the minimal redundancy is calculated as:

$$
\min R(\boldsymbol{X}),\ R = \frac{1}{J^2}\sum_{\boldsymbol{x}_i, \boldsymbol{x}_j \in \boldsymbol{X}} I(\boldsymbol{x}_i, \boldsymbol{x}_j) \tag{6.14}
$$

By combining D and R, an operator $\Psi(D, R)$ is designed to select the exergy-related variables:

$$
\max \Psi(\boldsymbol{X}, \boldsymbol{Y}),\ \Psi(\boldsymbol{X}, \boldsymbol{Y}) = D/R \tag{6.15}
$$

And the joint incremental search algorithm is used to obtain near-optimal variables[28]:

$$
\max \nabla\Psi(\boldsymbol{X}, \boldsymbol{Y}),\ \nabla\Psi(\boldsymbol{X}, \boldsymbol{Y}) = \max\left\{ I(\boldsymbol{x}_i, \boldsymbol{Y}) \Big/ \frac{1}{J'-1}\sum_{\boldsymbol{x}_i \in \boldsymbol{x}_{J'-1}} I(\boldsymbol{x}_i, \boldsymbol{x}_j) \right\} \tag{6.16}
$$

where, J' is the number of exergy-related variables.

6.2.3 Process monitoring based on ISTDD

Assume that $\boldsymbol{X}'$ is the exergy-related variables selected by mRMR. Then, considering the spatial information of the production equipment where each variable in $\boldsymbol{X}'$ is located, the third-order tensor $\boldsymbol{X}' \in R^{I\times J'\times K}$ is constructed, as shown in **Fig. 6.5**. Based on the $\boldsymbol{X}'$, the ISTDD model can be developed.

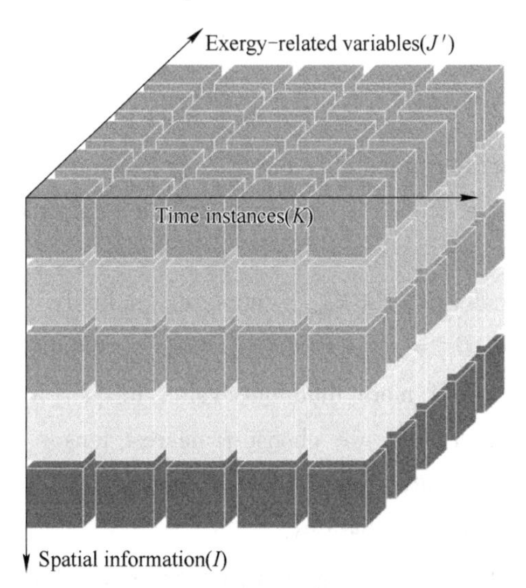

Fig. 6.5 Third-order tensor of exergy-related variables with spatial information

The center and radius of the ISTDD model are computed as:

$$A = \sum_i \alpha_i X_i' \tag{6.17}$$

$$R = \sqrt{1 - 2\sum_i \alpha_i K(X_i', \boldsymbol{X}_{\mathrm{ST}}') + \sum_{i,j} \alpha_i\alpha_j K(X_i', X_j')} \tag{6.18}$$

where, $\boldsymbol{X}_{\mathrm{ST}}'$ is one of support tensors, and α_i is the Lagrange multiplier.

For a test tensor sample $\boldsymbol{X}_{\mathrm{test}}'$, the distance between the center A and $\boldsymbol{X}_{\mathrm{test}}'$ can be calculated as:

$$D = \sqrt{1 - 2\sum_i \alpha_i K(X_i', \boldsymbol{X}_{\mathrm{test}}') + \sum_{i,j} \alpha_i\alpha_j K(X_i', X_j')} \tag{6.19}$$

If D is smaller than R, the test tensor sample $\boldsymbol{X}_{\mathrm{test}}'$ is located inside the hypersphere, which means that HSMP is under normal condition, otherwise, HSMP is in fault condition.

After the fault is detected, it is necessary to identify the fault type. Assume that the historical fault data $\boldsymbol{X}^f \in R^{I\times J'\times K}(f=1, 2, \cdots, F)$ include F kinds of faults. Based on the historical fault data, F fault hyperspheres can be constructed by the ISTDD model. Ideally, any two fault hyperspheres are independent of each other, and the test tensor sample $\boldsymbol{X}'_{\text{test}}$ can be easily identified. However, ISTDD only makes binary output for $\boldsymbol{X}'_{\text{test}}$. If $\boldsymbol{X}'_{\text{test}}$ is projected inside the overlap zone of several fault hyperspheres at the same time, it will be difficult to identify the fault type of $\boldsymbol{X}'_{\text{test}}$. In order to solve this thorny problem, the distance from $\boldsymbol{X}'_{\text{test}}$ to the center of the hypersphere is transformed into the probability of belonging to the hypersphere:

$$P_f(\boldsymbol{X}'_{\text{test}}) = \frac{1}{1 + \exp(D_f - R_f)(1 + 1/D_f)} \tag{6.20}$$

where, $P_f(\boldsymbol{X}'_{\text{test}})$ denotes the probability that $\boldsymbol{X}'_{\text{test}}$ belongs to fault f ; $D_f = \| \varphi(\boldsymbol{X}'_{\text{test}}) - A_f \|$ is the distance to the center of fault hypersphere f ; A_f and R_f are the center and radius of fault hypersphere f.

When D_f increases gradually, $P_f(\boldsymbol{X}'_{\text{test}})$ decreases gradually. When $D_f = 0$, $P_f(\boldsymbol{X}'_{\text{test}})$ approaches to 1. When $D_f = R_f$, $P_f(\boldsymbol{X}'_{\text{test}}) = 0.5$. When D_f approaches infinity, $P_f(\boldsymbol{X}'_{\text{test}})$ approaches to 0. If more than one value in $\{P_1(\boldsymbol{X}'_{\text{test}}), P_2(\boldsymbol{X}'_{\text{test}}), \cdots, P_F(\boldsymbol{X}'_{\text{test}})\}$ is greater than 0.5, we choose k nearest tensor samples around $\boldsymbol{X}'_{\text{test}}$ to calculate the average probability, and determine the fault type of $\boldsymbol{X}'_{\text{test}}$ according to the maximum value of the average probability.

The flowchart of ISTDD-based exergy-related process monitoring scheme for HSMP is presented in **Fig. 6.6**.

6.2.4 Robust version of ISTDD

When historical fault data are available, they can be integrated with historical normal data to enhance the monitoring performance of the ISTDD model. In this subsection, the enhanced version of ISTDD is called robust improved support tensor data description (RISTDD) . Assume that normal tensor samples (labeled 1) are enumerated by indices i , j , and fault tensor samples (labeled −1) are enumerated by indices l , m . Inspired by the successful use of SVDD with negative examples, the optimization function of STDD can be rewritten as:

$$\min R^2 + C_1 \sum_i \xi_i + C_2 \sum_i \xi_l$$

$$\text{s.t.} \ \| \phi(X_i^r) - A_r \|_F^2 \leqslant R_r^2 + \xi_i, \ \| \phi(X_l^f) - A_r \|_F^2 \geqslant R_r^2 - \xi_l, \ \xi_i \geqslant 0, \quad \xi_l \geqslant 0 \tag{6.21}$$

Similarly, the Lagrange multipliers α_i, α_l, γ_i, γ_l can be introduced to equation (6.21):

$$L(R_r, A_r, \xi_i, \xi_l, \alpha_i, \alpha_l, \gamma_i, \gamma_l) = R_r^2 + C_1 \sum_i \xi_i + C_2 \sum_l \xi_l - \sum_i \gamma_i \xi_i - \sum_l \gamma_l \xi_l - \sum_i \alpha_i (R_r^2 + \xi_i - \| \phi(X_i') - A_r \|_F^2) - \sum_l \alpha_l (\| \phi(X_l^f) - A_r \|_F^2 - R_r^2 + \xi_l) \tag{6.22}$$

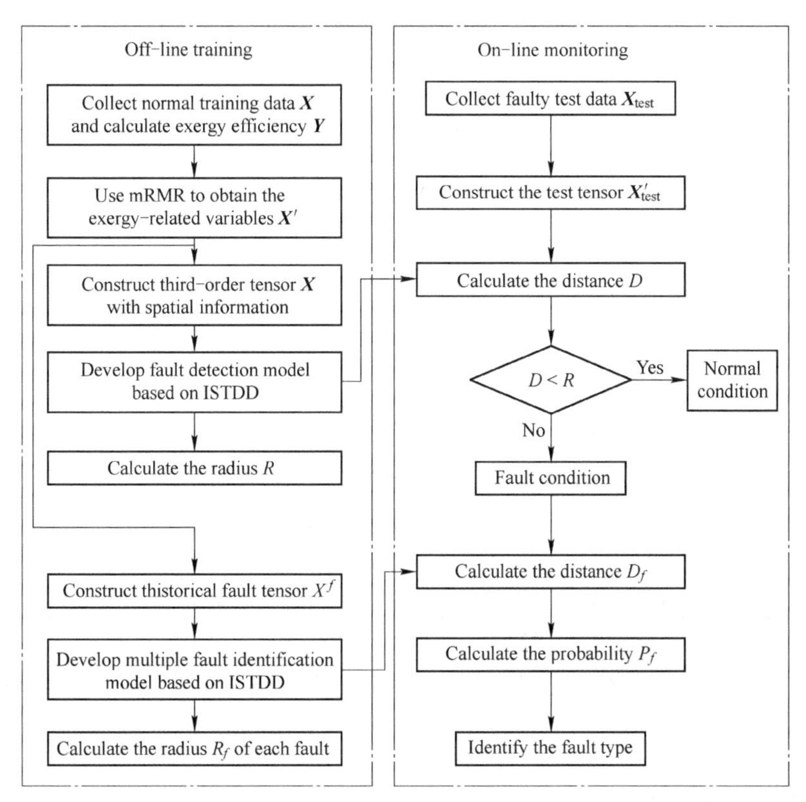

Fig. 6.6 Flowchart of ISTDD-based exergy-related process monitoring scheme

The partial derivatives of $L(R_r, A_r, \xi_i, \xi_l, \alpha_i, \alpha_l, \gamma_i, \gamma_l)$ are set to zero with respect to R_r, A_r, ξ_i and ξ_l as follows:

$$\begin{gathered} \frac{\partial L}{\partial R_r} = 0 \Rightarrow \sum_i \alpha_i - \sum_l \alpha_l = 1 \\ \frac{\partial L}{\partial A_r} = 0 \Rightarrow A_r = \sum_i \alpha_i \phi(X_i') - \sum_l \alpha_l \phi(X_l^f) \\ 0 \leqslant a_i \leqslant C_1, \quad 0 \leqslant a_l \leqslant C_2, \quad \forall i, l \end{gathered} \tag{6.23}$$

Thus, equation (6.22) can be rewritten as follows:

$$L = \sum_i \alpha_i K(X_i^{\prime}, X_i^{\prime}) - \sum_l \alpha_l K(X_l^f, X_l^f) - \sum_i \sum_j \alpha_i \alpha_j K(X_i^{\prime}, X_j^{\prime}) - \sum_l \sum_m \alpha_l \alpha_m K(X_l^f, X_m^f) + 2\sum_l \sum_j \alpha_l \alpha_j K(X_l^f, X_j^{\prime}) \tag{6.24}$$

Then, new variables $\alpha_i^* = \mathrm{Lab}_i \times \alpha_i$ ($\mathrm{Lab}_i = 1$ represents normal data, and $\mathrm{Lab}_i = -1$ represents fault data) are defined, the constraints in equation (6.23) are transformed into:

$$\begin{gathered} \sum_i \alpha_i^* = 1 \\ A_r^* = \sum_i \alpha_i^* \phi(X_i^*) \end{gathered} \tag{6.25}$$

where, $\boldsymbol{X}^* = [\boldsymbol{X}', \boldsymbol{X}^f]$ is the overall set of normal tensor samples and fault tensor samples.

The radius R_r of the hypersphere can be obtained as follows:

$$R_r = \sqrt{1 - 2\sum_i \alpha_i^* K(X_i^*, \boldsymbol{X}_{\mathrm{ST}}^*) + \sum_i \sum_j \alpha_i^* \alpha_j^* K(X_i^*, X_j^*)} \tag{6.26}$$

where, $\boldsymbol{X}_{\mathrm{ST}}^*$ represents one of the support tensors.

For a test tensor sample $\boldsymbol{X}_{\mathrm{test}}$, the distance to the center of the hypersphere can be calculated:

$$D_r = \sqrt{1 - 2\sum_i \alpha_i^* K(X_i^*, \boldsymbol{X}_{\mathrm{test}}) + \sum_i \sum_j \alpha_i^* \alpha_j^* K(X_i^*, X_j^*)} \tag{6.27}$$

If D_r is smaller than R_r, the HSMP is in normal condition, otherwise, HSMP is in fault condition.

6.3 Case study

HSMP is an important process in iron and steel production, which has the characteristics of complex mechanism, large-scale, multi working conditions. HSMP is an automatic production line running in accordance with the specified sequence. The production line is mainly composed of the furnace, roughing mill area, transfer table and shear, finishing mill area, laminar cooling and coiler. First, the slab is heated to 1200 ℃ in the furnace, and then enters the roughing mill area. The thickness and width of the slab is roughly processed to meet the expected requirements. After passing through the output mill table, the finishing mill can further accurately make it reach the preset thickness and width. Then, the 850 ℃-strip steel is cooled to the required temperature through laminar cooling equipment, and finally coiled by coiler to facilitate loading. The finishing mill area includes 7 stands, and the slab is continuously rolled to the required specification by these stands. Each stand is equipped with two types of rolls, including two supporting rolls on the outside and two working rolls on the inside. By applying rolling force and bending force to the roll, it is ensured that the most important quality index, namely the thickness of strip steel, can be reduced as expected, and the crown of strip steel can be quantified to an acceptable level. The four rolls are driven by the motor, so the strip steel can move forward smoothly. Meanwhile, the finishing mill exit temperature should be strictly controlled within a certain range to meet the required microstructure and properties inside the strip steel. In our previous work, more specific description of HSMP can be found[29]. In this section, the exergy-related process monitoring methods based on ISTDD and RISTDD are applied in a practical HSMP, and three representative faults are used for illustrating the monitoring performance of the proposed method.

6.3.1 Description of process data

For testing the validity of the proposed process monitoring scheme, normal and fault data were collected from production debugging of Ansteel Group in the people's Republic of China, including two strip products with thickness 2.70 mm and 3.95 mm. The process variables are listed in **Table 6.1**. 2000 normal samples and 1000 samples for each fault are used for ISTDD and RISTDD model training. Based on the 2000 normal samples, the exergy $E_{x,\mathrm{sup}}$, $E_{x,\mathrm{e}}$, $E_{x,\mathrm{mp}}$, $E_{x,\mathrm{bp}}$, $E_{xl,\mathrm{out}}$ are calculated. Then, according to the above five exergy values, the exergy efficiency can be obtained as:

$$\eta_{\mathrm{ex}} = \frac{E_{x,\mathrm{mp}} + E_{x,\mathrm{bp}}}{E_{x,\mathrm{sup}} + E_{x,\mathrm{e}}} \tag{6.28}$$

In order to verify process monitoring performance, the proposed monitoring scheme is compared with robust SVDD (RSVDD) and STDD. Three typical faults are listed in **Table 6.2**, which usually occur in the finishing mill area. 2000 test samples in each fault scenario are used for online monitoring, including 1000 fault samples and 1000 normal samples. The kernel parameter σ and penalty parameter C of the above algorithms can be optimized by genetic algorithm.

Table 6.1 Process variables in the HSMP

Variable	Description	Unit
1-7	Roll gap of the i_{th} stand, $i=1, \cdots, 7$	mm
8-14	Rolling force of the i_{th} stand, $i=1, \cdots, 7$	MN
15-20	Bending force of the i_{th} stand, $i=2, \cdots, 7$	MN

Continued Table 6. 1

Variable	Description	Unit
21-27	Rolling speed of the i_{th} stand, $i=1, \cdots, 7$	mm/s
28-34	Current of the i_{th} stand, $i=1, \cdots, 7$	A
35-41	Power of the i_{th} stand, $i=1, \cdots, 7$	kW
42-47	Tension of the i_{th} looper, $i=1, \cdots, 6$	MPa
48	Strip thickness of finishing mill exit	mm
49	Strip temperature of roughing mill exit	℃
50	Strip temperature of finishing mill exit	℃
51	Strip flatness of finishing mill exit	I
52	Strip crown of finishing mill exit	mm
53	Strip width of finishing mill exit	mm

Table 6. 2 Typical faults in the finishing mill area

Fault No.	Fault description	Fault durat
1 (3. 95 mm)	Failure of gap control loop in the 4_{th} stand	1001_{st} to 2000_{th}
2 (3. 95 mm)	stiction of the cooling valve between the 2_{nd} and the 3_{rd} stands	501_{st} to 1500_{th}
3 (2. 70 mm)	Malfunction of bending force measuring sensor in the 5_{th} stand	501_{st} to 1500_{th}

6. 3. 2 Process monitoring results and discussions

In this work, the fault detection rate (FDR) and false alarm rate (FAR) are used to assess the practicability of the RSVDD, STDD, ISTDD, and RISTDD, which are generally calculated as:

$$\mathrm{FDR} = \frac{\mathrm{TP}}{\mathrm{TP} + \mathrm{FN}} \tag{6.29}$$

$$\mathrm{FAR} = \frac{\mathrm{FP}}{\mathrm{FP} + \mathrm{TN}} \tag{6.30}$$

where, TP is the number of fault samples that are correctly detected; TP + FN is the total number of fault samples; FP is the number of normal samples that are incorrectly identified as faults; FP + TN is the total number of normal samples.

An effective monitoring strategy usually has higher FDR, lower FAR, and smaller detection time delay. The relationship between the number of exergy-related variables and FDR is shown in **Fig. 6. 7**. For the strip product with thickness 3. 95 mm, the FDRs of Fault 1 and Fault 2 have the maximum value when the number of exergy-related variables is 7. For the strip product with thickness 2.70 mm, the FDR of Fault 3 has the maximum value when the the number of exergy-related variables is 6. Variable selection based on mRMR is only a screening of process variables and does not extract the linear relationship between process variables like traditional multivariate statistical analysis. By properly selecting the number of exergy-related variables, exergy information can be effectively extracted, and redundant noise information can be eliminated at the same time. The FDRs and FARs of RSVDD, STDD and the proposed methods are quantified in **Table 6.3** For each fault scenario, the best monitoring results are highlighted in bold type. It is clear that RISTDD has a higher FDR and moderate FAR. Based on the monitoring results, it is not difficult to find that the proposed method performs best in all fault scenarios.

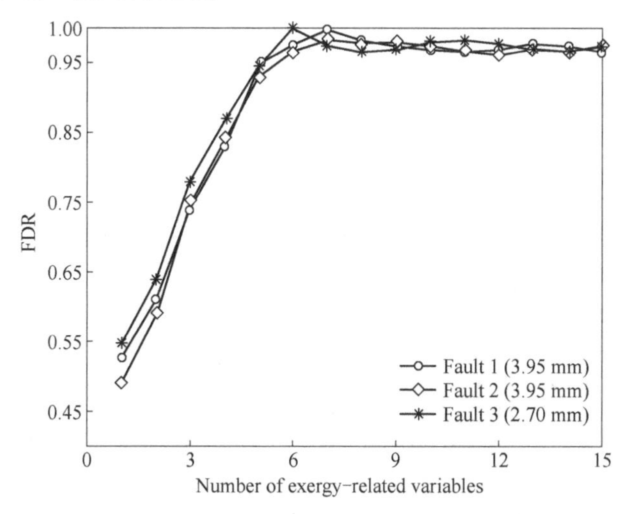

Fig. 6.7 The relationship between the number of exergy-related variables and FDR

Table 6.3 Comparison of different methods in terms of FDR and FAR

Fault No.	FDR				FAR			
	RSVDD	STDD	ISTDD	RISTDD	RSVDD	STDD	ISTDD	RISTDD
1	0.922	0.885	0.965	**0.998**	0.027	**0**	0.015	0.008
2	0.895	0.916	0.943	**0.984**	0.477	0.434	0.337	**0.273**
3	0.938	0.94	0.984	**1**	0.093	0.056	0.03	**0**

Fault 1 occurs at the hydraulic roll gap control loop in the 4_{th} stand from the 501_{st} test sample. The roll gap of the 4_{th} stand is directly affected, and the rolling force of the 4_{th} stand is also affected. Due to feedback control mechanism, the roll gaps and rolling forces of the following stands will be adjusted. Finally, the quality indexes of strip steel such as thickness and flatness are affected. The fault detection results of RSVDD, STDD, ISTDD, and RISTDD are shown in **Fig. 6. 8**(a)-(d) . It can be found that the D statistic of STDD has a large detection delay, which makes its FDR smaller than other methods. A lot of false alarms can be seen in **Fig. 6. 8**(a), which means the RSVDD has the largest FAR. Compared with RSVDD and STDD, ISTDD provides more acceptable performance. In addition, RISTDD marginally performs better than ISTDD as shown in **Fig. 6. 8** (d) . Thus, the proposed methods are effective to detect Fault 1.

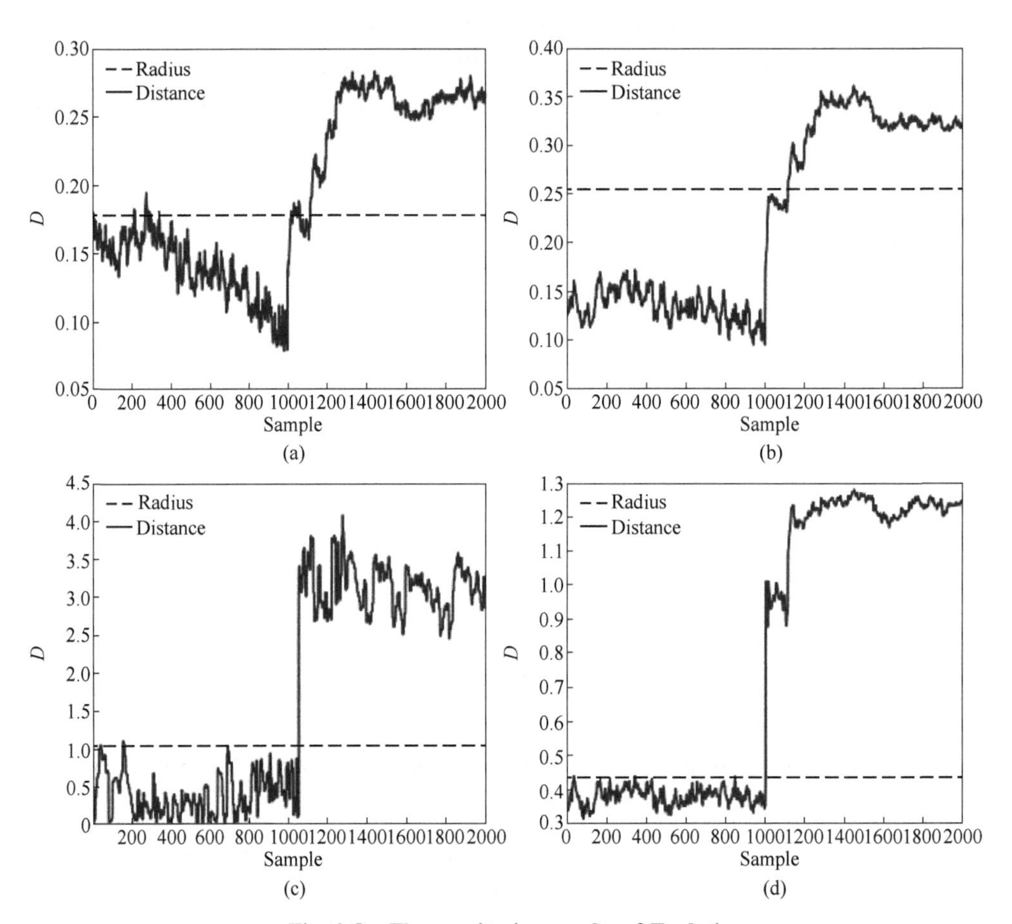

Fig. 6. 8 The monitoring results of Fault 1

(a) The assessment result of RSVDD; (b) The assessment result of STDD;
(c) The assessment result of ISTDD; (d) The assessment result of RISTDD

Fault 2 occurs in the cooling valve between the 2_{nd} stand and the 3_{rd} stand, which leads to abnormal strip temperature in the 3_{rd} stand. Then, the rolling force and bending force of the 3_{rd} stand, and the looper tension between the 3_{rd} stand and the 4_{th} stand are changed. As important quality indexes, the strip's flatness and temperature of finishing mill exit will be ultimately influenced. **Fig. 6.9** (a)-(d) shows the fault detection results of RSVDD, STDD, ISTDD, and RISTDD. For this fault, the FDRs of RSVDD and STDD are moderate, but there are many missing alarms in the beginning of the fault. Moreover, although the fault was manually removed after 1500 samples, plenty of false alarms exist in the final stage, which means the FARs of both methods are relatively large. The reason for this phenomenon can be explained as that it takes time for the following stands to recover from the fault condition after troubleshooting. Therefore, the measurement results of the 4_{th} and 5_{th} stand are still abnormal for a short time. Similar with RSVDD and STDD, ISTDD and RISTDD also have some false alarms in the initial

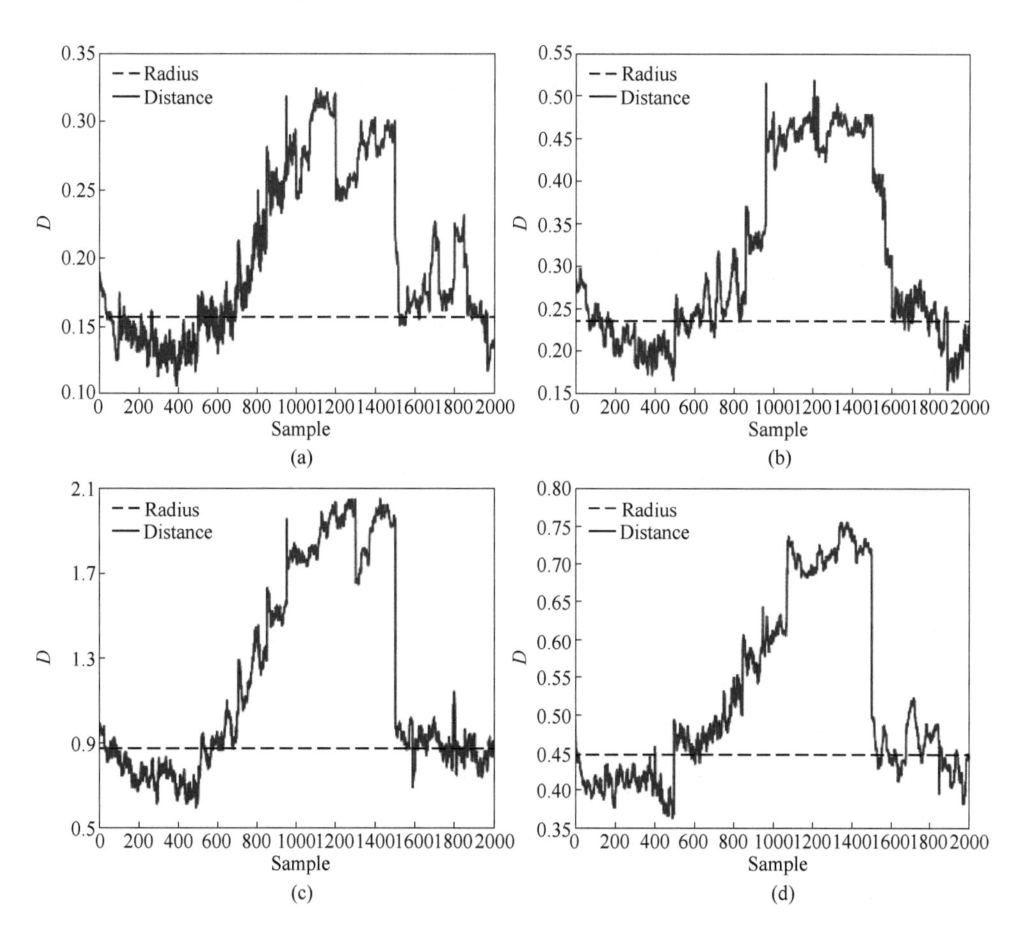

Fig. 6.9 The monitoring results of Fault 2

(a) The assessment result of RSVDD; (b) The assessment result of STDD;
(c) The assessment result of ISTDD; (d) The assessment result of RISTDD

and final stage. However, the monitoring results of the proposed methods have been improved, which means more fault samples can be detected. It is worth noting that the FAR of RISTDD is reduced by nearly 20% compared with traditional methods.

Fault 3 is caused by the malfunction of bending force measuring sensor equipped in the 5_{th} stand. When it occurs at the 501_{st} test sample, the bending force of the 5_{th} stand increases significantly. With feedback regulation of automatic control system, the bending forces of the 6_{th} stand and the 7_{th} stand are changed subsequently. Fault 3 belongs to a kind of step transition, which has little effect on the thickness. However, the crown and flatness of exit strip will be seriously influenced. The fault detection results of Fault 3 are illustrated in **Fig. 6. 10** (a)-(d) . RSVDD and STDD-based methods cannot completely detect this fault. When the process is under normal condition, some test samples are judged as fault samples. To be worse, many test

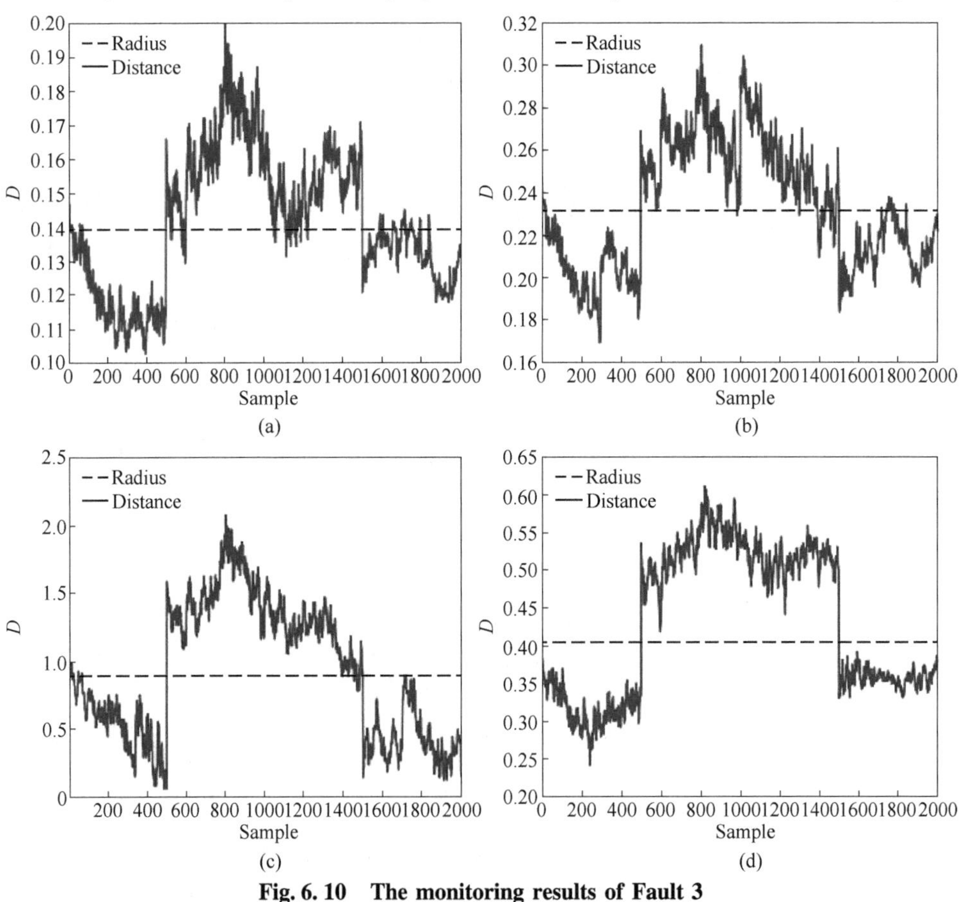

Fig. 6. 10 The monitoring results of Fault 3

(a) The assessment result of RSVDD; (b) The assessment result of STDD;
(c) The assessment result of ISTDD; (d) The assessment result of RISTDD

samples are regarded as normal samples in the fault condition. By contrast, RISTDD can perfectly detected this fault with no false alarms and missing alarms. The aforementioned monitoring results and analyses demonstrate that proposed method gives the best fault detection performance. After the fault is detected, the fault type can be identified according to the value of $P_f(\boldsymbol{X}'_{\text{test}})$.

6.4 Conclusions

In this study, an exergy-related STDD based process monitoring method has been proposed for HSMP. Energy flow and spatial information of HSMP were jointly considered in the developed monitoring model. First of all, mRMR was used to select exergy-related process variables. Then, third-order tensor was constructed to train the ISTDD model. Based on the historical fault tensor, the multiple fault identification model was developed. Moreover, the robust version of ISTDD was discussed. In order to accomplish the fault detection and identification task, the new scheme utilized the distance to the center of the hypersphere and the probability as the evaluating indicators for detecting and identifying faults. The effectiveness of the presented methods has been compared with other existing methods through a practical HSMP. Evidently, the monitoring performances of RSVDD and STDD are inferior to that of ISTDD and RISTDD, and the proposed methods have larger fault detection rates and smaller fault alarm rates. Future work considers topics with exergy-related operating performance assessment for HSMP.

References

[1] Ding S X, Yin S, Peng K X, et al. A novel scheme for key performance indicator prediction and diagnosis with application to an industrial hot strip mill [J]. IEEE Transactions on Industrial Informatics, 2013, 9 (4): 2239-2247.

[2] Jiao R H, Peng K X, Dong J. Remaining useful life prediction for a roller in a hot strip mill based on deep recurrent neural networks [J]. IEEE/CAA Journal of Automatica Sinica, 2021, 8 (7): 1345-1354.

[3] Liao Y B, Ragai I, Huang Z Y, et al. Manufacturing process monitoring using time-frequency representation and transfer learning of deep neural networks [J]. Journal of Manufacturing Processes, 2021, 68: 231-248.

[4] Qin S J, Chiang L. Advances and opportunities in machine learning for process data analytics [J]. Computers & Chemical Engineering, 2019, 126: 465-473.

[5] Wang Y Q, Si Y, Huang B, et al. Survey on the theoretical research and engineering applications of multivariate statistics process monitoring algorithms: 2008—2017 [J]. The Canadian Journal

of Chemical Engineering, 2018, 96 (10): 2073-2085.

[6] Ge Z Q, Song Z H, Gao F R. Review of recent research on data-based process monitoring [J]. Industrial & Engineering Chemistry Research, 2013, 52: 3543-3562.

[7] Lee J M, Yoo C K, Choi S W, et al. Nonlinear process monitoring using kernel principal component analysis [J]. Chemical Engineering Science, 2004, 59: 223-234.

[8] Li G, Qin S J, Zhou D H. Geometric properties of partial least squares for process monitoring [J]. Automatica, 2010, 46: 204-210.

[9] Kano M, Tanaka S, Hasebe S, et al. Monitoring independent components for fault detection [J]. AIChE Journal. 2003, 49 (4): 969-976.

[10] Zhang K, Dong J, Peng K X. A novel dynamic non-Gaussian approach for quality-related fault diagnosis with application to the hot strip mill process [J]. Journal of the Franklin Institute, 2017, 354: 702-721.

[11] Zhao C H, Wang W, Tian C, et al. Fine-scale modelling and monitoring of wide-range nonstationary batch processes with dynamic analytics [J]. IEEE Transactions on Industrial Electronics, 2021, 68 (9): 8808-8818.

[12] Ge Z Q, Song Z H, Ding S X, et al. Data mining and analytics in the process industry: The role of machine learning [J]. IEEE Access 2017, 5: 20590-20616.

[13] Zhang C F, Peng K X, Dong J. An incipient fault detection and self-learning identification method based on robust SVDD and RBM-PNN [J]. Journal of Process Control, 2020, 85: 173-183.

[14] Tax D, Duin D. Support vector data description [J]. Machine Learning, 2004, 54 (1): 45-66.

[15] Wang J L, Ma L Y, Liu W M, et al. Batch process monitoring by kernel similarity-based support vector data description [J]. CIESC Journal, 2017, 68 (9): 3494-3500.

[16] Lv Z M, Yan X F, Jiang Q C, et al. Just-in-time learning multiple subspace support vector data description used for non-Gaussian dynamic batch process monitoring [J]. Journal of Chemometrics, 2019, 33: 1-13.

[17] Zhang Y F, Li X S. Two-step support vector data description for dynamic, non-linear, and non-Gaussian processes monitoring [J]. The Canadian Journal of Chemical Engineering, 2020, 98 (10): 2109-2124.

[18] Liu C, Gryllias K. A semi-supervised support vector data description-based fault detection method for rolling element bearings based on cyclic spectral analysis [J]. Mechanical Systems & Signal Processing, 2020, 140: 1-24.

[19] Yuan P, Mao Z Z, Wang B. A pruned support vector data description-based outlier detection method: Applied to robust process monitoring [J]. Transactions of the Institute of Measurement and Control, 2020, 42 (11): 2113-2126.

[20] Marais H J, Schoor G V, Uren K R. Energy-based fault detection for an autothermal reformer [J]. IFAC-PapersOnLine, 2016, 49 (7): 353-358.

[21] Durand C P, Schoor G V, Nieuwoudt C. Enthalpy-entropy graph approach for the classification of faults in the main power system of a closed Brayton cycle HTGR [J]. Annals of Nuclear Energy, 2009, 36 (6): 703-711.

[22] Marais H J, Schoor G V, Uren K R. The merits of exergy-based fault detection in petrochemical processes [J]. Journal of Process Control, 2017, 74 (2): 110-119.

[23] Zhou M F, Liu Z H, Cai Y J, et al. Incipient fault detection based on exergy efficiency and support vector data description [J]. Journal of Chemical Engineering of Japan, 2019, 52 (6): 562-569.

[24] Deng X W, Jiang P, Peng X N. Support high-order tensor data description for outlier detection in high-dimensional big sensor data [J]. Future Generation Computer Systems, 2018, 81: 177-187.

[25] Signoretto M, Lathauwer L D, Suykens J. A kernel-based framework to tensorial data analysis [J]. Neural Networks, 2011, 24 (8): 861-874.

[26] Kraskov A, Stogbauer H, Andrzejak R G, et al. Hierarchical clustering using mutual information [J]. Europhysics Letters, 2005, 70 (2): 278-284.

[27] Xu C, Zhao S, Liu F. Distributed plant-wide process monitoring based on PCA with minimal redundancy maximal relevance [J]. Chemometrics & Intelligent Laboratory Systems, 2017, 169: 53-63.

[28] Ma L, Dong J, Hu C J, et al. A novel decentralized detection framework for quality-related faults in manufacturing industrial processes [J]. Neurocomputing, 2021, 428 (7): 30-41.

[29] Zhang C F, Peng K X, Dong J. An extensible quality-related fault isolation framework based on dual broad partial least squares with application to the hot rolling process [J]. Expert System with Application, 2020, 167 (1): 114166.

Chapter 7 Exergy-Related Fault Detection and Diagnosis Based on TransCGAN

Fault detection and diagnosis (FDD) plays a vital role in preventing industrial process accidents and ensuring system safety[1]. Reliable process monitoring mechanism can greatly reduce the risk of system faults or unexpected downtime, thereby improving production efficiency and reducing production costs[2-3]. With the development of Internet of things, artificial intelligence and other technologies, sensor technology is widely used in the process industries. However, as the complexity and dimensionality of the data collected by sensors increase, the effectiveness of statistical and shallow learning-based methods is limited and difficult to meet the actual needs of safety production.

In recent years, the rapid development of deep learning (DL) technology has brought new opportunities to FDD. By multi-layer nonlinear function mapping, DL-based methods can make full use of data and effectively mine the complex interrelations between data, so as to achieve more abstract data expression. Dai et al. proposed a novel power transformer fault diagnosis method with deep belief networks (DBN) and tested the diagnostic model using various oil chromatographic datasets[4]. Shao et al. developed an improved convolutional deep belief network (CDBN) for feature learning and fault diagnosis of rolling bearing[5]. Viola et al. developed a deep convolutional generative adversarial network (DCGAN) to generate image information of rotating bearing to improve the sample balance of the training set, and convolutional neural network (CNN) was used to distinguish the sample categories[6]. Qi et al. designed a novel stacked sparse autoencoder (SSAE) for machine fault diagnosis, which can extract high-level features and have good fault diagnostic performance[7].

Although current research has shown the effectiveness of DL, there are still some issues that deserve further investigation. Firstly, the above methods seldom consider the concept of energy. However, in the actual process industries, the material flow will be driven and acted on by the energy flow, according to the set program, and run along the specific process network dynamically and orderly. Marais et al. proposed an energy-based fault detection method using exergy information[8]. As the concept combining the first and second laws of thermodynamics, exergy can be used to better quantify the direction of low efficiency and distinguish energy quality. Secondly, the historical data of process industries show complex spatiotemporal characteristic. Current research rarely focuses on the analysis of spatiotemporal relationship of process data. The effective use and extraction of spatiotemporal information is the key to ensure the accuracy of FDD. Thirdly, most of the current research are based on the assumption that the distribution of faults is approximately balanced. They rely on large amounts of data and show slow convergence. In other words, the diagnosis accuracy of DL-based methods depend on the number of training samples. In the actual process industries, it is difficult to obtain enough fault samples because normal operating condition is more common than fault operating condition. Based on the above analysis, it is necessary to develop a novel FDD framework that takes the above three issues into account. The main contributions of this chapter are summarized as follows:

(1) Based on Spearman's rank correlation coefficient (SRCC) and maximal information coefficient (MIC), a new mixed correlation coefficient (MCC) is developed to select the exergy-related variables.

(2) Gramian angular field (GAF) is used to convert original process data to two-dimensional (2D) image data for representation, which can fully retain the spatiotemporal information of original process data.

(3) A novel conditional generative adversarial networks (CGAN) is developed for FDD, whose basic unit is the transformer encoder, also called TransCGAN in this chapter.

7.1 Review of CGAN and transformer

7.1.1 Conditional generative adversarial networks

Conditional generation adversarial network (CGAN) is a variant network that imposes conditional constraints on GAN[9], and its structure is shown in **Fig. 7.1**. Structurally, CGAN introduces a conditional variable $\boldsymbol{y}$ (representing the desired labels) in the generator G and discriminator D, respectively. G will generate samples that meet the condition $\boldsymbol{y}$, while D will judge the generated samples to determine whether they meet the condition $\boldsymbol{y}$ and whether they are authentic, and make corresponding optimization improvements to the objective function. By feeding the conditional variable $\boldsymbol{y}$ into G and D as the additional input layer, GAN changes from an unsupervised network to a supervised network. $\boldsymbol{y}$ can effectively guide the training process of G and D. The objective function of CGAN is a two-player minimax game:

$$\min_G \max_D V(D, G) = E_{x \sim p_{data(x)}}[\lg D(x|y)] + E_{z \sim p_{z(z)}}[\lg(1 - D(G(z|y)))] \quad (7.1)$$

where, $p_{z(z)}$ and $p_{data(x)}$ are the prior distribution of random noise z and real data x, respectively.

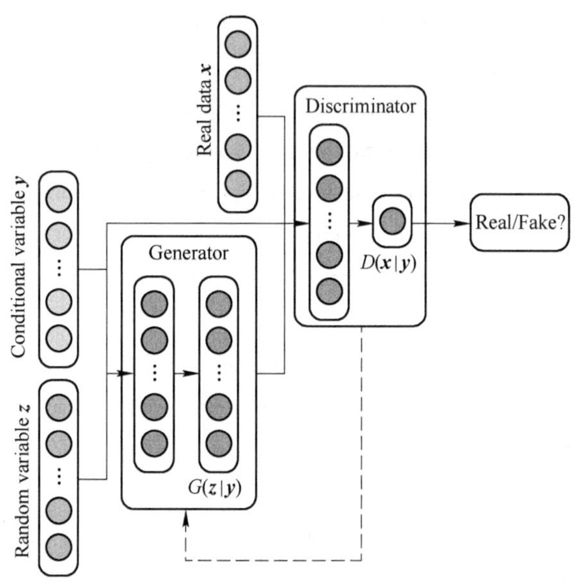

Fig. 7.1 Architecture of CGAN

7.1.2 Transformer model

Transformer model, also known as attention mechanism model, is widely used in the field of natural language processing[10]. Transformer model includes an encoder and a decoder, and its structure is shown in **Fig. 7.2**. The encoder is composed of 6 layers with the same structure. Each layer contains 2 sub-layers: a multi-head attention mechanism and a position-wise fully connected feed-forward network. A residual connection is employed around each group of sub-layers, and then processed by layer normalization. Thus, the output of each sub-layer can be calculated as follows:

$$\text{Output} = \text{LayerNorm}(x + \text{Sublayer}(x)) \quad (7.2)$$

where, x is the input sequence.

The decoder has a similar structure to the encoder, but with an extra sub-layer (masked multi-head attention). This sub-layer is used to mask unpredicted information, ensuring that each prediction is based on known information only. The core of transformer is the attention mechanism. Usually, we describe an attention function as mapping a query and a set of key-value pairs to an output. The self-attention function and multi-head self-attention function can be computed as follows:

$$\mathrm{Attention}(\boldsymbol{Q},\ \boldsymbol{K},\ \boldsymbol{V}) = \mathrm{softmax}\left(\frac{\boldsymbol{Q}\boldsymbol{K}^{\mathrm{T}}}{\sqrt{d_k}}\right)\boldsymbol{V} \tag{7.3}$$

$$\mathrm{MultiHead}(\boldsymbol{Q},\ \boldsymbol{K},\ \boldsymbol{V}) = \mathrm{Concat}(\mathrm{head}_1,\ \cdots,\ \mathrm{head}_h)\boldsymbol{W}^O \tag{7.4}$$

where, the queries, keys, and values are packed together into matrices $\boldsymbol{Q}$, $\boldsymbol{K}$ and $\boldsymbol{V}$, $\boldsymbol{W}_i^Q \in R^{d_{\mathrm{model}} \times d_q}$, $\boldsymbol{W}_i^K \in R^{d_{\mathrm{model}} \times d_k}$, $\boldsymbol{W}_i^V \in R^{d_{\mathrm{model}} \times d_v}$, and $\boldsymbol{W}_i^O \in R^{d_v \times d_{\mathrm{model}}}$ are parameter matrices, and $\mathrm{head}_i = \mathrm{Attention}(\boldsymbol{Q}\boldsymbol{W}_i^Q;\ \boldsymbol{K}\boldsymbol{W}_i^K;\ \boldsymbol{V}\boldsymbol{W}_i^V)$.

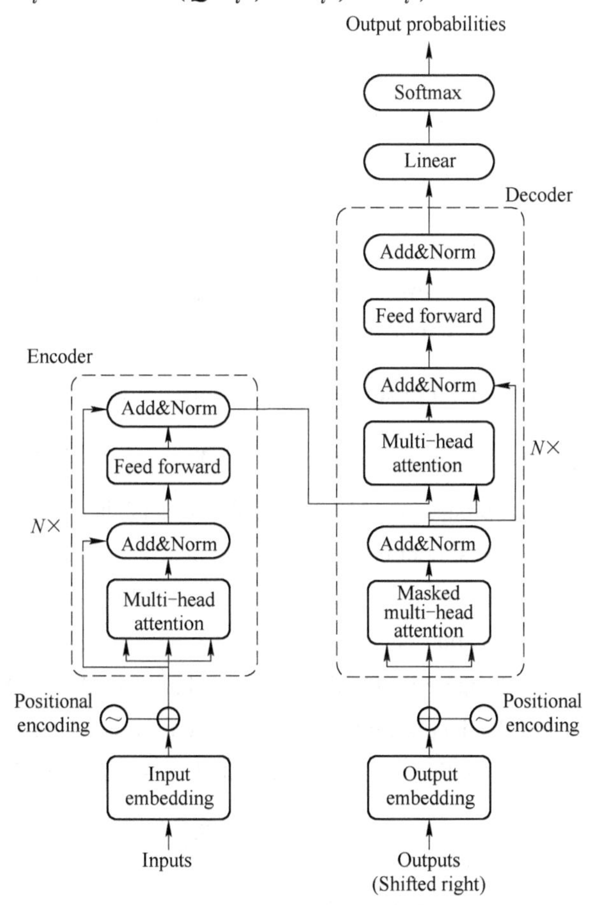

Fig. 7.2 Architecture of transformer model

7.2 Exergy-related FDD framework

HSMP plays an crucial role in modern iron and steel industry. Effective fault detection and diagnosis for HSMP can ensure process safety and product quality. However, HSMP has high energy consumption, spatio-temporal characteristic, and unbalanced data samples (normal samples far outnumber the faulty ones), which greatly increases the difficulty of fault detection and diagnosis. As the unity of quality and quantity reflecting energy, exergy contains the performance change information of the process, and can explain the fault change information better than the general attribute variable. Therefore, it is necessary to consider the energy flow information in developing FDD model. Exergy efficiency is not only the specific form of the exergy information in HSMP, but also the evaluation index of the overall energy consumption. However, the calculation of exergy efficiency is complex, which is usually difficult to obtain online. Meanwhile, there are measurement errors in the practical process. In this work, a new mixed correlation coefficient (MCC) is developed to select the exergy-related variables. On this basis, the original one-dimensional data are converted into two-dimensional images by Gramian angular fields (GAF), which can fully retain the time-dependent information of the data. Then, transformer and CGAN are combined to extract the spatio-temporal feature of image data and generate more high-quality fault samples, so as to realize effective FDD for HSMP with complex characteristics.

7.2.1 Exergy-related variable selection and image data conversion

Process data are usually collected as a matrix $\boldsymbol{X} = [\boldsymbol{X}_1, \boldsymbol{X}_2, \cdots, \boldsymbol{X}_J] \in R^{N \times J}$, where J refers to the number of process variables, and N refers to the number of samples. And the exergy efficiency $\boldsymbol{Y}$ was obtained by off-line analysis. Both $\boldsymbol{X}$ and $\boldsymbol{Y}$ have been normalized. The correlations between $\boldsymbol{X}$ and $\boldsymbol{Y}$ are evaluated by MCC, which is based on Spearman's rank correlation coefficient (SRCC)[11] and maximal information coefficient (MIC)[12]. Compared with Pearson correlation analysis (PCA), Spearman correlation analysis (SCA) has advantages such as being unaffected by data dimensions and being able to handle nonlinear data. The SRCC between $\boldsymbol{X}_i$ and $\boldsymbol{Y}$ is defined as[13]:

$$\mathrm{SRCC}(\boldsymbol{X}_i, \boldsymbol{Y}) = \frac{\mathrm{cov}(\boldsymbol{X}_i, \boldsymbol{Y})}{\sqrt{\mathrm{var}(\boldsymbol{X}_i)\mathrm{var}(\boldsymbol{Y})}} = \frac{\sum_{k=1}^{N} (X_i^k - \overline{X}_i)(Y^k - \overline{Y})}{\sqrt{\sum_{k=1}^{N} (X_i^k - \overline{X}_i)^2 \sum_{k=1}^{N} (Y^k - \overline{Y})^2}} \tag{7.5}$$

The SRCC ranges from −1 to 1. The closer SRCC is to 0, the less the correlation between $\boldsymbol{X}_i$ and $\boldsymbol{Y}$. MIC is another frequently used correlation measure proposed by Reshef et al., which is independent of data assumptions about distribution. The MIC between $\boldsymbol{X}_i$ and $\boldsymbol{Y}$ is calculated as follows[14]:

$$\mathrm{MIC}(\boldsymbol{X}_i, \boldsymbol{Y}) = \max\{I(\boldsymbol{X}_i, \boldsymbol{Y}) / \log_2 \min\{n_{\boldsymbol{X}_i}, n_{\boldsymbol{Y}}\} \tag{7.6}$$

where

$$\begin{aligned} I(\boldsymbol{X}_i, \boldsymbol{Y}) &= H(\boldsymbol{X}_i) + H(\boldsymbol{Y}) - H(\boldsymbol{X}_i, \boldsymbol{Y}) \\ &= \sum_{k=1}^{n_{\boldsymbol{X}_i}} p(X_i^k) \log_2 \frac{1}{p(X_i^k)} + \sum_{l=1}^{n_{\boldsymbol{Y}}} p(Y^l) \log_2 \frac{1}{p(Y^l)} - \\ &\quad \sum_{k=1}^{n_{\boldsymbol{X}_i}} \sum_{l=1}^{n_{\boldsymbol{Y}}} p(X_i^k Y^l) \log_2 \frac{1}{p(X_i^k Y^l)} \end{aligned} \tag{7.7}$$

$n_{\boldsymbol{X}_i} \cdot n_{\boldsymbol{Y}} < B(n)$, $B(n) = n^{0.6}$, $n_{\boldsymbol{X}_i}$ and $n_{\boldsymbol{Y}}$ are the number of bins of the partition of the $\boldsymbol{X}_i$- and $\boldsymbol{Y}$- axis, respectively. $p(X_i^k)$ represents the probability of X_i^k. The MIC ranges from 0 to 1. The larger the MIC is, the higher correlation between $\boldsymbol{X}_i$ and $\boldsymbol{Y}$. To ensure the stability and reliability of the correlation analysis[15], a mixed correlation coefficient is designed as follows:

$$\mathrm{MCC}(\boldsymbol{X}_i, \boldsymbol{Y}) = \omega\,|\mathrm{SRCC}(\boldsymbol{X}_i, \boldsymbol{Y})| + (1 - \omega)\mathrm{MIC}(\boldsymbol{X}_i, \boldsymbol{Y}) \tag{7.8}$$

where, $0 < \omega < 1$ is an equilibrium parameter. In this work, ω is set as 0.5. It is noted that $\mathrm{MCC}(\boldsymbol{X}_i, \boldsymbol{Y})$ ranges from 0 to 1. In order to select exergy-related variables, we denote η as a threshold for MCC. If $\mathrm{MIC}(\boldsymbol{X}_i, \boldsymbol{Y})$ is larger than η, $\boldsymbol{X}_i$ will be regarded as an exergy-related variable.

Assmue that $\boldsymbol{X}_{\mathrm{Exergy}} = [\boldsymbol{X}_1, \boldsymbol{X}_2, \cdots, \boldsymbol{X}_R] \in R^{N\times R}$ are exergy-related variables. Based on GAF, we represent the $\boldsymbol{X}_{\mathrm{Exergy}}$ in a polar coordinate system instead of the typical Cartesian coordinates[16]. GAF can fully preserve the signal information and time dependent information of process data, which can be more convenient for spatiotemporal information extraction after being converted into image data. Given an exergy-related variable $\boldsymbol{X}_i = [x_{i,1}, x_{i,2}, \cdots, x_{i,N}]^{\mathrm{T}}$, we can represent it in a polar coordinate system instead of the typical Cartesian coordinates by encoding the value as the angular cosine as follows:

$$\varphi_{i,j} = \arccos x_{i,j}, \quad -1 \leqslant x_{i,j} \leqslant 1, \quad x_{i,j} \in \boldsymbol{X}_i \tag{7.9}$$

Then, the Gramian angular summation field (GASF) matrix and Gramian angular difference field (GADF) matrix of $\boldsymbol{X}_i$ are defined as follows:

$$\mathbf{GASF}_i = \begin{bmatrix} \cos(\varphi_{i,1} + \varphi_{i,1}) & \cdots & \cos(\varphi_{i,1} + \varphi_{i,N}) \\ \cos(\varphi_{i,2} + \varphi_{i,1}) & \cdots & \cos(\varphi_{i,2} + \varphi_{i,N}) \\ \vdots & \ddots & \vdots \\ \cos(\varphi_{i,N} + \varphi_{i,1}) & \cdots & \cos(\varphi_{i,N} + \varphi_{i,N}) \end{bmatrix} = \boldsymbol{X}_i^{\mathrm{T}}\boldsymbol{X}_i - \sqrt{\boldsymbol{I} - \boldsymbol{X}_i^2}^{\mathrm{T}} \sqrt{\boldsymbol{I} - \boldsymbol{X}_i^2} \tag{7.10}$$

$$\mathbf{GADF}_i = \begin{bmatrix} \sin(\varphi_{i,1} + \varphi_{i,1}) & \cdots & \sin(\varphi_{i,1} + \varphi_{i,N}) \\ \sin(\varphi_{i,2} + \varphi_{i,1}) & \cdots & \sin(\varphi_{i,2} + \varphi_{i,N}) \\ \vdots & \ddots & \vdots \\ \sin(\varphi_{i,N} + \varphi_{i,1}) & \cdots & \sin(\varphi_{i,N} + \varphi_{i,N}) \end{bmatrix} = \sqrt{\boldsymbol{I} - \boldsymbol{X}_i^2}^{\mathrm{T}} \boldsymbol{X}_i - \boldsymbol{X}_i^{\mathrm{T}} \sqrt{\boldsymbol{I} - \boldsymbol{X}_i^2} \tag{7.11}$$

where, $\boldsymbol{I}$ denotes a unit vector $[1, 1, \cdots, 1] \in R^{1 \times N}$.

Fig. 7.3 shows the GASF and GADF conversion process of an exergy-related variable (Rolling force).

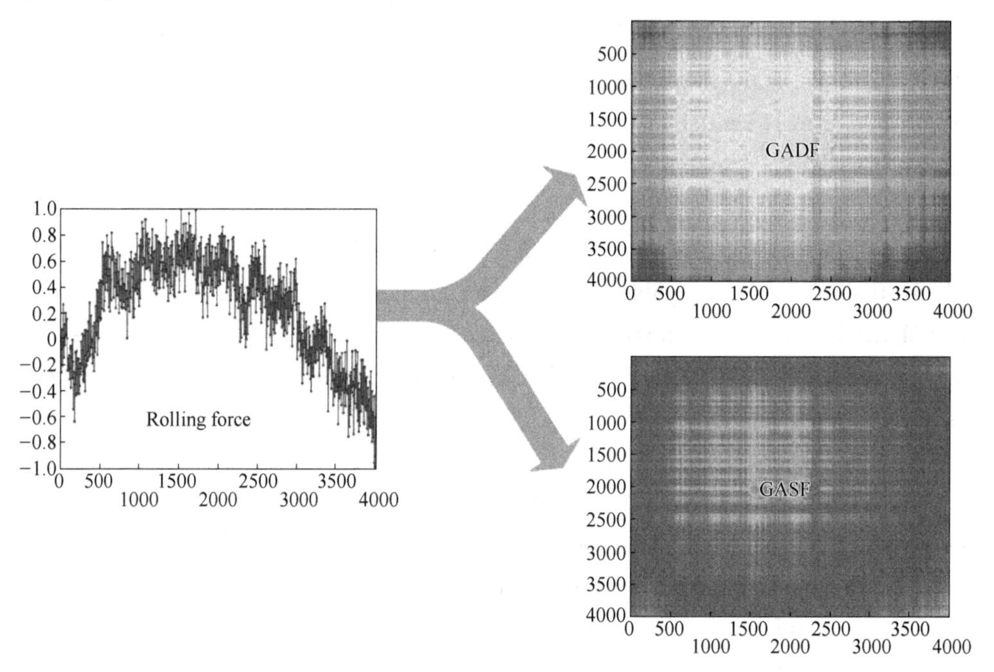

Fig. 7.3 GASF and GADF conversion process

(Scan the QR code on the front of the book for color picture)

7.2.2 Exergy-related FDD based on TransCGAN

Inspired by previous works [9] and [10], we combine the advantages of transformer model and CGAN to build a new transformer-based conditional generative adversarial networks (TransCGAN) completely free of convolutions as shown in the **Fig. 7.4**. As the basic block, the transformer encoder consists of two parts, namely the multi-head self-attention module and the feed-forward multiple layer perceptron (MLP) as shown in the bottom right corner of **Fig. 7.4**. And layer normalization is applied before these two parts[17]. The memory-friendly transformer-based generator has three stages, and each stage consists of several transformer encoders (4, 2, and 2 by default) . As the input of the generator, the random noise is processed to form a vector by MLP. And the vector is reshaped into a feature map. By stages, it can gradually increase the feature map resolution while reducing the size of the input images. After each stage, an upsampling module is inserted to synthesize higher resolution images, which consisting of a reshaping and pixelshuffle module[18] . Unlike the generator that require precise synthesis of each pixel, the transformer-based discriminator can capture the global relationship of the input images through a self-attention mechanism to distinguish the authenticity of the input images.

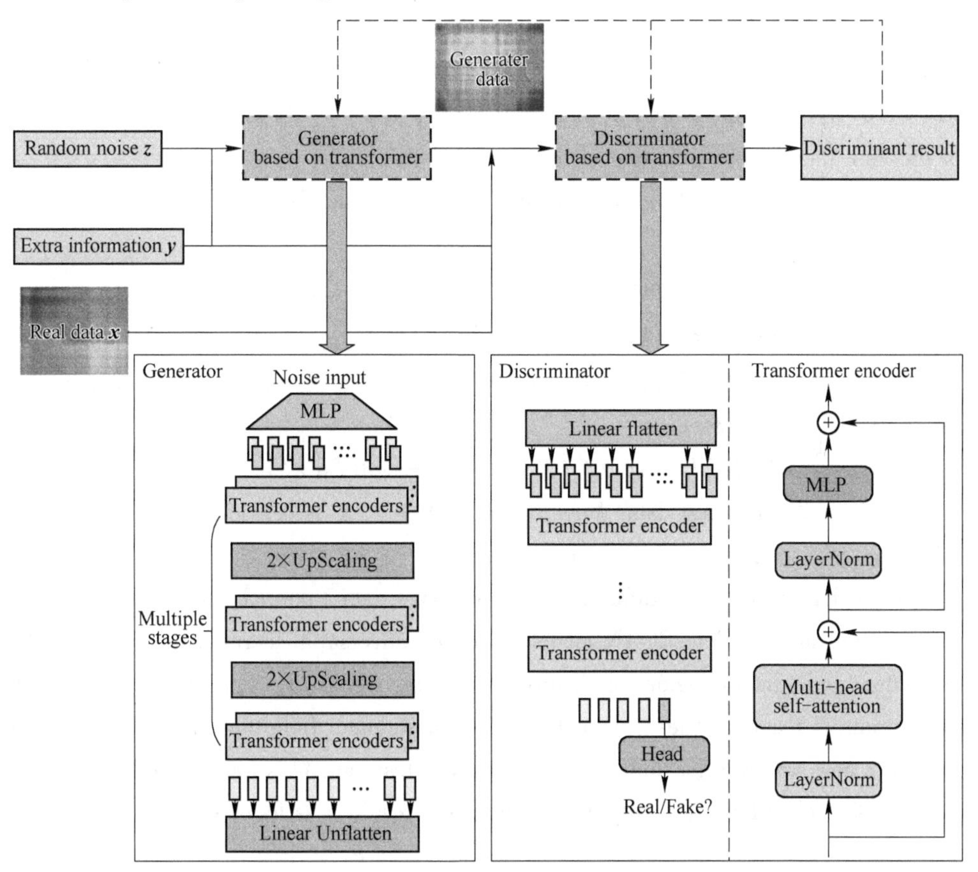

Fig. 7.4 TransCGAN

(Scan the QR code on the front of the book for color picture)

In fault detection stage, GASF and GADF conversion of normal data are carried out, and then, the TransCGAN model is trained and tested by the image data. The generator

G and discriminator D are updated alternately until $V(D, G)$ converges. We find the optimal random noise $\boldsymbol{Z}_i$ to minimize the reconstruction error of generator G for a given image data $\boldsymbol{X}_i$, and the generator fault detection index (G-FDI) is defined as:

$$F_G(\boldsymbol{X}_{\text{Exergy}}) = \min \sum_{i=1}^{R} w_i \ \| \boldsymbol{X}_i - G(\boldsymbol{Z}_i | \boldsymbol{y}) \|^2 \tag{7.12}$$

where, $w_i = \dfrac{\text{MIC}(\boldsymbol{X}_i, \boldsymbol{Y})}{\sum_{i=1}^{R} \text{MIC}(\boldsymbol{X}_i, \boldsymbol{Y})}$ is the weight coefficient of the i_{th} TransCGAN model.

For the trained discriminator D , a discriminator fault detection index (D-FDI) can be calculated as:

$$F_D(\boldsymbol{X}_{\text{Exergy}}) = - \sum_{i=1}^{R} w_i D(\boldsymbol{X}_i | \boldsymbol{y}) \tag{7.13}$$

where, $D(\boldsymbol{X}_i | \boldsymbol{y})$ is the output of discriminator D ; the "–" is used to make the D-FDI of the fault data higher than that of the normal data.

When the significance level α is set to 5%, the thresholds of G-FDI and D-FDI are defined as[19]:

$$\begin{cases} T_G = 95 \text{ quantile of } \{f_G(\boldsymbol{X}_{\text{Exergy}}) \,|\, \boldsymbol{X}_{\text{Exergy}} \in \boldsymbol{X}_{\text{train}}\} \\ T_D = 95 \text{ quantile of } \{f_D(\boldsymbol{X}_{\text{Exergy}}) \,|\, \boldsymbol{X}_{\text{Exergy}} \in \boldsymbol{X}_{\text{train}}\} \end{cases} \tag{7.14}$$

Based on the G-FDI, D-FDI, and their thresholds, a comprehensive fault detection index (CFDI) is designed as:

$$\text{CFDI}(\boldsymbol{X}') = \text{sgn}(f_G(\boldsymbol{X}') - T_G) + \text{sgn}(f_D(\boldsymbol{X}') - T_D) \tag{7.15}$$

where, $\boldsymbol{X}' \in \boldsymbol{X}_{\text{test}}$ is a test data. When $\text{CFDI}(\boldsymbol{X}') \geqslant 0$, $\boldsymbol{X}'$ is considered as a fault data; otherwise, $\boldsymbol{X}'$ is regarded a normal one.

In fault diagnosis stage, different kinds of image data (normal and different faults) are used to train the TransCGAN model. Based on adversarial learning mechanism, the generator and discriminator of TransCGAN are iteratively optimized until Nash equilibrium is reached. Then, we extract the discriminator from the trained TransCGAN to form a new structure for fault classification. The test data $\boldsymbol{X}'$ is input into the network, for a C-class classifier, the output fault diagnosis result is as follows:

$$H(\boldsymbol{X}') = \sum_{i=1}^{R} w_i D(\boldsymbol{X}_i') = \begin{bmatrix} P_1 \\ P_2 \\ \vdots \\ P_C \end{bmatrix} \tag{7.16}$$

$$\text{Faul ttype} = \text{argmax}\{P_c, \ c = 1, 2, \cdots, C\} \tag{7.17}$$

7.2.3 Further discussions and comments

7.2.3.1 Hyperparameter selection of TransCGAN

In the process of fault detection and diagnosis, the model hyperparameters are crucial to the training efficiency and stability of TransCGAN. The structure of generator G and discriminator D can influence the training procedure. To handle this issue, the following hyperparameter optimization tricks are used:

(1) Since the G and D are learning from each other, layer normalization (LN) is used in each transformer encoder to prevent the problem of gradient vanishing during training. Adding LN layer solves the problem of poor initialization, while ensuring gradient propagation to each layer, and also prevents the G from converging to the same point.

(2) Learning rate l is an important hyperparameter in network optimization. If l is too large, the TransCGAN will constantly fluctuate during the training process, or even fail to converge. If l is too small, the convergence speed is too slow. We adopt Adam stochastic optimization for parameter optimization, which combines momentum and RMSProp. It can adaptively adjust l and use momentum to guide parameter updating[20]. In this work, the initial l is 1×10^{-4} for both G and D.

(3) The role of activation function is to perform nonlinear conversion in the network and enhance the representation and learning ability of the network. The selection of activation function is important for TransCGAN. Gaussian error linear unit (GELU) is selected as the high-performing activation function of MLP in this work[21].

7.2.3.2 Improved technique for training TransCGAN

As a drawback of GAN, the training of GAN is extremely unstable. Generally speaking, when optimizing G, the default assumption is that the discriminative ability of D is better than the current generation ability of G, so that D can guide G to optimize in a better direction. We can first update the parameters of D once or multiple times, and then update the parameters of G. When the discriminator loss decreases, the generator loss increases (and vice versa), which makes gradient descent fail to converge. Recently, Heusel et al. proposed a two time-scale update rule (TTUR) for GAN training[22]. TTUR sets individual learning rates for both G and D to make D converge faster than G. Usually, G has slower update rules, while D has faster update rules. Based on TTUR, TransCGAN can converge to a stationary local Nash equilibrium, G and D can be updated at a rate of 1 : 1, and better results can be obtained in the same amount of time only by modifying the update rate.

7.3 Case study

In this section, the exergy-related fault diagnosis method is applied to a real HSMP. First, an overview of the HSMP is presented. Then, three state-of-the-art (SOTA) methods are compared with the proposed method. 41 process variables are selected for experiments, as shown in **Table 7.1.**

Table 7.1 Variable descriptions in the FMZ

Variable	Description	Unit
1-7	Average roll gap of the i_{th} stand, $i=1, \cdots, 7$	mm
8-14	Rolling force of the i_{th} stand, $i=1, \cdots, 7$	MN
15-21	Rolling speed of the i_{th} stand, $i=1, \cdots, 7$	mm/s
22-27	Bending force of the i_{th} stand, $i=2, \cdots, 7$	MN
28-34	Current of the i_{th} stand, $i=1, \cdots, 7$	A
35-41	Power of the i_{th} stand, $i=1, \cdots, 7$	kW

7.3.1 Fault detection results and analysis

In order to verify the fault detection performance of the TransCGAN, it is compared with the conditional-deep convolutional generative adversarial network (C-DCGAN)[23]. We use 4000 normal samples to train TransCGAN and C-DCGAN model, respectively. Three types of faults are provided in **Table 7.2**, which usually occur in the HSMP. There are 1000 testing samples for each fault. In the test dataset of Fault 1, the first 500 samples are normal, and Fault 1 is introduced to the hydraulic roll gap control loop from the 501_{st} test sample. In the test datasets of Fault 2 and Fault 3, the first 250 samples and the last 250 samples are normal, Fault 2 occurs in the cooling valve between the 2_{nd} stand and the 3_{rd} stand from the 251_{st} test sample, and Fault 3 is introduced to the bending force measuring sensor from the 251_{st} test sample. Fault detection rate (FDR) and false alarm rate (FAR) are used to evaluate the detection performance of TransCGAN and *C*-DCGAN. The FDR and FAR can be calculated as:

$$\mathrm{FDR} = \frac{\mathrm{TP}}{\mathrm{TP} + \mathrm{FN}} \tag{7.18}$$

$$\mathrm{FAR} = \frac{\mathrm{FP}}{\mathrm{FP} + \mathrm{TN}} \tag{7.19}$$

where TP represents correctly detected fault samples, TP + FN denotes the total of fault samples. FP is the normal samples incorrectly identified as fault samples, FP + TN denotes the total of normal samples. An effective fault detection usually has high FDR and low FAR.

Table 7.2 Fault types in the HSMP

Fault	Fault description
1	Failure of gap control loop in the 4_{th} stand
2	Stiction of cooling valve between the 2_{nd} and the 3_{rd} stands
3	Malfunction of bending force measuring sensor in the 5_{th} stand

Fig. 7.5-Fig. 7.7 show the fault detection results of the TransCGAN and C-DCGAN for Fault 1, Fault 2, and Fault 3, respectively. And the FDRs and FARs of the TransCGAN and C-DCGAN are summarized in **Table 7.3**, where the higher FDRs and lower FARs are bolded. Obviously, TransCGAN has fewer missing alarms and false alarms than C-DCGAN, which means TransCGAN provides more acceptable detection performance. Compared with Fault 1, it is more difficult to detect Fault 2 and Fault 3. For Fault 2, the FDRs of C-DCGAN is moderate, but there are many missing alarms during the occurrence of this fault. In addition, C-DCGAN has more false alarms than TransCGAN before Fault 3 is introduced and after it disappears. It is worth noting that the FAR of TransCGAN is reduced by approximately 8% compared with C-DCGAN. Moreover, the proposed network can detect more fault samples than C-DCGAN in the duration of Fault 3. According to the aforementioned detection results and analyses, the TransCGAN has better fault detection performance than C-DCGAN.

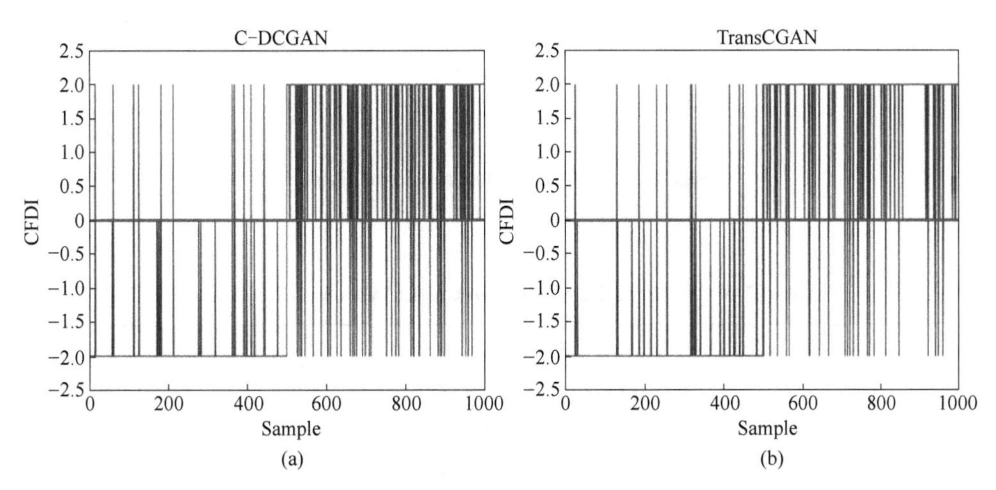

Fig. 7.5 Fault detection results of Fault 1

(a) C-DCGAN; (b) TransCGAN

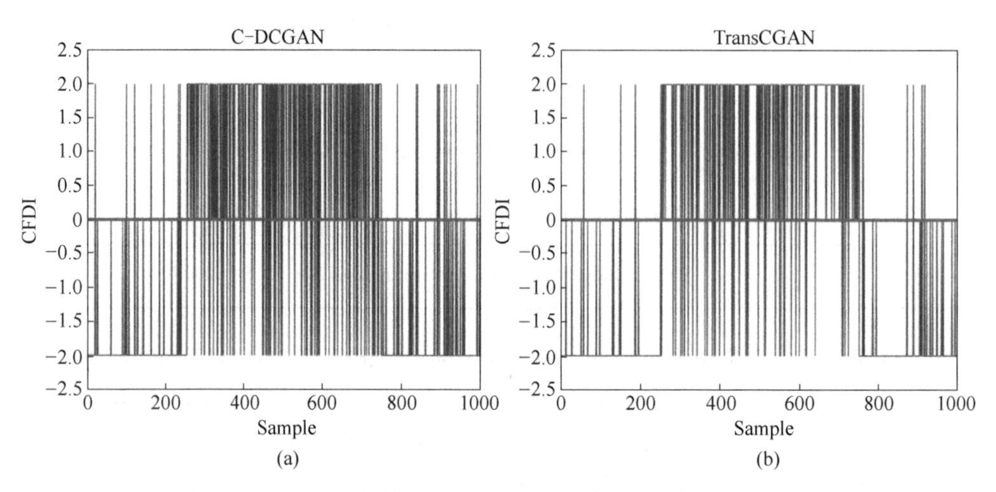

Fig. 7. 6 Fault detection results of Fault 2

(a) C-DCGAN; (b) TransCGAN

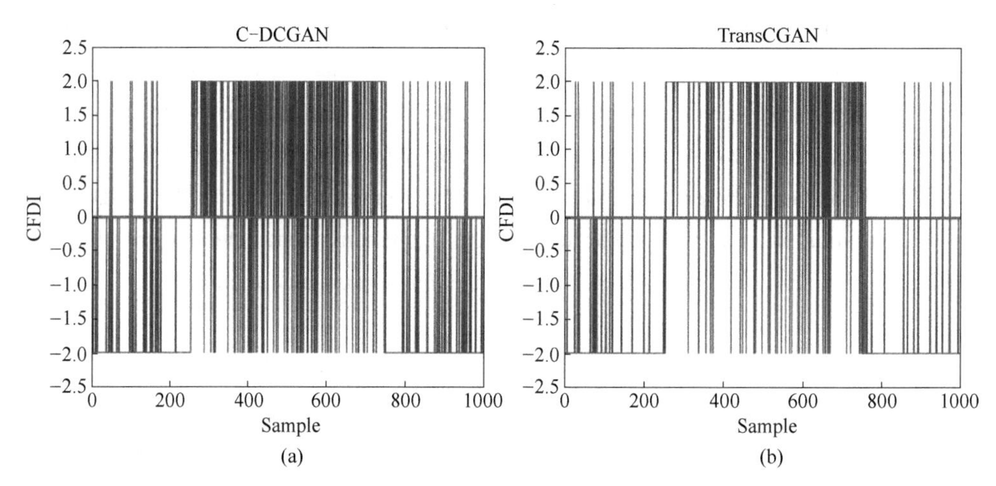

Fig. 7. 7 Fault detection results of Fault 3

(a) C-DCGAN; (b) TransCGAN

Table 7. 3 FDRs and FARs of the C-DCGAN and TransCGAN

Fault No.	FDR		FAR	
	C-DCGAN	TransCGAN	C-DCGAN	TransCGAN
1	92. 4%	**95. 6%**	**4. 6%**	5. 0%
2	84. 6%	**91. 2%**	9. 8%	**6. 4%**
3	83. 2%	**91. 0%**	11. 2%	**6. 8%**
Average	86. 7%	**92. 6%**	8. 5%	**6. 1%**

7.3.2 Fault diagnosis results and analysis

GAN is used to generate high-quality data samples for data enhancement. Therefore, it is important to evaluate the similarity between the generated data and the original data, and we conduct relevant experiments to evaluate whether the generated samples are suitable as training data sets in fault diagnosis tasks. First, we use training data to train the generator and generate new samples. Then, the mixed data of the original training data and the generated data is used to train the classifier. The classifier is tested on a given labeled test dataset, and the classification accuracy demonstrates the quality of the generated data and the fault diagnosis performance of the models. Normal samples are labelled as 0, the samples of Fault 1, Fault 2, Fault 3 are labeled as 1, 2, and 3, respectively. In this experiment, a total of 4000 samples are collected for training networks, including 2500 normal samples, 500 Fault 1 samples, 500 Fault 2 samples, and 500 Fault 3 samples. Besides, 500 samples are generated for each process condition. We also collect 2000 testing samples, including 500 samples of each process condition. In this work, a classification accuracy (CA) of each process condition is used to evaluate the diagnosis performance, which can be calculated as:

$$CA_s = \frac{\text{samples(correctly identified)}}{\text{total samples}} \quad s = 0,\ 1,\ 2,\ 3 \tag{7.20}$$

In order to investigate the impact of adding generated training samples to the classifier model, four scenarios are considered in this work, including 0 generated samples, 100 generated samples, 300 generated samples, and 500 generated samples. Based on the results in **Table 7.4**, it can be found that the CA_s are improved to a certain extent, indicating that the generated samples provide auxiliary and useful information for model training, which is helpful for fault diagnosis in most scenarios. It should be noted that when the training data is sufficient, there is only a small improvement for CA_s.

Table 7.4 The CA_s of TransCGAN after adding different number of generated samples

Adding samples	0	100	300	500
CA_0	98.8%	98.2%	98.4%	**99.0%**
CA_1	96.2%	96.0%	96.6%	**97.4%**
CA_2	95.2%	95.4	**95.8%**	**95.8%**
CA_3	96.0%	95.8%	96.4%	**96.6%**
Average	96.3%	96.4%	96.8%	**97.2%**

For comparison, the fault diagnosis models are also developed by CDBN, SSAE, and Wasserstein generative adversarial networks (WGAN)[24]. **Table 7.5** shows the CA_s of the above five models with 500 generated samples added. **Fig. 7.8** (a)-(e) show the confusion matrices of C-DCGAN, CDBN, SSAE, WGAN, and the proposed method, respectively. For most faults, it is obvious that the diagnosis performance of proposed method is better than the other four methods. Based on the above simulation results and analysis, it can be concluded that the proposed framework not only works well in fault detection, but also achieves good results in fault diagnosis, which will provide theoretical basis and technical support for process safety and product quality of the HSMP.

Table 7.5 The CA_s of different model

Method	CA_0	CA_1	CA_2	CA_3	Average
C-DCGAN	96.6%	95.8%	95.0%	95.4%	95.7%
CDBN	97.0%	96.2%	95.4%	95.0%	95.9%
SSAE	98.2%	964%	**96.0%**	95.6%	96.6%
WGAN	93.4%	94.0%	92.6%	92.2%	93.1%
TransCGAN	**99.0%**	**97.4%**	95.8%	**96.6%**	**97.2%**

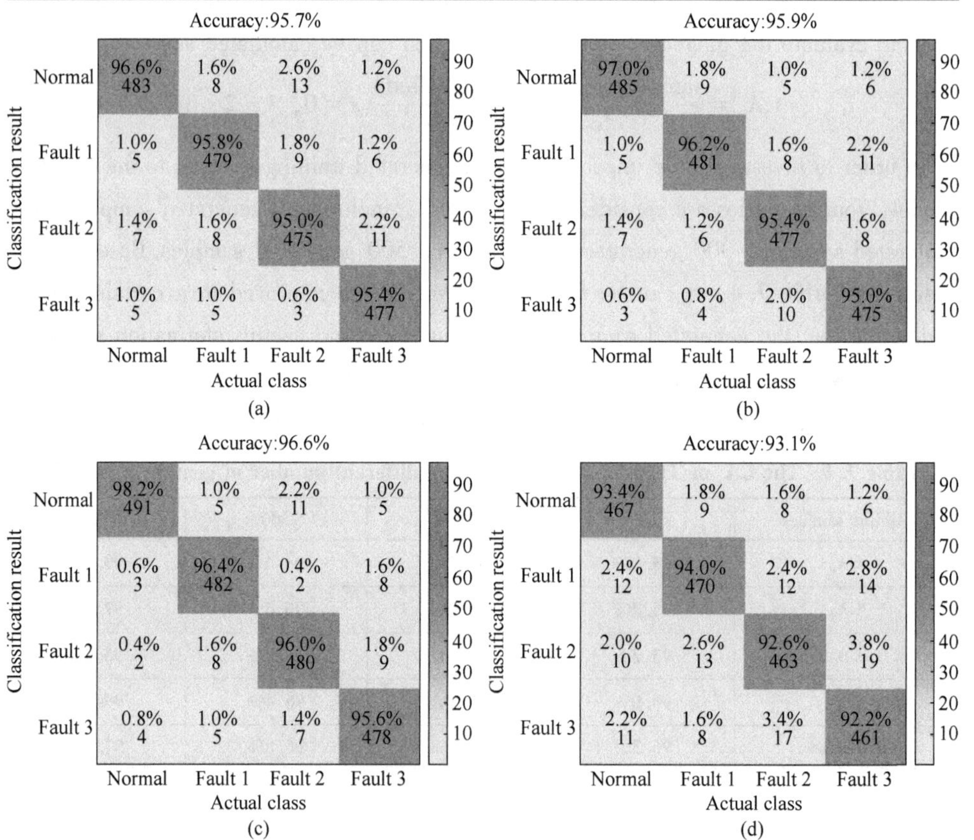

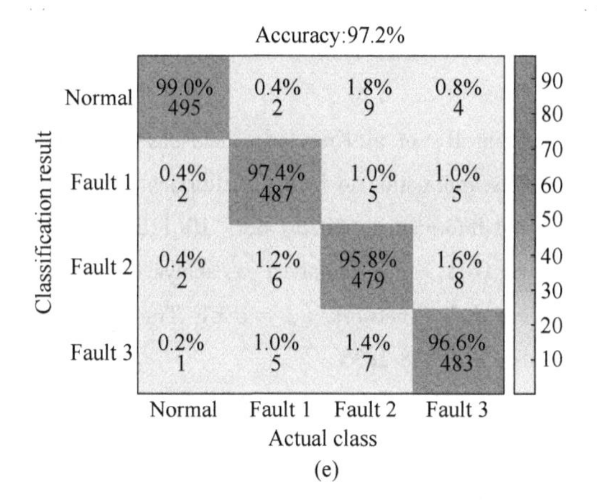

Fig. 7. 8 Fault diagnosis results of different methods

(a) C-DCGAN; (b) CDBN; (c) SSAE; (d) WGAN; (e) TransCGAN

(Scan the QR code on the front of the book for color picture)

7. 4 Conclusions

In this chapter, a novel exergy-related fault detection and diagnosis framework has been proposed for the hot rolling process. Different from existing methods, the proposed framework considers energy consumption, spatio-temporal correlation, and unbalanced data characteristics. First, a new correlation coefficient is designed to select the exergy-related variables. Then, GAF converts the original one-dimensional data into image data. After that, a transformer encoder-based CGAN is developed to extract the spatio-temporal feature of the exergy-related variables and generate high-quality samples. Finally, the fault detection and diagnosis strategy is proposed to realize effective process monitoring. The experiment results indicate that compared to other SOTA methods, the proposed framework can effectively detect and identify the faults in HRP. For outlooks, this work only considers normal and several fault conditions of HRP, does not provide a more detailed division of the operating performance under normal condition. Future works will consider the operating performance assessment issue under normal condition, which will directly influence economic benefits of the enterprise.

References

[1] Liu Z Y, Zhang J F, He X, et al. Fault diagnosis of rotating machinery with limited expert interaction: A multicriteria active learning approach based on broad learning system [J]. IEEE Transactions on Control Systems Technology, 2023, 31 (2): 953-960.

[2] Zhang S Q, Wang R J, Si Y P, et al. An improved convolutional neural network for three-phase inverter fault diagnosis [J]. IEEE Transactions on Instrumentation & Measurement, 2021, 71: 3510915.

[3] Ren H, Yang C H, Sun B, et al. Knowledge-data-based synchronization states analysis for process monitoring and its application to hydrometallurgical zinc purification process [J]. IEEE Transactions on Industrial Informatics, 2023, doi: 10.1109/TII.2023.3268411.

[4] Dai J J, Song H, Sheng G H, et al. Dissolved gas analysis of insulating oil for power transformer fault diagnosis with deep belief network [J]. IEEE Transactions on Dielectrics & Electrical Insulation, 2017, 24 (5): 2828-2835.

[5] Shao H, Jiang H, Zhang H, et al. Rolling bearing fault feature learning using improved convolutional deep belief network with compressed sensing [J]. Mechanical Systems & Signal Processing, 2018, 100: 743-765.

[6] Viola J, Chen Y Q, Wang J. FaultFace: Deep convolutional generative adversarial network (DCGAN) based ball-bearing failure detection method [J]. Information Sciences, 2021, 542: 195-211.

[7] Qi Y, Shen C, Wang D, et al. Stacked sparse autoencoder-based deep network for fault diagnosis of rotating machinery [J]. IEEE Access, 2017, 5: 15066-15079.

[8] Marais H J, Schoor G V, Uren K R. The merits of exergy-based fault detection in petrochemical processes [J]. Journal of Process Control, 2017, 74 (2): 110-119.

[9] Mirza M, Osindero S. Conditional generative adversarial nets [J]. Computer Science, 2014, 10: 2672-2680.

[10] Vaswani A, Shazeer N, Parmar N, et al. Attention is all you need [J]. Neural Information Processing System (NIPS), 2017.

[11] Spearman C. The proof and measurement of association between two things [J]. International Journal of Epidemiology, 1904, 15 (1): 72-101.

[12] Reshef D N, Reshef Y A, Finucane H K, et al. Detecting novel associations in large data sets [J]. Science, 2011, 334 (6062): 1518-1524.

[13] Zhang W J, Li X. General correlation and partial correlation analysis in finding interactions: With Spearman rank correlation and proportion correlation as correlation measures [J]. Network Biology, 2015, 5 (4): 163-168.

[14] Zhang Y, Jia S L, Huang H Y, et al. A novel algorithm for the precise calculation of the maximal information coefficient [J]. Science Report, 2014, 4: 6662.

[15] Li S, Zhou X F, Pan F C, et al. Correlated and weakly correlated fault detection based on variable division and ICA [J]. Computers & Industrial Engineering, 2017, 112: 320-335.

[16] Sun S Y, Ren J. GASF-MSNN: A new fault diagnosis model for spatiotemporal information extraction [J]. Industrial Engineering Chemical Research, 2021, 60: 6235-6248.

[17] Ba J L, Kiros J R, Hinton G E. Layer normalization [J]. 2016, arXiv: 1607.06450.

[18] Shi W, Caballero J, Totz J, et al. Real-time single image and video super-resolution using an

efficient sub-pixel convolutional neural network [J]. Computer Vision Pattern Recognition (CVPR), 2016.

[19] Wang H G, Li X, Zhang T. Generative adversarialnetwork based novelty detection using minimized reconstruction error [J]. Frontiers of Information Technology & Electronic Engineering, 2018, 19 (1): 116-125.

[20] Khan A H, Cao X, Li S, et al. BAS-ADAM: An ADAM based approach to improve the performance of beetle antennae search optimizer [J]. IEEE/CAA Journal of Automatica Sinica, 2020, 7 (2): 461-471.

[21] Hendrycks D, Gimpel K. Gaussian Error Linear Units (GELUs) [J]. 2016, arXiv: 1606.08415.

[22] Heusel M, Ramsauer H, Unterthiner T, et al. GANs trained by a two time-scale update rule converge to a local Nash equilibrium [J]. 2017, arXiv: 1706.08500.

[23] Luo J, Huang J Y, Huang H M. A case study of conditional deep convolutional generative adversarial networks in machine faultdiagnosis [J]. Journal of Intelligent Manufacturing, 2021, 32: 407-425.

[24] Arjovsky M, Chintala S, Bottou L. Wasserstein generative adversarial networks [J]. 2017, arXiv: 1701.07875.

Chapter 8 Distributed CVRAE-Based Spatio-Temporal Process Monitoring

Process monitoring is a critical means to ensure the safety and stability of industrial processes[1-5]. As industrial processes become more complex than before, research on plant-wide process monitoring techniques has received much attention[6-7]. A plant-wide process is typically a process with multiple operating units, workshops or even plants. The large-scale and highly complex characteristic of plant-wide processes poses significant challenges for process modeling and monitoring.

Since mechanism analysis of large-scale systems is laborious and even unavailable, model-based monitoring methods, which are employed for small-scale systems, are rarely used for large-scale systems. Data-driven methods, which require less system knowledge, have become mainstream for large-scale process modeling and monitoring. Different from traditional centralized monitoring techniques for small-scale processes, plant-wide process monitoring techniques usually divide process into multiple blocks according to empirical knowledge or data-based approaches. As a result, multiple monitoring models are established for individual blocks, and then the monitoring results of multiple blocks can be fused by information fusion techniques to obtain global monitoring results. The monitoring methods based on multi-block division have the following advantages. Firstly, the modeling complexity is reduced. Secondly, the divided blocks can better describe the local characteristic of large-scale systems. Thirdly, the most responsible local system can be easily located once a fault occurs.

Based on the divided blocks by experiential knowledge, MacGregor proposed the multi-block PLS (MBPLS) method to establish monitoring charts for each local block and the entire process[8]. The unified multi-block analyses for both PCA and PLS were provided by Qin et al. [9]. Zhang et al. considered the nonlinear characteristic of plant-wide processes, and proposed multi-block kernel PCA (MBKPCA)[10] and multi-block kernel PLS (MBKPLS)[11] methods for fault diagnosis. Recently, more and more data-based methods have been used for multi-block division. Ge proposed a distributed PCA (DPCA) method and reserved variables to the corresponding blocks by contribution indices[12]. Jiang and Yan introduced a multi-block PCA method that employed mutual information to divide blocks automatically[13]. Zhang et al. divided the plant-wide processes by mixed similarity measure and then proposed a distributed gap support vector data description (Gap-SVDD) method for sub-block monitoring[14].

In practical industrial processes, different operation units are physically connected to each other. However, the methods mentioned above only decomposed the entire process into several blocks and used the local measurements of each block to detect faults, ignoring the interactions between different blocks. It would be more effective to consider the correlation information of neighboring blocks for local block monitoring. Chen et al. proposed a distributed CCA (DCCA) method to detect local faults by considering the correlation information measured from other blocks[15]. Zhang et al. applied DCCA to the monitoring of the HSMP, which reduced the transmission and computational costs compared with the centralized CCA method[16]. Peng et al. used the partly-connected topology to select the neighboring variables, then the residuals generated by CCA model is used for process monitoring[17]. Jiang et al. used LASSO to determine communication variables, then developed a latent variable correlation analysis (LVCA) framework to achieve distributed monitoring for plantwide processes[18].

Spatially, multiple subsystems are distributed in different locations in tandem. Information is transmitted between subsystems through material flow, energy flow and information flow. Temporally, the process variables present the time-dependent auto-correlation and variable dependent cross-correlation, due to the influence of material balance, energy balance, close loop control or other factors. And due to the different times of processing materials arriving at different subsystems, the transfers of information between different subsystems have a high delay, which makes systems exhibit a large time-delay characteristic. Thosespatio-temporal characteristics require that the influence of both the correlation between subsystems and the dynamic and time delay that existed in the system should be considered in the modeling and monitoring of tandem plant-wide processes.

Although few articles have considered the correlations between sub-systems or sub-processes, these methods are based on statistical learning and cannot extract deep features between and within sub-processes, especially temporal related information[19-20]. In addition, there is little research on how to describe the dependencies between the preceding and following subsystems for tandem industrial processes. Therefore, this chapter proposes a novelspatio-temporal monitoring model based on distributed conditional variational recurrent autoencoder (CVRAE) to model and monitor the tandem industrial processes. The main contributions can be described in two aspects:

(1) A distributed VRAE modeling method is proposed to extractspatio-temporal features of tandem processes with the characteristics of correlation, dynamics and nonlinearity between and within subsystems.

(2) Two-level monitoring statistics are established for detecting process faults and locating the subsystem where the fault occurred.

8.1 Review of LSTM, CVAE, and DSVDD

8.1.1 Long Short-Term Memory

As a special kind of recurrent neural network (RNN), Lony-Short-Term Memory (LSTM) network is mainly used to process time series data, which can use past information to calculate the current output. It is proposed to address the vanishing gradient problem that the standard RNN easily forgets the information from a long time ago[21]. The structure of LSTM unit constructed by Graves[22] is shown in **Fig. 8.1**. The cell state $\boldsymbol{c}_t$ and the hidden state $\boldsymbol{h}_t$ record the information in the time series. The updating of $\boldsymbol{c}_t$ decides what information will be added and how much information from previous states will be forgotten. The hidden state $\boldsymbol{h}_t$ decides the final output. The forward calculation formulas are shown as below:

$$\boldsymbol{f}_t = \sigma(\boldsymbol{W}_{\mathrm{f}} \cdot [\boldsymbol{h}_{t-1}, \boldsymbol{x}_t] + \boldsymbol{b}_{\mathrm{f}}) \tag{8.1}$$

$$\boldsymbol{i}_t = \sigma(\boldsymbol{W}_i \cdot [\boldsymbol{h}_{t-1}, \boldsymbol{x}_t] + \boldsymbol{b}_i) \tag{8.2}$$

$$\tilde{\boldsymbol{c}}_t = \sigma(\boldsymbol{W}_c \cdot [\boldsymbol{h}_{t-1}, \boldsymbol{x}_t] + \boldsymbol{b}_{\mathrm{c}}) \tag{8.3}$$

$$\boldsymbol{c}_t = \boldsymbol{f}_t * \boldsymbol{c}_{t-1} + \boldsymbol{i}_t * \tilde{\boldsymbol{c}}_t \tag{8.4}$$

$$\boldsymbol{o}_t = \sigma(\boldsymbol{W}_{\mathrm{o}} \cdot [\boldsymbol{h}_{t-1}, \boldsymbol{x}_t] + \boldsymbol{b}_{\mathrm{o}}) \tag{8.5}$$

$$\boldsymbol{h}_t = \boldsymbol{o}_t * \tanh(\boldsymbol{c}_t) \tag{8.6}$$

where, $\boldsymbol{W}$ and $\boldsymbol{b}$ refer to the network weight matrix and bias vector, respectively. $\sigma(\ )$ is sigmoid function, whose output is between 0 and 1. The output of function tanh( ) is between −1 and 1.

The forget gate $\boldsymbol{f}_t$ in equation (8.1) decides which information of cell state $\boldsymbol{c}_{t-1}$ will be forgot. The input gate it in equation (8.2) decides how much candidate cell information $\tilde{\boldsymbol{c}}_t$ in equation (8.3) is added to the cell state $\boldsymbol{c}_t$. Then, the new cell state $\boldsymbol{c}_t$ is updated by equation (8.4). Finally, the output state $\boldsymbol{h}_t$ is calculated by the output gate $\boldsymbol{o}_t$ and the cell state $\boldsymbol{c}_t$.

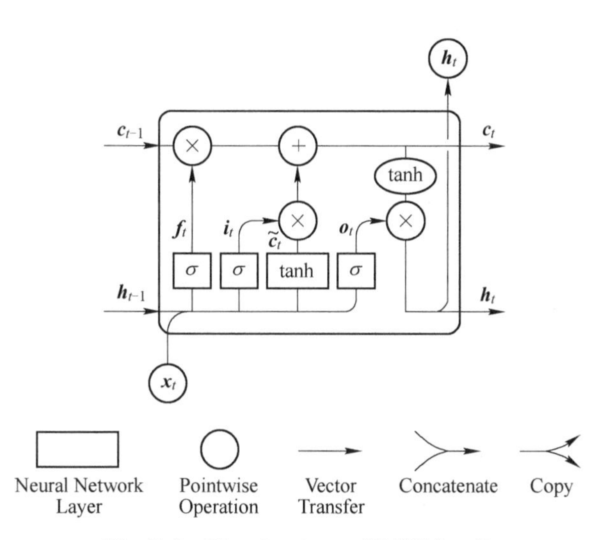

Fig. 8.1 The structure of LSTM unit

8.1.2 Conditional variational autoencoder

As a special autoencoder model, variational autoencoder (VAE)[23-24] is generally used for data generation. While guaranteeing the minimization of reconstruction error, VAE projects the observation data to a lowdimensional latent space and makes them obey a prior distribution as far as possible. In general, the standard normal distribution is selected as the prior distribution. Then the distribution of the data generated by prior distribution can approximate the true distribution of observation data. However, the simple prior assumption will lead to poor performance in modeling complex systems.

To better learn the representation of more complex distribution, conditional VAE (CVAE) introduces conditional constraints into VAE model that make the prior distribution determined by input conditions[24]. Similar to VAE, CVAE comprises two processes: recognition and generation. From the perspective of a probabilistic graphical model, CVAE is shown in **Fig. 8.2**.

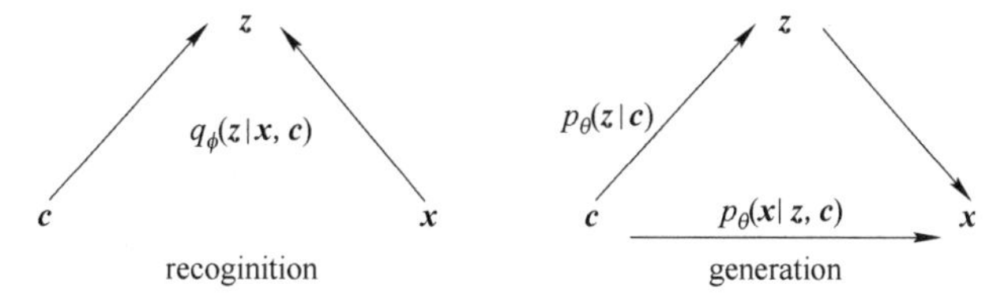

Fig. 8.2 The probabilistic graphical model of CVAE

The CVAE model is composed of three parts: recognition network, prior network and generation network. The network framework of CVAE for calculating and training process is shown in **Fig. 8. 3**, where $\boldsymbol{x}$ and $\boldsymbol{c}$ represent the process variables and condition variables, respectively, and $\boldsymbol{z}$ represent the latent variables. The dashed lines in **Fig. 8. 3** denote the loss terms of objective function for model training.

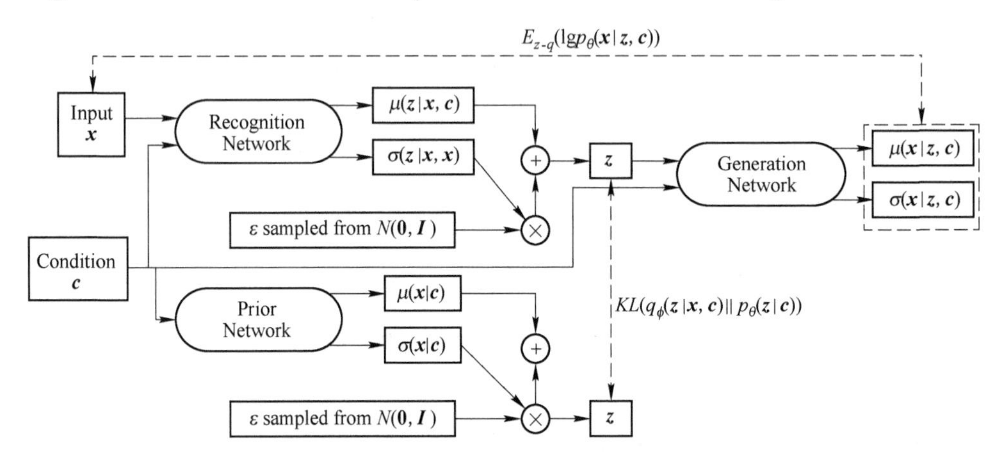

Fig. 8. 3　The network framework of CVAE

In the recognition process, the posterior distribution $q_\phi(\boldsymbol{z}|\boldsymbol{x}, \boldsymbol{c})$ is determined by recognition network, whose inputs are the observation $\boldsymbol{x}$ and the condition $\boldsymbol{c}$. During generation process, the prior distribution $p_\theta(\boldsymbol{z}|\boldsymbol{c})$ is determined by the prior network with the input $\boldsymbol{c}$. The observation $\boldsymbol{x}$ are generated by the generation network with the latent variables $\boldsymbol{z}$ and the condition $\boldsymbol{c}$ as input. If CVAE is used for data generation, the latent variables $\boldsymbol{z}$ are sampled from distribution $p_\theta(\boldsymbol{z}|\boldsymbol{c})$. If CVAE is used for data reconstruction, the latent variables $\boldsymbol{z}$ are sampled from distribution $q_\phi(\boldsymbol{z}|\boldsymbol{x}, \boldsymbol{c})$.

The training target of CVAE is to maximize the conditional log likelihood:

$$\max \lg p(\boldsymbol{x} \mid \boldsymbol{c}) = \max \sum_{i=1}^{N} \lg p(\boldsymbol{x}_i|\boldsymbol{c}_i) \tag{8.7}$$

Given a train sample $(\boldsymbol{x}_i, \boldsymbol{c}_i)$, the Kullback-Leibler (KL) divergence measures the difference between $q_\phi(\boldsymbol{z}|\boldsymbol{x}_i, \boldsymbol{c}_i)$ and $p_\theta(\boldsymbol{z}|\boldsymbol{x}_i, \boldsymbol{c}_i)$, can be formulated as:

$$\begin{aligned} D_{\mathrm{KL}}[q_\phi(\boldsymbol{z}|\boldsymbol{x}_i, \boldsymbol{c}_i)||p_\theta(\boldsymbol{z}|\boldsymbol{x}_i, \boldsymbol{c}_i)] &= E_{z\text{-}q}[\lg q_\phi(\boldsymbol{z}|\boldsymbol{x}_i, \boldsymbol{c}_i) - \lg p_\theta(\boldsymbol{x}_i|\boldsymbol{z}, \boldsymbol{c}_i) - \\ &\quad \lg p_\theta(\boldsymbol{x}_i|\boldsymbol{c}_i)] + \lg p_\theta(\boldsymbol{x}_i) \\ &= -E_{z\text{-}q}[\lg p_\theta(\boldsymbol{x}_i|\boldsymbol{z}, \boldsymbol{c}_i)] + \\ &\quad D_{\mathrm{KL}}[q_\phi(\boldsymbol{z}|\boldsymbol{x}_i, \boldsymbol{c}_i)||p_\theta(\boldsymbol{z}|\boldsymbol{x}_i)] + \\ &\quad \lg p_\theta(\boldsymbol{x}_i|\boldsymbol{c}_i) \end{aligned} \tag{8.8}$$

The posterior distribution$q_\phi(\boldsymbol{z}|\boldsymbol{x}_i, \boldsymbol{c}_i)$ is the variational approximation of the true distribution $p_\theta(z|\boldsymbol{x}_i, \boldsymbol{c}_i)$. When the posterior distribution is close enough to the true distribution, the KL divergence is close to 0. Define the objective function of VAE as:

$$L(\theta, \phi; \boldsymbol{x}_i, \boldsymbol{c}_i) = \lg p_\theta(\boldsymbol{x}_i|\boldsymbol{c}_i) - \mathrm{KL}[q_\phi(z|\boldsymbol{x}_i, \boldsymbol{c}_i) || p_\theta(z|\boldsymbol{x}_i, \boldsymbol{c}_i)] \tag{8.9}$$

The training target of CVAE is translated to maximize $L(\theta, \phi; \boldsymbol{x}_i, \boldsymbol{c}_i)$. The objective function of CVAE is expressed as:

$$L(\theta, \phi; \boldsymbol{x}_i, \boldsymbol{c}_i) = E_{z\text{-}q}[\lg p_\theta(\boldsymbol{x}_i|z, \boldsymbol{c}_i)] - D_{\mathrm{KL}}[q_\phi(z|\boldsymbol{x}_i, \boldsymbol{c}_i) \| p_\theta(z|\boldsymbol{c}_i)] \tag{8.10}$$

8.1.3 Deep support vector data description

SVDD is a technique to find the smallest hyper-sphere that encloses the majority of the data in the feature space[25]. Usually, SVDD is used for anomaly detection, where those data that fall outside the hyper-sphere are considered anomalous. Let ϕ_k be a feature mapping, which projects data into feature space $\boldsymbol{F}_k$. $c_k \in \boldsymbol{F}_k$ and $R_k > \boldsymbol{0}$ are the center and the radius of the hyper-sphere, respectively. The SVDD primal problem is given by:

$$\begin{aligned} &\min_{R_k, c_k, \xi} R_k^2 + \frac{1}{vn}\sum_i \xi_i \\ &\text{s.t. } \|\phi_k(\boldsymbol{x}_i) - \boldsymbol{c}_k\|_{\mathrm{F}_k}^2 \leqslant R^2 + \xi_i,\ \xi_i \geqslant 0,\ \forall i \end{aligned} \tag{8.11}$$

Here, nonnegative slack variable ξ_i allows the margin to be soft and hyperparameter $v \in (\boldsymbol{0}, \boldsymbol{1}]$ controls the tradeoff between penalty ξ_i and the volume of the sphere.

The most commonly used method of SVDD is the one based on kernel machine. Due to its poor computational performance for large-scale data, Ruff et al. proposed deep SVDD algorithm[20], which trains a neural network to minimize the volume of a hyper-sphere that encloses the network representations of the data. The idea of deep SVDD is shown in **Fig. 8.4**. The objective function of deep SVDD is defined as below:

$$\min_{W} \frac{1}{n}\sum_{i=1}^{n} \|\phi(\boldsymbol{x}_i; \boldsymbol{W}) - \boldsymbol{c}\|^2 + \frac{\lambda}{2}\sum_{l=1}^{L} \|\boldsymbol{W}^l\|_F^2 \tag{8.12}$$

where, $\phi(\boldsymbol{x}_i; \boldsymbol{W})$ is a neural network with L hidden layers and weights $\boldsymbol{W} = \{\boldsymbol{w}^1, \boldsymbol{w}^2, \cdots, \boldsymbol{w}^L\}$. The first term of the function is a quadratic loss for penalizing the distance of every network representation $\phi(\boldsymbol{x}_i; \boldsymbol{W})$ to the center $\boldsymbol{c}$. The last term is a weight decay regularizer on parameters $\boldsymbol{W}$ with hyperparameter λ, where $\| \ \|_F$ denotes the Frobenius norm.

In order to avoid the hyper-sphere collapse, that means the radius R^* of the optimal solution equals to 0, there are three points to be noted here. Firstly, the hypersphere center $\boldsymbol{c}$ is fixed as the mean of the initial network representations, and can not be 0. Secondly, there are no bias terms in neural networks of deep SVDD. Finally, the unbounded activation function such as ReLU should be used in deep SVDD.

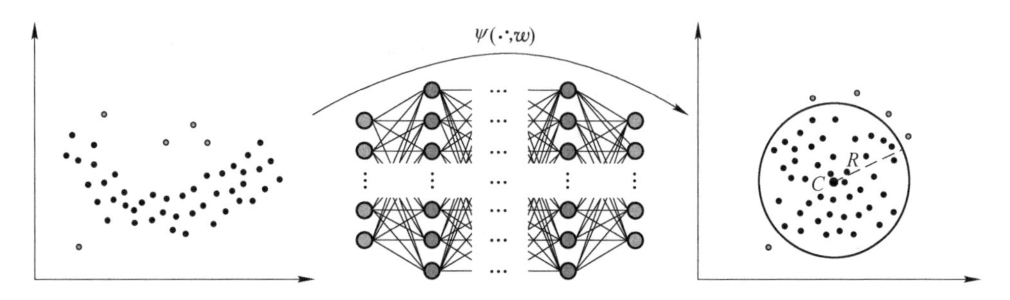

Fig. 8. 4 The network presentation of deep SVDD

(Scan the QR code on the front of the book for color picture)

8. 2 System description and methodology

8. 2. 1 System description

Consider the dynamic collaborative stochastic distribution (CSD) system which consists of N subsystems connected in tandem, which is shown in **Fig. 8. 5** (a). Denote $\boldsymbol{x}_i(t)$ and $z_i(t)$ as the state and the output variables of the i_{th} subsystem at time t, respectively. Due to the dynamic nature, the structure expanded in the time domain is presented in **Fig. 8. 5** (b), where n is the order of the process.

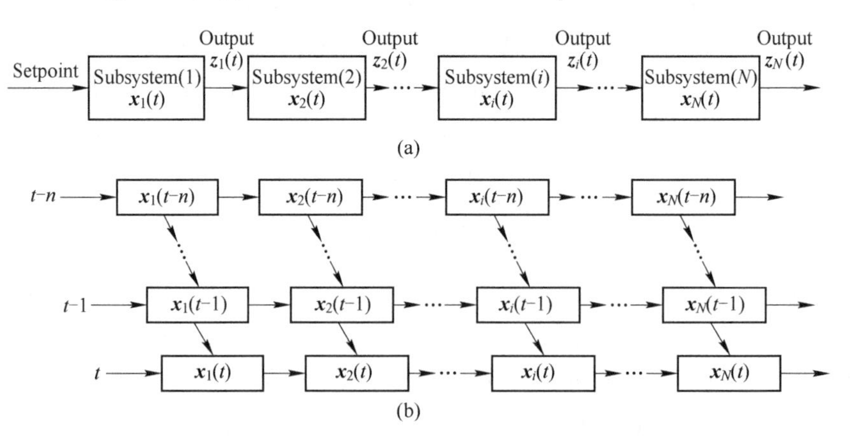

Fig. 8. 5 Structure of collaborative stochastic distribution system

(a) The tandem structure of CSD; (b) The structure expanded in the time domain

For the i_{th} subsystem, the output $z_{i-1}(t)$ of the previous subsystem is a part of its input variables. Considering the effect of unknown disturbance $\boldsymbol{o}_i(t)$, the process of the i_{th} subsystem is presented as below

$$\begin{cases} \boldsymbol{x}_i(t) = g_i(\boldsymbol{X}_i(t-1), \boldsymbol{Z}_{i-1}(t), \boldsymbol{O}_i(t)) \\ z_i(t) = h_i(\boldsymbol{X}_i(t-1), \boldsymbol{Z}_{i-1}(t), \boldsymbol{O}_i(t)) \end{cases} \tag{8.13}$$

where, $g(\ )$ and $h(\ )$ denote nonlinear functions, $\boldsymbol{Z}_i(t) = [z_i(t-n), z_i(t-n+\mathbf{1}), \cdots, z_i(t)]$, and $\boldsymbol{X}_i(t) = [\boldsymbol{x}_i(t-n), \boldsymbol{x}_i(t-n+\mathbf{1}), \cdots, \boldsymbol{x}_i(t)]$. It can be found that the state variables $\boldsymbol{x}_i(t)$ of the i_{th} subsystem are determined by joint effects of the state of the past time and the output of the previous subsystems. In the future, $f_i(t)$ represents the unknown fault in the i_{th} subsystem. The current state and the output are directly affected by the fault $f_i(t)$. Then the change of the output further influences the subsequent subsystem. Therefore, the fault can be propagated through the series path of the process.

Furthermore, the time-delay d should be considered in the CSD system since most of the practical complex industrial processes are nonlinear macrohysteretic systems, such as the steel rolling process and chemical process. In general, d is considered as a known constant in order to guarantee the stability of the system.

8.2.2 Distributed conditional variational recurrent autoencoder

In order to model the dynamic CSD system, it needs to consider not only the effect of the previous subsystem, but also the effect of early states. In this section, the CVRAE model is used to model the subsystems, where the condition features extracted from the previous subsystem by LSTM network but related to the current one is used as the constrained conditions of CVRAE model. Another LSTM network is introduced into CVRAE model to extract the dynamic features. The structure of CVRAE model for the i_{th} subsystem is shown in **Fig. 8.6**.

The distributed model is constructed by connecting all CVRAE models of each subsystem in series according to the path of CDS system. For each subsystem, the condition features are extracted from the hidden dynamic feature $\boldsymbol{h}(t)$ of the previous subsystem. The final product quality can be viewed as the output of the last subsystem, which can be represented by the posterior probability $p_\theta(\boldsymbol{y}(t) | z_N(t), \boldsymbol{c}_N(t))$. The predicted model is constructed similarly to the recognition model of CVRAE. Its model parameters are learned by maximizing log likelihood estimation, whose objective function is represented as $\boldsymbol{E}_{z\text{-}q}[p_\theta(\boldsymbol{y}(t) | z_N(t), \boldsymbol{c}_N(t))]$. The structure of the integrated model is shown in **Fig. 8.7**.

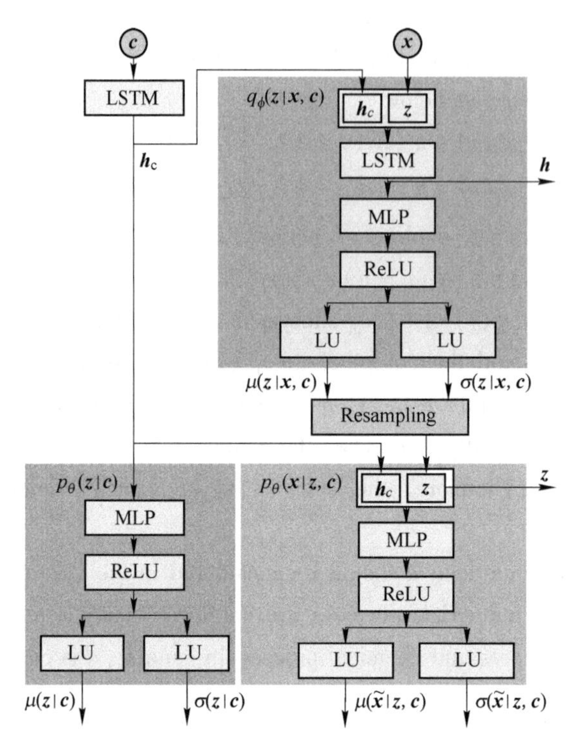

Fig. 8. 6 The network framework of CVRAE

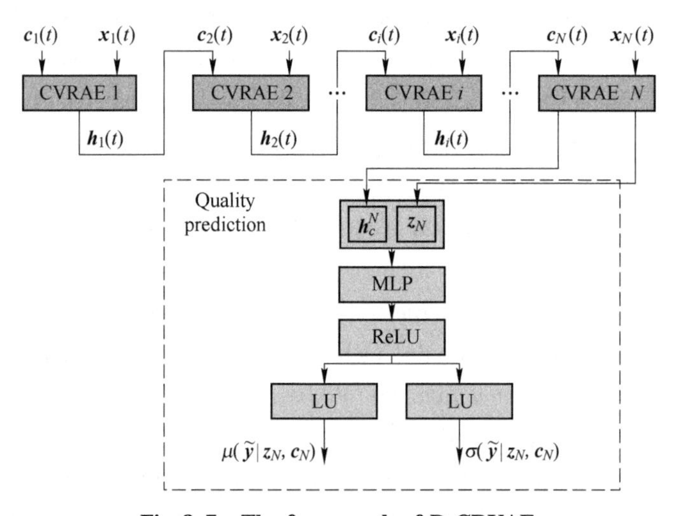

Fig. 8. 7 The framework of D-CRVAE

For a tandem industrial process, the main factors that affect the first subsystem are the parameters of the raw material or the measurement data of the previous processes. Therefore, the input conditions of the first CVRAE sub-model are the raw material characteristics that are not affected by the first monitored process or the operation data of the previous process, etc. For example, in the finishing process of HSM, the parameters before the first subsystem include the force and gap of the roughing mill stand and the finish rolling entry temperature, etc. These parameters are used as input conditions for the first sub-model to participate in model training and calculation.

Merge all objective functions of the subsystems, then an integrated objective function of the D-CVRAE model is formularized as:

$$L(\theta, \phi) = \sum_{i=1}^{N} [E_{z\text{-}q}(\lg p_\theta(\boldsymbol{x}_i | z, \boldsymbol{c}_i)) - D_{\mathrm{KL}}(q_\phi(z | \boldsymbol{x}_i, \boldsymbol{c}_i) || p_\theta(z | \boldsymbol{x}_i))] + E_{z\text{-}q}[p_\theta(\boldsymbol{y}(t) \mid z_N(t), \boldsymbol{c}_N(t))] \tag{8.14}$$

In the offline training phase, the normal historical data are first collected. Then the historical data are divided into several parts according to the physical distribution of equipment and the experiential knowledge. Due to the existence of time-delay between subsystems, the effect of time-delay should be eliminated before training the model. The time-delay d between subsystems can be determined based on the experience or mechanism model. The historical data of each subsystem are aligned according to the estimated time-delay d. Finally, the normalized data are fed into the designed distributed model for model training.

8.2.3 Hierarchical framework of process monitoring

In this section, a global-local modeling and monitoring framework is proposed. Firstly, a D-CVRAE model is established to learn the distributed representations of each local unit. Then a joint representation for all local units is constructed.

8.2.3.1 Local Unit Monitoring

A D-CVRAE model is established for all N operation units. This section focuses on establishing the local monitoring statistics for each operation unit. For the time t, the observation data are $\boldsymbol{x}(t) = \{\boldsymbol{x}_1(t), \boldsymbol{x}_2(t), \cdots, \boldsymbol{x}_N(t)\}$, where N denotes the number of operation units. Considering the i_{th} operation unit, the estimation value of latent variables $z_i(t)$ is the output mean value $\boldsymbol{\mu}(z_i(t) | \boldsymbol{x}_i(t), \boldsymbol{c}_i(t))$ of the recognition network. Because the prior probability is changed with the condition $\boldsymbol{c}_i(t)$, which is given by the previous operation unit, the estimation value should be firstly normalized by the prior probability$p(z_i(t) | \boldsymbol{c}_i(t))$, and is formulated as:

$$\underline{z}_i(t) = \frac{\boldsymbol{\mu}(z_i(t) \mid \boldsymbol{x}_i(t),\ \boldsymbol{c}_i(t)) - \boldsymbol{\mu}(z_i(t) \mid \boldsymbol{c}_i(t))}{\boldsymbol{\sigma}(z_i(t) \mid \boldsymbol{c}_i(t))} \tag{8.15}$$

where, $\boldsymbol{\mu}(z_i(t) \mid \boldsymbol{c}_i(t))$ and $\boldsymbol{\sigma}(z_i(t) \mid \boldsymbol{c}_i(t))$ are the mean and the variance calculated by prior network $p_\theta(z_i \mid \boldsymbol{c}_i)$, respectively.

The KL divergence term of CVRAE makes the posterior distribution $q_\phi(z_i \mid \boldsymbol{c}_i)$ approximate to the true prior distribution $p(z_i \mid \boldsymbol{c}_i)$, but not equal to it. Therefore, the normalized representation $\underline{z}_i(t)$ does not completely follow the standard normal distribution $N(\boldsymbol{0},\ \boldsymbol{I})$. Let $\overline{\boldsymbol{Z}}_i$ and $\Sigma_{\boldsymbol{Z}_i}$ denote the mean vector and covariance matrix of the normalized representation $z_i(t)$ of the historical training dataset, respectively. The monitoring statistic T^2 for local operation unit is established as:

$$T_i^2(t) = (\underline{z}_i(t) - \bar{z})^{\mathrm{T}} \Sigma_{\underline{z}}^{-1}(\underline{z}_i(t) - \bar{z}) \tag{8.16}$$

T_i^2 detects the fault occurrence in the i_{th} local unit with the condition information from the previous unit. The control limit of T_i^2 is determined as follows:

$$T_{\text{lim},\ i}^2 = \lambda_\alpha^2(m_i) \tag{8.17}$$

where, m_i is the latent variable dimension of the i_{th} local unit; α is the confidence level.

The probability of observation $\boldsymbol{x}_i(t)$ with the reconstructed probability distribution $q_\phi(\hat{\boldsymbol{x}}_i(t) \mid z_i(t),\ \boldsymbol{c}_i(t))$ denotes the reconstruction error of $\boldsymbol{x}_i(t)$. S samples are drawn from the distribution $q_\phi(z_i(t) \mid \boldsymbol{x}_i(t),\ \boldsymbol{c}_i(t))$. For each sample $z_i^s(t)$, the squared Mahalanobis distance between $\boldsymbol{x}_i(t)$ and the center of the probability distribution $p_\theta(\hat{\boldsymbol{x}}_i^s(t) \mid z_i^s(t),\ \boldsymbol{c}_i(t))$ follows $\chi^2(n_i)$, where n_i is the dimension of $\boldsymbol{x}_i(t)$. That is

$$\begin{aligned} D(\boldsymbol{x}_i(t),\ \hat{\boldsymbol{x}}_i^s(t)) = (\boldsymbol{x}_i(t) - \boldsymbol{\mu}(\hat{\boldsymbol{x}}_i^s(t)))^{\mathrm{T}} \Lambda^{-1}(\hat{\boldsymbol{x}}_i^s(t)) \\ (\boldsymbol{x}_i(t) - \boldsymbol{\mu}(\hat{\boldsymbol{x}}_i^s(t))) : \chi^2(n_i) \end{aligned} \tag{8.18}$$

where, $\Lambda(\hat{\boldsymbol{x}}_i^s(t))$ is a diagonal matrix whose diagonal values are the output variance of the generation network $p_\theta(\boldsymbol{x}_i \mid z_i,\ \boldsymbol{c}_i)$; $\boldsymbol{\mu}(\hat{\boldsymbol{x}}_i^s(t))$ denotes the output mean vector of the generation network.

Define a local Mahalanobis distance-based probability index as:

$$P_i^s(\boldsymbol{x}_i(t)) = Pr\{D < D(\boldsymbol{x}_i(t),\ \hat{\boldsymbol{x}}_i^s(t))\} \tag{8.19}$$

Then the probabilistic statistic for the reconstruction error, combining the local Mahalanobis distance across all samples of $z_i^s(t)$, is given by:

$$Q_i(\boldsymbol{x}_i(t)) = \frac{1}{S}\sum_{s=1}^{S} P_i^s(\boldsymbol{x}_i(t)) \tag{8.20}$$

The control limit of the monitoring statistic Q_i is directly equal to the user-specified confidence level α . The two statistics T_i^2 and Q_i can effectively monitor the local unit.

8.2.3.2 Global Monitoring

Even though the local monitoring statistics provide effective monitoring performance for the local unit, global monitoring statistic of all units is also important in the plant-wide process. For the i_{th} local unit, the representation $z_i(t)$ is sampled from posterior distribution $q_\phi(z_i(t)|\boldsymbol{x}_i(t), \boldsymbol{c}_i(t))$ through D-CVRAE model. The sampled representation $z_i(t)$ is firstly normalized by the prior probability $p_\theta(z_i(t)|\boldsymbol{c}_i(t))$, and the normalized representation is denoted as $\tilde{z}_i(t)$. The joint representation vector $\tilde{z}(t) = (\tilde{z}_1^{\mathrm{T}}(t), \tilde{z}_2^{\mathrm{T}}(t), \cdots, \tilde{z}_N^{\mathrm{T}}(t))^{\mathrm{T}}$ can be constructed. Then a deep SVDD model is established based on the joint representations for all training data. For the joint representations $\tilde{z}(t)$, the monitoring statistic can be defined as the distance between the mapping point and the center of the hyper-sphere, that is:

$$D(\tilde{z}(t)) = \| \phi(\tilde{z}(t); w) - \boldsymbol{c}_k \|_2 \tag{8.21}$$

The control limit of the statistic is determined by kernel density estimation method[26] with the confidence level α, which is denoted D_{lim}. In summary, the flowchart of the proposed plant-wide process monitoring framework is shown in **Fig. 8.8**.

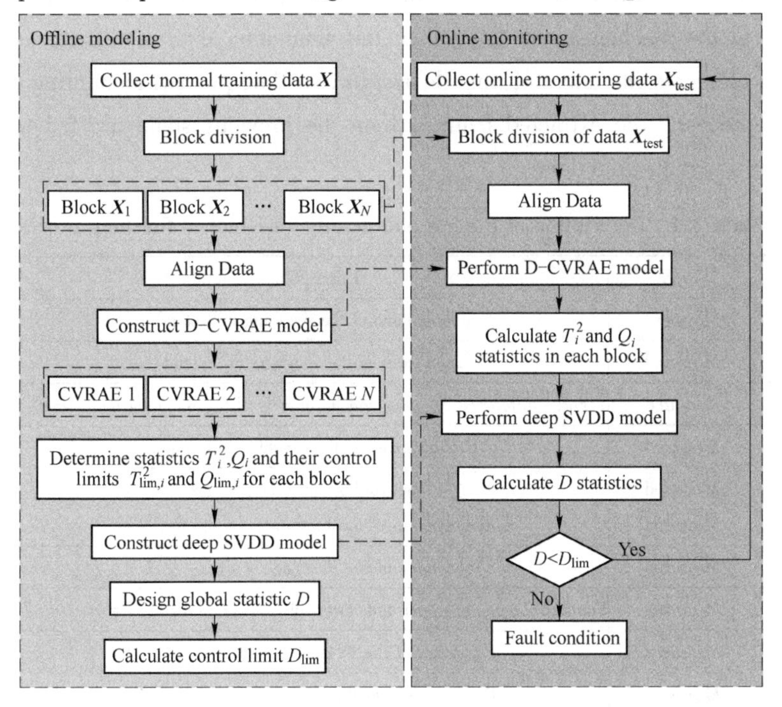

Fig. 8.8 Flowchart of the proposed plant-wide process monitoring framework

8.3 Case study

The proposed process monitoring framework is applied to the finishing process, which is the key part of HSMP. For this purpose, the normal operating data and two types of abnormal operation data collected from the finishing process are used to validate the applicability of the distributed CVRAE.

8.3.1 Description of process variables and faults

In this section, one class of manufacturing process is taken for the algorithm verification, whose target thickness is 3.75 mm. The real measurement data of 10 production batches are collected as training data, and each batch contains 5000 samples. The sampling period is 0.01 s. The selected process and quality variables include 5 measured variables before the entrance of the finishing roll, 53 process variables in 7 mill stands, and one finishing mill exit thickness, which are listed in **Table 8.1**. Two kinds of frequently occurring faults are considered, which are listed in **Table 8.2**. Two batches of measurement data corresponding to the two faults are collected to verify the fault detection performance. Because there is a certain time delay for steel plates to arrive at different stands, the time delays are firstly calculated according to the mechanism models, then the simulation data of different stands are aligned to eliminate the effect of time delays. In real-time process monitoring of the real industrial process, the extracted features from the previous stand are fed to the next stand after a certain time delay.

Table 8.1 Description of process and quality variables in finishing process

Variable	Type	Description	Unit
Gr	Measured	Average gap of the roughing mill stand	mm
TFr	Measured	Total force of the roughing mill stand	MN
DFr	Measured	Force difference of upper and lower rolls of the roughing mill stand	MN
Sr	Measured	Rolling speed of the roughing mill stand	m/s
FET G1-G7	Measured Measured	Finish rolling entry temperature Average gap of the Fi stand, $i=1, \cdots, 7$	℃ mm
TF1-TF7	Measured	Total force of the Fi stand, $i=1, \cdots, 7$	MN
DF1-DF7	Measured	Force difference of upper and lower rolls of the Fi stand, $i=1, \cdots, 7$	MN
P1-P7	Measured	Power of the Fi stand, $i=1, \cdots, 7$	kW
S1-S7	Measured	Rolling speed of the Fi stand, $i=1, \cdots, 7$	m/s

Contrnued Table 8.1

Variable	Type	Description	Unit
Q2-Q7	Measured	Work roll shifting force of the Fi stand, $i=1$, ⋯, 7	MN
B2-B7	Measured	Work roll bending force of the Fi stand, $i=1$, ⋯, 7	MN
T1-T6	Measured	Looper tension between the Fi and F $i+1$ stands, $i=1$, ⋯, 6	MN
y	Quality	Finishing mill exit thickness of the strip	mm

Table 8.2 Typical faults in finishing mill

Fault No.	Description	Occurrence time (s)
1	Malfunction of gap control loop in F4 stand	20
2	Fault of cooling valve between F3 and F4	20

8.3.2 Monitoring results and analyses

For the proposed process monitoring framework, the historical normal operating condition (NOC) dataset is first normalized. Then the 53 process variables are divided into 7 parts according to stands, whose variable numbers of each operation unit are {6, 8, 8, 8, 8, 8, 7}, and the 5 measured variables before the entrance of the finishing roll are selected as the condition input c1 for the first sub-model. For each local unit of D-CVRAE, the dimension of all hidden layers is set as 30, and the MLP is designed as a two-layer structure. The latent and conditional representation dimensions are set as 5 and 3, respectively. The neural network of the deep SVDD model is designed with 2 hidden layers, whose dimensions of each hidden features are {50, 7}, and the hyperparameter λ is set as 0.01.

Firstly, one batch of normal operating data is used to validate the model's accuracy in predicting the finishing mill exist thickness, **Fig. 8.9** shows the predicted result. In **Fig. 8.9** (a), the actual values are represented by blue lines, the predicted values are represented by red lines, and the yellow area denotes the confidence intervals of prediction with 95% confidence level. Since the thickness prediction results are presented as a probability distribution, real time online monitoring of finishing mill exit thickness can be achieved by judging whether the actual thickness is within the confidence interval of the thickness prediction. It can be seen that the predicted values have a good tracking of the actual thickness variation, and the actual values are always within the confidence interval. **Fig. 8.9** (b) shows the scatter plot of predicted value distribution. The color going from blue to red represents the magnitude of absolute error. According to the statistics, the prediction error for 95% of samples is less than 6.6 μm, and the mean absolute error (MAE) and the root mean squared error (RMSE) are 1.8 μm and 3 μm, respectively. This test performance indicates that it is suitable for practical application.

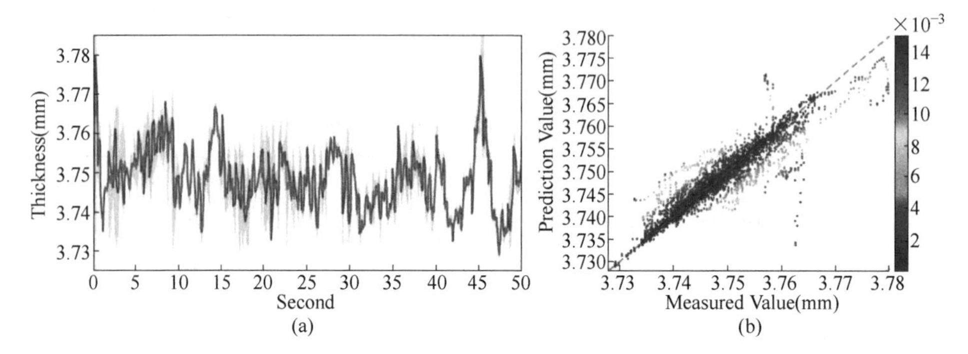

Fig. 8.9 The prediction of finishing mill exit thickness

(a) thickness predictions; (b) prediction performance

(Scan the QR code on the front of the book for color picture)

Then the performance of fault detection based on DCVRAE is compared with distributed variational recurrent autoencoder (D-VRAE), global variational recurrent autoencoder (G-VRAE), and global dynamic kernel PCA (G-DKPCA)[27] based methods. All three compared methods are used for nonlinear dynamic process monitoring. G-VRAE models the whole system by variational recurrent autoencoder (VRAE), and the monitoring indicators are similar to the local monitoring statistics of the proposed method. D-VRAE is a distributed modeling approach, which models the local subsystems by VRAE, not considering the effect of adjacent subsystems, and the other steps are consistent with the proposed method. The modeling and monitoring method based on VRAE refers to the literature[28].

The fault detection rate (FDR) and false alarm rate (FAR) are used to evaluate the performance of the two methods, which measure the proportion of faults that are correctly detected after they have occurred and the proportion of normal samples that are incorrectly identified as faults before they occur, respectively.

The FDRs and FARs of compared methods for local and global monitoring are shown in **Table 8.3** and **Table 8.4**, respectively. As can be seen from the comparison results in **Table 8.3**, due to the use of the condition information of the previous subsystem, the two faults were effectively detected on their counterparts and their subsequent stands with FDRs above 90%. In contrast, D-VRAE does not consider the influence of the previous subsystem, and the farther away from the subsystem where the fault occurs, the worse the fault monitoring effect is. It shows that the local monitoring performance of the proposed method is better than the method based on D-VRAE. The comparison results in **Table 8.4** show that for both faults, the FDRs of D-CVRAE methods is 100%, which is higher than the D-VRAE method, while the average FAR of D-CVRAE method is only 0.96%, which is lower than the two global monitoring methods, G-DKPCA and G-VRAE. It shows that the proposed method has more accurate fault detection results than other methods in global monitoring.

Table 8.3 The FDRs and FARs of D-VRAE, D-CVRAE for local monitoring

Fault				F1	F2	F3	F4	F5	F6	F7
FDR	1	D-VARE	T^2	0.97	1	0.5	100	8.67	0.8	2.93
			Q	0.53	0.83	0.7	100	97.77	0.57	26.23
		D-CVRAE	T^2	1.3	1.03	0.87	100	86.67	61.3	90.97
			Q	0.7	0.9	0.9	100	100	98.03	95.73
	2	D-VARE	T^2	0.97	1	100	80.83	1.37	2.27	9
			Q	0.53	0.63	100	100	0.57	2.83	96.97
		D-CVRAE	T^2	1.3	1.03	100	100	53.27	49.67	85.37
			Q	0.77	0.83	100	100	99.83	94.2	99.47
FAR	Average	D-VRAE	T^2	0.97	0.87	1.03	0.67	0.71	0.71	1.03
			Q	1.62	1.12	1.02	0.81	0.46	0.41	1.47
		D-CVRAE	T^2	0.86	1.07	0.96	0.71	1.03	0.91	0.81
			Q	0.86	1.47	0.91	0.56	0.41	0.56	0.46

Table 8.4 The FDRs and FARs of different methods for global monitoring

Fault		G-DKPCA		G-VRAE		D-VRAE	D-CVRAE
		T^2	*SPE*	T^2	Q	D	D
FDR	1	99.87	100	79.6	100	100	100
	2	25.4	99.97	31.23	100	98.23	100
FAR	Average	12.39	7.92	0.39	1.88	0.42	0.96

Fault 1 introduces a malfunction of the hydraulic gap control loop in F4 stand and directly increases the roll gap (G4). This reduces the rolling force and increases the slab exit thickness of F4 stand. Since the slab becomes thicker, the roll gap of the subsequent stands increases, and these stands need to increase the rolling force to counteract the effect of the increasing slab thickness. This feed-forward control will make the further away from stand F4, the less noticeable the impact from this fault will be, and the final finishing exit thickness will be controlled within a reasonable range. The global monitoring results of four methods are sketched in **Fig. 8. 10**. The yellow lines denote statistics which the process is faulty. It can be seen that all four methods give timely detection. However, the GDKPCA method has high false alarm rates (FARs) at the early process, and the statistic T^2 of G-DKPCA has 0. 05 s time delay for Fault 1. The statistic T^2 of G-VRAE has the lowest fault detection rates (FDRs), which means the statistic built by latent variables is insensitive to the fault. Though the

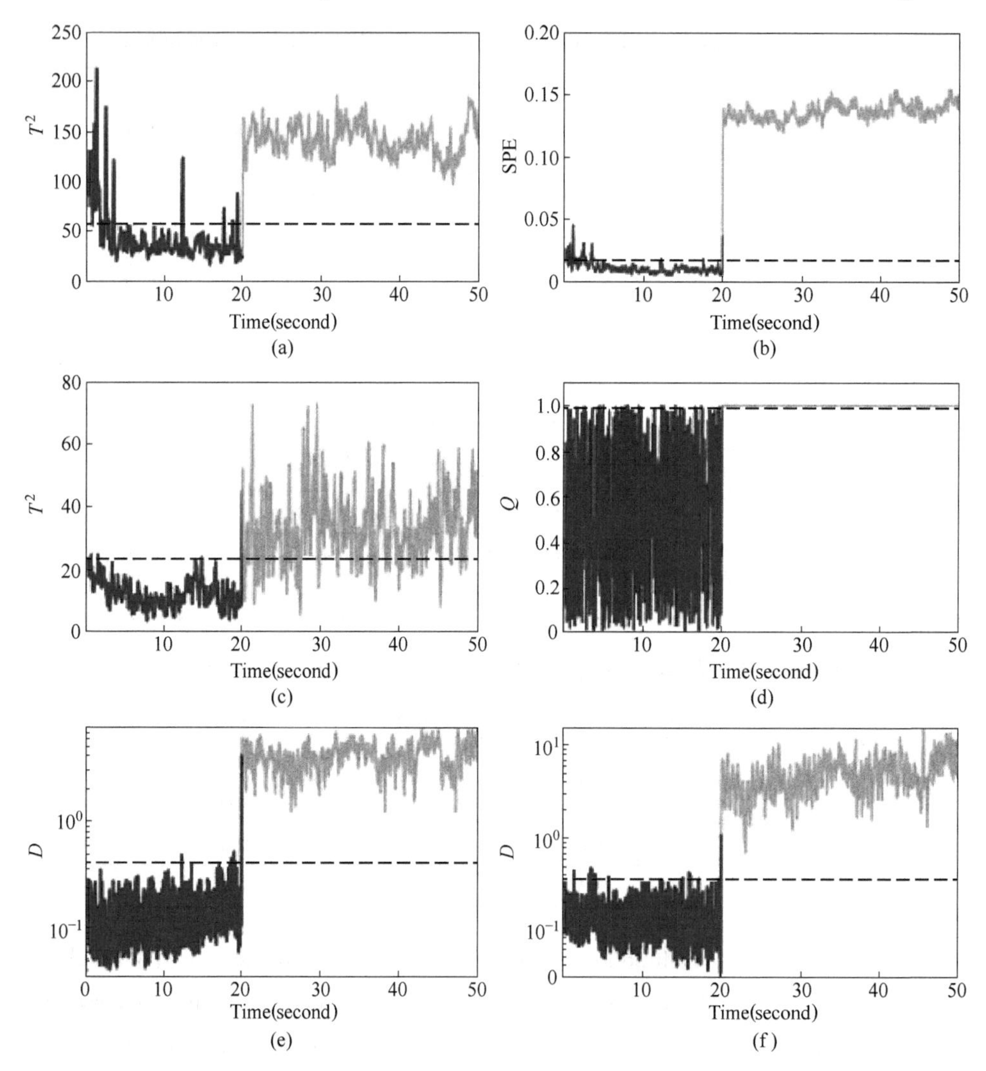

Fig. 8. 10 Global process monitoring results of Fault 1

(a) (b) G-DKPCA; (c) (d) G-VRAE; (e) D-VRAE; (f) D-CVRAE

statistics of D-VRAE and D-CVRAE are based on latent variables, both of them can detect the fault timely. The distributed methods can promote detection performance compared with global methods. The local monitoring result of proposed method is shown in **Fig. 8.11**. It can be seen that the local monitoring statistics of F4-F7 stands detect this fault effectively. According to the series structures of FMP, F4 stand is identified as the part occurring fault first. Due to the actions of material and energy flow, the subsequent stands are also affected by this fault. But the feedback regulation existing in the subsequent stands can reduce the impact of this fault, which are reflected in the statistics T^2 of F5-F7 stands.

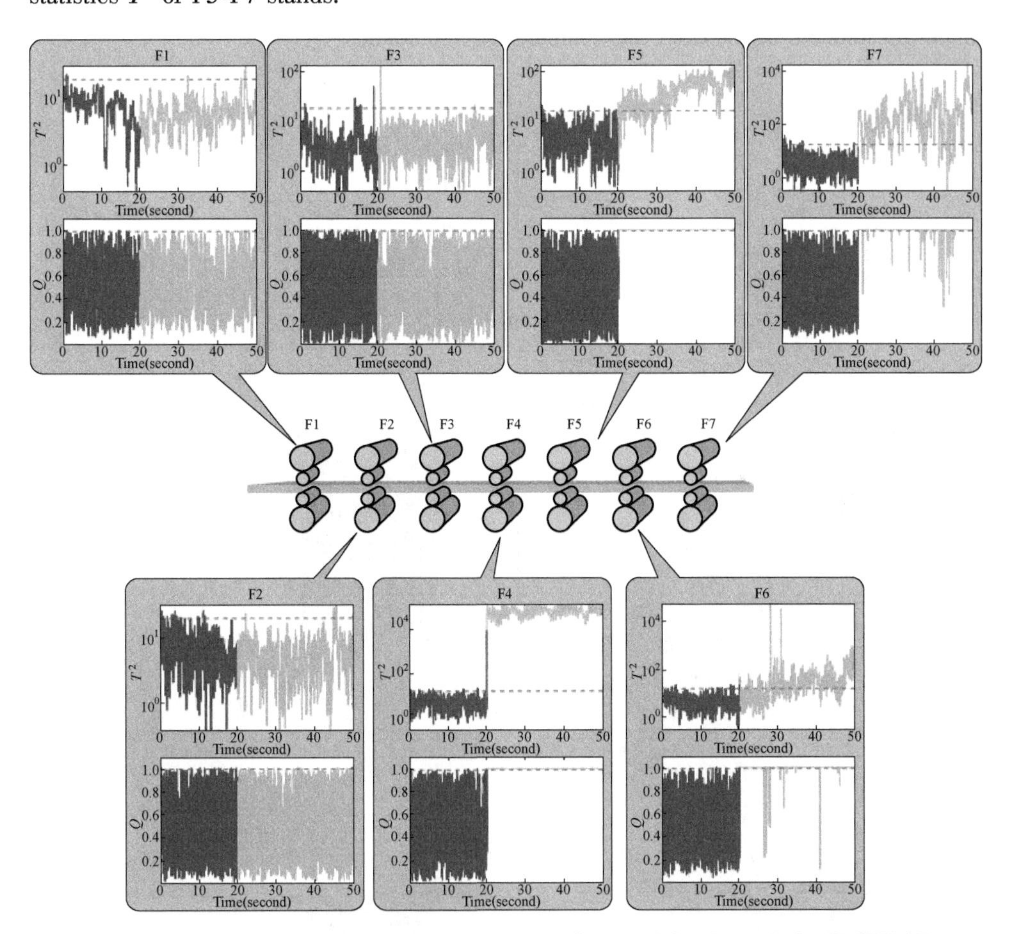

Fig. 8.11 The local process monitoring results of Fault 1 for 7 stands by D-CVRAE

Fault 2 is a blockage failure of the cooling valve between F2 and F3 stand. This failure causes the temperature of the slab entering F3 stand to rise, which directly affects the slab's stress factor. And the slab entry temperature of subsequent stands is higher than expected, resulting in the deviation of the rolling force and roll gap (TF3-TF7 and G3-G7) of each stand, which in turn affects the final finishing exit thickness. The global monitoring results of four methods are shown in **Fig. 8.12**. The four methods alarm without delay. But the statistics T^2 of both GDKPCA and G-VRAE fail to detect the fault effectively. The method based on D-CVRAE gives a better detection result than that based on D-VRAE. The local monitoring result of our proposed method is presented in **Fig. 8.13**. It is indicated that the fault occurs in F3 stand.

Based on the above experiment results and analysis, it can be concluded that the proposed method is capable of handling process modeling and monitoring tandem plantwide process with spatial and temporal characteristics.

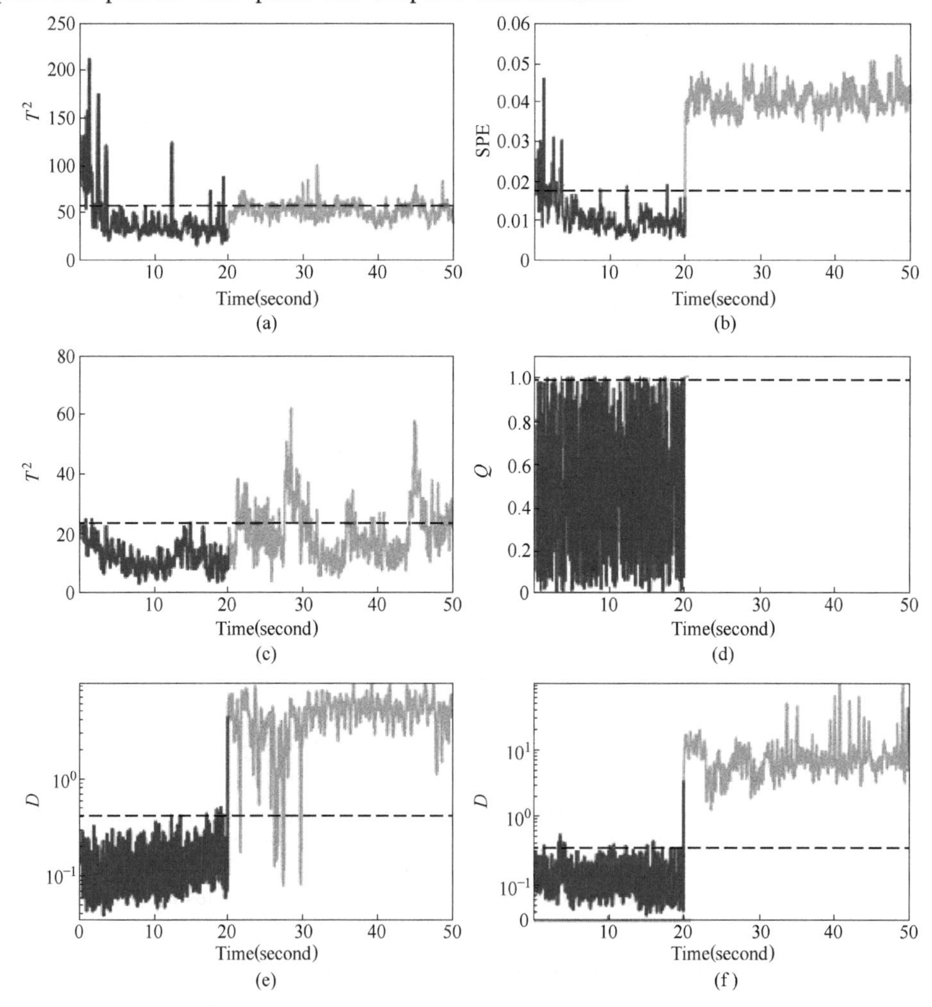

Fig. 8.12 Global process monitoring results of Fault 2

(a) (b) G-DKPCA; (c) (d) G-VRAE; (e) D-VRAE; (f) D-CVRAE

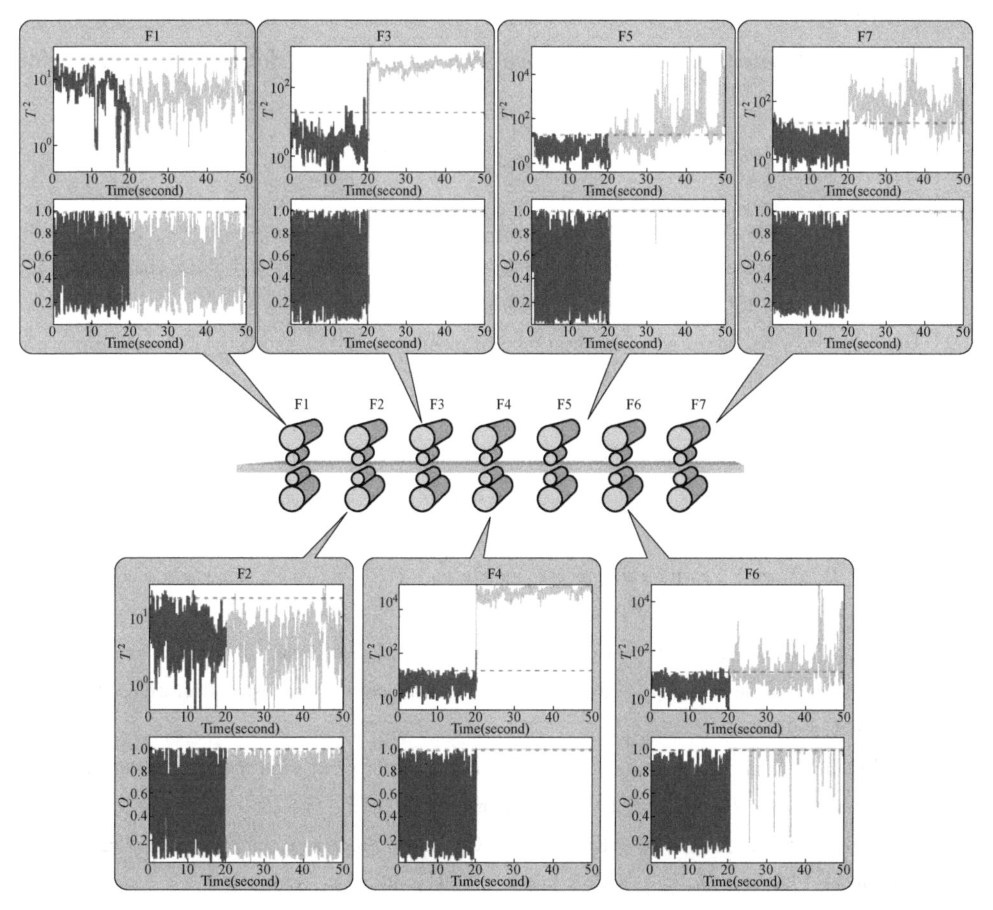

Fig. 8.13 The local process monitoring results of Fault 2 for 7 stands by D-CVRAE

8.4 Conclusions

In this chapter, a novel D-CVRAE algorithm is proposed for the tandem plant-wide process modeling with spatiotemporal constraints. The related features extracted from the adjacent subsystem with a certain time delay are used as the constrained condition of the local CRVAE sub-model. Then the multiple local sub-models are combined to build a distributed model. A quality prediction model is also developed by D-CRVAE. The local monitoring statistics are built through the CVRAE sub-model. A global monitoring statistic based on deep SVDD is calculated to determine whether the whole process is faulty. The local monitoring results can help quickly determine which subsystem is the cause of the fault. Finally, the proposed method is applied to the HSM process. Both theoretical analysis and experiment results show better performance of D-CVRAE than those of G-DKPCA, G-VRAE and DVRAE. The utilization of related information from the adjacent subsystem effectively improves the monitoring performance for plant-wide processes.

References

[1] Tidriri K, Chatti N, Verron S, et al. Bridging datadriven and model-based approaches for process fault diagnosis and health monitoring: A review of researches and future challenges [J]. Annual Reviews in Control, 2016, 42: 63-81.

[2] Qin S J. Survey on data-driven industrial process monitoring and diagnosis [J]. Annual reviews in control, 2012, 36 (2): 220-234.

[3] Yin S, Ding S X, Xie X, et al. A review on basic data driven approaches for industrial process monitoring [J]. IEEE Transactions on Industrial electronics, 2014, 61 (11): 6418- 6428.

[4] Peng K X, Ma L, Zhang K. Review of quality-related fault detection and diagnosis techniques for complex industrial processes [J]. Acta Automatica Sinica, 2017, 43 (3): 349-365.

[5] He Q P, Wang J. Statistical process monitoring as a big data analytics tool for smart manufacturing [J]. Journal of Process Control, 2018, 67: 35-43.

[6] Jiang Q, Yan X, Huang B. Review and perspectives of data-driven distributed monitoring for industrial plantwide processes [J]. Industrial & Engineering Chemistry Research, 2019, 58 (29): 12899-12912.

[7] Ge Z Q. Review on data-driven modeling and monitoring for plant-wide industrial processes [J]. Chemometrics and Intelligent Laboratory Systems, 2017, 171: 16-25.

[8] MacGregor J F, Jaeckle C, Kiparissides C, et al. Process monitoring and diagnosis by multiblock PLS methods [J]. AIChE Journal, 1994, 40 (5): 826-838.

[9] Qin S J, Valle S, Piovoso M J. On unifying multiblock analysis with application to decentralized process monitoring [J]. Journal of Chemometrics: A Journal of the Chemometrics Society, 2001, 15 (9): 715-742.

[10] Zhang Y W, Hong Z, Qin S J. Decentralized fault diagnosis of large-scale processes using multiblock kernel principal component analysis [J]. Acta Automatica Sinica, 2010, 36 (4): 593-597.

[11] Zhang Y W, Zhou H, Qin S J, et al. Decentralized fault diagnosis of large-scale processes using multiblock kernel partial least squares [J]. IEEE Transactions on Industrial Informatics, 2009, 6 (1): 3-10.

[12] Ge Z Q, Song Z H. Distributed PCA model for plant-wide process monitoring [J]. Industrial & Engineering Chemistry Research, 2013, 52 (5): 1947-1957.

[13] Jiang Q, Yan X. Plant-wide process monitoring based on mutual information-multiblock principal component analysis [J]. ISA transactions, 2014, 53 (5): 1516-1527.

[14] Zhang C F, Peng K X, Dong J. A novel plant-wide process monitoring framework based on distributed Gap-SVDD with adaptive radius [J]. Neurocomputing, 2019, 350: 1-12.

[15] Chen Z, Cao Y, Ding S X, et al. A distributed canonical correlation analysis-based fault detection method for plant-wide process monitoring [J]. IEEE Transactions on Industrial Informatics, 2019, 15 (5): 2710-2720.

[16] Zhang K, Peng K X, Ding S X, et al. A correlation-based distributed fault detection method and its application to a hot tandem rolling mill process [J]. IEEE Transactions on Industrial Electronics, 2019, 67 (3): 2380-2390.

[17] Peng X, Ding S X, Du W, et al. Distributed process monitoring based on canonical correlation analysis with partly-connected topology [J]. Control Engineering Practice, 2020, 101: 104500.

[18] Jiang Q, Chen S, Yan X, et al. Data driven communication efficient distributed monitoring for multiunit industrial plant-wide processes [J]. IEEE Transactions on Automation Science and Engineering, 2022, 19 (3): 1913-1923.

[19] Tang P, Peng K X, Dong J. Nonlinear quality-related fault detection using combined deep variational information bottleneck and variational autoencoder [J]. ISA transactions, 2021, 114: 444-454.

[20] Ruff L, Vandermeulen R, Goernitz N, et al. Deep one-class classification in International conference on machine learning [J]. PMLR, 2018: 4393-4402.

[21] Hochreiter S, Schmidhuber J. Long short-term memory [J]. Neural computation, 1997, 9 (8): 1735-1780.

[22] Graves A. Generating sequences with recurrent neural networks [J]. 2013, arXiv: 1308.0850.

[23] Kingma D P, Welling M. Auto-encoding variational Bayes [J]. 2013, arXiv: 1312.6114.

[24] Doersch C. Tutorial on variational autoencoders [J]. 2016, arXiv: 1606.05908.

[25] Tax D M, Duin R P. Support vector data description [J]. Machine learning, 2004, 54 (1): 45-66.

[26] Gonzalez,R, Huang B, Lau E. Process monitoring using kernel density estimation and Bayesian networking with an industrial case study [J]. ISA Transactions, 2015, 58: 330-347.

[27] Choi S W, Lee I B. Nonlinear dynamic process monitoring based on dynamic kernel PCA [J]. Chemical engineering science, 2004, 59 (24): 5897-5908.

[28] Tang P, Peng K X, Dong J, et al. A variational autoencoders approach for process monitoring and fault diagnosis [J]. International Journal of System Control and Information Processing, 2021, 3 (3): 229-245.

Chapter 9 VAE-BAB-based Process Monitoring and Fault Isolation Framework

Fault detection and diagnosis technology is an essential approach to ensure process safety and product quality, which has been widely used in industrial productions. With the rapid development of automation and informatization in industrial processes, immense amounts of data has been accumulated in industrial productions. Data-driven process monitoring technologies have subsequently developed rapidly. One of the most popular technologies is based on multivariate statistical process monitoring (MSPM), such as PCA, PLS, CCA[1-5]. Due to the high complexity of modern industries, process variables are highly nonlinear, and highly correlated to each other. Traditional MSPM methods can not handle these characteristics very well, so the monitoring performance for complex nonlinear processes is limited.

To resolve the problem of nonlinear process monitoring, different types of monitoring methods have been proposed. The most popular method is kernel method, such as KPCA[6-7], KPLS[8-10], SVDD[11-13]. Because of the limited of kernel technique, these methods are just suitable for modeling with small and middle-sized samples. Deep learning, who can handle largescale data and extract deep features, has achieved great success in industrial applications[14-15].

Recently, some scholars have applied deep learning to the field of process monitoring, such as autoencoder (AE), denosising AE, sparse AE and restricted Bolzmann machines[16-19]. As one efficient deep learning algorithm, Variational autoencoder (VAE) learns representations from data in nonlinear manner, combining the variational Bayesian approach with neural networks[20-21]. The probabilistic graphical structure of VAE offers a probabilistic interpretation of the process data. This characteristic makes VAE is available to describe model uncertainties. Some researchers have applied VAE algorithm in process monitoring[22-24].

After one fault is detected, the variables related to the fault should be isolated to diagnose the root causes of the fault-by-fault isolation approach. Contribution plot is an efficient method for fault diagnosis and widely used, in which the faulty variables are considered to have high contributions to the fault detection indexes. Qin and Alcala analyzed the diagnosability of contribution plot, though it would involve fault "smearing", and lead to misdiagnosis[25]. Another popular fault isolation approach is reconstruction-based contribution (RBC). The core concept of RBC is that each variable is reconstructed based on the other variables by minimizing fault index. The variables whose monitoring index reduce most significantly are considered to be critical to the fault. However, when reconstructing a non-faulty variable, some of the remaining variables are still faulty. The fault information used for reconstruction will lead to "smearing effects"[26]. Reconstruction based approach can identify the true fault from a set of candidate faults without ambiguity, but the prior knowledge of the fault direction is needed[27].

In recent years, some multivariate fault isolation approaches are developed using variable selection strategies, such as the least absolute shrinkage and selection operators (LASSO)[28-29] and branch and bound (BAB) search[30-31]. The variable selection strategies, such as sparse loading-based contribution[32], are also applied in performance assessment and diagnosis. LASSO adds an L1 penalty term to the reconstructed monitoring statistic. The variables corresponding to nonzero regression coefficients are believed to be faulty. LASSO can isolate faults more effectively. However, LASSO may not identify all of faulty variables, especially when existing groups of highly correlated variables. BAB is an efficient tool to resolve combinatorial optimization problems. The key idea is that the subset of candidate solutions can be removed by examining the bound of the objective function. Compared with exhaustive search, BAB can reduce the number of searches dramatically. However, these approaches mentioned above are applied in linear systems, which limited the ability to process nonlinear systems. Because of the nonlinear coupling interactions among variables, even few researchers have explored the multivariate fault isolation problem for nonlinear systems[33-34], there are still many challenges.

In this chapter, a process monitoring and fault isolation framework is proposed based on VAE to address the difficult of fault diagnosis in nonlinear processes. The critical issue of the proposed method is that how to construct a proper monitoring index, and how to accurately isolate the faulty variables. In the proposed method, the VAE model is trained using the date under normal operating conditions. Due to the probability directed graph structure of VAE, even though the complex distribution of observation data is unknown, its probability density can be still inferred by integrating the posterior probability. A probabilistic monitoring index can be designed based on the idea. If the fault variables are known, the predicted value calculated by normal variables replace the fault variables, the process should go back to normal. The key of fault isolation is the reconstruction of presumptive fault variables. However, the nonlinear structure of multi-layer neural networks leads to the difficult of variable reconstruction. Because VAE is one kind of probabilistic models, the missing variable can be estimated by probabilistic derivation. Therefore, the variable reconstruction is implemented by a proposed missing variable estimation method. Then a BAB strategy combined with missing variable estimation method is designed to search the optimal fault variable set. The contributions can be summarized as follows:

(1) A novel process monitoring and fault isolation framework is proposed;

(2) A uniform probabilistic monitoring index based on VAE is derived for process monitoring;

(3) A BAB strategy missing and a novel missing value estimation method are combined for multivariate fault isolation.

9.1 Principle of variational autoencoder

VAE is a directed probabilistic graphical model (DPGM) realized by variational Bayes with an autoenconder architecture[20]. The structure of VAE is shown in **Fig. 9.1**, who is consist of generative model $p_\theta(\boldsymbol{x}|\boldsymbol{z})$ and recognition model $q_\phi(\boldsymbol{z}|\boldsymbol{x})$, where $\boldsymbol{x} \in R^n$ and $\boldsymbol{z} \in R^m$ represent the process variables and latent variables, respectively. $\boldsymbol{\theta}$ and $\boldsymbol{\phi}$ are the corresponding parameters of models.

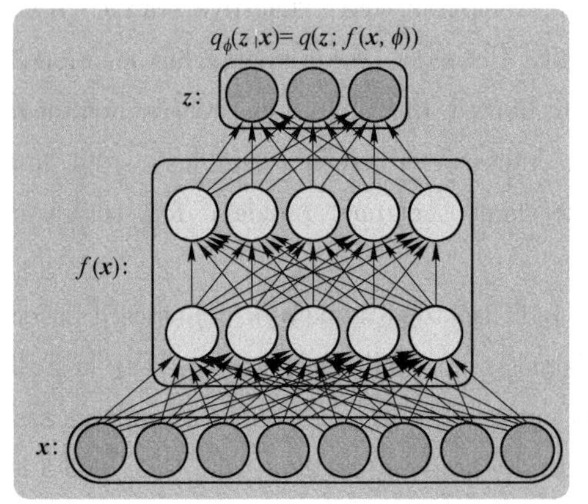

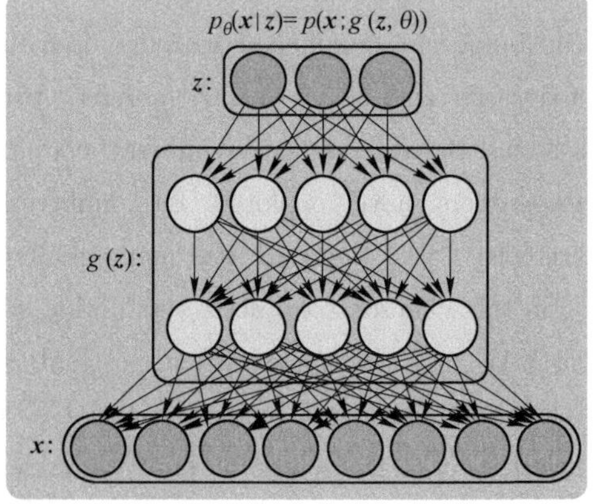

Fig. 9.1　The structure of VAE

Generative model $p_\theta(\boldsymbol{x}|\boldsymbol{z})$ generates high-dimensional process variables $\boldsymbol{x}$ by latent variables $\boldsymbol{z}$, which subjects to a predefined prior distribution. In general, the prior distribution of $\boldsymbol{z}$ is chosen to be standard normal distribution $p(\boldsymbol{z})=N(\boldsymbol{0}, \boldsymbol{I})$. Generative model can be described as a multivariate normal distribution:

$$p_{\theta}(\boldsymbol{x} \mid z) = N(\boldsymbol{\mu}(\boldsymbol{x} \mid z;\ \theta),\ \mathrm{diag}(\boldsymbol{\sigma}^2(\boldsymbol{x} \mid z;\ \theta))) \tag{9.1}$$

where, $\boldsymbol{\mu}(z;\ \boldsymbol{\theta})$ and $\mathrm{diag}(\boldsymbol{\sigma}^2(z;\ \boldsymbol{\theta}))$ denote mean vector and covariance matrix of normal distribution, respectively. $\mathrm{diag}(\boldsymbol{\sigma}^2(z;\ \boldsymbol{\theta}))$ denotes a diagonal covariance matrix with an assumption that different variables are independent. Generative model is developed by a multi-layer neural networks with input data $\boldsymbol{x}$ and outputs $(\boldsymbol{\mu}(z),\ \mathrm{diag}(\boldsymbol{\sigma}^2(z)))$.

A nonlinear dimensionality reduction is performed by recognition model to map process variables to a low-dimensional space of the latent variables. Recognition model $q_{\phi}(z \mid \boldsymbol{x})$ is an approximation of the intractable true posterior distribution $p_{\theta}(z \mid \boldsymbol{x})$. Similar to the generative model, its distribution in observation $\boldsymbol{x}_i$ is presented as follows:

$$q_{\phi}(z \mid \boldsymbol{x}_i) = N(\boldsymbol{\mu}(\boldsymbol{x}_i;\ \phi),\ \mathrm{diag}(\boldsymbol{\sigma}^2(\boldsymbol{x}_i;\ \phi))) \tag{9.2}$$

In order to estimate unknown parameters $\{\boldsymbol{\theta},\ \boldsymbol{\phi}\}$, the variational lower bound $L(\boldsymbol{\theta},\ \boldsymbol{\phi};\ \boldsymbol{x}_i)$ of $\lg p(\boldsymbol{x}_i)$ is given as the cost function of VAE:

$$L(\boldsymbol{\theta},\ \boldsymbol{\phi};\ \boldsymbol{x}_i) = E_{q_{\varphi}(z \mid \boldsymbol{x}_i)}[\lg p_{\theta}(\boldsymbol{x}_i \mid z)] - D_{\mathrm{KL}}[q_{\phi}(z \mid \boldsymbol{x}_i) \parallel p_{\theta}(z)] \tag{9.3}$$

where, $E_{q_{\phi}(z \mid \boldsymbol{x}_i)}[\lg p_{\theta}(\boldsymbol{x}_i \mid z)]$ denotes the expectation of $\lg p_{\theta}(\boldsymbol{x}_i \mid z)$ w. r. t $q_{\phi}(z \mid \boldsymbol{x}_i)$. $D_{\mathrm{KL}}(p \| q)$ is the Kullback-Leibler (KL) divergence, which measures the difference between two probability distributions. The calculation process of $L(\boldsymbol{\theta},\ \boldsymbol{\phi};\ \boldsymbol{x}_i)$ is shown in **Fig. 9.2**.

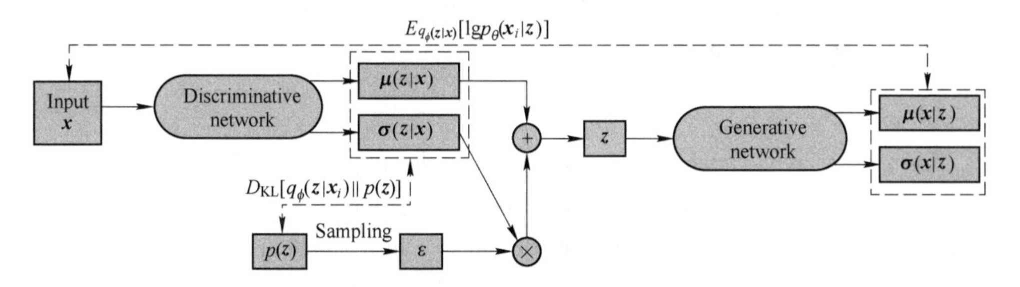

Fig. 9.2 The calculation process of VAE

Mini-batch stochastic gradient descent is used to train VAE model. The cost function for a mini-batch with N_m samples is written as:

$$L(\boldsymbol{\theta},\ \boldsymbol{\phi}) = \frac{1}{S \cdot N_m} \sum_{i=1}^{N_m} \sum_{s=1}^{S} \lg p_{\theta}(\boldsymbol{x}_i \mid \boldsymbol{z}^{(i,\ S)}) - \frac{1}{N_m} \sum_{i=1}^{N_m} D_{\mathrm{KL}}[q_{\phi}(z \mid \boldsymbol{x}_i) \| p_{\theta}(z)] \tag{9.4}$$

where, S samples are drawn from $q_{\phi}(z \mid \boldsymbol{x}_i)$, denoted as $\boldsymbol{x}^{(i,1)}$, $\boldsymbol{x}^{(i,2)}$, $\cdots$, $\boldsymbol{x}^{(i,S)}$. The more detailed introduction of VAE is shown in references [20-21].

9.2 Process monitoring and fault isolation based on VAE and BAB

In this section, a process monitoring and fault isolation framework is constructed based on VAE. Firstly, a single statistic index and the corresponded control limit are derived for nonlinear process monitoring. Then, a variable reconstruction strategy and BAB are integrated for multivariate fault isolation. In order to reconstruct unknown variables by the given observed variables, a missing value estimation method is derived. The detailed procedures is presented in the rest of this section.

9.2.1 VAE-based process monitoring

The proposed process monitoring method makes use of VAE model introduced in the previous section. The process data under NOC is collected and normalized to zero mean and unit variance as training dataset. After the network structure is designed, the VAE model is trained by mini-batch stochastic gradient descent. Once the VAE model is built, the marginal probability density of observation variables can be estimated by the DPGM of VAE, which can be a single statistic index to detect process fault. Furthermore, it provides the foundation of subsequent fault isolation. The control limit for process monitoring can be determined by Monte Carlo sampling.

Based on generative model of VAE, the marginal probability density of measured variables $\boldsymbol{x}$ is expressed as:

$$p(\boldsymbol{x}) = \int p_\theta(\boldsymbol{x} | z) p(z) \mathrm{d}z \tag{9.5}$$

Given measured variables $\boldsymbol{x}$, the large probability value of $p_\theta(\boldsymbol{x} | \boldsymbol{z})$ usually concentrates in a small area. If the samplers from $p(z)$ don't fall into the area, the efficient number of samplers is far smaller than actual number. Therefore, based on importance sampling, the posterior probability $q_\phi(z | \boldsymbol{x})$, whose distribution concentrates in the area, is used to sample. Equation (9.5) is rewritten as:

$$p(\boldsymbol{x}) = \int \frac{p_\theta(\boldsymbol{x} | z) p(z)}{q_\phi(z | \boldsymbol{x})} q_\phi(z | \boldsymbol{x}) \mathrm{d}z = E_{z\text{-}q}\left[\frac{p_\theta(\boldsymbol{x} | z) p(z)}{q_\phi(z | \boldsymbol{x})}\right] \tag{9.6}$$

where, $p(\boldsymbol{x})$ can be expressed by expectation with probability $q_\phi(z | \boldsymbol{x})$, and be approximated by the mean value of K samples from $q_\phi(z | \boldsymbol{x})$.

As a result, $p(\boldsymbol{x})$ is estimated as:

$$p(\boldsymbol{x}) \simeq \sum_{k=1}^{K} \frac{p_\theta(\boldsymbol{x} | z_k) p(z_k)}{q_\phi(z_k | \boldsymbol{x})} \tag{9.7}$$

The smaller the probability $p(\boldsymbol{x})$ is, the more possible the fault occurs. Therefore, $p(\boldsymbol{x})$ can be used to detect the fault. Then a control limit is required to identify any abnormal behavior. Given the β confidence bound, the likelihood threshold h is designed that satisfies the following integral[35]:

$$\int_{x:\ p(x)>h} p(\boldsymbol{x})\mathrm{d}\boldsymbol{x} = \beta \tag{9.8}$$

For VAE model, the calculation of integral is intractable and therefore it's impossible to obtain the threshold h directly. The integral can be approximated by Markov chain Monte Carlo (MCMC) sampling method.

DPGM constructed by recognition model $q_\phi(\boldsymbol{z}|\boldsymbol{x})$ and generative model $p_\theta(\boldsymbol{x}|\boldsymbol{z})$ is a Markov Chain (MC). Given an initial state, after finite sate transitions, the probability distribution $p(\boldsymbol{x})$ converges to a stationary distribution. Therefore, large number of samples subjected to complex distribution can be generated by MCMC sampling. Then the control limit h is calculated by integral approximation:

(1) Generate N_S samples, $\{\boldsymbol{x}^j, j = 1, 2, \cdots, N_S\}$, from $p(\boldsymbol{x})$ by MCMC sampling.

(2) Caculate the likelihood $p(\boldsymbol{x}_j)$ of samples $\boldsymbol{x}^j$.

(3) Sort $p(\boldsymbol{x}_j)$, $j = 1, 2, \cdots, N_S$ in descending order.

(4) The control limit is determined by $h = p(\boldsymbol{x}_l)$, where $l = N_S\beta$.

$p(\boldsymbol{x})$ is a number that approaches to 0, especially when $\boldsymbol{x}$ deviate from normal operating area. In order to show monitoring results better, the negative logarithm $-\lg p(\boldsymbol{x})$ is used as monitoring index, which is expressed as:

$$\mathrm{LP}(\boldsymbol{x}) = -\lg p(\boldsymbol{x}) \tag{9.9}$$

The control limit is redefined as $h = -\lg p(\boldsymbol{x}_l)$. Then data $\boldsymbol{x}_i$ is considered to be faulty if $\mathrm{LP}(\boldsymbol{x}_i) > h$.

9.2.2 Missing variable estimation

The measured data $\boldsymbol{x}$ are divided into observed variables $\boldsymbol{x}_\mathrm{o}$ and missing variables $\boldsymbol{x}_\mathrm{m}$, which is denoted as $\boldsymbol{x} = [\boldsymbol{x}_\mathrm{o}, \boldsymbol{x}_\mathrm{m}]$. Missing variables $\boldsymbol{x}_\mathrm{m}$ can be estimated by maximizing the likelihood $p(\boldsymbol{x}_\mathrm{m}, \boldsymbol{x}_\mathrm{o})$. The maximum likelihood estimation is expressed as:

$$p(\boldsymbol{x}_\mathrm{m}^*, \boldsymbol{x}_\mathrm{o}) = \max_{\boldsymbol{x}_\mathrm{m} \in R^d} p(\boldsymbol{x}_\mathrm{m}, \boldsymbol{x}_\mathrm{o}) \tag{9.10}$$

Because observed variables $\boldsymbol{x}_\mathrm{o}$ are known, $p(\boldsymbol{x}_\mathrm{o})$ is constant. $p(\boldsymbol{x}_\mathrm{m}|\boldsymbol{x}_\mathrm{o})$ is proportional to $p(\boldsymbol{x}_\mathrm{m}, \boldsymbol{x}_\mathrm{o})$, where $p(\boldsymbol{x}_\mathrm{m}, \boldsymbol{x}_\mathrm{o}) = p(\boldsymbol{x}_\mathrm{m} | \boldsymbol{x}_\mathrm{o})p(\boldsymbol{x}_\mathrm{o})$. Then the missing variables $\boldsymbol{x}_\mathrm{m}$ can be estimated by calculating $\max_{\boldsymbol{x}_\mathrm{m} \in R^d} p(\boldsymbol{x}_\mathrm{m} | \boldsymbol{x}_\mathrm{o})$. The expectation $E_{\boldsymbol{x}_\mathrm{o}}(\boldsymbol{x}_\mathrm{m})$ is the optimal estimation of missing variables $\boldsymbol{x}_{\mathrm{m}}$.

The prior probability distribution $p(\boldsymbol{x})$ is unknown. $p(\boldsymbol{x}_{\mathrm{m}}, \boldsymbol{x}_{\mathrm{o}})$ can be expressed with the prior probability distribution $p(z)$:

$$p(\boldsymbol{x}_{\mathrm{m}} | \boldsymbol{x}_{\mathrm{o}}) = \int p_{\theta}(\boldsymbol{x}_{\mathrm{m}}, z | \boldsymbol{x}_{\mathrm{o}}) \mathrm{d}z = \int p_{\theta}(\boldsymbol{x}_{\mathrm{m}} | z) p_{\theta}(z | \boldsymbol{x}_{\mathrm{o}}) \mathrm{d}z \tag{9.11}$$

By defining the conditional distribution:

$$p(z | \boldsymbol{x}_{\mathrm{o}}) = \frac{p_{\theta}(\boldsymbol{x}_{\mathrm{o}} | z) p(z)}{p(\boldsymbol{x}_{\mathrm{o}})} = \frac{p_{\theta}(\boldsymbol{x}_{\mathrm{o}} | z) p(z)}{\int p_{\theta}(\boldsymbol{x}_{\mathrm{o}} \mid z) p(z) \mathrm{d}z} \tag{9.12}$$

$p(\boldsymbol{x}_{\mathrm{m}} | x_{\mathrm{o}})$ is rewritten as:

$$p(x_{\mathrm{m}} | x_{\mathrm{o}}) = \frac{\int p_{\theta}(\boldsymbol{x}_{\mathrm{m}} | z) p_{\theta}(\boldsymbol{x}_{\mathrm{o}} | z) p(z) \mathrm{d}z}{\int p_{\theta}(\boldsymbol{x}_{\mathrm{o}} | z) p(z) \mathrm{d}z} \tag{9.13}$$

For the purpose of missing variables estimation, the expectation $E_{\boldsymbol{x}_{\mathrm{o}}}(\boldsymbol{x}_{\mathrm{m}})$ can be expressed as:

$$\begin{aligned} E_{\boldsymbol{x}_{\mathrm{o}}}(\boldsymbol{x}_{\mathrm{m}}) &= \int \boldsymbol{x}_{\mathrm{m}} p(\boldsymbol{x}_{\mathrm{m}} | \boldsymbol{x}_{\mathrm{o}}) \mathrm{d}\boldsymbol{x}_{\mathrm{m}} \\ &= \frac{\iint \boldsymbol{x}_{\mathrm{m}} p_{\theta}(\boldsymbol{x}_{\mathrm{m}} | z) p_{\theta}(\boldsymbol{x}_{\mathrm{o}} | z) p(z) \mathrm{d}x_{\mathrm{m}} \mathrm{d}z}{\int p_{\theta}(\boldsymbol{x}_{\mathrm{o}} | z) p(z) \mathrm{d}z} \\ &= \frac{\int \boldsymbol{u}_{\boldsymbol{x}_{\mathrm{o}} | z} p_{\theta}(\boldsymbol{x}_{\mathrm{o}} | z) p(z) \mathrm{d}z}{\int p_{\theta}(\boldsymbol{x}_{\mathrm{o}} | z) p(z) \mathrm{d}z} \end{aligned} \tag{9.14}$$

where, $\boldsymbol{u}_{\boldsymbol{x}_{\mathrm{o}} | z}$ is the mean of distribution $p_{\theta}(\boldsymbol{x}_{\mathrm{o}} | z)$, which can be calculated by generative model $p_{\theta}(\boldsymbol{x} \mid z)$.

According to importance sampling, the posterior probability $q_{\phi}(z | \boldsymbol{x})$ is introduced into equation (9.14) . The formula is rewritten as:

$$\begin{aligned} E_{\boldsymbol{x}_{\mathrm{o}}}(\boldsymbol{x}_{\mathrm{m}}) &= \frac{\int \frac{\boldsymbol{u}_{\boldsymbol{x}_{\mathrm{o}} | z} p_{\theta}(\boldsymbol{x}_{\mathrm{o}} | z) p(z)}{p(z | \boldsymbol{x}_{\mathrm{m}}, \boldsymbol{x}_{\mathrm{o}})} p(z | \boldsymbol{x}_{\mathrm{m}}, \boldsymbol{x}_{\mathrm{o}}) \mathrm{d}z}{\int \frac{p_{\theta}(\boldsymbol{x}_{\mathrm{o}} | z) p(z)}{q_{\phi}(z | \boldsymbol{x}_{\mathrm{m}} | \boldsymbol{x}_{\mathrm{o}})} q_{\phi}(z | \boldsymbol{x}_{\mathrm{m}}, \boldsymbol{x}_{\mathrm{o}}) \mathrm{d}z} \\ &= \frac{\sum_{s=1}^{S} \frac{\boldsymbol{u}_{\boldsymbol{x}_{\mathrm{o}} | z} p_{\theta}(\boldsymbol{x}_{\mathrm{o}} | z_{\mathrm{s}}) p(z_{\mathrm{s}})}{q_{\phi}(z_{\mathrm{s}} | \boldsymbol{x}_{\mathrm{m}}, \boldsymbol{x}_{\mathrm{o}})}}{\sum_{s=1}^{S} \frac{p_{\theta}(\boldsymbol{x}_{\mathrm{o}} | z_{\mathrm{s}}) p(z_{\mathrm{s}})}{q_{\phi}(z_{\mathrm{s}} | \boldsymbol{x}_{\mathrm{m}}, \boldsymbol{x}_{\mathrm{o}})}} \end{aligned} \tag{9.15}$$

where, the two integral terms are approximated by the mean value of S samples generated from $q_{\phi}(z|\boldsymbol{x}_{\mathrm{m}}, \boldsymbol{x}_{\mathrm{o}})$.

The missing variable estimation is an iterative process, that $E_{\boldsymbol{x}_{\mathrm{o}}}(\boldsymbol{x}_{\mathrm{m}})$ replaces previous $\boldsymbol{x}_{\mathrm{m}}$ to perform the next iteration operation until it converges.

9.2.3 Multivariate fault isolation using BAB

Once a fault is successfully detected, the root cause of the fault should be found by fault isolation. The task of fault isolation method is to identify which variables are the most responsible for the occurrence of the fault. If the combination of fault variables is known, the monitoring index will be below the threshold h after the fault variables is reconstructed. The variable reconstruction is implemented by missing variable estimation method, and BAB searches the optimal combination of fault variables.

Let $\boldsymbol{x}_n = \{x_1, x_2, \cdots, x_n\}$ denote the set of all n variables. The objective of BAB is to find the optimal subset $\boldsymbol{x}_{\mathrm{r}}^*$ with the selection criterion $\phi(\boldsymbol{x}_{\mathrm{r}}^*)$ is minimal, that

$$\phi(\boldsymbol{x}_{\mathrm{r}}^*) = \min_{x_{\mathrm{r}} \in R^n} \phi(\boldsymbol{x}_n) \tag{9.16}$$

If using exhaustive search, the number of candidate subsets to be searched is 2^n, which grows exponentially with variable size. BAB can provide a global optimal solution, while greatly reduces the search times by efficient bounding and pruning strategy.

Consider an example where process data has 5 variables, the tree structure adopted by BAB algorithm is shown in **Fig. 9.3**. Each node represents a candidate set, whose observed variable set $\boldsymbol{x}_{\mathrm{o}}$ is determined by the node number of current node and its parent nodes. The algorithm starts from the root node, corresponding to all variables being missing, and gradually makes some missing variables observed through bottom-up search. For example, take the case of node with label "1" in level L1, which is the sub-node of the root, the missing variable set is $\{x_2, x_3, x_4, x_5\}$. Its sub-node with label "2" in level L2 has missing variable set $\{x_3, x_4, x_5\}$, where variable x_2 is removed. Branches of each node are non-overlapping sub-problems (sub-nodes) of original problem (parent nodes). Bound of the objective criterion is determined from nodes that have been examined. If any node is out of the bound, its sub-nodes cannot be optimal, and its branch is pruned; otherwise continue search for its sub-nodes. Compare with exhaustive search, the strategy of bounding and pruning can be more efficient to search optimal solution. In the meanwhile, the globally optimal solution is guaranteed.

For one node in BAB tree with label j, $\boldsymbol{x}_{\mathrm{m}}$ and $\boldsymbol{x}_{\mathrm{o}}$ denote missing variable set and observe variable, respectively. Define the candidate set $\boldsymbol{x}_{\mathrm{c}}$ of observed variables as

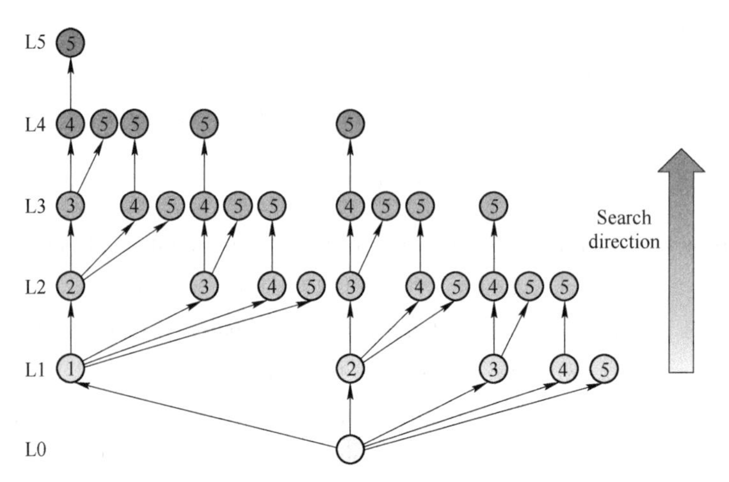

Fig. 9. 3 Branch and bound search tree for 5 process variables

$\boldsymbol{x}_c = \{x_{j+1}, x_{j+2}, \cdots, x_n\} \subset x_m$. Because of the uniqueness of missing variables for each node, $\boldsymbol{x}_m$ represents each node. The sub-node of node $\boldsymbol{x}_m$ is represented as:

$$\Gamma = \{\boldsymbol{x}_o \setminus x_f | x_f \subset \boldsymbol{x}_c\} \tag{9.17}$$

In order to solve fault isolation problems by BAB, the objective function $\phi(\boldsymbol{x}_m)$ should satisfy the condition:

$$\forall x_l \in \boldsymbol{x}_c, \ \phi(\boldsymbol{x}_m \setminus x_l) \geqslant \phi(\boldsymbol{x}_m) \tag{9.18}$$

In this chapter, the monitoring index $\mathrm{LP}(\boldsymbol{x}_m^*, \boldsymbol{x}_o)$ is chosen as objective criterion $\phi(\boldsymbol{x}_m)$, where $\boldsymbol{x}_m^*$ is the estimation by missing variable estimation method. Divide $\boldsymbol{x}_m$ into two part $\{\boldsymbol{x}_{m1}, \boldsymbol{x}_{m2}\}$. According to equation (9.9), it can be inferred that $p(\boldsymbol{x}_{m1}^*, \boldsymbol{x}_{m2}^*, \boldsymbol{x}_o) \geqslant p(\boldsymbol{x}_{m1}^*, \boldsymbol{x}_{m2}, \boldsymbol{x}_o)$. As a result, the monitoring index $\mathrm{LP}(\boldsymbol{x})$ satisfies equation (9.18) .

For the optimal fault variable set $\boldsymbol{x}_m$, the observe variable set is $\boldsymbol{x}_o = \boldsymbol{x} \setminus \boldsymbol{x}_m$, which denotes the rest variables of $\boldsymbol{x}$ depart from $\boldsymbol{x}_m$. $\boldsymbol{x}_m$ should satisfy the following conditions:

(1) $\phi(\boldsymbol{x}_m) < h$.

(2) $\forall x_l \in \boldsymbol{x}_m, \ \phi(\boldsymbol{x}_m \setminus x_l) > h$.

(3) $\forall x_l \in \boldsymbol{x}_m$ and $x_j \in \boldsymbol{x}_o$, $\phi(\boldsymbol{x}_m) \leqslant \phi(\boldsymbol{x}_m \setminus x_l, x_j)$.

Based on the above conditions, it is important to guarantee $\phi(\boldsymbol{x}_m) < h$ with the number of missing variables as little as possible. In the meanwhile, $\phi(\boldsymbol{x}_m)$ should be the smallest at the same level, where the size of missing variables is the same.

Define the level count and objective function of optimal node $\boldsymbol{x}_m^*$ as $\{L^*, \phi^*\}$. For one node $\boldsymbol{x}_m$, the end node is $\boldsymbol{x}_m \setminus \boldsymbol{x}_c$. The pruning rules are shown in **Table 9.1**, where if condition of any row is satisfied, the node and its branch is pruned. And the optimal node information is updated by the rules list in **Table 9.2**.

Table 9.1 Pruning rules for BAB

Rule No.	Condition	Pruning node
1	$\phi(\boldsymbol{x}_{\mathrm{m}})>h$	$\boldsymbol{x}_{\mathrm{m}}$
2	$\phi(\boldsymbol{x}_{\mathrm{m}})>\phi^{*}$, $L(\boldsymbol{x}_{\mathrm{m}})=L^{*}$	$\boldsymbol{x}_{\mathrm{m}}$
3	$L(\boldsymbol{x}_{\mathrm{m}}\setminus\boldsymbol{x}_{\mathrm{c}})\leqslant L^{*}$	$\boldsymbol{x}_{\mathrm{m}}$

Table 9.2 optimal node information update rules for BAB

Rule No.	Condition	Optimal node information
1	$\phi(\boldsymbol{x}_{\mathrm{m}})<\phi^{*},L(\boldsymbol{x}_{\mathrm{m}})=L^{*}$	$\boldsymbol{x}_{\mathrm{m}},\{\phi(\boldsymbol{x}_{\mathrm{m}}),L(\boldsymbol{x}_{\mathrm{m}})\}$
2	$\phi(\boldsymbol{x}_{\mathrm{m}})<h,L(\boldsymbol{x}_{\mathrm{m}})>L^{*}$	$\boldsymbol{x}_{\mathrm{m}},\{\phi(\boldsymbol{x}_{\mathrm{m}}),L(\boldsymbol{x}_{\mathrm{m}})\}$
3	$\phi(\boldsymbol{x}_{\mathrm{m}}\setminus\boldsymbol{x}_{\mathrm{c}})<h,L(\boldsymbol{x}_{\mathrm{m}}\setminus\boldsymbol{x}_{\mathrm{c}})>L^{*}$	$\boldsymbol{x}_{\mathrm{m}}\setminus\boldsymbol{x}_{\mathrm{c}},\{\phi(\boldsymbol{x}_{\mathrm{m}}\setminus\boldsymbol{x}_{\mathrm{c}}),L(\boldsymbol{x}_{\mathrm{m}}\setminus\boldsymbol{x}_{\mathrm{c}})\}$
4	$\phi(\boldsymbol{x}_{\mathrm{m}}\setminus\boldsymbol{x}_{\mathrm{c}})<\phi^{*},L(\boldsymbol{x}_{\mathrm{m}}\setminus\boldsymbol{x}_{\mathrm{c}})=L^{*}$	$\boldsymbol{x}_{\mathrm{m}}\setminus\boldsymbol{x}_{\mathrm{c}},\{\phi(\boldsymbol{x}_{\mathrm{m}}\setminus\boldsymbol{x}_{\mathrm{c}}),L(\boldsymbol{x}_{\mathrm{m}}\setminus\boldsymbol{x}_{\mathrm{c}})\}$

Some additional effective skills are adopted, which are list as follow:

(1) For variable $x_i \in \boldsymbol{x}_{\mathrm{c}}$, sort the sub-nodes in descending order according to the objective function $\phi(\boldsymbol{x}_{\mathrm{m}} \setminus \boldsymbol{x}_i)$.

(2) If the branch isn't pruned by $\phi(\boldsymbol{x}_{\mathrm{m}} \setminus \boldsymbol{x}_{\mathrm{c}})$, for any variable $x_i \in \boldsymbol{x}_{\mathrm{c}}$, calculate $\phi(\boldsymbol{x}_{\mathrm{m}} \setminus \boldsymbol{x}_{\mathrm{c}}, \boldsymbol{x}_i)$ to determine pruning or information updating.

(3) Use multi-thread and GPU to accelerate the computation.

In summary, the complete fault isolation algorithm based on BAB runs as follows:

(1) Initialize root node with $\boldsymbol{x}_{\mathrm{c}} = \boldsymbol{x}_n$ and $\boldsymbol{x}_{\mathrm{m}} = \boldsymbol{x}_n$. Set the optimal information $L^* = 0$ and $\boldsymbol{x}_{\mathrm{m}} = \{x_1, x_2, \cdots, x_{\mathrm{m}}\}$.

(2) For each node in current level, perform the following operation:

1) $\forall x_l \in \boldsymbol{x}_{\mathrm{c}}$, calculate $\phi(\boldsymbol{x}_{\mathrm{m}} \setminus x_l)$. Sort sub-nodes in descending order.

2) If $\phi(\boldsymbol{x}_{\mathrm{m}} \setminus x_l) > h$, prune the sub-node and its branch.

3) For sub-node $\boldsymbol{x}_{\mathrm{m}}^s$, performe the optimal node update rules and pruning rules. If $\boldsymbol{x}_{\mathrm{m}}^s$ and its branch is pruned, skip next step. Else, go to next step.

4) $\forall x_i \in \boldsymbol{x}_{\mathrm{c}}$, if $\phi(\boldsymbol{x}_{\mathrm{m}} \setminus \boldsymbol{x}_{\mathrm{c}}, \boldsymbol{x}_i) < h$, update the optimal node information by rule 4 in **Table 9.2**, where $\boldsymbol{x}_{\mathrm{m}} \setminus \boldsymbol{x}_{\mathrm{c}}$ is replaced by $\{\boldsymbol{x}_{\mathrm{m}}^s \setminus \boldsymbol{x}_{\mathrm{c}}, \boldsymbol{x}_i\}$, and prune the node $\boldsymbol{x}_{\mathrm{m}}^s$ and its branch.

(3) For all the rest sub-nodes, prune the sub-node, where $L(\boldsymbol{x}_{\mathrm{m}}^s \setminus \boldsymbol{x}_{\mathrm{c}}^s) < L^*$. If the number of the rest sub-nodes is zero, end of the algorithm.

(4) The count of level is increased by one and skip to step 4 to start the next iteration.

In online monitoring, the monitoring index is calculated for each process data, and compared to the control limit h. When the index is beyond the limit, the process is considered out of control, and the fault isolation procedure is performed to identify the fault variables. Based on the fault isolation result, the fault magnitude of each variable can be estimated. Because of the correlation between variables, many variables would be effected by faulty. But only few variables have larger contribution to faulty. In order to make sure which variables are critical to the fault, the fault isolation procedure is re-performed with the control limit $(\mathrm{LP}(\boldsymbol{x}) - h) \times 10\% + h$ to find these variables who can reduce monitoring index by more than 90%. The flow diagram of fault diagnosis is shown in **Fig. 9. 4**.

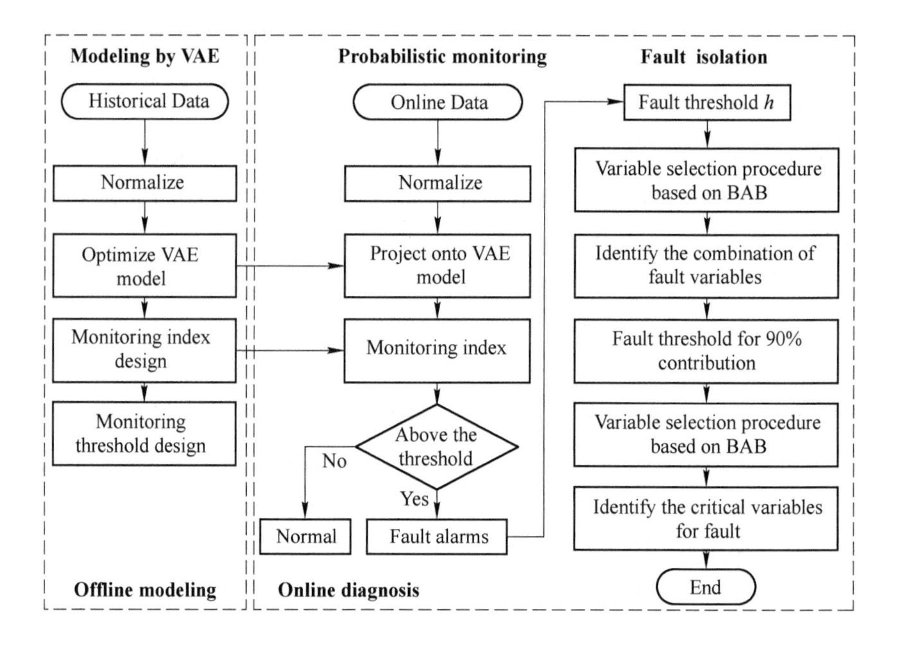

Fig. 9. 4 Branch and bound search tree for 5 process variables

9. 3 Case study

In this section, the performance of the proposed fault diagnosis framework is verified by one nonlinear numerical instance and one real industrial process case.

9.3.1 Numerical case

In this subsection, a numerical case of nonlinear process consisting of nine measured variables is considered, which is presented as following formula:

$$\begin{cases} x_1 = t_1 + e_1 \\ x_2 = t_1^2 - t_1 + e_2 \\ x_3 = - t_1^3 + 3t_2 + e_3 \\ x_4 = t_2 + e_4 \\ x_5 = t_2^2 - 2t_3 - 1 + e_5 \\ x_6 = 4\sin t_2 + t_1 + e_6 \\ x_7 = t_3 + e_7 \\ x_8 = \sin(0.5t_3) + 2.5t_1 + e_8 \\ x_9 = - t_3^3 + 3t_2 t_3 + e_9 \end{cases} \tag{9.19}$$

where, $t_k \in [0.01, 1]$, $k = 1, 2, 3$ are the latent factors, and $e_j \sim N(0, 0.01)$, $j = 1, 2, \cdots, 9$ denote the Gaussian distributed noise. The process is a typical nonlinear complex process, whose measured variables are interacted with each other. In this case study, 10000 normal samples are firstly generated and collected as training samples according to equation (9.19). Then two different faults are considering. For each fault, 400 test samples are generated, whose first 200 samples are normal and the rest are faulty samples. These faults are described as follows:

Fault 1: set deviation in x_3: $x_3 = x_3^* + f$

Fault 2: set deviation in t_3: $t_3 = t_3^* + f$

where, t_3^* and x_3^* are normal values; f is the magnitude of introduced fault, whose value is set to 1.

For the proposed fault diagnosis framework, the historical dataset under normal operating condition (NOC) is normalized firstly. Then a VAE model with 9-30-3-30-9 network structure is built, which denotes the size of the latent variable layer and hidden layer are 3 and 30, respectively. The dimension of latent variable is determined by 5-fold cross validation. During model training, Adam optimizer with a learning rate of 0.0001 is used, and the batch size is set as 64. The activation function of hidden layer is Tanh. For fault detection, the confidence limit h for 95% confidence level is determined by MCMC sampling, whose relationship between confidence bound α and threshold h is shown in **Fig. 9.5**.

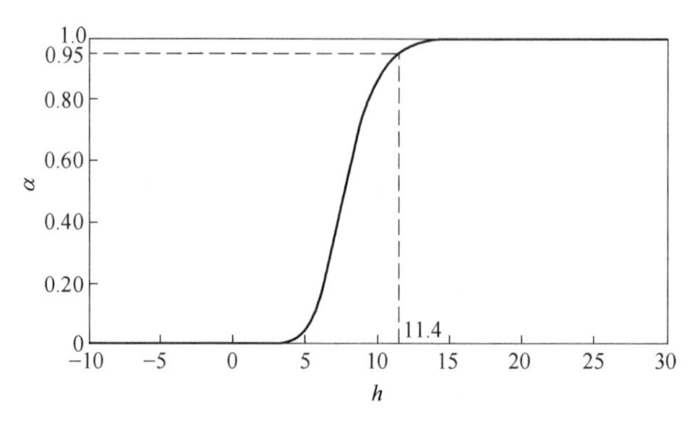

Fig. 9. 5 The relationship between *α* and *h* for numerical case

The process data under NOC is used to validate the efficiency of missing variable estimation method. For each variable, using the rest variables to reconstruct the current variable by missing variable estimation. **Fig. 9. 6** shows results of reconstruction. The original variables are represented by blue lines, and the reconstruction variables are represented by red lines. It can be seen that the reconstructed variable can follow the change of the original variable well.

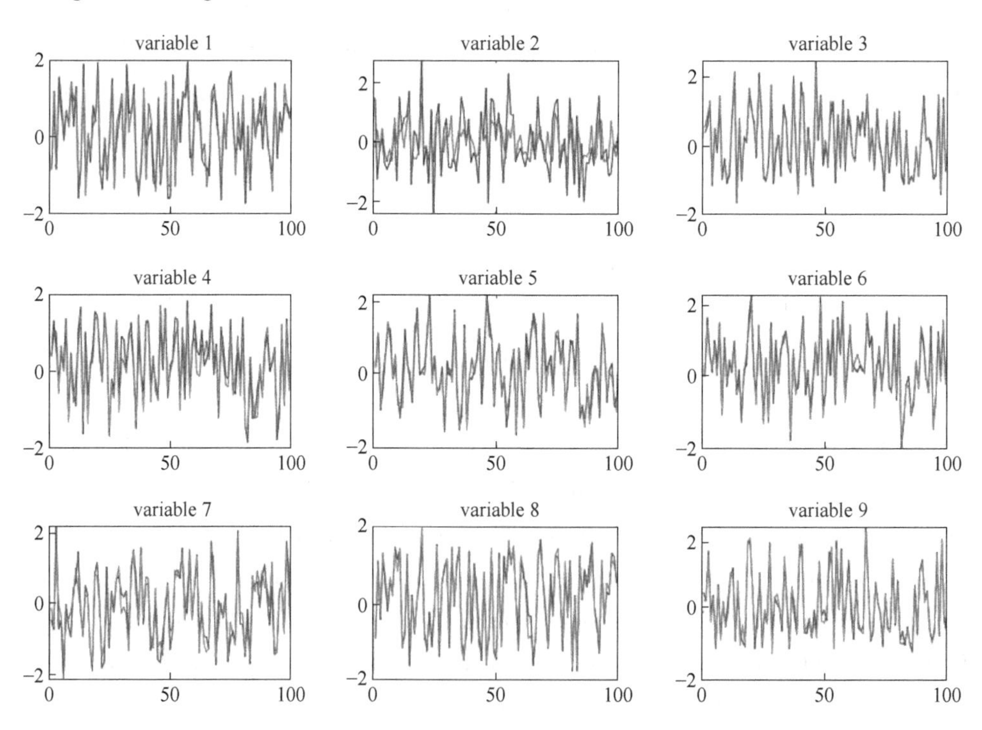

Fig. 9. 6 Original and reconstructed variables under normal operation for numerical case

(Scan the QR code on the front of the book for color picture)

The performance of fault detection based on VAE is compared with several conventional data-driven fault detection methods, including probabilistic PCA (PPCA)[36], KPCA[6] and DBN[37]. PPCA can be seen as a linear particular case of VAE. KPCA and DBN are a popular kernel method and a deep learning method for monitoring nonlinear process, respectively. For the sake of fairness, DBN has the same parameters as VAE.

The fault detection rates (FDRs) and false alarm rates (FARs) of four methods for two faults are listed in **Table 9.3** and **Table 9.4**, respectively. It can be seen that the proposed VAE method achieves the best results for the two fault cases. The performance of VAE-based process monitoring method is verified.

Fault 1 is a step change occurred in the measured variable x_3, which not affects other measured variables. **Fig. 9.7** shows the monitoring index before and after fault isolation. The blue line represents monitoring index, and the orange dots represent the reconstructed monitoring index calculated by the reconstruction variables, which is estimated by fault isolation. It can be seen that the fault is detected in time without delay in fault detection procedure. If fault is detected, the reconstructed indexes by fault isolation reduce to the control limit.

Table 9.3 FDRs of two faults for numerical case

Fault No.	PPCA		KPCA		DBN		VAE
	SPE	T^2	SPE	T^2	SPE	T^2	PL
1	0.485	0.02	0.58	0.16	0.68	0.23	0.84
2	0.855	0.595	0.87	0.615	0.92	0.74	0.955

Table 9.4 Average FARs for numerical case

PPCA		KPCA		DBN		VAE
SPE	T^2	SPE	T^2	SPE	T^2	PL
0.018	0.075	0.018	0.073	0.02	0.035	0.02

As comparison, the RBC analysis is carried out for fault isolation. For variable x_l, $l = 1, 2, \cdots, m$, the contribution rate of x_l is defined as $(\mathrm{LP}(\boldsymbol{x}) - \phi(x_l))/\mathrm{LP}(\boldsymbol{x}) - h$, whose control limit is 0.1. LASSO method proposed in reference [30] is also compared with BAB method. The fault isolation results at 203_{rd} sample by these fault isolation methods are shown in **Fig. 9.8**. **Fig. 9.8** (a) shows the contribution plots corresponding to RBC. The variables x_1, x_2, x_3, x_6, x_8 and x_9 are identified as fault variables. However, all the variables except x_3 are normal. So RBC method could lead to false diagnosis. Both LASSO and BAB methods identify variables x_3 as the only fault variable, which is consistent with the analysis of fault mechanism.

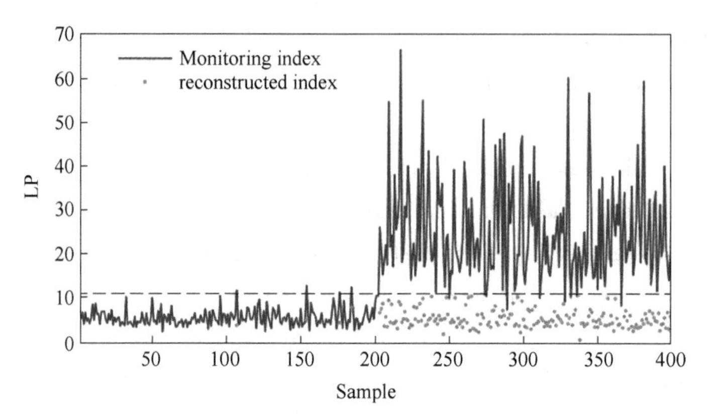

Fig. 9.7 The monitoring indexes before and after fault isolation for Fault 1

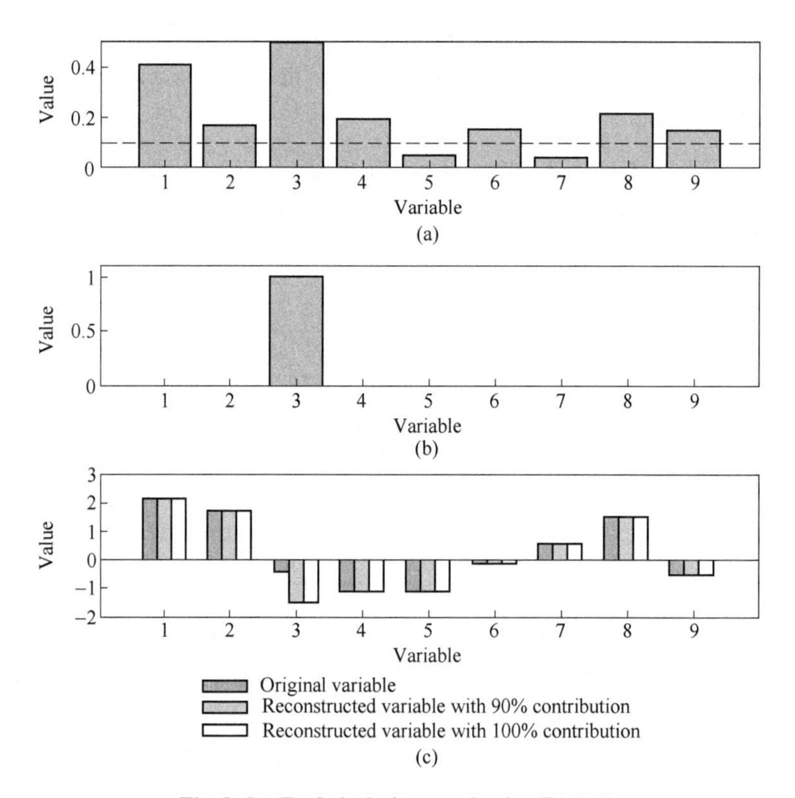

Fig. 9.8 Fault isolation results for Fault 1

(a) RBC; (b) LASSO; (c) BAB

Fig. 9.9 shows the fault isolation results for the whole process. The method based on RBC indicates that variable x_3 is of the greatest contribution to the fault. But variables x_6, x_8 and x_9 are also considered as fault variables. This is because of "smearing effects". The fault isolation result based on LASSO is shown in **Fig. 9.9** (b). The variable x_3 is identified as fault variable. The variable x_8 is also considered to be affected by the fault occasionally. Because the LASSO method is based on PCA arithmetic, it cannot handle the nonlinear data well. Part of fault samples are not detected as fault. The fault isolation result based on BAB is shown in **Fig. 9.9** (c), the variables who are less effected by the fault are green, and the variables contribution exceed 90% are yellow. Almost all the diagnostic results of these fault samples are that variable x_3 is the unique fault variable. The fault isolation based on BAB has better accuracy than the method based on RBC and LASSO.

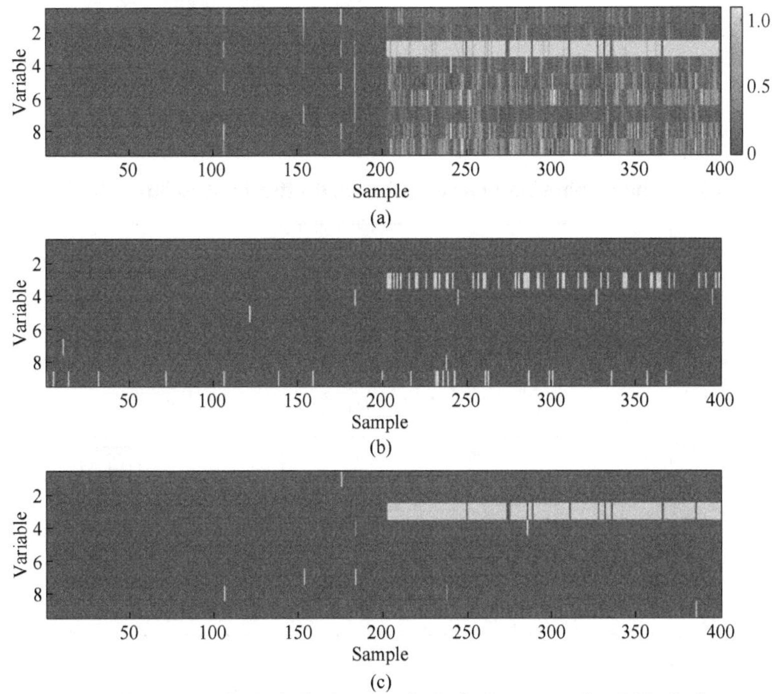

Fig. 9.9 The fault isolation result (whole process) of Fault 1

(a) RBC; (b) LASSO; (c) BAB

(Scan the QR code on the front of the book for color picture)

Fault 2 is a step change occurred in latent factor t_3, which affects the measured variables x_5, x_7, x_8 and x_9. **Fig. 9.10** shows the monitoring index before and after fault isolation. The fault is detected after the 201_{st} sample timely. After fault is detected, the reconstructed indexes by fault isolation drop below the control limit.

The fault isolation results at 202_{nd} sample by these fault isolation methods are shown in **Fig. 9. 11**. The RBC analysis identifies variables x_3, x_5 and x_9 as fault variables, where x_3 is normal variable in real situation, and the change of variables x_7 and x_8 are not reflected. The LASSO method only identifies variable x_9 as fault variables. The proposed fault isolation method based on BAB indicates that variables x_5, x_7 and x_9 are critical to the fault. Compared with other two method, BAB method provides more accurate fault information, and not misdiagnose the normal variable as fault one.

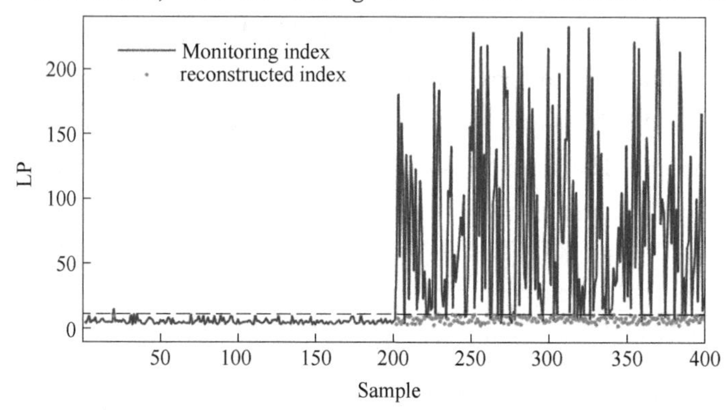

Fig. 9. 10 The monitoring indexes before and after fault isolation for Fault 2

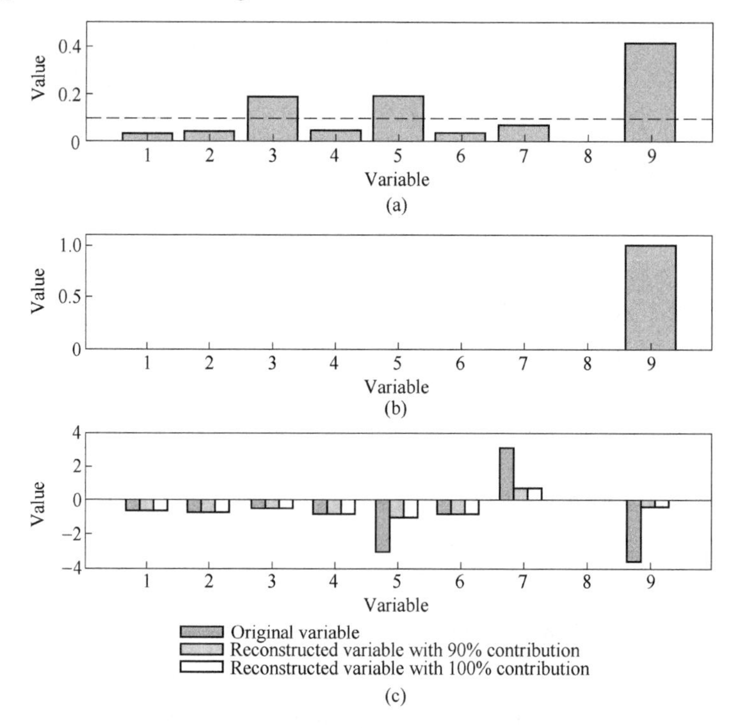

Fig. 9. 11 Fault isolation results for Fault 2

(a) RBC; (b) LASSO; (c) BAB

Fig. 9. 12 shows the fault isolation results for the whole process. According to fault isolation result based on RBC shown in **Fig. 9. 12** (a), the primary effected variables by fault are x_3, x_5 and x_9. The primary fault variables identified by LASSO and BAB methods are x_5, x_7 and x_9, which are more accurate than RBC method. Because of the global optimal search characteristic of BAB algorithm and the nonlinear characteristic of VAE, the misdiagnosis of BAB method are apparently less than LASSO.

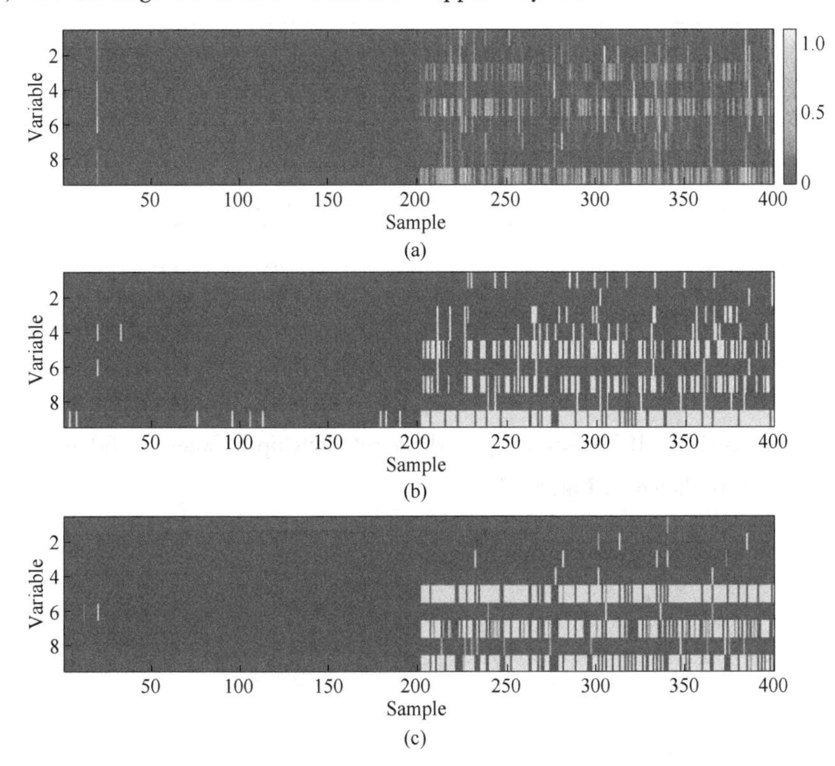

Fig. 9. 12 The fault isolation result (whole process) of Fault 2

(a) RBC; (b) LASSO; (c) BAB

(Scan the QR code on the front of the book for color picture)

9. 3. 2 Finishing mill process case

In this subsection, one class of strip's manufacturing process are taken for this simulation, whose thickness target is 3. 95 mm. The gap, rolling force and bending force in each mill stand and the final thickness in exit of finishing mill are selected as process variables. The 21 variables are listed in **Table 9. 5**. Two kinds of frequently occurring faults are used for verifying the performance of fault detection, which are listed in **Table 9. 6**.

Table 9.5 Description of process and quality variables in FMP

Variable	Type	Description	Unit
G1-G7	Measured	Average gap of the i_{th}, $i=1, \cdots, 7$	mm
F1-F7	Measured	Total force of the i_{th}, $i=1, \cdots, 7$	MN
B2-B7	Measured	Work roll bending force the i_{th}, $i=2, \cdots, 7$	MN
y	Quality	Finishing mill exit thickness of strip	mm

Table 9.6 Typical faults in finishing mill

Fault No.	Description	Occurrence time (s)
1	Sensor fault of bending force in F5 stand of strip1	10
2	Fault of cooling valve between F2 and F3 in strip1	10

The historical dataset under NOC is normalized firstly. Then a VAE model with 21-30-7-30-21 network structure is built. During model training, Adam optimizer with a learning rate of 0.0001 is used, and the batch size is set as 64. The activation function of hidden layer is Tanh. For fault detection, the confidence limit h for 95% confidence level is determined by MCMC sampling, whose relationship between confidence bound α and threshold h is shown in **Fig. 9.13**.

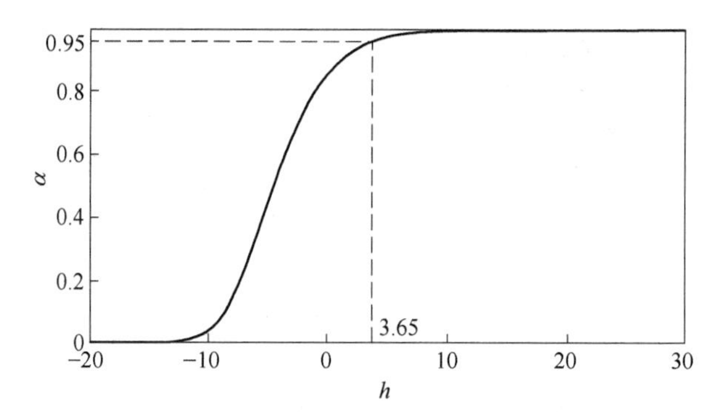

Fig. 9.13 The relationship between α and h for FMP case

The part reconstructed results of process data under NOC are shown in **Fig. 9.14**. The reconstructed variable can follow the change of the original variable, and the noises are restrained.

The performance of fault detection based on VAE is compared with PPCA, KPCA and DBN. The FDRs and FARs of four methods for two faults are listed in **Table 9.7** and **Table 9.8**, respectively. The proposed VAE method achieves the best results for the two fault cases.

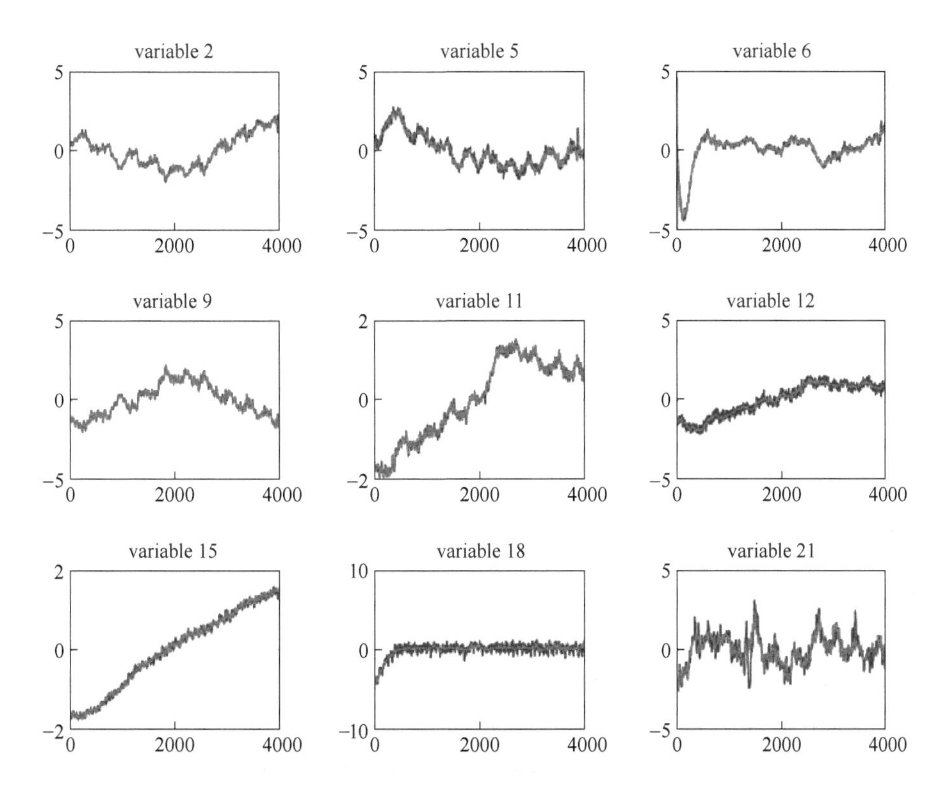

Fig. 9.14 Original and reconstructed variables under normal operation for FMP case

(Scan the QR code on the front of the book for color picture)

Table 9.7 FDRs of two faults in FMP case

Fault No.	PPCA		KPCA		DBN		VAE
	SPE	T^2	SPE	T^2	SPE	T^2	PL
1	0.998	0.263	1.000	0.784	1.000	0.571	1.000
2	0.956	0.562	0.998	0.573	0.998	0.967	1.000

Table 9.8 Average FARs in FMP case

PPCA		KPCA		DBN		VAE
SPE	T^2	SPE	T^2	SPE	T^2	PL
0.03	0.116	0.038	0.113	0.006	0.03	0.001

Fault 1 involves a step change in the measurement of bending force on F5 stand. When the fault occurs, there is a sudden increase in x_{18}. Subsequently, the close loop control reacts to drop the bending force in F6 (x_{19}) as well as F7 (x_{20}). **Fig. 9.15** shows the monitoring index before and after fault isolation, which can be found that the fault is detected in time. Once fault is detected, the reconstructed indexes can be dropped below the control limit by fault isolation.

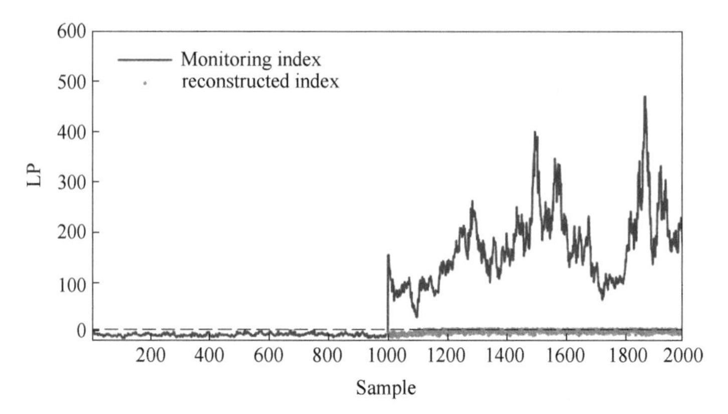

Fig. 9. 15 The monitoring indexes before and after fault isolation for Fault 1 in FMP case

The fault isolation method based on BAB is compared with RBC and LASSO, whose fault isolation results at 1380_{th} sample are shown in **Fig. 9. 16**. **Fig. 9. 16** (a) shows the contribution plots corresponding to RBC. It can be inferred that x_{15} , x_{18} and x_{19} have the biggest contribution. However, the variable x_{15} , who is not affected by fault, is mistaken as fault variables by RBC analysis. In **Fig. 9. 16** (b) and (c), the fault isolation based on both LASSO and BAB identify variables x_{18} , x_{19} and x_{20} as fault variables. And variable x_{18} is identified as critical fault variable by BAB method. This is consistent with the analysis of fault mechanism.

Fig. 9. 17 shows the fault isolation results for the whole process. The method based on RBC indicates that variable x_{18} is of the greatest contribution to the fault. But x_{19} and x_{20} are not identified, because of " smearing effects" . **Fig. 9. 17** (b) shows the fault isolation results based on LASSO. Because the LASSO method is based linear model, in the initial stages of fault detection, some samples are false detected as fault. Although variables x_{18}, x_{19} are diagnosed as fault variables, x_{20} is diagnosed just in the previous period of fault impact, and normal variables x_5, x_{12} are mistaken for fault variable in part of time. The fault isolation result based on BAB is shown in **Fig. 9. 17** (c) . In the beginning of the fault, variables x_{18} is firstly affected by the fault. After a while, variables x_{19} and x_{20} are gradually effected. But because of the close loop control, the value changes of the two variables are much small than the change of x_{18} . Therefore, x_{18} is considered as the primary variable effected by the fault. The fault isolation based on BAB has better accuracy than the method based on RBC and LASSO, and offers more detailed information.

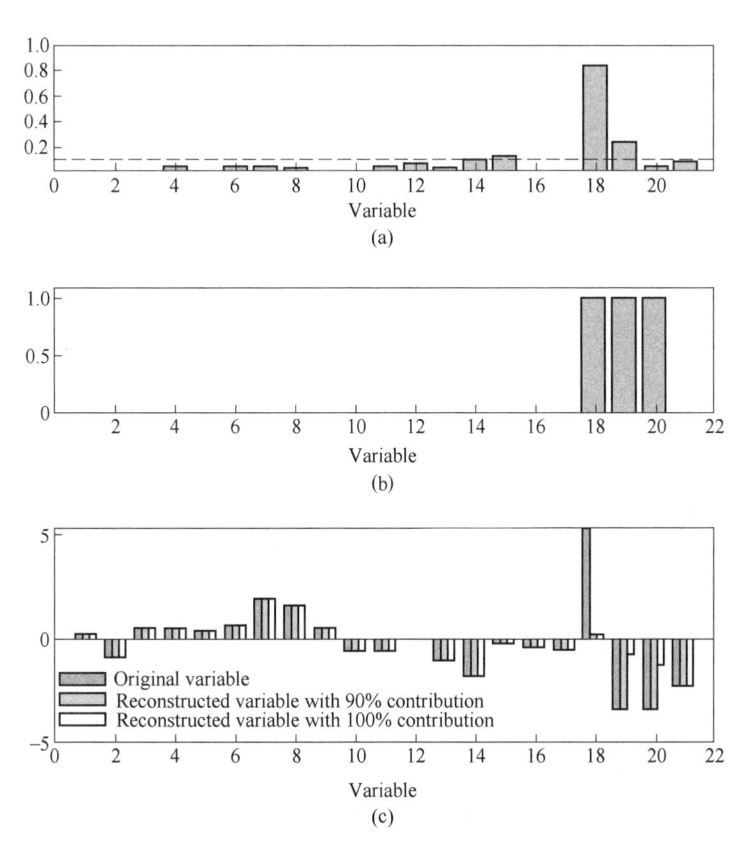

Fig. 9. 16 Fault isolation results for Fault 1 in FMP case

(a) RBC; (b) LASSO; (c) BAB

Fault 2 is an actuator fault of cooling valve between F2 and F3 stand, which is prevalent in finishing mill process. Its occurrence causes severe variations both in stand total force and gap ($x_3 - x_7$ and $x_{10} - x_{14}$) from F3 to F7 stand sequentially. Exit thickness (x_{21}) is influenced ultimately.

Fig. 9. 18 shows the monitoring indexes before and after fault isolation. The fault is detected after the 1001_{th} sample timely. After the fault occurred, the reconstructed indexes by fault isolation drops below the control limit. The fault isolation results at the 1800_{th} sample by the three fault isolation methods are shown in **Fig. 9. 19**. The total force and gap from F3 to F7 stand and exit thickness had been influenced at the 1800_{th} sample. The BAB method identifies variables x_5, x_6, x_7, x_{10} , x_{11} and x_{21} as primary fault variables. And the fault have weak influence on variables x_3, x_4, x_{12} , x_{13} , x_{14}. All the relevant variables are isolated successfully by BAB method. The RBC analysis mistakenly identifies variables x_2, x_{16} , x_{18} as fault variables, which would provide

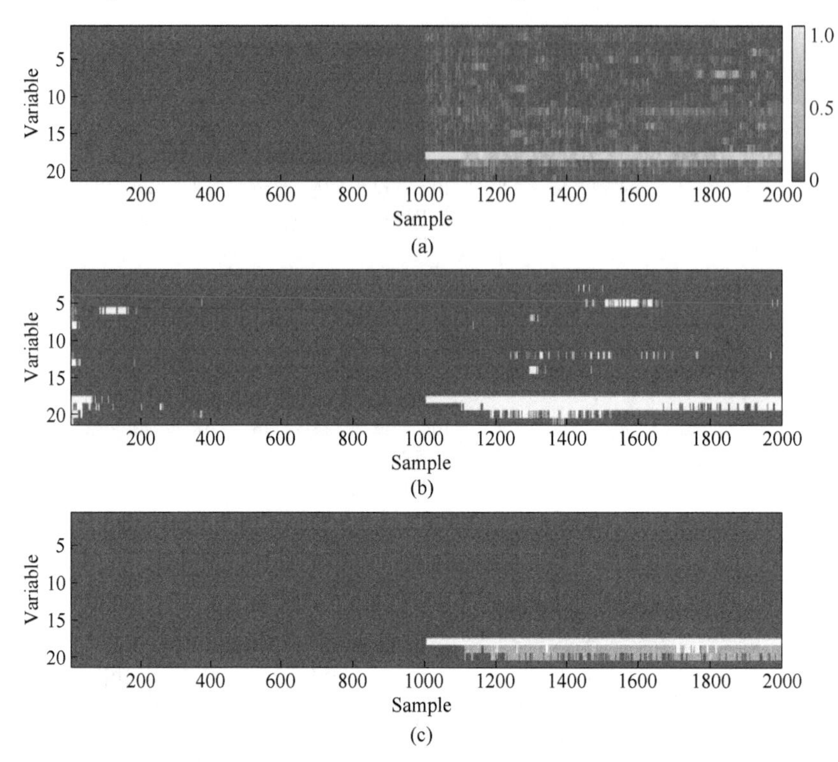

Fig. 9. 17 The fault isolation result (whole process) of Fault 1 in FMP case

(a) RBC; (b) LASSO; (c) BAB

(Scan the QR code on the front of the book for color picture)

incorrect information for maintenance personnel. For the fault isolation result of LASSO, fault variable x_4 is not identified.

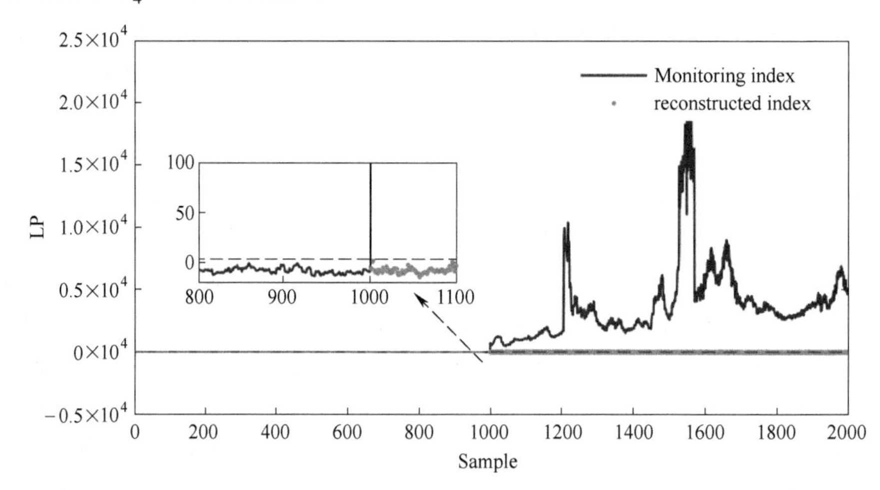

Fig. 9. 18 The monitoring indexes before and after fault isolation for Fault 2 in FMP case

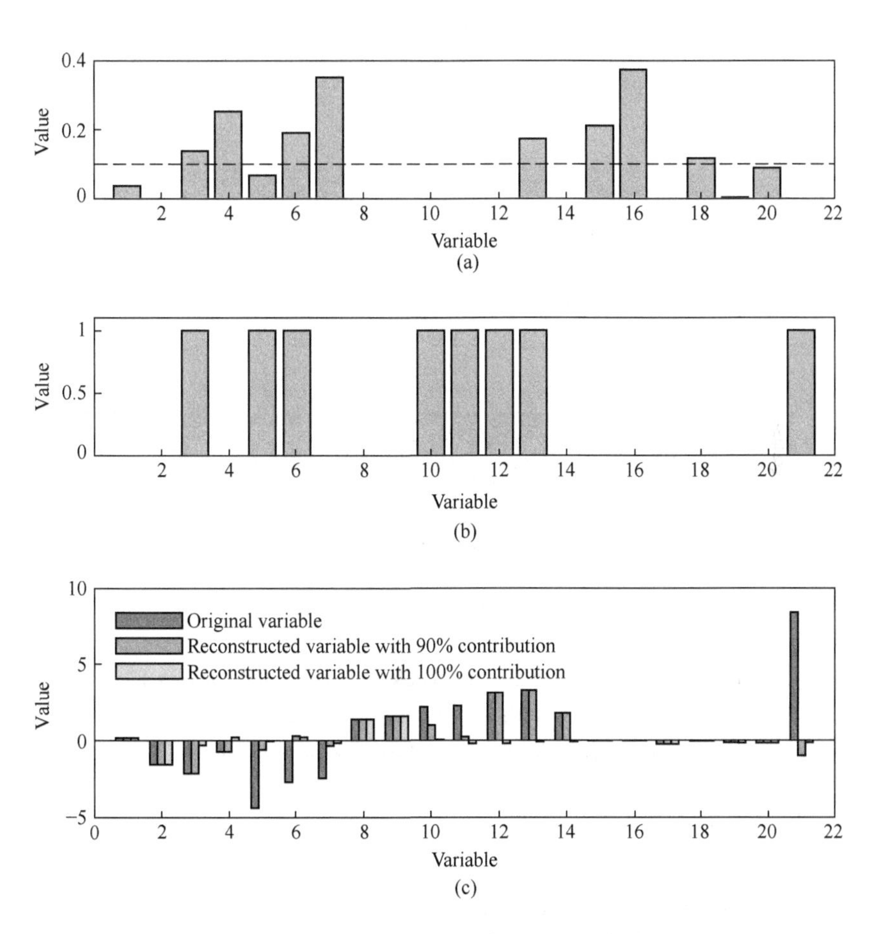

Fig. 9.19 Fault isolation results for Fault 2 in FMP case

(a) RBC; (b) LASSO; (c) BAB

Fig. 9.20 shows the fault isolation results for the whole process. The fault directly effects total force and gap (x_3 and x_7), but RBC method fails to identify x_3. And variable x_{16} is not affected by the fault, is also identified as fault variable by RBC method. So RBC method can't offer much useful information for maintenance. For the LASSO method, there are some kind of problem that variables x_4, x_7 and x_{14} are missed isolation, and variables x_6 and x_{20} are wrongly isolated. The BAB method effectively identifies all the fault variables. Based on the fault isolation result, the propagation of this fault can be easily inferred with the route from F3 stand to exit thickness.

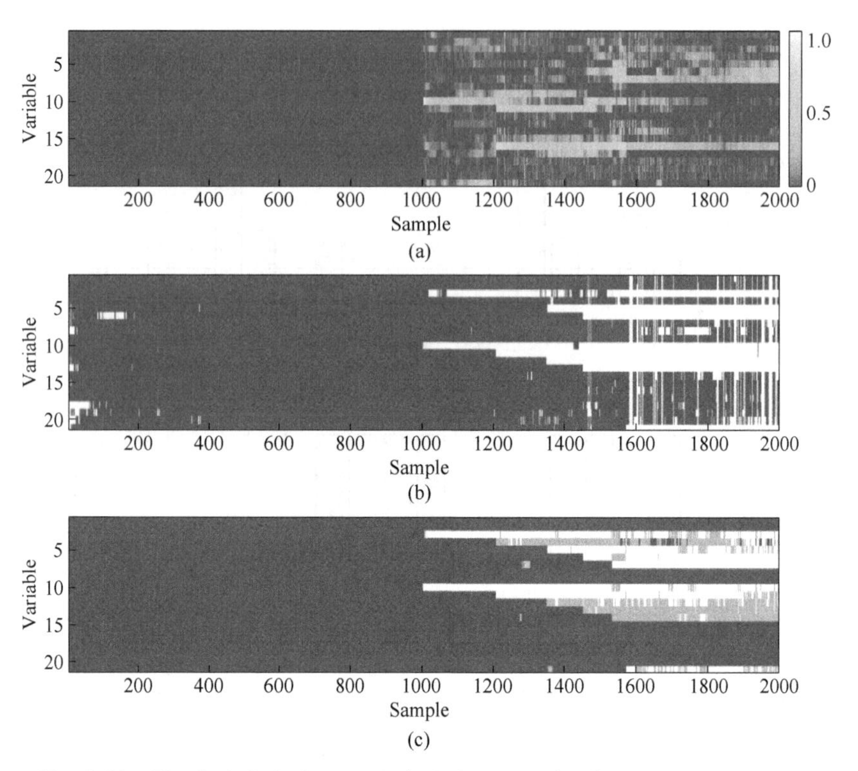

Fig. 9.20 The fault isolation result (whole process) of Fault 2 in FMP case

(a) RBC; (b) LASSO; (c) BAB

(Scan the QR code on the front of the book for color picture)

9.4 Conclusions

In this chapter, a process monitoring and fault isolation framework based on VAE is proposed. In fault detection procedure, a monitoring index based on probability density is derived to detect process faults. In fault isolation procedure, a missing value estimation method and BAB are integrated to isolate multivariate related to fault. To show the effectiveness, the proposed framework is applied to a numerical case and an HSMP case. It can be concluded from analysis and simulation that the monitoring index is proper for fault detection, and the fault isolation method based on BAB is more effective than other existing methods for nonlinear fault diagnosis.

References

[1] Venkatasubramanian V, Rengaswamy R, Kavuri S N, et al. A review of process fault detection and diagnosis: Part Ⅲ: Process history based methods [J]. Computers & chemical engineering,

2003, 27 (3): 327-346.

[2] Tidriri K, Chatti N, Verron S, et al. Bridging data-driven and model-based approaches for process fault diagnosis and health monitoring: A review of researches and future challenges [J]. Annual Reviews in Control, 2016, 42: 63-81.

[3] Qin S J. Survey on data-driven industrial process monitoring and diagnosis [J]. Annual reviews in control, 2012, 36 (2): 220-234.

[4] Yin S, Ding S X, Xie X, et al. A review on basic data-driven approaches for industrial process monitoring [J]. IEEE Transactions on Industrial Electronics, 2014, 61 (11): 6418-6428.

[5] Severson K, Chaiwatanodom P, Braatz R D. Perspectives on process monitoring of industrial systems [J]. Annual Reviews in Control, 2016, 42: 190-200.

[6] Lee J M, Yoo C, Choi S W, et al. Nonlinear process monitoring using kernel principal component analysis [J]. Chemical engineering science, 2004, 59 (1): 223-234.

[7] Choi S W, Lee C, Lee J M, et al. Fault detection and identification of nonlinear processes based on kernel PCA [J]. Chemometrics & intelligent laboratory systems, 2005, 75 (1): 55-67.

[8] Peng K, Zhang K, Li G. Quality-related process monitoring based on total kernel PLS model and its industrial application [J]. Mathematical Problems in Engineering, 2013: 707953.

[9] Jiang Q, Yan X. Quality-driven kernel projection to latent structure model for nonlinear process monitoring [J]. IEEE Access, 2019, 7: 74450-74458.

[10] Jia Q, Zhang Y. Quality-related fault detection approach based on dynamic kernel partial least squares [J]. Chemical Engineering Research and Design, 2016, 106: 242-252.

[11] Ge Z, Gao F, Song Z. Batch process monitoring based on support vector data description method [J]. Journal of Process Control, 2011, 21 (6): 949-959.

[12] Ge Z, Song Z. Bagging support vector data description model for batch process monitoring [J]. Journal of Process Control, 2013, 23 (8): 1090-1096.

[13] Yin L, Wang H, Fan W. Active learning based support vector data description method for robust novelty detection [J]. Knowledge-Based Systems, 2018, 153: 40-52.

[14] Sun Q, Ge Z. Probabilistic sequential network for deep learning of complex process data and soft sensor application [J]. IEEE Transactions on Industrial Informatics, 2018, 15 (5): 2700-2709.

[15] Yao L, Ge Z. Deep learning of semi supervised process data with hierarchical extreme learning machine and soft sensor application [J]. IEEE Transactions on Industrial Electronics, 2017, 65 (2): 1490-1498.

[16] Shaheryar A, Yin X C, Hao H W, et al. A denoising based auto associative model for robust sensor monitoring in nuclear power plants [J]. Science and Technology of Nuclear Installations, 2016: 9746948.

[17] Zhang Z, Jiang T, Li S, et al. Automated feature learning for nonlinear process monitoring: an approach using stacked denoising autoencoder and k-nearest neighbor rule [J]. Journal of Process Control, 2018, 64: 49-61.

[18] Mao T, Zhang Y, Zhou H, et al. Data driven injection molding process monitoring using sparse auto encoder technique [C] //2015 IEEE International Conference on Advanced Intelligent Mechatronics (AIM), 2015: 524-528.

[19] Hornberger K, Vacchini B. Monitoring derivation of the quantum linear Boltzmann equation [J]. Physical Review A, 2008, 77 (2): 022112.

[20] Kingma D P, Welling M. Auto-encoding variational bayes [J]. 2013, arXiv: 1312.6114.

[21] Doersch C. Tutorial on variational autoencoders [J]. 2016, arXiv: 1606.05908.

[22] Lee S, Kwak M, Tsui K L, et al. Process monitoring using variational autoencoder for high-dimensional nonlinear processes [J]. Engineering Applications of Artificial Intelligence, 2019, 83: 13-27.

[23] Zhang Z, Jiang T, Zhan C, et al. Gaussian feature learning based on variational autoencoder for improving nonlinear process monitoring [J]. Journal of Process Control, 2019, 75: 136-155.

[24] Wang K, Forbes M G, Gopaluni B, et al. Systematic development of a new variational autoencoder model based on uncertain data for monitoring nonlinear processes [J]. IEEE Access, 2019, 7: 22554-22565.

[25] Alcala C F, Qin S J. Analysis and generalization of fault diagnosis methods for process monitoring [J]. Journal of Process Control, 2011, 21 (3): 322-330.

[26] Kerkhof P V D, Vanlaer J, Ginset G, et al. Analysis of smearing-out in contribution plot based fault isolation for statistical process control [J]. Chemical Engineering Science, 2013, 104: 285-293.

[27] Alcala C F, Qin S J. Reconstruction-based contribution for process monitoring [J]. Automatica, 2009, 45 (7): 1593-1600.

[28] Kariwala V, Odiowei P E, Cao Y, et al. A branch and bound method for isolation of faulty variables through missing variable analysis [J]. Journal of Process Control, 2010, 20 (10): 1198-1206.

[29] He B, Yang X, Chen T, et al. Reconstruction-based multivariate contribution analysis for fault isolation: A branch and bound approach [J]. Journal of Process Control, 2012, 22 (7): 1228-1236.

[30] Yan Z, Yao Y. Variable selection method for fault isolation using least absolute shrinkage and selection operator (lasso) [J]. Chemometrics & Intelligent Laboratory Systems, 2015, 146: 136-146.

[31] Yan Z, Yao Y, Huang T B, et al. Reconstruction-based multivariate process fault isolation using Bayesian lasso [J]. Industrial & Engineering Chemistry Research, 2018, 57 (30): 9779-9787.

[32] Wang K, Chen J, Song Z. A sparse loading-based contribution method for multivariate control performance diagnosis [J]. Journal of Process Control, 2020, 85: 199-213.

[33] Ren S, Si F, Zhou J, et al. A new reconstruction-based auto-associative neural network for

fault diagnosis in nonlinear systems [J]. Chemometrics & Intelligent Laboratory Systems, 2018, 172: 118-128.

[34] Yu W K, Zhao C H. Robust monitoring and fault isolation of nonlinear industrial processes using denoising autoencoder and elastic net [J]. IEEE Transactions on Control Systems Technology, 2020, 28 (3): 1083-1091.

[35] Chen T, Sun Y. Probabilistic contribution analysis for statistical process monitoring: A missing variable approach [J]. Control Engineering Practice, 2009, 17 (4): 469-477.

[36] Kim D, Lee I B. Process monitoring based on probabilistic PCA [J]. Chemometrics & intelligent laboratory systems, 2003, 67 (2): 109-123.

[37] Tang P, Peng K X, Zhang K, et al. A deep belief network-based fault detection method for nonlinear processes [J]. IFAC-PapersOnLine, 2018, 51 (24): 9-14.

Chapter 10 Robust Fault Classification Based on WGAN and SLN

Process industry is developing towards intelligence, integration and automation. In order to ensure its normal operation, more and more variables need to be monitored[1-3]. Meanwhile, with the development of sensor technology, the frequency of data collection is increasing. Recently, streaming industrial data has become a new terminology, whose status information changes dynamically with time[4]. As an important carrier of information and knowledge, datadriven based process monitoring methods has received more and more attention from academia and industry[5-7]. When a fault is detected, the fault data needs to be further analyzed to know the fault type for equipment maintenance[8-10]. Therefore, when process industry is abnormal, the fault classification plays a core role in the resumption of production.

Fault classification is essentially a multi-class classification problem, so many classification methods in the field of artificial intelligence can be used as effective tools to handle this industrial problem. Guglielmi et al. applied the multilayer feedforward and radial basis function (RBF) neural networks to the fault diagnosis system of high-pressure feedwater heaters in thermal power plants[11]. Yu and Qin proposed a monitoring method based on Gaussian mixture model (GMM) under multiple operating conditions[12]. Feng et al. developed a fault classification method based joint Fisher discriminant analysis (JFDA) to deal with the non-Gaussianity and nonlinearity of Tennessee Eastman process (TEP)[13]. Lee et al. applied the fault detection and classification convolutional neural network (FDC-CNN) model to a chemical vapor deposition process[14]. Pan et al. proposed a novel deep learning network called LiftingNet to learn features adaptively from raw mechanical data without prior knowledge[15]. These methods are mainly based on supervised learning or unsupervised learning, but the complexity of streaming industrial data is far from ideal.

Compared with normal operating condition, faults occur infrequently and unexpectedly. Under low sampling frequency, fault data are difficult to collect. In addition, it is impossible to judge whether the collected data is fault data and what kind of fault data it is without adding the fault diagnosis system. Lack of fault labels may lead to overfitting issue in supervised learning. Labeling collected data manually is usually time-consuming, costly, and difficult to achieve. In fact, unlabeled data may contain some important process information, which helps to achieve better fault classification results. Thus, semi-supervised learning is more applicable to real industrial data by integrating the advantages of supervised and unsupervised learning. Some scholars have applied semi-supervised learning to the field of fault classification. Li et al. proposed a feature extraction method based on semi supervised kernel marginal Fisher analysis (SSKMFA) and applied it to bearing fault classification[16]. Ko and Kim developed a novel semi-supervised deep generative model to make effective use of massive unlabeled data for fault classification[17]. Wang et al. proposed a semi-supervised fault classification method based on inter-relational Mahalanobis stacked autoencoder (IRMSAE) for industrial hydrocracking process[18].

Although these semi-supervised methods have achieved good classification results in related fields, most of them only focus on modeling task with complete data. Due to the broken of measurement devices, the errors in data transmission and even willful damages, streaming industrial data frequently has missing values[19]. However, traditional case deletion methods and statistical imputation methods rarely take into account the temporal information of streaming industrial data[20]. In recent years, Luo et al. have successfully applied gated recurrent unit (GRU) and generative adversarial network (GAN) to time series imputation. The experimental results on Physionet dataset and KDD CUP 2018 Dataset show the superiority of GRU-GAN compared to other imputation algorithms in terms of imputation accuracy[21]. In addition, Rasmus and Valpola proposed a semi-supervised ladder network (SLN) by adding lateral connections between the encoder and the decoder on the basis of denoising autoencoder (DAE)[25]. Compared with the DAE, the SLN has better classification performance, stronger robustness and adaptability.

Inspired by the success of GAN in time series imputation and SLN in semi-supervised learning, this paper proposes a robust semi-supervised fault classification method for streaming industrial data to deal with the lack of labeled data and missing values simultaneously. The main contributions are summarized as follows:

(1) Inspired by the effectiveness and simplicity of minimal gated unit (MGU), we propose an enhanced MGU (EMGU) to deal with irregular time delay and obtain the latent information from streaming industrial data.

(2) Based on Wasserstein generative adversarial network (WGAN) and EMGU, we propose the EMGU-based WGAN to impute the initial incomplete unlabeled

streaming industrial data.

(3) After missing data imputation, SLN is used for semi-supervised fault classification and then the proposed EMGUWGAN-SLN method is applied to a real HSMP.

10.1 Principle of WGAN, MGU and SLN

In this section, the algorithm of WGAN, MGU and SLN are revisited briefly.

10.1.1 Wasserstein generative adversarial network

To solve the problem of gradient vanishing and instability in GAN training, Arjovsky et al. proposed WGAN using Wasserstein distance instead of Jensen-Shannon (JS) divergence[22]. The advantage of Wasserstein distance over JS divergence is that if there is no overlap between the two distributions, the Wasserstein distance still reflects the distance of the distribution. In addition, the JS divergence do not change continuously, but the Wasserstein distance changes smoothly. When using gradient descent to optimize model parameters, the JS divergence cannot be derived, but the Wasserstein distance can. The Wasserstein distance measures the distance between distributions of input data $\boldsymbol{x}$ and generated data $\boldsymbol{y}$ as follows:

$$W(P_{\mathrm{r}}, P_{\mathrm{g}}) = \inf_{\gamma \sim \Pi(P_{\mathrm{r}}, P_{\mathrm{g}})} \mathrm{E}_{(\boldsymbol{x},\boldsymbol{y}) \sim \gamma}[\|\boldsymbol{x} - \boldsymbol{y}\|] \tag{10.1}$$

where, $\Pi(P_{\mathrm{r}}, P_{\mathrm{g}})$ represents all joint distributions $\gamma(\boldsymbol{x}, \boldsymbol{y})$ whose marginals are P_{r} and P_{g}, respectively.

Since the lower bound of Wasserstein distance cannot be solved directly, equation (10.1) can be transformed into a dual problem (consider K-Lipschitz for some constant K):

$$\begin{aligned} K \times W(P_{\mathrm{r}}, P_{\mathrm{g}}) &\approx \max_{\|D\|_{\mathrm{L}} \leqslant K} E_{\boldsymbol{x}\text{-}P_{\mathrm{r}}}[D(\boldsymbol{x})] - E_{\boldsymbol{y}\text{-}P_{\mathrm{g}}}[D(\boldsymbol{y})] \\ &= \max_{\|D\|_{\mathrm{L}} \leqslant K} E_{\boldsymbol{x}\text{-}P_{\mathrm{r}}}[D(\boldsymbol{x})] - E_{\boldsymbol{z}\text{-}P_{\mathrm{g}}}[D(G(\boldsymbol{z}))] \end{aligned} \tag{10.2}$$

where, function $D(\cdot)$ is the mapping of discriminator. $G(\boldsymbol{z})$ maps the random noise $\boldsymbol{z}$ to $\boldsymbol{y}$. The value of K does not affect the direction of the gradient. Thus, the objective functions of generator (G) and discriminator (D) in WGAN are defined as follows:

$$V(G) = \min - E_{\boldsymbol{z}\text{-}P_{\mathrm{g}}}[D(G(\boldsymbol{z}))] \tag{10.3}$$

$$V(D) = \min E_{\boldsymbol{z}\text{-}P_{\mathrm{g}}}[D(G(\boldsymbol{z}))] - E_{\boldsymbol{x}\text{-}P_{\mathrm{r}}}[D(\boldsymbol{x})] \tag{10.4}$$

10.1.2 Minimal gated unit

Inspired by the effectiveness of GRU, Zhou et al. proposed a novel simplified gated unit for RNN with a single gate, as shown in **Fig. 10.1**. By analyzing the importance of each gate of LSTM and GRU, it is found that the forget gate $\boldsymbol{f}_t$ is critical. On the basis of GRU, MGU further combines the reset gate $\boldsymbol{r}_t$ and update gate z_t as forgetting gate $\boldsymbol{f}_t$ [23]. Thus, the update rule of hidden state in MGU is as follows:

$$\boldsymbol{f}_t = \sigma(\boldsymbol{W}_{\mathrm{f}}[\boldsymbol{h}_{t-1}, \boldsymbol{x}_t] + \boldsymbol{b}_{\mathrm{f}}) \tag{10.5}$$

$$\widetilde{\boldsymbol{h}}_t = \tanh(\boldsymbol{W}_{\mathrm{h}}[\boldsymbol{f}_t \odot \boldsymbol{h}_{t-1}, \boldsymbol{x}_t] + \boldsymbol{b}_{\mathrm{h}}) \tag{10.6}$$

$$\boldsymbol{h}_t = (1 - \boldsymbol{f}_t) \odot \boldsymbol{h}_{t-1} + \boldsymbol{f}_t \odot \widetilde{\boldsymbol{h}}_t \tag{10.7}$$

In the MGU network, only the hidden state $\boldsymbol{h}_t$ is transmitted between units.

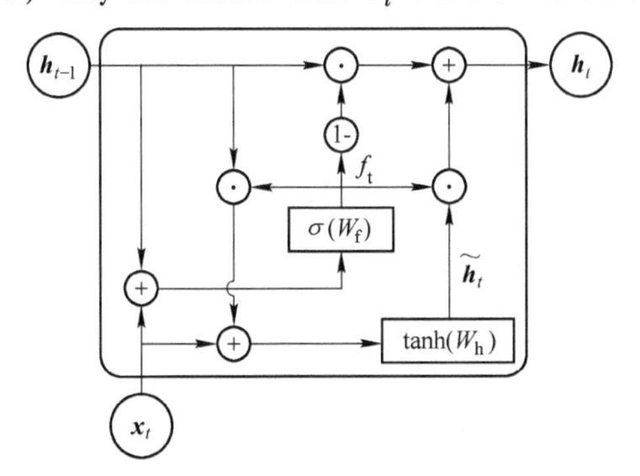

Fig. 10.1 Illustration of minimal gated unit

10.1.3 Semi-supervised ladder network

By integrating supervised learning and ladder network, Valpola and Rasmus proposed the SLN[24-25], as shown in **Fig. 10.2**. SLN has two encoders (corrupted encoder and clean encoder) and one decoder. Gaussian noise is added to each layer of the corrupted encoder. The decoder is connected to the corrupted encoder. The clean encoder shares the weights with the corrupted encoder. Compared with denoising autoencoders, the upper layers of SLN can discard some secondary features, focusing on the key features that play a decisive role in classification, while the lower layers can reconstruct the discarded secondary features through lateral connections.

Consider a training set consists of N labeled data $\{\boldsymbol{x}(n), t(n) \mid 1 \leqslant n \leqslant N\}$ with targets t and M unlabeled data $\{\boldsymbol{x}(n) \mid N+1 \leqslant n \leqslant N+M\}$. Normally, the unlabeled data are plentiful whereas the labeled data are scarce, that is $N \ll M$. The total cost function of SLN consists of the supervised cost function C_{S} and the unsupervised denoising cost function C_{U}. The supervised cost C_{S} is the cross entropy of the noisy output $\tilde{y}$ matching the label $t(n)$ given the input $\boldsymbol{x}(n)$:

$$C_{\mathrm{S}} = -\frac{1}{N}\sum_{n=1}^{N}\lg P(\tilde{y} = t(n) \mid \boldsymbol{x}(n)) \tag{10.8}$$

The unsupervised denoising cost function C_{U} is the mean squared reconstruction error of each hidden layer:

$$\begin{aligned} C_{\mathrm{U}} &= \sum_{l=0}^{L}\lambda_l C_{\mathrm{U}}^{(l)} \\ &= \sum_{l=0}^{L}\frac{\lambda_l}{(N+M)m_l}\sum_{n=1}^{N+M} \| \boldsymbol{z}^{(l)}(n) - \hat{\boldsymbol{z}}_{\mathrm{BN}}^{(l)}(n) \|^2 \end{aligned} \tag{10.9}$$

$$\hat{\boldsymbol{z}}_{\mathrm{BN}}^{(l)} = \frac{\hat{\boldsymbol{z}}^{(l)} - \mathrm{mean}(\boldsymbol{z}_{\mathrm{pre}}^{(l)})}{\sqrt{\mathrm{var}(\boldsymbol{z}_{\mathrm{pre}}^{(l)})}} \tag{10.10}$$

where, the projection $\boldsymbol{z}_{\mathrm{pre}}^{(l)}$ is selected as the target for denoising; L is number of latent variable layers; m_l is the width of layer l ; the hyperparameter λ_l a layerwise multiplier determining the importance of the denoising cost.

Thus, the total cost function of SLN is:

$$C = C_{\mathrm{S}} + C_{\mathrm{U}} \tag{10.11}$$

All parameters of the model can be trained through back propagation. After the training, input test data into the clean encoder to get the prediction results.

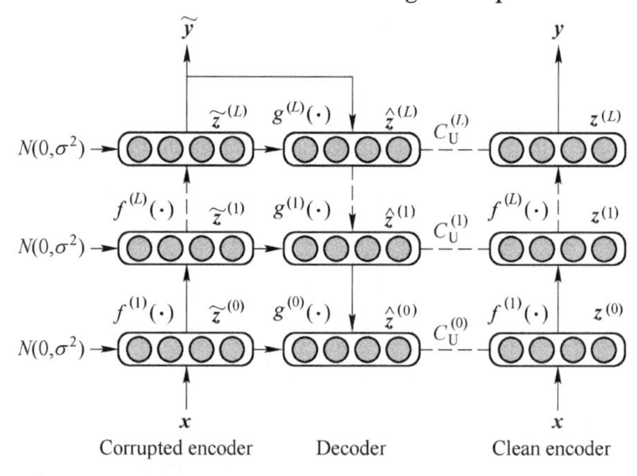

Fig. 10.2 Illustration of semi-supervised ladder network

10.2 Fault classification based on WGAN-SLN

HSMP is gradually becoming large-scale, diversified, and integrated. More and more sensors are equipped to the manufacturing facilities. The speed of data generation is faster, and the data dimension becomes higher. Although the streaming industrial data is growing explosively, the density of data value is low, and the data quality is uneven. Streaming industrial data often have non-ideal characteristics such as unbalanced, missing values, outliers, etc. In the real HRP, there are often some missing values in the massive data collected by sensors. Meanwhile, it is usually the case that labeled samples are scarce whereas unlabeled samples are numerous owing to the high cost of labeling data. Adequate labeled data is hard to acquire from industrial applications. Once a fault occurs, the production line needs to be adjusted and maintained immediately. However, conventional fault classification methods usually aim at ideal process data, and seldom consider the non-ideal characteristics of streaming industrial data, which is difficult to achieve good classification results. Data quality largely determines the accuracy and robustness of fault classification. Therefore, this paper focuses on missing values and lack of labels of streaming industrial data, selects EMGU as the basic network of WGAN to handle the missing values of unlabeled incomplete data. And then, labeled complete data and unlabeled complete data are utilized to train the SLN model, so as to realize the effective fault classification for HSMP.

10. 2. 1 Missing data imputation based on EMGU and WGAN

When designing the detailed structure of the WGAN, MGU is adopted as the basic network of generator and discriminator. However, due to the incompleteness of the data, the time delay between two consecutive complete observations varies greatly, which makes it difficult for MGU to achieve acceptable application results. Therefore, the EMGU is proposed to deal with irregular time delay and obtain the latent information from them, whose structure is shown in **Fig. 10. 3**.

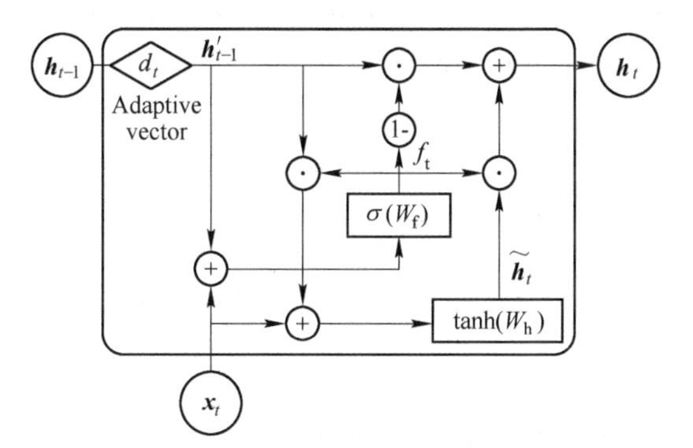

Fig. 10. 3 Illustration of EMGU

Given an incomplete data set $\boldsymbol{X} = [\boldsymbol{x}_{t_0}, \cdots, \boldsymbol{x}_{t_{n-1}}]^{\mathrm{T}} \in R^{n\times m}$ observed in $\boldsymbol{T} = [t_0, \cdots, t_{n-1}]^{\mathrm{T}}$. The structure matrix $\boldsymbol{S} \in R^{n\times m}$ is introduced to describe whether the values of $\boldsymbol{X}$ exist or not. If $x_{t_i}^j$ exists, $S_{t_i}^j$ is equal to 1, otherwise $S_{t_i}^j = 0$. In order to record the time lag between current value and last complete data, the time-delay matrix $\boldsymbol{\theta} \in R^{n\times m}$ can be calculated according to the following rules:

$$\theta_{t_i}^j = \begin{cases} t_i - t_{i-1}, & S_{t_{i-1}}^j == 1 \\ \theta_{t_{i-1}}^j + t_i - t_{i-1}, & S_{t_{i-1}}^j == 0 \text{ and } i > 0 \\ 0, & i == 0 \end{cases} \tag{10.12}$$

This chapter gives a simple example to show how to calculate the time-delay matrix $\boldsymbol{\theta}$. Assume that $n = 3$, $m = 4$, and "×" represents the missing data, $\boldsymbol{X}$ and $\boldsymbol{T}$ are as follows:

$$\boldsymbol{X} = \begin{bmatrix} 3762 & \times & 2394 & 2518 \\ 3745 & 2800 & \times & \times \\ \times & 2766 & \times & 2568 \end{bmatrix}, \ \boldsymbol{T} = \begin{bmatrix} 0 \\ 4 \\ 12 \end{bmatrix} \tag{10.13}$$

Then, the time-delay matrix $\boldsymbol{\theta}$ can be calculated by $\boldsymbol{S}$ and $\boldsymbol{T}$:

$$\boldsymbol{S} = \begin{bmatrix} 1 & 0 & 1 & 1 \\ 1 & 1 & 0 & 0 \\ 0 & 1 & 0 & 1 \end{bmatrix}, \ \boldsymbol{\theta} = \begin{bmatrix} 0 & 0 & 0 & 0 \\ 4 & 4 & 4 & 4 \\ 8 & 8 & 12 & 12 \end{bmatrix} \tag{10.14}$$

Based on the time-delay matrix $\boldsymbol{\theta}$, an adaptive vector $\boldsymbol{d}$ is designed to control the influence of the past observations:

$$d_{t_i} = \mathrm{e}^{-\max(0,\ \boldsymbol{W}_{\mathrm{d}}\theta_{t_i} + b_{\mathrm{d}})} \tag{10.15}$$

where, W_{d} and $\boldsymbol{b}_{\mathrm{d}}$ are training parameters. Negative exponential formulation is used to make sure that $d_{t_i} \in (0, 1]$. The larger the time-delay matrix $\boldsymbol{\theta}$, the smaller the adaptive vector $\boldsymbol{d}$. In addition, full weight matrix is used for $\boldsymbol{W}_{\mathrm{d}}$ to capture the interactions of the $\boldsymbol{\theta}$' s variables. After obtaining the adaptive vector, we update the MGU hidden state by element-wise multiplying the adaptive vector. The detailed update functions of EMGU are as follows:

$$\boldsymbol{h}'_{t_{i-1}} = d_{t_i} \odot \boldsymbol{h}_{t_{i-1}} \tag{10.16}$$

$$\boldsymbol{f}_{t_i} = \sigma(\boldsymbol{W}_{\mathrm{f}}[\boldsymbol{h}'_{t_{i-1}}, \boldsymbol{x}_{t_i}] + \boldsymbol{b}_{\mathrm{f}}) \tag{10.17}$$

$$\widetilde{\boldsymbol{h}}_{t_i} = \tanh(\boldsymbol{W}_{\mathrm{h}}[\boldsymbol{f}_{t_i} \odot \boldsymbol{h}'_{t-1}, \boldsymbol{x}] + \boldsymbol{b}_{\mathrm{h}}) \tag{10.18}$$

$$\boldsymbol{h}_{t_i} = (1 - \boldsymbol{f}_{t_i}) \odot \boldsymbol{h}_{t_{i-1}} + \boldsymbol{f}_{t_i} \odot \widetilde{\boldsymbol{h}}_{t_i} \tag{10.19}$$

The $\boldsymbol{G}$ and $\boldsymbol{D}$ in WGAN are both composed of a EMGU layer and a full-connection layer. In order to avoid overfitting, dropout is used in the full connection layer[26].

Because generated samples can change a lot with the changing of the input random noise vector z, a two-part cost function is designed to evaluate the imputation fitness:

$$C_{\mathrm{WGAN}}(z) = -\lambda D(G(z)) + \| \boldsymbol{X} \odot \boldsymbol{S} - G(z) \odot \boldsymbol{S} \|_2 \tag{10.20}$$

where, the first part is the discrimination loss; the second part is the reconstruction loss; λ is the hyper-parameter that controls the trade-off between them.

For any unlabeled incomplete data $\boldsymbol{x}^{\mathrm{UI}}$, the random noise vector z are fed into $\boldsymbol{G}$ to get $G(z)$. When $C_{\mathrm{WGAN}}(z)$ converging to the optimal solution, the unlabeled complete $\boldsymbol{x}^{\mathrm{UC}}$ can be imputed by $G(z)$ as follows:

$$\boldsymbol{x}^{\mathrm{UC}} = \boldsymbol{x}^{\mathrm{UI}} \odot \boldsymbol{S} + G(z) \odot (1 - \boldsymbol{S}) \tag{10.21}$$

10.2.2 Semi-supervised fault classification based on SLN

Assuming that $\{\boldsymbol{x}^{\mathrm{LC}}(n), t(n) \mid 1 \leqslant n \leqslant N\}$ is the labeled complete data, and $\{\boldsymbol{x}^{\mathrm{UC}}(n) \mid N+1 \leqslant n \leqslant M\}$ is imputed by unlabeled incomplete data in subsection B. The update functions for individual layers of the corrupted encoder are as follows:

$$\tilde{\boldsymbol{x}}^{\mathrm{LC}} = \boldsymbol{x}^{\mathrm{LC}} + \boldsymbol{\varepsilon} \tag{10.22}$$

$$\tilde{\boldsymbol{h}}^{(0)} = \tilde{\boldsymbol{z}}^{(0)} = \tilde{\boldsymbol{x}}^{\mathrm{LC}} \tag{10.23}$$

$$\tilde{\boldsymbol{z}}_{\mathrm{pre}}^{(l)} = \boldsymbol{W}^{(l)} \tilde{\boldsymbol{h}}^{(l-1)}, \quad 1 \leqslant l \leqslant L \tag{10.24}$$

$$\tilde{\boldsymbol{z}}^{(l)} = \mathrm{BN}(\tilde{\boldsymbol{z}}_{\mathrm{pre}}^{(l)}) + \boldsymbol{\varepsilon}, \quad 1 \leqslant l \leqslant L \tag{10.25}$$

$$\tilde{\boldsymbol{h}}^{(l)} = \varphi(\gamma^{(l)}(\tilde{\boldsymbol{z}}^{(l)} + \beta^{(l)})), \quad 1 \leqslant l \leqslant L \tag{10.26}$$

$$\tilde{\boldsymbol{y}} = \tilde{\boldsymbol{h}}^{(L)} \tag{10.27}$$

where, $\boldsymbol{\varepsilon}$ is the Gaussian noise; $\boldsymbol{W}^{(l)}$ is the weight from layer $l-1$ to layer l; $\tilde{\boldsymbol{h}}^{(l-1)}$ is the post-activation at layer $l-1$; $\tilde{\boldsymbol{z}}_{\mathrm{pre}}^{(l)}$ is the pre-normalization; BN is a component-wise batch normalization[27]; $\gamma^{(l)}$ and $\beta^{(l)}$ are used for shifting and scaling after batch normalization; $\varphi(\cdot)$ is the activation function.

In this work, the exponential linear unit (ELU) is selected as the activation function for layer l to layer $l-1$. Layer l is the classification layer, and large-margin softmax is selected as the activation function for prediction output[28]. Considering that the clean encoder is similar to corrupted encoder in structure, if $\tilde{\boldsymbol{h}}$ and $\tilde{\boldsymbol{z}}$ are replaced with $\boldsymbol{h}$ and $\boldsymbol{z}$, and the Gaussian noise $\boldsymbol{\varepsilon}$ is removed, the update functions of clean encoder will be obtained.

Compared with DAE, the biggest difference of SLN is that the lateral connections are added in the decoding process, that is, SLN can not only obtain information from the upper layer, but also from the same layer of corrupted encoder, as shown in the second column in **Fig. 10. 2**. The update functions for individual layers of the decoder are as follows:

$$\hat{\boldsymbol{u}}^{(l)} = \mathrm{BN}(\tilde{\boldsymbol{y}}), \quad l = L \tag{10.28}$$

$$\hat{\boldsymbol{u}}^{(l)} = \mathrm{BN}(\boldsymbol{V}^{(l+1)}\hat{\boldsymbol{z}}^{(l+1)}), \quad 0 \leqslant l \leqslant L-1 \tag{10.29}$$

$$\hat{z}_i^{(l)} = g_i(\tilde{z}_i^{(l)}, \hat{u}_i^{(l)}), \quad 0 \leqslant l \leqslant L \tag{10.30}$$

$$\hat{\boldsymbol{x}}^{LC} = \hat{\boldsymbol{z}}^{(0)} \tag{10.31}$$

where, $\boldsymbol{V}^{(l+1)}$ is the weight from layer $l+1$ to layer l; $\tilde{z}_i^{(l)}$ represents the value of the i_{th} neuron of $\tilde{z}^{(l)}$; $\hat{u}_i^{(l)}$ represents the value of the i_{th} neuron of $\hat{u}^{(l)}$; $\hat{z}_i^{(l)}$ represents the output of the i_{th} neuron in decoding layer l.

$g_i(\cdot, \cdot)$ is the nonlinear mapping function, which can operate each neuron as follows:

$$g_i(\tilde{z}_i^{(l)}, \hat{u}_i^{(l)}) = (\tilde{z}_i^{(l)} - \mu_i(\hat{u}_i^{(l)}))v_i(\hat{u}_i^{(l)}) + \mu_i(\hat{u}_i^{(l)}) \tag{10.32}$$

$$\mu_i(\hat{u}_i^{(l)}) = \eta_{1,i}^{(l)}\mathrm{SIG}(\eta_{2,i}^{(l)}\hat{u}_i^{(l)} + \eta_{3,i}^{(l)}) + \eta_{4,i}^{(l)}\hat{u}_i^{(l)} + \eta_{5,i}^{(l)} \tag{10.33}$$

$$v_i(\hat{u}_i^{(l)}) = \eta_{6,i}^{(l)}\mathrm{SIG}(\eta_{7,i}^{(l)}\hat{u}_i^{(l)} + \eta_{8,i}^{(l)}) + \eta_{9,i}^{(l)}\hat{u}_i^{(l)} + \eta_{10,i}^{(l)} \tag{10.34}$$

where, $\eta_{1,i}^{(l)}, \cdots, \eta_{10,i}^{(l)}$ are parameters of $g_i(\cdot, \cdot)$ corresponding to the neuron i at layer l; $\mathrm{SIG}(\cdot)$ is the sigmoid function.

In fact, the output $\hat{z}_i^{(l)}$ only depends on $\tilde{z}_i^{(l)}$, not all $\tilde{z}^{(l)}$.

All parameters of SLN can be obtained by minimizing the total cost function in equation (10. 11). Assume that the process contains K types of faults, the fault classification result for a test data $\boldsymbol{x}_{\mathrm{test}}$ is the prediction output of the clean encoder, namely, $F(\boldsymbol{x}_{\mathrm{test}}) = \max\{y_1, \cdots, y_K\}$. The network structure of the proposed method based on EMGU-WGAN-SLN is shown in **Fig. 10. 4**.

10. 3 Case study

10. 3. 1 Experimental setup

In this paper, 50 process and quality variables are selected for fault classification of the HSMP, as shown in **Table 10. 1**[29-30]. To verify the performance of imputation and semi-supervised classification of the EMGU-WGAN-SLN, it is compared with the traditional SLN[25], virtual adversarial training (VAT)[31], semi-supervised encoder

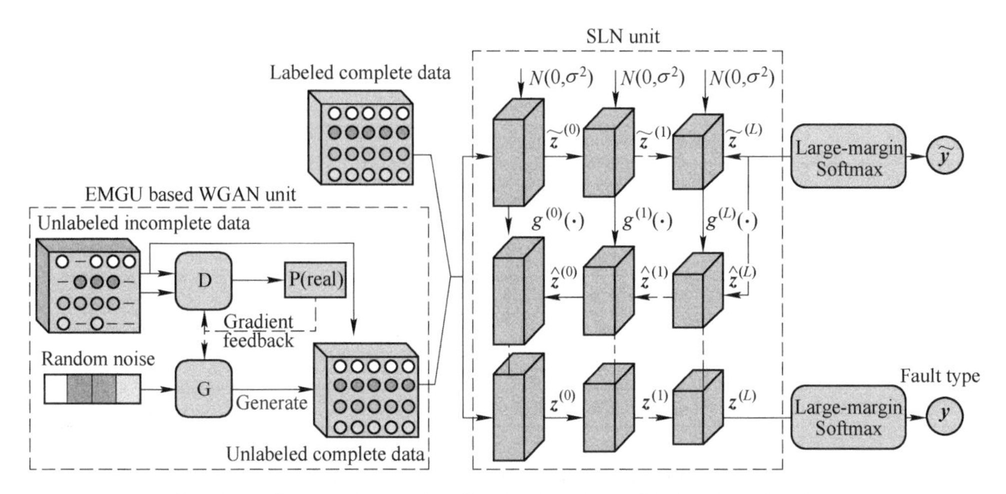

Fig. 10.4 Robust fault classification based on EMGU-WGAN-SLN

generative adversarial network (SSE-GAN)[32], categorical generative adversarial network (CAT-GAN)[33], and graph-based activity regularization (GAR)[34]. For SLN, VAT, SSE-GAN, CATGAN and GAR, multiple Imputation (MI) is used to deal with the missing data[35]. Three typical faults occurring in the HSMP are selected in this work, which are shown in **Table 10.2**. Compared with normal production data, fault data is relatively scarce. Therefore, we collected multiple batches of production data of the same product from the historical database in a hot rolling plant of Angang Steel, and sorted out the fault samples to form the simulation data used in this work. For each fault, 1000 samples are used for offline modeling. Considering the missing data cases, 5%, 10%, and 15% missing data are randomly generated. Meanwhile, N = 30 and 300 are randomly chosen in each fault for the supervised cost. The number of hidden units in EMGU of G and D is 50, the dimension of random noise is 50, the layer sizes of SLN unit is 50-100-50-25-3, the training epochs is 50, the batch size is 100. Adam optimization algorithm is used to optimize the network[36]. It can adjust the learning rate more flexibly by calculating the first moment estimation and the second moment estimation of gradient, which is suitable for parameter optimization of complex networks. For the parameter settings of the compared models, please refer to the corresponding papers. Generally, we first adjust the parameters that have great impact on the experimental results, and fix other parameters at the same time, adjust the parameters from coarse to fine stages. After getting a relatively good experimental result, we adjust other parameters in the same way. For the testing set, 500 samples are collected for each fault, totaling 1500 samples.

Table 10. 1 Description of variables in the HSMP

Variable	Description	Unit
1-7	Working-side gap of the i_{th} stand, $i=1, \cdots, 7$	mm
8-14	Driving-side gap of the i_{th} stand, $i=1, \cdots, 7$	mm
15-21	Working-side rolling force of the i_{th} stand, $i=1, \cdots, 7$	MN
22-28	Driving-side rolling force of the i_{th} stand, $i=1, \cdots, 7$	MN
29-35	Rolling speed of the i_{th} stand, $i=1, \cdots, 7$	mm/s
36-41	Bending force of the the i_{th} stand, $i=2, \cdots, 7$	MN
42-48	Current value of the i_{th} stand, $i=1, \cdots, 7$	MN
49	Finishing mill exit strip thickness	mm
50	Finshing mill exit strip flatness	I

Table 10. 2 Typical faults in the HSMP

Fault No.	Description
Fault 1	Malfunction of gap control loop in the 4_{th} stand
Fault 2	Sensor fault of bending force in the 5_{th} stand
Fault 3	Fault of cooling valve between the 2_{nd} and 3_{rd} stands

10. 3. 2 Analysis of fault classification results

For each fault, the classification accuracy is used to evaluate the fault classification performance of these four methods, which can be calculated as follows:

$$\mathrm{Acc}_i = \frac{\text{samples(correctly identified)}}{\text{total samples of Fault } i}, \quad i = 1, 2, 3 \tag{10.35}$$

where, i represents the typical faults in **Table 10. 2**; Acc is the average accuracy of all faults.

Comparable classification results have been observed through the above methods, it is easy to find out from **Table 10. 3** that the proposed method is more effective and has better performance than the other methods. The detailed analysis of each case is as follows.

Table 10. 3 Acc of different methods at different missing data

Missing data	Model	$N=30$	$N=300$	Average
5% missing data	SLN	83. 2%	91. 3%	87. 3%
	VAT	86. 3%	94. 5%	90. 4%
	SSE-GAN	86. 6%	95. 5%	91. 1%
	CAT-GAN	84. 2%	90. 9%	87. 6%
	GAR	87. 8%	95. 1%	91. 5%
	Proposed	92. 1%	97. 1%	94. 6%

Continued Table 10. 3

Missing data	Model	N=30	N=300	Average
10% missing data	SLN	79. 1%	87. 9%	83. 5%
	VAT	80. 7%	88. 8%	84. 8%
	SSE-GAN	80. 4%	89. 7%	85. 1%
	CAT-GAN	80. 8%	88. 3%	84. 6%
	GAR	81. 6%	90. 5%	86. 1%
	Proposed	90. 9%	95. 7%	93. 3%
15% missing data	SLN	74. 1%	83. 0%	78. 6%
	VAT	75. 5%	84. 2%	79. 9%
	SSE-GAN	75. 7%	83. 9%	79. 8%
	CAT-GAN	74. 6%	82. 7%	78. 7%
	GAR	77. 3%	84. 6%	81. 0%
	Proposed	91. 1%	96. 1%	93. 6%

10. 3. 2. 1 Case 1 (N=30)

In this scenario, the training data contains 50 labeled samples for each fault, and the rest is unlabeled data. The classification results and confusion matrices of the four methods under different missing data (5%, 10%, and 15%) modeling environment are shown in **Fig. 10. 5**(a)-(f) and **Fig. 10. 7**(a)-(c), respectively. Obviously, when the proportion of missing data is 5%, the classification accuracies of the four methods are the highest. With the increase of missing data, the classification accuracies of MI-based SLN, VAT, SSE-GAN, CAT-GAN and GAR decrease significantly. While the classification accuracy of the EMGU-WGAN-SLN does not change significantly, which clearly verifies the robustness for the EMGU-based WGAN unit. When the proportion of missing data is 15%, the accuracy of the EMGU-WGAN-SLN is about 15% higher than that of the SLN and CAT-GAN.

10. 3. 2. 2 Case 2 (N=300)

As we increase the labeled samples to ten times of Case 1, the classification accuracies of the four methods have been improved to some extent. Detailed classification results and confusion matrices under 5-15% missing data modeling conditions are shown in **Fig. 10. 6**(a)-(f) and **Fig. 10. 8**(a)-(c). Same as Case 1, the EMGU-WGAN- SLN method performs well under the missing data modeling environment. As the proportion of missing data increases from 5% to 15%, there is also no significant change in classification performance, which demonstrates the robustness of the proposed method.

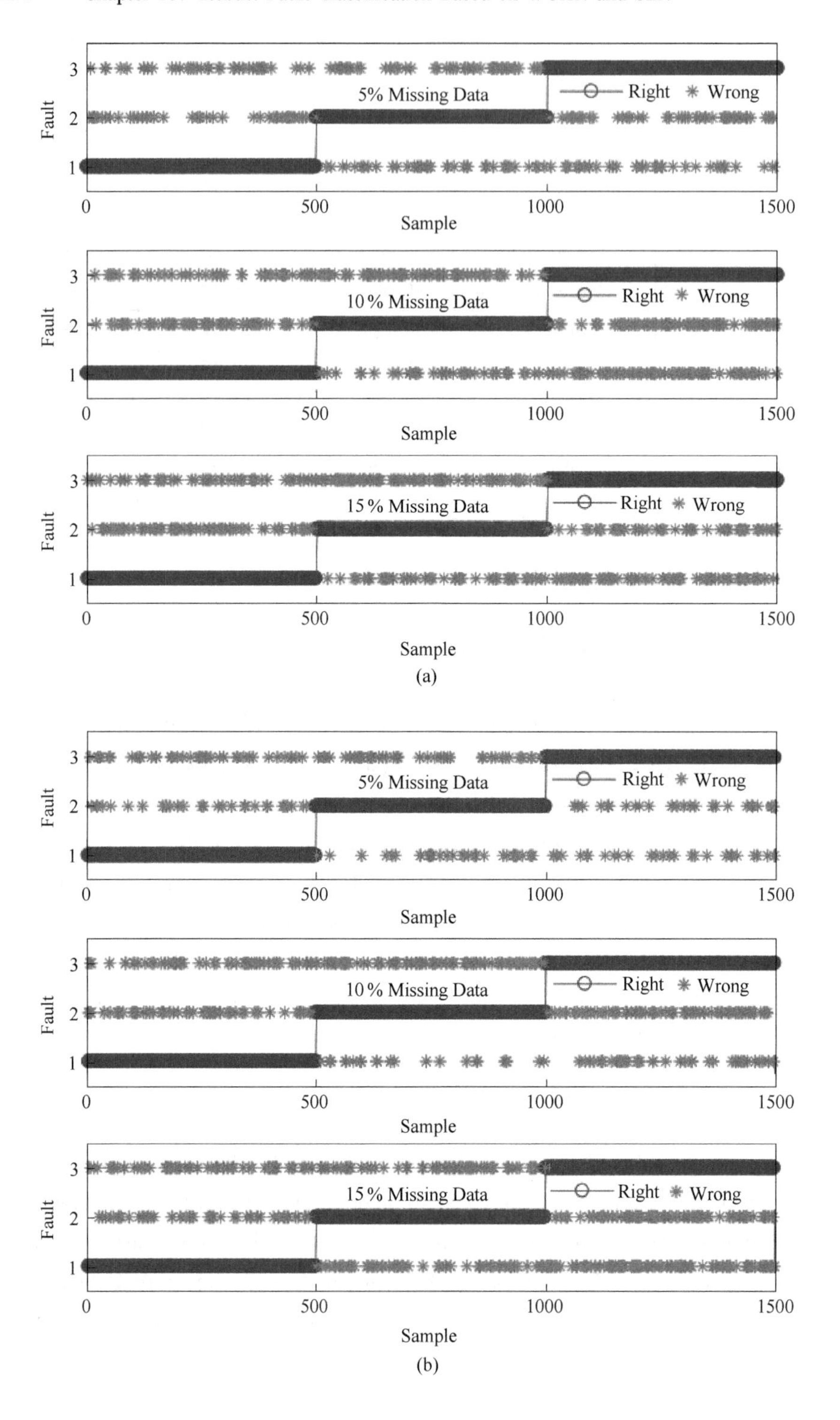

Fault
3
2
1
5% Missing Data
Right
Wrong
0
500
1000
1500
Sample
10% Missing Data
15% Missing Data

(a)

(b)

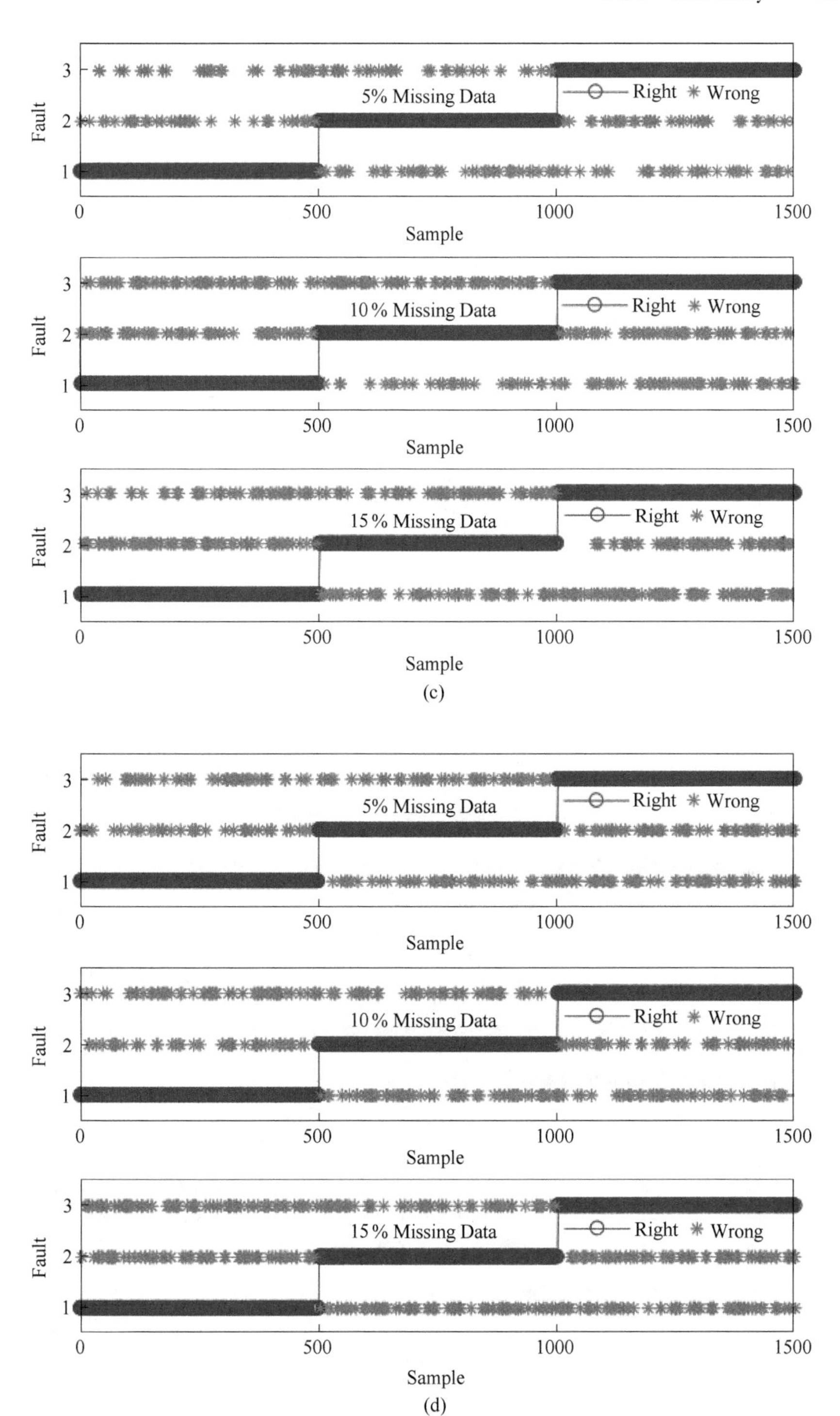

3
2
1
Fault
5% Missing Data
Right
Wrong
0
500
1000
1500
Sample
10% Missing Data
15% Missing Data

(c)

(d)

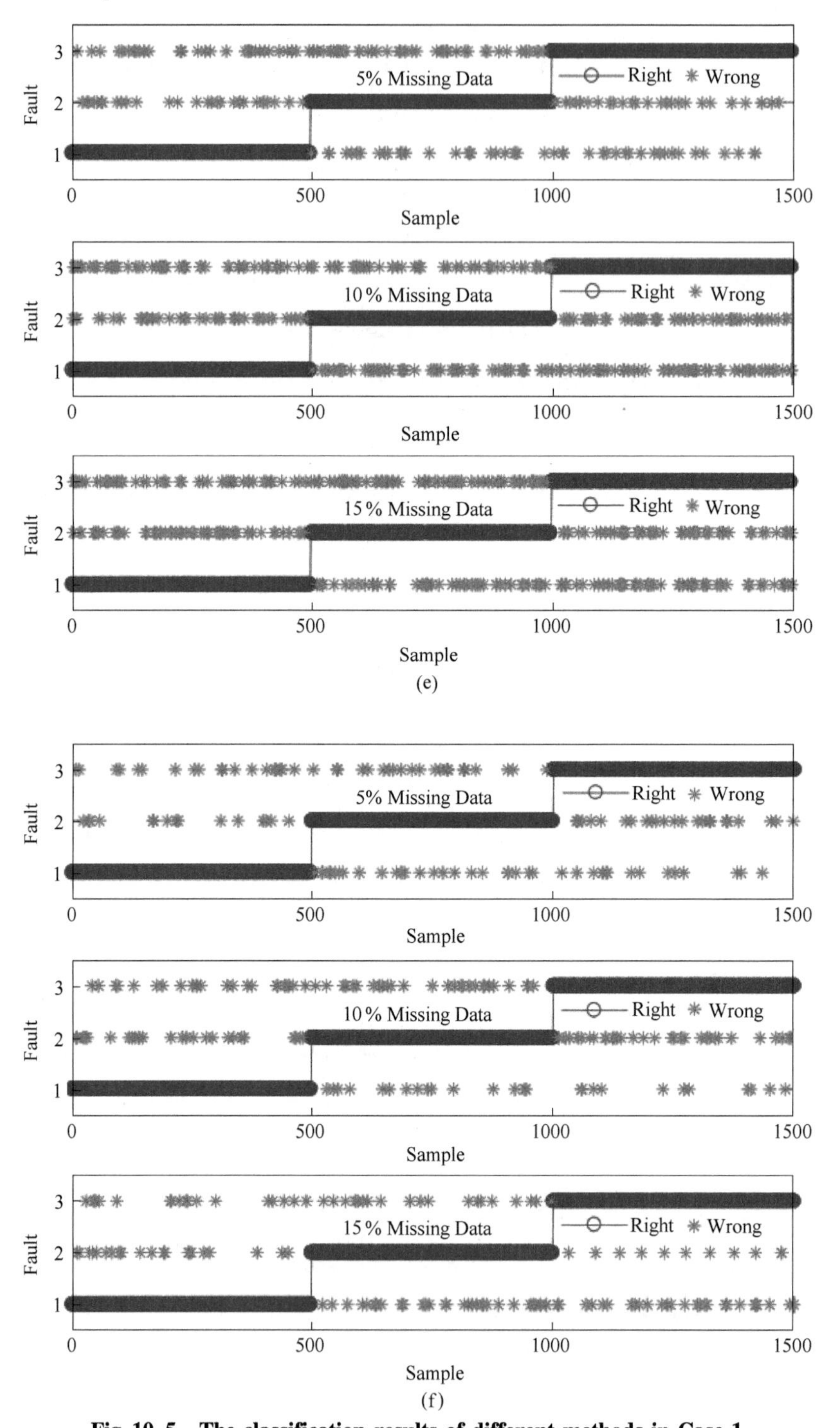

Fig. 10. 5 The classification results of different methods in Case 1

(a) SLN; (b) VAT; (c) SSE-GAN; (d) CAT-GAN; (e) GAR; (f) EMGU-WGAN-SLN

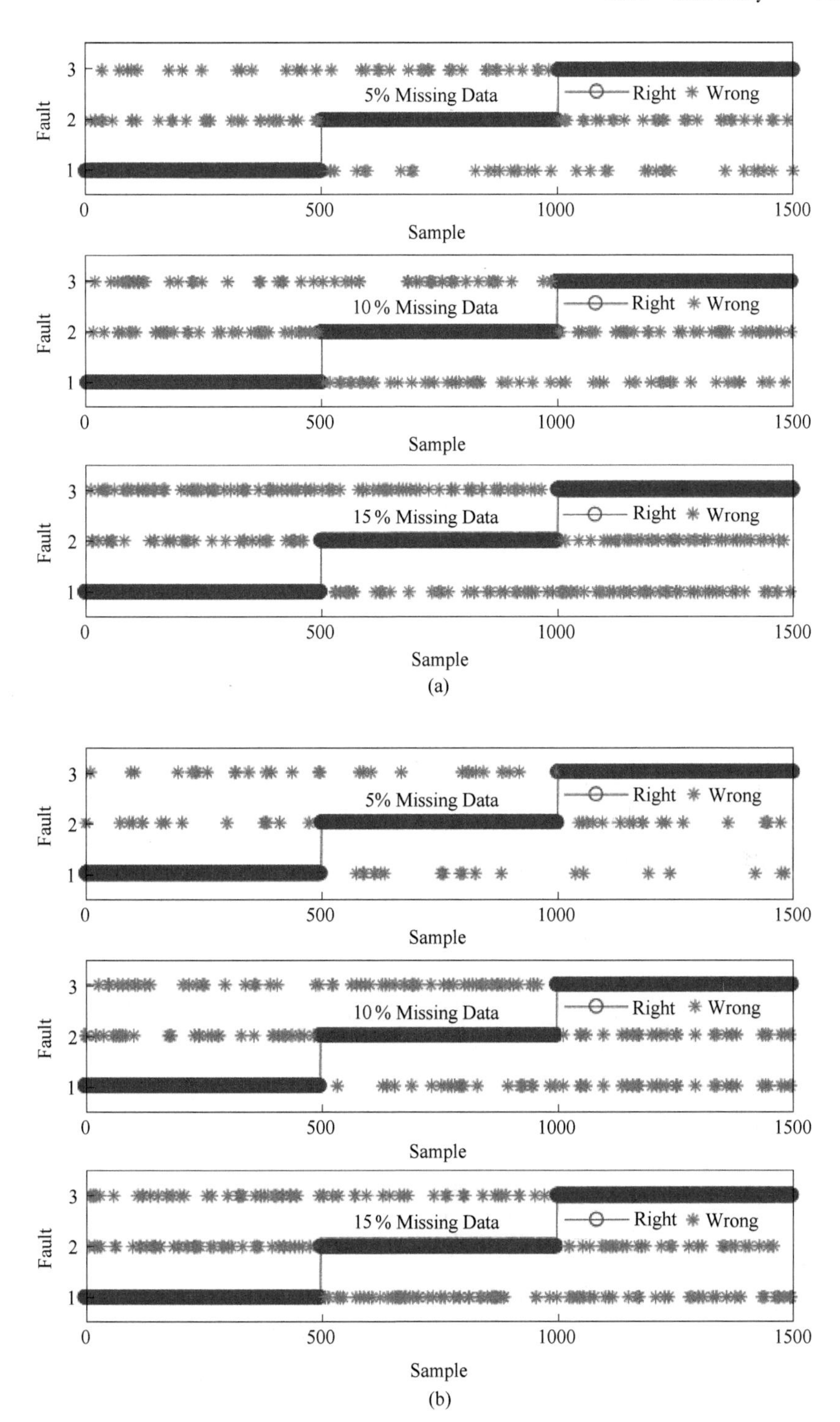
5% Missing Data
10% Missing Data
15% Missing Data
Right
Wrong
Fault
Sample
0
500
1000
1500
1
2
3

(a)

(b)

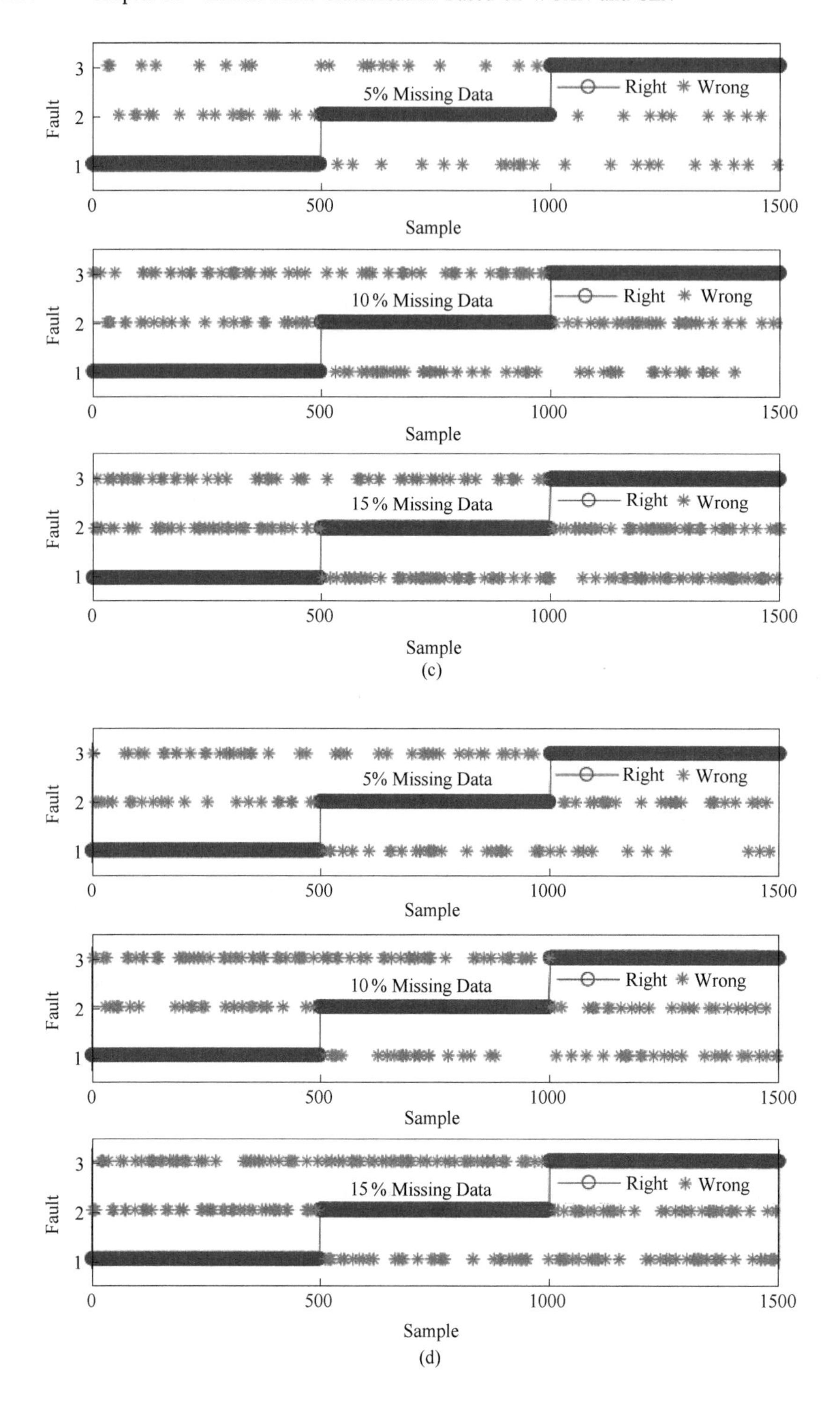

Fault
Sample
5% Missing Data
10% Missing Data
15% Missing Data
Right
Wrong
0
500
1000
1500
1
2
3

(c)

(d)

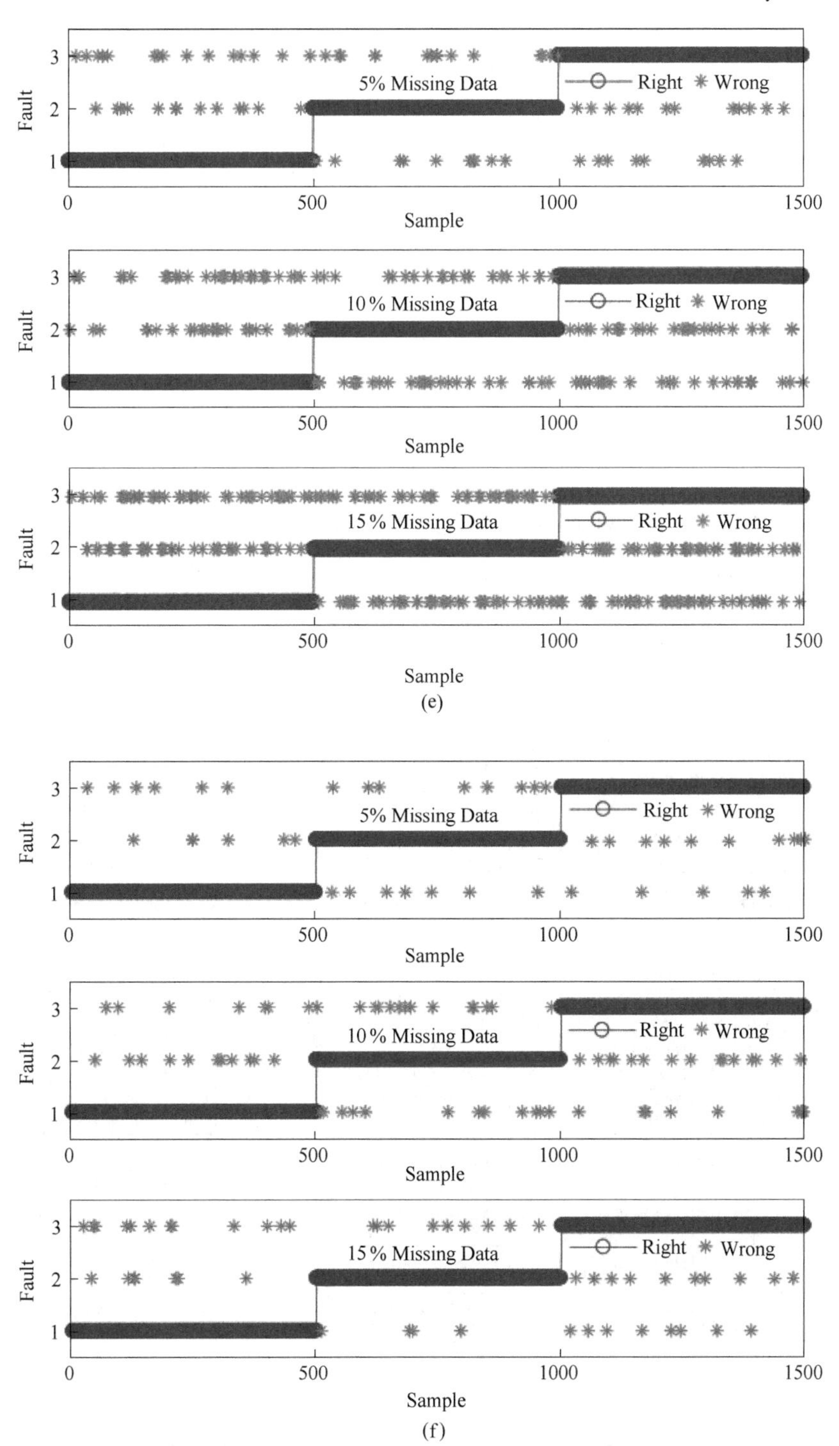

Fig. 10.6 The classification results of different methods in Case 2

(a) SLN; (b) VAT; (c) SSE-GAN; (d) CAT-GAN; (e) GAR; (f) EMGU-WGAN-SLN

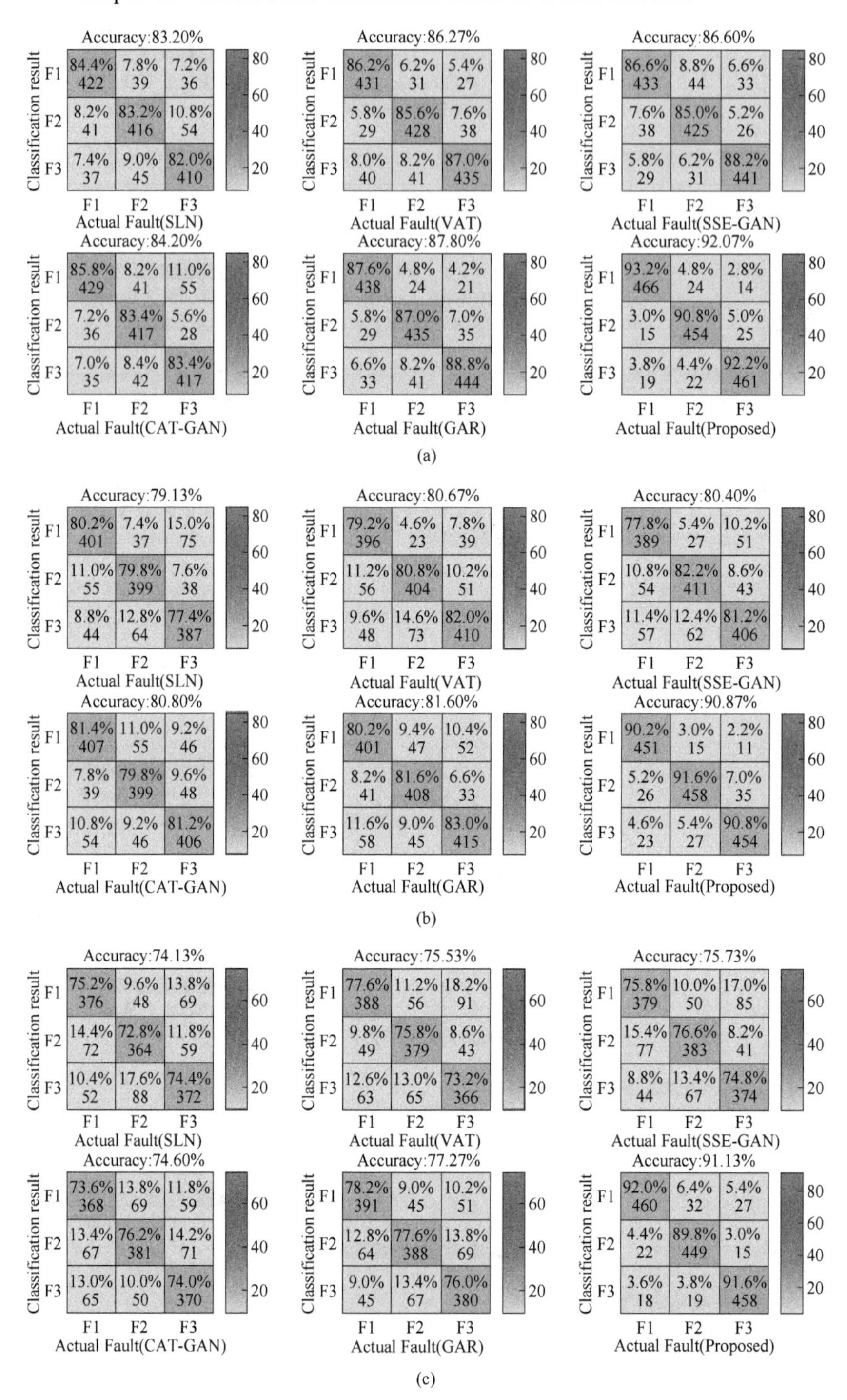

Fig. 10.7 Confusion matrices of different methods in Case 1

(a) 5% missing data; (b) 10% missing data; (c) 15% missing data

(Scan the QR code on the front of the book for color picture)

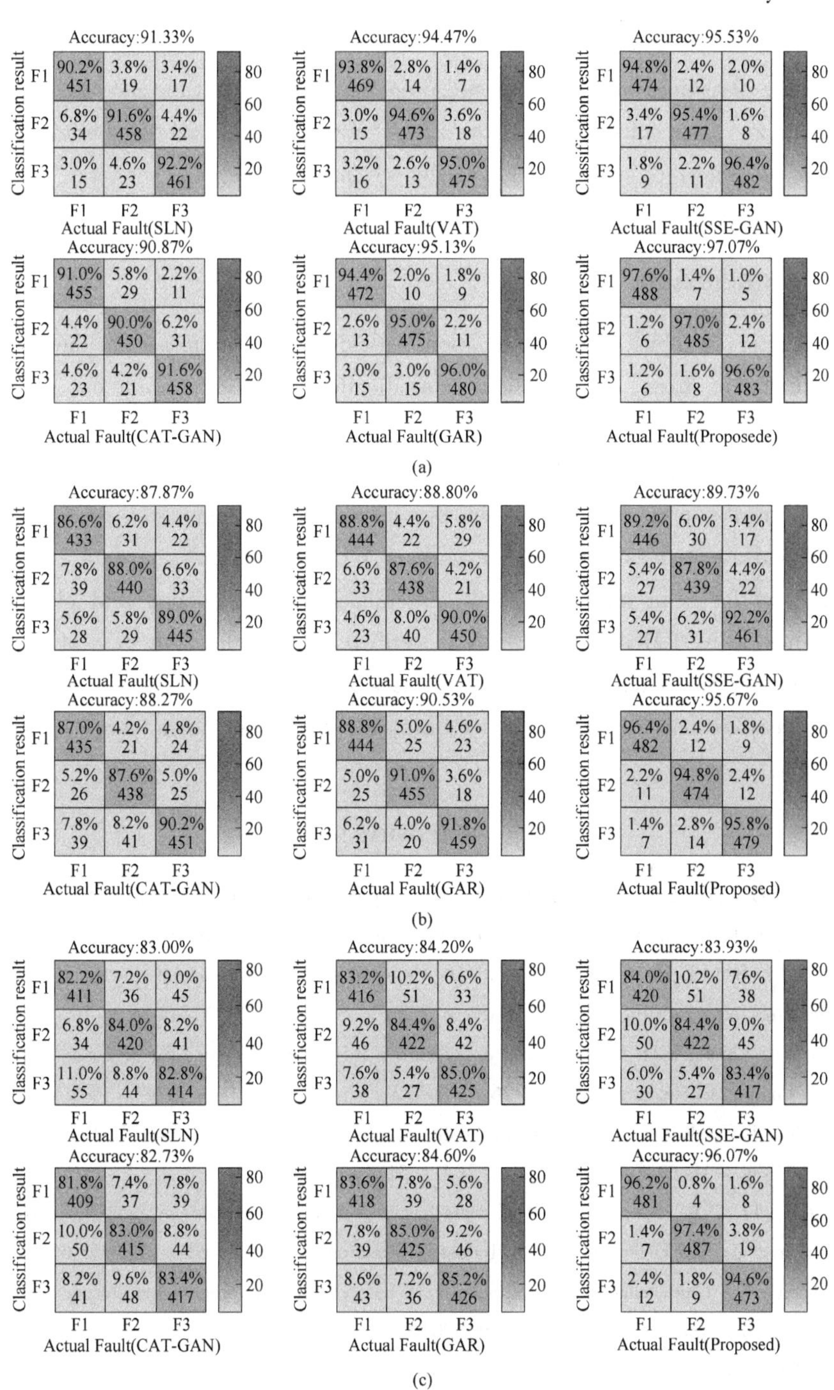

Fig. 10.8 Confusion matrices of different methods in Case 2

(a) 5% missing data; (b) 10% missing data; (c) 15% missing data

(Scan the QR code on the front of the book for color picture)

Although the classification accuracies of the other three methods are higher than that of Case 1, the classification accuracies still drop significantly with the increase of missing data. Anyhow, the results show that the EMGU-WGAN-SLN is more reliable under the modeling conditions of missing data existence and labeled data shortage, so it can improve the performance of the diagnosis phase.

10.4 Conclusions

In this chapter, a robust fault classification method has been proposed for the hot rolling process. Different from existing fault classification methods based on ideal data modeling, the EMGU-WGAN-SLN method works under non-ideal modeling conditions of missing data existence and lack of labels. The well-known WGAN and SLN are combined in the proposed method. In the missing data imputation stage, MGU is modified and adopted as the basic network of WGAN. After missing data is imputed, SLN is used for semi-supervised fault classification. The modeling and classification performance of the EMGU-WGAN-SLN is verified through the fault data of a real hot rolling process. Simulation results of two cases show that the proposed method have higher classification accuracy and robustness than the other three competitive approaches.

References

[1] Jiang Q C, Yan X F, Huang B. Review and perspectives of datadriven distributed monitoring for industrial plant-wide processes [J]. Ind. Eng. Chem. Res., 2019, 58 (29): 12899-12912.

[2] Ge Z Q, Song Z H, Ding S X, et al. Data mining and analytics in the process industry: The role of machine learning [J]. IEEE Access, 2017, 5 (99): 20590-20616.

[3] Qin S J, Chiang L H. Advances and opportunities in machine learning for process data analytics [J]. Comput. Chem. Eng, 2019, 126: 465-473.

[4] Bezdek J C, Keller J M. Streaming data analysis: Clustering or classification [J]. IEEE Trans. Syst., Man, Cybern., Syst., 2021, 51 (1): 91-102.

[5] Yin S, Li X W, Gao H J, et al. Data-based techniques focused on modern industry: An overview [J]. IEEE Transactions on Industrial Electronics, 2015, 62 (1): 657-667.

[6] Severson K, Chaiwatanodom P, Braatz R D. Perspectives on process monitoring of industrial systems [J]. Annual Review of Control, 2016, 42: 190-200.

[7] Ge Z Q. Process data analytics via probabilistic latent variable models: a tutorial review [J]. Industrial Engineering Chemical Research, 2018, 57: 12646-12661.

[8] Wang N, Yang F, Zhang R, et al. Intelligent fault diagnosis for chemical processes using deep learning multi model fusion [J]. IEEE Transactions on Cybernetics, 2022, 52 (7):

7121-7135.

[9] Chai, Z, Zhao C H. Fault-prototypical adapted network for cross domain industrial intelligent diagnosis [J]. IEEE Transactions on Automation Science & Engineering, 2022, 19 (4) 3649-3658.

[10] Zhang C F, Peng K X, Dong J. An extensible quality-related fault isolation framework based on dual broad partial least squares with application to the hot rolling process [J]. Expert System with Application, 2020, 167: 114166.

[11] Guglielmi G, Parisini T, Rossi G. Keynote paper: Fault diagnosis and neural networks: Apowerplant application [J]. Control Engineering Practice, 1995, 3 (5): 601-620.

[12] Yu J, Qin S J. Multimode process monitoring with Bayesian inference-based finite Gaussian mixture models [J]. AIChE Journal, 2008, 54 (7): 1811-1829.

[13] Feng J, Wang J, Zhang H W, et al. Fault diagnosis method of joint fisher discriminant analysis based on local and global manifold learning and its kernel version [J]. IEEE Transactions on Automation Science & Engineering, 2016, 13 (1): 122-133.

[14] Lee K B, Cheon S, Kim C O. A convolutional neural network for fault classification and diagnosis in semiconductor manufacturing processes [J]. IEEE Transactions on Semiconductor Manufacturing, 2017, 30 (2): 135-142.

[15] Pan J, Zi Y Y, Chen J L, et al. LiftingNet: a novel deep learning network with layerwise feature learning from noisy mechanical data for fault classification [J]. IEEE Transactions on Industrial Electronics, 2018, 65 (6): 4973-4982.

[16] Li J, Xuan J, Shi T. Feature extraction based on semi-supervised kernel Marginal Fisher analysis and its application in bearing fault diagnosis [J]. Mechanical Systems & Signal Processing, 2013, 41 (1/2): 113-126.

[17] Ko T, Kim H. Fault classification in high-dimensional complex processes using semi-supervised deep convolutional generative models [J]. IEEE Transactions on Industrial Informatics, 2020, 16 (4): 2868-2877.

[18] Wang Y L, Aman A M, Liu C L, et al. Inter-relational Mahalanobis SAE with semi-supervised strategy for fault classification in chemical processes [J]. Chemometrics & Intelligent Laboratory Systems, 2022, 228: 104624.

[19] Zhu J L, Ge Z Q, Song Z H, et al. Review and big data perspectives on robust data mining approaches for industrial process modeling with outliers and missing data [J]. Annual Review of Control, 2018, 46: 107-133.

[20] Silva L O, Zarate L E. A brief review of the main approaches for, treatment of missing data [J]. Intelligent Data Analysis, 2014, 18 (6): 1177-1198.

[21] Luo Y H, Cai X R, Zhang Y, et al. Multivariate time series imputation with generative adversarial networks [J]. NeurIPS, 2018.

[22] Arjovsky M, Chintala S, Bottou L. Wasserstein generative adversarial networks [J]. 2017, arXiv: 1701. 07875.

[23] Zhou G B, Wu J X, Zhang C L, et al. Minimal gated unit for recurrent neural networks [J]. International Journal of Automation & Computing, 2016, 13 (3): 226-234.

[24] Valpola H. From neural PCA to deep unsupervised learning [J]. ICALM, 2015.

[25] Rasmus A, Valpola H, Honkala M, et al. Semi-supervised learning with ladder network [J]. Computer Science, 2015, 9 (1): 1-9.

[26] Srivastava N, Hinton G, Krizhevsky A, et al. Dropout: a simple way to prevent neural networks from overfitting [J]. Journal of Machine Learning Research, 2014, 15 (1): 1929-1958.

[27] Ioffe S, Szegedy C. Batch normalization: Accelerating deep network training by reducing internal covariate shift [J]. ICML, 2015.

[28] Liu W Y, Wen Y D, Yu Z D, et al. Large margin softmax loss for convolutional neural networks [J]. ICML, 2016.

[29] Ma A, Dong J, Peng K X, et al. Hierarchical monitoring and root-cause diagnosis framework for key performance indicator related multiple faults in process industries [J]. IEEE Transactions on Industrial Informatics, 2019, 15 (4): 2091-2100.

[30] Zhang C F, Peng K X, Dong J, et al. KPI-related operating performance assessment based on distributed ImRMR-KOCTA for hot strip mill process [J]. Expert System with Application, 2022, 209: 118273.

[31] Miyato T, Maeda S, Koyama M, et al. Virtual adversarial training: A regularization method for supervised and semi-supervised learning [J]. IEEE Transactions on Pattern Analysis & Machine Intelligence, 2017, 7 (23): 99-112.

[32] Fu X, Shen Y T, Li H W, et al. A semi-supervised encoder generative adversarial networks model for image classification [J]. Acta Automatica Sinica, 2020, 46 (3): 531-539.

[33] Springenberg J T. Unsupervised and semi-supervised learning with categorical generative adversarial networks [J]. ICLR, 2015.

[34] Kilinc O, Uysal I. GAR: an efficient and scalable graph-based activity regularization for semi-supervised learning [J]. Neurocomputing, 2018, 296 (28): 46-54.

[35] Murray J S. Multiple imputation: a review of practical and theoretical findings [J]. Statistical Science, 2018, 33 (2): 142-159.

[36] Kingma D, Ba J. Adam: A method for stochastic optimization [J]. ICML, 2015.

Chapter 11 Extensible Quality-Related Fault Isolation Framework Based on DBPLS

In order to meet the market demands for multi-specification and high value-added steels, the modern iron and steel industry is developing towards the direction of high efficiency and large-scale integration[1-3]. As production scales up and complexity level increases, adopting reasonable quality related monitoring methods to ensure production safety and product quality has become a top priority in automation of the steel industry[4-5]. For instance, there is a strong link between the thickness of strip steels and rolling forces in HSMP. Once a fault occurs in production, the thickness will be affected to varying degrees. However, unlike process variables, quality ones usually cannot be measured in real time, which causes delay in real-time monitoring and evaluation of product quality. Thus, how to effectively deal with such faults deserves attention[6-9].

In the last two decades, many data-driven methods have been proposed to implement process monitoring. Compared with model-based methods, data-driven ones are developed without a known physical model, which makes it more suitable for complex industrial processes. Partial least squares (PLS) is one of the most attractive quality-related methods. MacGregor et al. showed its power to process monitoring and fault diagnosis[10]. Zhou et al. [11] developed a total PLS model, which have improved the fault detectability of PLS. Qin and Zheng[12] proposed a concurrent PLS monitoring method, which offers complete monitoring of faults that happen in different subspaces. Yin et al. proposed an improved PLS approach to decompose the measurable process variables into the key-performance-indicator-related and unrelated parts for diagnosing key-performance-indicator-related faults[13].

From the perspective of classification, fault detection is a one-class classification (OCC) problem, and fault isolation can be regarded as a multiclass classification (MCC) problem[14], which is shown in **Fig. 11. 1**. By assigning abnormal samples to the most relevant fault types, the root causes can be effectively determined. Fisher discriminant analysis (FDA) is a well-known linear classification method, optimal in terms of maximizing the separation between different faults[15-16]. FDA can perform well under linear conditions, but it cannot be effective to deal with nonlinear data. As a representative nonlinear classification method, support vector machine (SVM) is designed based on the Vapnik-Chervonenkis theory and has superior performance in nonlinear classification[17]. In addition, SVM also shows good generalization ability in practice. However, the classification accuracy of SVM is affected by the kernel functions and hyperparameters. Chiang et al.[18] investigated the proficiencies of FDA and SVM for fault diagnosis in Tennessee Eastman process (TEP). Artificial neural network (ANN) has also been successfully used for process monitoring[19-20]. However, the learning process is time-consuming and the network structures are decided empirically.

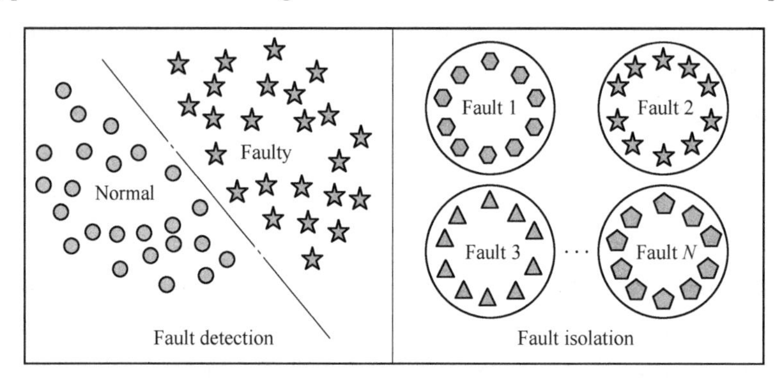

Fig. 11. 1 The difference between fault detection and fault isolation

With the rapid development of machine learning, deep neural network has shown better performance than the above methods in different fields. Several deep learning methods have been used for fault isolation, such as deep belief networks (DBN)[21], stack auto-encoders (SAE)[22], and convolutional neural networks (CNN)[23]. Zhang and Zhao[24] proposed an extensible DBN-based fault isolation method for complex chemical processes. Shao et al.[25] developed a new optimization DBN model for rolling bearing fault isolation. Wen et al.[26] developed a deep transfer learning approach based on a three-layer SAE to isolate motor bearing faults. Yu and Zhao[27] proposed a broad CNN based fault classification method with incremental learning capability for a three-phase flow (TPF) facility. Although the above deep learning based methods are powerful for fault isolation, they also have some drawbacks:

(1) The network structures are complex and have many hyperparameters, resulting in a time-consuming training process.

(2) If a new fault occurs, the models need to be retrained completely, which increases the difficulty of real-time fault isolation.

(3) Since quality variables are often difficult to be measured or are unavailable,

most of the above models, only use process variables when modeling and ignore the correlation between process and quality variables.

As a result, it is difficult to effectively deal with quality-related faults. As an effective classification method, a broad learning system (BLS) is a form of flat network that seeks to offer an alternative approach to deep structure neural networks[28]. BLS has incremental learning capability, which means the network structure can be updated without a complete retraining process for new fault samples. Generally, quality variables are usually difficult to be sampled online, and do not have real-time characteristic. As a result, quality data cannot be used as the input of BLS. This means that BLS is not suitable for quality-related fault isolation. Considering the important role of quality variables in fault isolation, PLS can extract the correlations between process variables and quality variables[29]. Meanwhile, as a useful regression algorithm, PLS can be used for soft sensor modeling and estimate the difficult-to-measure quality variables[30]. Therefore, this chapter incorporates the merits of BLS and PLS, and puts forward a dual broad partial least squares (DBPLS) with extensible capability for quality-related fault isolation in HSMP. Based on above motivations, the main contributions of this chapter are:

(1) A quality-related fault isolation framework is developed by combining BLS and PLS models. It can improve the classification performance by extracting the correlations between process variables and quality ones, and has extensible capability for new coming fault samples.

(2) A new weighted similarity index is proposed for just-in-time-learning PLS. Based on that, quality variables can be estimated online and used as the quality input of the fault isolation model.

11.1 Review of BLS and PLS

In this section, the BLS and PLS algorithms are reviewed briefly.

11.1.1 Broad learning system

BLS is an effective incremental learning system based on a random vector functional link neural network (RVFLNN) proposed by C. L. Philip Chen. The network structure of BLS is shown in **Fig. 11.2**. There are three main differences between traditional RVFLNN and BLS:

(1) RVFLNN takes inputs directly and establishes the enhancement nodes, while BLS first maps inputs to construct a set of features.

(2) The initial weights of RVFLNN are randomly generated, while those of BLS are fine-tuned using a sparse auto-encoder.

(3) BLS has incremental learning algorithms that can update the system dynamically. Considering that feedforward neural networks usually suffer from slow convergence and can be trapped in a local minimum, BLS uses the pseudo-inverse and ridge regression learning algorithms to directly obtain the output connecting weights.

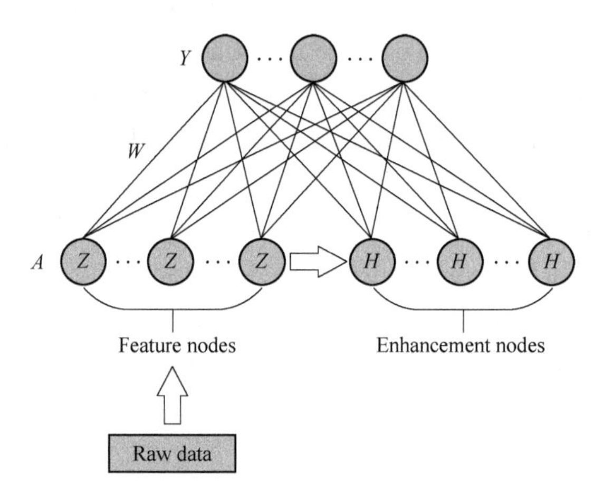

Fig. 11. 2 Basic structure of BLS

Denote the raw data as $\boldsymbol{X} \in R^{n\times m}$, where n is the number of samples, m is the number of variables, and $\boldsymbol{Y} \in R^{n\times c}$ is the output matrix, where c is the class number. $\boldsymbol{X}$ is usually preprocessed to have zero mean and unit variance. Assume that the number of feature nodes is \$b\$, the features can be represented as follows:

$$\boldsymbol{Z} = \phi(\boldsymbol{X}\boldsymbol{W}_{\mathrm{e}} + \boldsymbol{E}_{\mathrm{e}}) \tag{11.1}$$

where, $\boldsymbol{Z} \in R^{n\times b}$ is the feature matrix, while $\boldsymbol{W}_{\mathrm{e}} \in R^{m\times b}$ and $\boldsymbol{E}_{\mathrm{e}} \in R^{n\times b}$ are randomly generated. Taking advantage of a sparse auto-encoder, $\boldsymbol{W}_{\mathrm{e}}$ can be fine-tuned to obtain better features. Function $\phi(\cdot)$ has no explicit restrictions, which means that the common choices, such as kernel mappings, nonlinear transformations, or convolutional functions, are acceptable.

Similarly, assume that the number of enhancement nodes is d, the enhancement components can be represented as follows:

$$\boldsymbol{H} = \xi(\boldsymbol{Z}\boldsymbol{W}_{\mathrm{h}} + \boldsymbol{E}_{\mathrm{h}}) \tag{11.2}$$

where, $\boldsymbol{H} \in R^{n\times d}$ is the enhancement matrix; $\boldsymbol{W}_{\mathrm{h}} \in R^{b\times d}$ and $\boldsymbol{E}_{\mathrm{h}} \in R^{n\times d}$ are randomly generated; $\xi(\cdot)$ is an activation function similar to $\phi(\cdot)$. Then, $\boldsymbol{Z}$ is combined with $\boldsymbol{H}$

as the actual inputs of BLS. Hence, the BLS model is:

$$Y = [Z|H]W = AW \tag{11.3}$$

where, $A \in R^{n\times(b+d)}$ is the actual inputs of BLS; $W = [Z|H]^{+} Y = A^{+} Y$ is the connecting weight matrix and can be computed through the following ridge regression approximation:

$$A^{+} = \lim_{\lambda \to 0} (\lambda I + AA^{\mathrm{T}})^{-1} A^{\mathrm{T}} \tag{11.4}$$

where, λ is the parameter of L_2-norm in ridge regression, which is used for the bias-variance trade-off.

11.1.2 Partial least squares

In real industrial processes, the raw data X can be further divided into two groups, namely process data $X_P \in R^{n\times p}$ and quality data $X_Q \in R^{n\times q}$, where p and q are the number of process variables and quality ones, respectively. By extracting the latent variables from X_P and X_Q, PLS can not only maximize variation and correlation between X_P and X_Q, but also predict X_Q with X_P. The decomposition of X_Q with X_P can be represented as follows:

$$\begin{aligned} X_P &= TP^{\mathrm{T}} + E_P \\ X_Q &= TQ^{\mathrm{T}} + E_Q \end{aligned} \tag{11.5}$$

where, $T = [t_1, \cdots, t_a]$ is the score matrix, a is the number of latent variables, $P = [p_1, \cdots, p_a]$ and $Q = [q_1, \cdots, q_a]$ are load matrices of X_P and X_Q, E_P and E_Q are corresponding residual matrices. The above-mentioned parameters can be calculated by several methods, but the most instructive one is noniterative partial least squares (NIPALS)[30-31]. NIPALS calculates score, load matrices, and an additional set of vectors known as weights $D = [d_1, \cdots, d_a]$[32]. T cannot be directly computed from X with D, hence the regression coefficients matrix $R = [r_1, \cdots, r_a]$ is introduced by:

$$r_1 = d_1,\ r_i = \prod_{j=1}^{i-1} (I_p - d_j p_j^{\mathrm{T}}) d_i,\ i \neq 1 \tag{11.6}$$

Then, T can be calculated from X as:

$$T = XR \tag{11.7}$$

The relationship between P, R, and R can be represented as follows[33]:

$$\begin{aligned} & R = D\,(P^{\mathrm{T}}D)^{-1} \\ & \text{s.t. } P^{\mathrm{T}}R = R^{\mathrm{T}}P = D^{\mathrm{T}}D = I_a \end{aligned} \tag{11.8}$$

11.2 DBPLS-based fault isolation framework

11.2.1 Offline modeling of DBPLS

The DBPLS model is consisted of two BLS units and one PLS unit. First, the BLS units are used to extract features from the process data $\boldsymbol{X}_P$ and the quality data $\boldsymbol{X}_Q$ using the corresponding feature and enhanced nodes. Then, the PLS unit deals with the features and generates fusion nodes. Finally, all the feature nodes, enhanced nodes and fusion nodes are combined as the expanded input matrix to connect the output layer for classification. The detailed procedures for offline modeling are provided as follows:

(1) Step 1: Assume that there are N_{PF} feature nodes and N_{PE} enhanced nodes in BLS_P. They can be represented as follows:

$$\begin{aligned} \boldsymbol{Z}_{PF} &= [\boldsymbol{z}_{PF,1}, \cdots, \boldsymbol{z}_{PF,n}]^{T} \\ \boldsymbol{H}_{PE} &= [\boldsymbol{h}_{PE,1}, \cdots, \boldsymbol{h}_{PE,n}]^{T} \\ \boldsymbol{A}_P &= [\boldsymbol{Z}_{PF} | \boldsymbol{H}_{PE}] \end{aligned} \tag{11.9}$$

where, $\boldsymbol{Z}_{PF}$ is the mapped feature matrix; $\boldsymbol{H}_{PE}$ is the enhanced feature matrix; $\boldsymbol{A}_P \in R^{n\times(N_{PF}+N_{PE})}$ is the augmented feature matrix of process data $\boldsymbol{X}_P$. $\boldsymbol{Z}_{PF}$ and $\boldsymbol{H}_{PE}$ can be calculated by equations (11.1) and (11.2), respectively.

(2) Step 2: Similarly, assume that there are N_{QF} feature nodes and N_{QE} enhanced nodes in $\boldsymbol{BLS}_Q$. They can be represented as:

$$\begin{aligned} \boldsymbol{Z}_{QF} &= [\boldsymbol{z}_{QF,1}, \cdots, \boldsymbol{z}_{QF,n}]^{T} \\ \boldsymbol{H}_{QE} &= [\boldsymbol{h}_{QE,1}, \cdots, \boldsymbol{h}_{QE,n}]^{T} \\ \boldsymbol{A}_Q &= [\boldsymbol{Z}_{QF} | \boldsymbol{H}_{QE}] \end{aligned} \tag{11.10}$$

where, $\boldsymbol{Z}_{QF}$ is the mapped feature matrix; $\boldsymbol{H}_{QE}$ is the enhanced feature matrix; $\boldsymbol{A}_Q \in R^{n\times(N_{QF}+N_{QE})}$ is the augmented feature matrix of quality data $\boldsymbol{X}_Q$.

(3) Step 3: Denote the fusion input as $\boldsymbol{T}_{IN} \in R^{n\times N_a}$, where N_a is the number of latent variables. Perfom PLS operation on $\boldsymbol{A}_P$ and $\boldsymbol{A}_Q$ to obtain $\boldsymbol{T}_{IN}$ as:

$$\begin{aligned} \boldsymbol{A}_P &= \boldsymbol{T}_{IN}\boldsymbol{P}^{T} + \boldsymbol{E}_P \\ \boldsymbol{A}_Q &= \boldsymbol{T}_{IN}\boldsymbol{Q}^{T} + \boldsymbol{E}_Q \end{aligned} \tag{11.11}$$

where, $\boldsymbol{P}$ and $\boldsymbol{Q}$ are the load matrices; $\boldsymbol{E}_P$ and $\boldsymbol{E}_Q$ are the residual matrices.

(4) Step 4: Assume that the number of fusion nodes is N_F. Then, the fusion nodes can be calculated as:

$$\boldsymbol{A}_F = \phi(\boldsymbol{T}_{IN}\boldsymbol{W}_F + \boldsymbol{E}_F) \tag{11.12}$$

where, $\boldsymbol{A}_F \in R^{n\times N_F}$ is the fusion feature matrix; $\boldsymbol{W}_F$ and $\boldsymbol{E}_F$ are randomly generated;

$\phi(\cdot)$ is a nonlinear activation function.

(5) Step 5: Combine all the feature, enhanced and fusion nodes as the expanded input matrix $A_{IN} = [A_P | A_Q | A_F] \in R^{n \times N_{IN}}$, where $N_{IN} = N_{PF} + N_{PE} + N_{QF} + N_{QE} + N_F$, and calculate the connecting weight matrix W based on the optimization function as:

$$\arg\min_{W} \| A_{IN} W - Y_F \|_2^2 + \lambda \| W \|_2^2 \tag{11.13}$$

where, Y_F is the corresponding fault type matrix of A_{IN}; λ is the further constraints on the sum of the squared weights. The parameter λ can be changed to avoid the overfitting problem in DBPLS. Similar to BLS, the solution of DBPLS can be calculated as:

$$W = (\lambda I + A_{IN}^{T} A_{IN})^{-1} A_{IN}^{T} Y \tag{11.14}$$

11.2.2 JITL-PLS-based soft sensor for quality variable estimation

After offline modeling of DBPLS, online fault isolation can be implemented. However, quality variables cannot be measured efficiently online. Therefore, it is important to use soft sensor to estimate these variables. In this subsection, just-in-time-learning (JITL) strategy is integrated with PLS for creating a soft sensor. One way to improve the performance of JITL is to make the selected local dataset and the query sample as similar as possible. Because of its simplicity, Euclidean distance (ED) is one of the most popular metrics for similarity measure. However, ED enlarges the role of absolute errors in similarity measure and does not fully consider relative errors, which is not suitable for highly biased data. Thus, a new weighted similarity index is proposed to highlight the data with small absolute variations and large relative changes. Assume that a data log $\{x_{P_i}, x_{Q_i}\}_{i=1,2,\cdots,N}$ is used as the database of JITL. For a new sample $x_{P_{new}}$, the weighted similarity index between $x_{P_{new}}$ and x_{P_i} is:

$$S_i = \frac{2\sum_{j=1}^{p}(p-j+1)e^{-z_i^j}}{(p+1)p}, \quad i = 1, 2, \cdots, N \tag{11.15}$$

where, $z_i^j = \left| \frac{x_{P_{new}}^j - x_{P_i}^j}{x_{P_i}^j} \right|$ is the relative error distance of the jth variable between $x_{P_{new}}$ and x_{P_i}. To be noted, all z_i^j have been re-arranged in descending order and $\{z_i^1, \cdots, z_i^p\}$ are used for calculating S_i. Assume that l samples $\begin{cases} X_{P_l} = [x_{P_1}, \cdots, x_{P_l}]^T \\ X_{Q_l} = [x_{Q_1}, \cdots, x_{Q_l}]^T \end{cases}$ with the first l largest similarity factors are used for local modeling. Then, the JITL-PLS soft sensor can be represented as:

$$\begin{aligned} X_{P_l} &= T_L P_l^{\mathrm{T}} + E_{P_l} \\ X_{Q_l} &= T_L Q_l^{\mathrm{T}} + E_{Q_l} \end{aligned} \tag{11.16}$$

where, P_l and Q_l are load matrices; E_{P_l} and E_{Q_l} are residual matrices.

Then, through the relationship in equations (11.7) and (11.8), the prediction of $x_{Q_{\text{new}}}$ is:

$$\hat{x}_{Q_{\text{new}}} = x_{P_{\text{new}}} D_l \left(P_l^{\mathrm{T}} D_l\right)^{-1} Q_l^{\mathrm{T}} \tag{11.17}$$

where, D_l is the weight matrix.

Finally, $x_{P_{\text{new}}}$ and $\hat{x}_{Q_{\text{new}}}$ are used as the inputs to DBPLS for online fault isolation.

11.2.3 DBPLS model with extensible capability

Based on the results presented so far, **Fig. 11.3** shows the proposed quality related fault isolation framework. When process variables are sampled, quality variables will be first estimated by the soft sensor. Then, both process and quality variables are input into the DBPLS model for fault isolation. Finally, the output of DBPLS is the fault type. However, DBPLS may not have a good classification performance if new faults occur. In order to achieve a better classification performance, DBPLS should be updated to reflect such faults. Without an entire training procedure, model extensibility is proposed to update the weights readily.

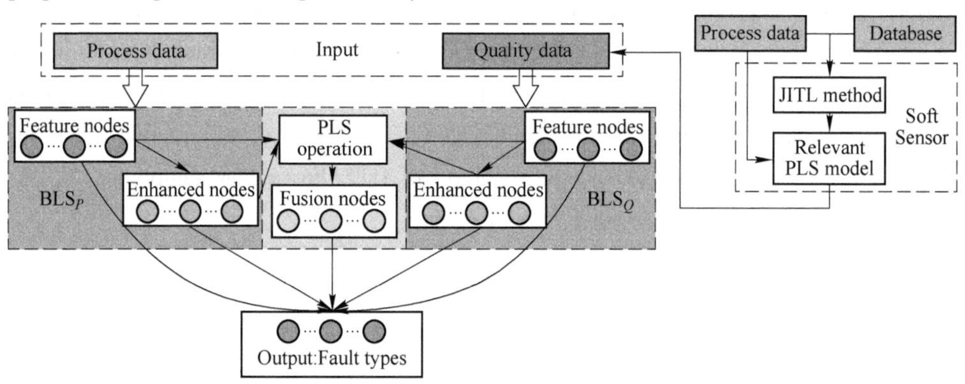

Fig. 11.3 Quality-related fault isolation framework based on DBPLS

Assume that there are C fault types in the initial training dataset, and C_{new} types of fault data are newly collected. Denote $X_{\text{new}} \in R^{n_{\text{new}} \times m}$ as the new fault data, where n_{new} is the number of samples. The increment of input matrix A_{new} can be represented as follows:

$$A_{\text{new}} = \left[A_{P_{\text{new}}} \mid A_{Q_{\text{new}}} \mid A_{F_{\text{new}}}\right] \tag{11.18}$$

where, $A_{P_{\text{new}}}$ and $A_{Q_{\text{new}}}$ are the augmented feature matrices; $A_{F_{\text{new}}}$ is the fusion feature matrix.

Then, the input matrix A_{IN} can be updated as $A_{IN_{\text{new}}} = \begin{bmatrix} A_{IN} \\ A_{\text{new}} \end{bmatrix}$. The pseudo-inverse of $A_{IN_{\text{new}}}$ is:

$$(A_{IN_{\text{new}}})^{+} = [(A_{IN})^{+} - BG^{T} | B] \tag{11.19}$$

where, $G^{T} = A_{\text{new}}(A_{IN})^{+}$, $V^{T} = A_{\text{new}} - G^{T}A_{IN}$, and if V equals to 0, $B = (A_{IN})^{+}G(I + G^{T}G)^{-1}$, otherwise $B = (V^{T})^{+}$. Hence, the updated weight matrix W_{new} can be obtained as follows:

$$\begin{aligned} W_{\text{new}} &= (A_{IN_{\text{new}}})^{+} Y_{F_{\text{new}}} \\ &= [(A_{IN})^{+} - BG^{T} | B] \begin{bmatrix} Y_{F} | \Phi \\ Y_{\text{new}} \end{bmatrix} \\ &= [W | \Phi] + (Y_{\text{new}} - A_{\text{new}}[W | \Phi])B \end{aligned} \tag{11.20}$$

where, $Y_{\text{new}} \in R^{n_{\text{new}} \times (C + C_{\text{new}})}$ is the expanded fault type matrix corresponding to the new faults; Φ represents the zero matrix.

Fig. 11.4 shows the ability of DBPLS to handle new faults.

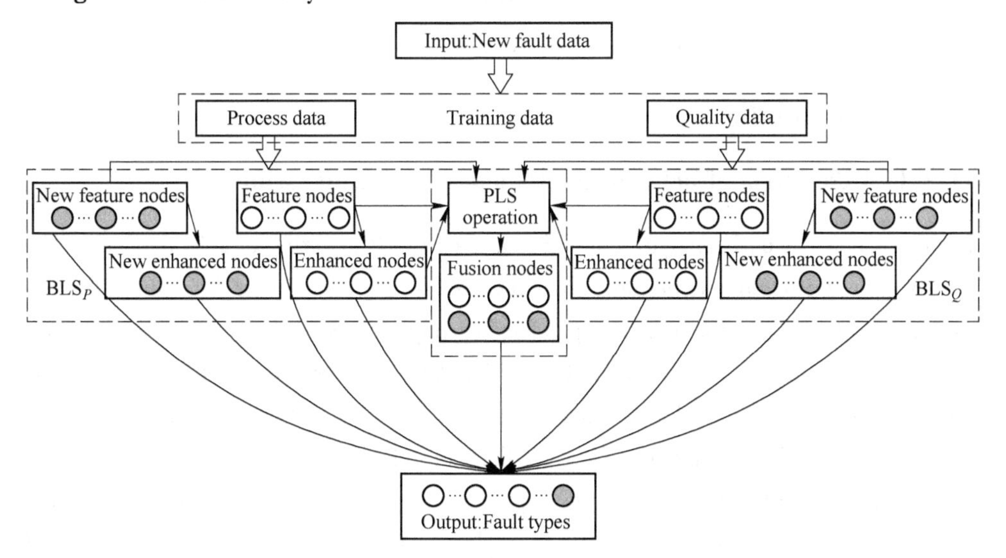

Fig. 11.4 The extensible capability of DBPLS

11.2.4 Further discussion on model optimization of DBPLS

The solution of weight matrix W is actually based on ridge regression (i.e., equation

(11.13)). However, ridge regression cannot produce a reduced-form model, for it keeps all variables in the model. If the dimension of input N_{IN} is very large, variable selection is needed to produce a sparse model. Owing to the nature of the L_1-penalty, the least absolute shrinkage and selection operator (LASSO) simultaneously shrinks and selects the variables. Compared with the existing fault samples, the sample number of new fault type may be much less in real industrial processes. In the case $N_{IN} > n$, LASSO selects at most n nodes before saturating, which seems to be a limiting feature for LASSO. As a combination of LASSO and ridge regression, the elastic net is particularly useful when N_{IN} is much larger than n. Thus, the weight matrix $\boldsymbol{W}$ of DBPLS can be calculated by the elastic net as:

$$\arg\min_{\boldsymbol{W}} \| \boldsymbol{A}_{IN}\boldsymbol{W} - \boldsymbol{Y}_{\mathrm{F}} \|_2^2 + \lambda_2 \| \boldsymbol{W} \|_2^2 + \lambda_1 \| \boldsymbol{W}_1 \tag{11.21}$$

where, $\lambda_2 \| \boldsymbol{W} \|_2^2$ is the L_2-penalty; $\lambda_1 \| \boldsymbol{W} \|_1$ is the LASSO penalty.

Let $\alpha = \dfrac{\lambda_2}{\lambda_1 + \lambda_2}$, equation (11.21) can be reformulated as:

$$\begin{aligned} &\arg\min_{\boldsymbol{W}} \| \boldsymbol{A}_{IN}\boldsymbol{W} - \boldsymbol{Y}_{\mathrm{F}} \|_2^2 \\ &\text{s. t.}\quad \alpha \| \boldsymbol{W} \|_2^2 + (1 - \alpha) \| \boldsymbol{W} \|_1 \leqslant r \text{ for some } r \end{aligned} \tag{11.22}$$

where, $\alpha \| \boldsymbol{W} \|_2^2 + (1 - \alpha) \| \boldsymbol{W} \|_1$ is the elastic net penalty.

11.3　Case study

11.3.1　Description of hot rolling CPS and experimental setup

In this section, experiments are applied to a real HSMP to verify the effectiveness of the proposed fault isolation framework. From the viewpoint of system architecture, the hot rolling manufacturing system is a typical cyber physical system (CPS). The quality-related fault isolation framework for hot rolling CPS is presented in **Fig. 11.5**. In the physical system layer, real-time data acquisition (e.g., values of rolling force) is implemented by the sensors installed on the hot rolling equipment. With the help of short-range communication technology, process data can be transmitted to wireless gateway or host computer systems. In the cyber-system layer, data from the physical system layer are first transmitted to each system server using the LAN, and then transmitted to the fault isolation platform through the Ethernet. The main tasks of the cyber-system layer include real-time monitoring, fault isolation model optimization, and knowledge base update. In the top layer, humans are the creators and operators of the physical and cyber-systems.

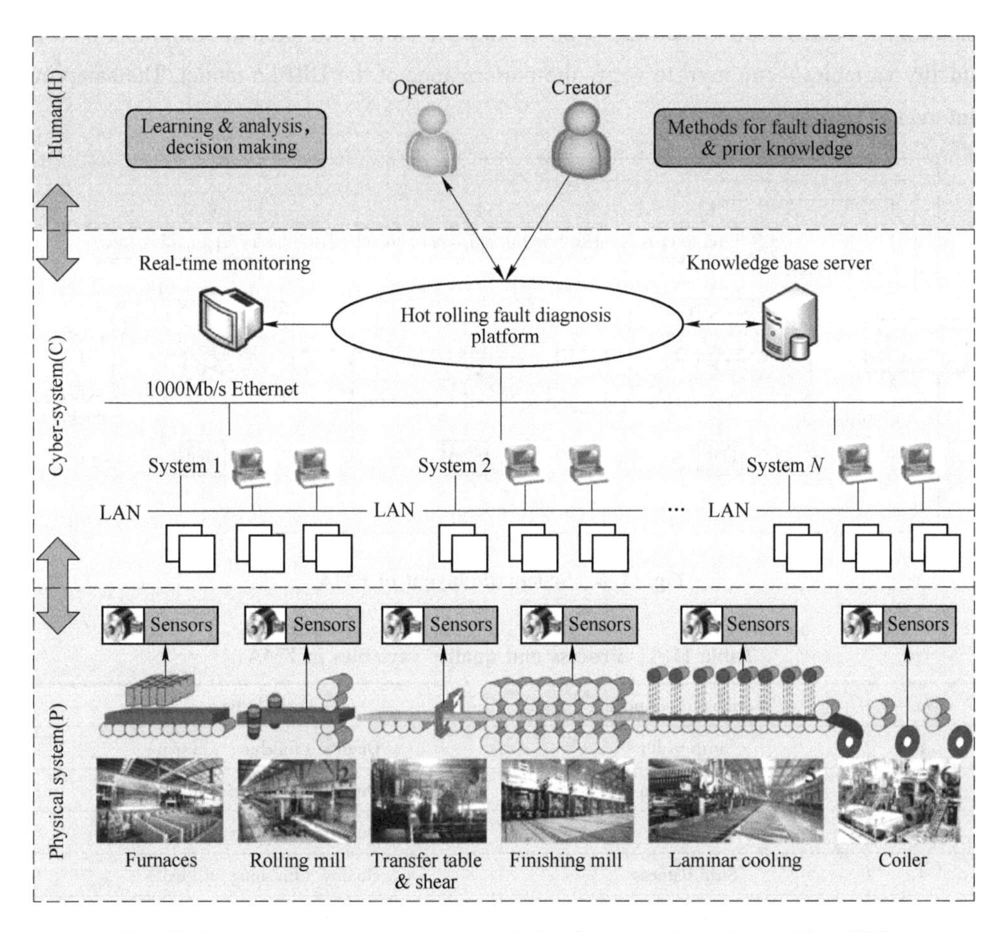

Fig. 11.5 The quality-related fault isolation framework for Hot rolling CPS

As the most important section of the HSMP, the finishing mill area (FMA) is a complex cascade subsystem consisting of seven roll stands, which is fully automated to ensure continuous production, stability, and high precision of the final products[34-35]. In each stand, a hydraulic cylinder is used for strip gauge control, and an electromechanical system rotates the rolls to make strip steels move forward smoothly, as shown in **Fig. 11.6**. The variables of the FMA are shown in **Table 11.1**, of which the first four are quality variables and the rest are process ones. Three typical quality-related faults of FMA are used in this chapter, which are given in **Table 11.2**. Fault 1 can lead to abnormal setting of the gap in the 4th stand and the output thickness can be instantly affected. Fault 2 can severely impact the strip surface temperature, which will further affect the output flatness. Fault 3 is caused by the malfunction of bending force sensor in

5th stand. After it occurs, the crown can be immediately influenced. We use 500 samples in each fault to train the DBPLS model, and 1000 testing samples (without quality variables) are used to verify the performance of the DBPLS model. The sampling interval is 0. 01 s.

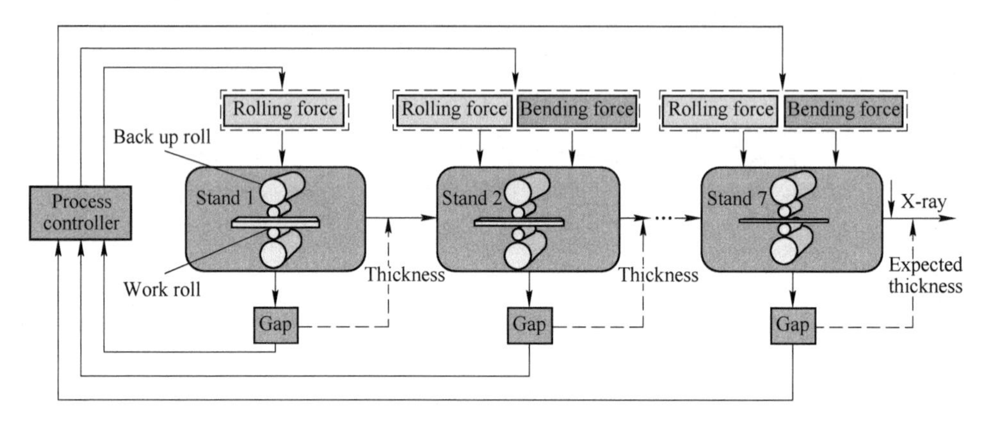

Fig. 11. 6 Schematic layout of FMA

Table 11. 1 Process and quality variables in FMA

No.	Variable description	Variable type
1	Strip width	Quality (finishing mill exit)
2	Strip crown	Quality (finishing mill exit)
3	Strip thickness	Quality (finishing mill exit)
4	Strip flatness	Quality (finishing mill exit)
5-11	Roll speeds of 7 stands	Process (each stand)
12-17	Bending forces of the last 6 stands	Process (each stand)
18-24	Work side roll gaps of 7 stands	Process (each stand)
25-31	Drive side roll gaps of 7 stands	Process (each stand)
32-38	Work side rolling force of 7 stands	Process (each stand)
39-45	Drive side rolling force of 7 stands	Process (each stand)

Table 11. 2 Typical quality-related faults in FMA

Fault No.	Fault description
1	Deterioration of the gap control loop in the 4_{th} stand
2	Actuator fault of the cooling valve between the 2_{nd} and 3_{rd} stand
3	Sensor fault of the bending force in the 5_{th} stand

11.3.2 Analysis of fault isolation results

When performing online fault isolation, soft sensor modeling is required to estimate quality variables. For comparing the proposed soft sensor with ED-based one, the root mean square error (RMSE) is defined as follow:

$$\mathrm{RMSE}=\sqrt{\frac{\sum_{i=1}^{n_{\mathrm{test}}}(\hat{\boldsymbol{x}}_{Q_i}-\boldsymbol{x}_{Q_i})^2}{n_{\mathrm{test}}}}\quad i=1,\ 2,\ \cdots,\ n_{\mathrm{test}} \tag{11.23}$$

where, n_{test} is the number of test samples, $\hat{\boldsymbol{x}}_{Q_i}$ and $\boldsymbol{x}_{Q_i}$ are the predicted and real values of the i_{th} test sample. However, this study does not focus on the estimation of quality variables. The strip thickness Q_{T} is selected for soft sensor development for simplicity.

In the modeling of JITL-PLS, two parameters should be chosen. Based on the cumulative percent variance (CPV) criterion, the number of latent variables in JITL-PLS is chosen as 11. Another key parameter is the number of samples for local modeling, which is highly related to the performance of JITL-PLS. Therefore, using different values of l from 20 to 200, the prediction results of test data are shown in **Fig. 11.7**(a). When the l value is chosen as 60, the best performance is obtained. The prediction results of these two soft sensors are shown in **Fig. 11.7** (b). In comparison with ED-based method, the results in **Fig. 11.7** demonstrate the effectiveness of the the proposed JITL-PLS in this thickness example.

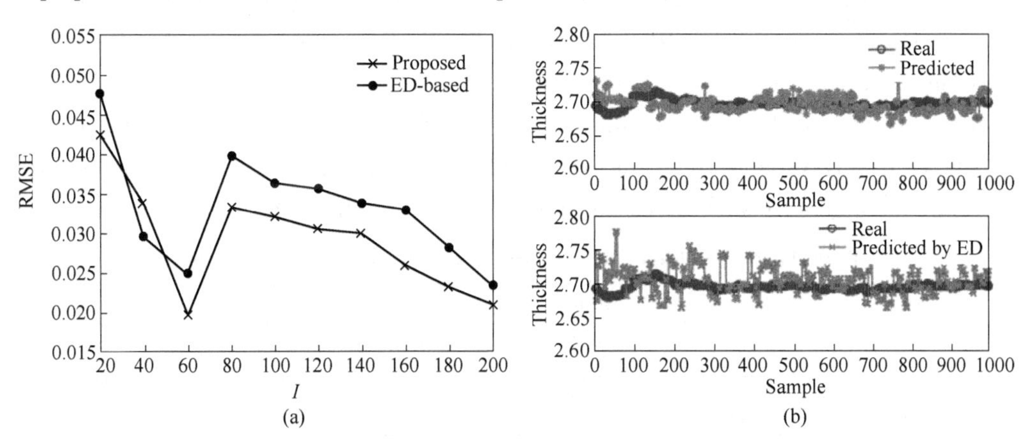

Fig. 11.7 Comparison between the proposed and ED-based

(a) RMSE under different l values; (b) Prediction results

Next, for verifying the effectiveness of the fault isolation, DBPLS is compared with BLS, XGBoost, DBN and SAE. The max depth, learning rate, and number of estimators in the XGBoost are set to be 10, 0.3, 200, respectively. The number of feature and enhancement nodes in BLS are 200 and 2000. The hidden-layer structure of DBN and SAE are both set as 50-50-40-40. For the proposed method, the number of feature nodes in $\mathbf{BLS}_P$ is 300 and 30 in $\mathbf{BLS}_Q$, while the number of enhanced nodes in BLS_P is 2000 and 200 in BLS_Q. The number of fusion nodes in DBPLS is set to be 2500, which gives an excellent classification performance for comparison. In this work, Precision, Recall, and Accuracy are used to evaluate the classification performance of different methods, which are defined as:

$$\text{Precision}_i = \frac{\text{TP}_i}{\text{TP}_i + \text{FP}_i} \tag{11.24}$$

$$\text{Recall}_i = \frac{\text{TP}_i}{\text{TP}_i + \text{FN}_i} \tag{11.25}$$

$$\text{Accuracy} = \frac{\text{samples(correctly isolated)}}{\text{total samples}} \tag{11.26}$$

where, i represents the fault type; TP_i, FP_i, and FN_i are the scale of true positive samples, false positive samples, and false negative samples, respectively.

The fault isolation results of testing samples for the quality-related faults are summarized in **Table 11.3**. From **Table 11.3**, it can be seen that the Accuracy of all methods are larger than 90%. Compared with the other four methods, DBPLS achieves the best classification performance for each quality-related fault. The Accuracy of DBPLS is 2.5 percentage points larger than that of the best one among BLS, XGBoost, DBN and SAE. The reason why DBPLS has better classification performance is that the fault isolation model not only contains process variables but also introduces quality ones. Quality variables are first estimated by the soft sensor, and then the correlations between quality variables and process variables are obtained in fusion layer to better isolate the quality-related faults. For different faults, **Fig. 11.8** shows the confusion

Table 11.3 Precision, Recall, and Accuracy of different algorithms

Index	BLS	XGBoost	DBN	SAE	DBPLS
Precision_1	0.915	0.923	0.926	0.906	0.949
Recall_1	0.952	0.955	0.957	0.959	0.984
Precision_2	0.847	0.952	0.952	0.948	0.973
Recall_2	0.895	0.908	0.891	0.902	0.932
Precision_3	0.925	0.938	0.917	0.950	0.964
Recall_3	0.938	0.948	0.945	0.941	0.969
Accuracy	0.928	0.937	0.931	0.934	0.962

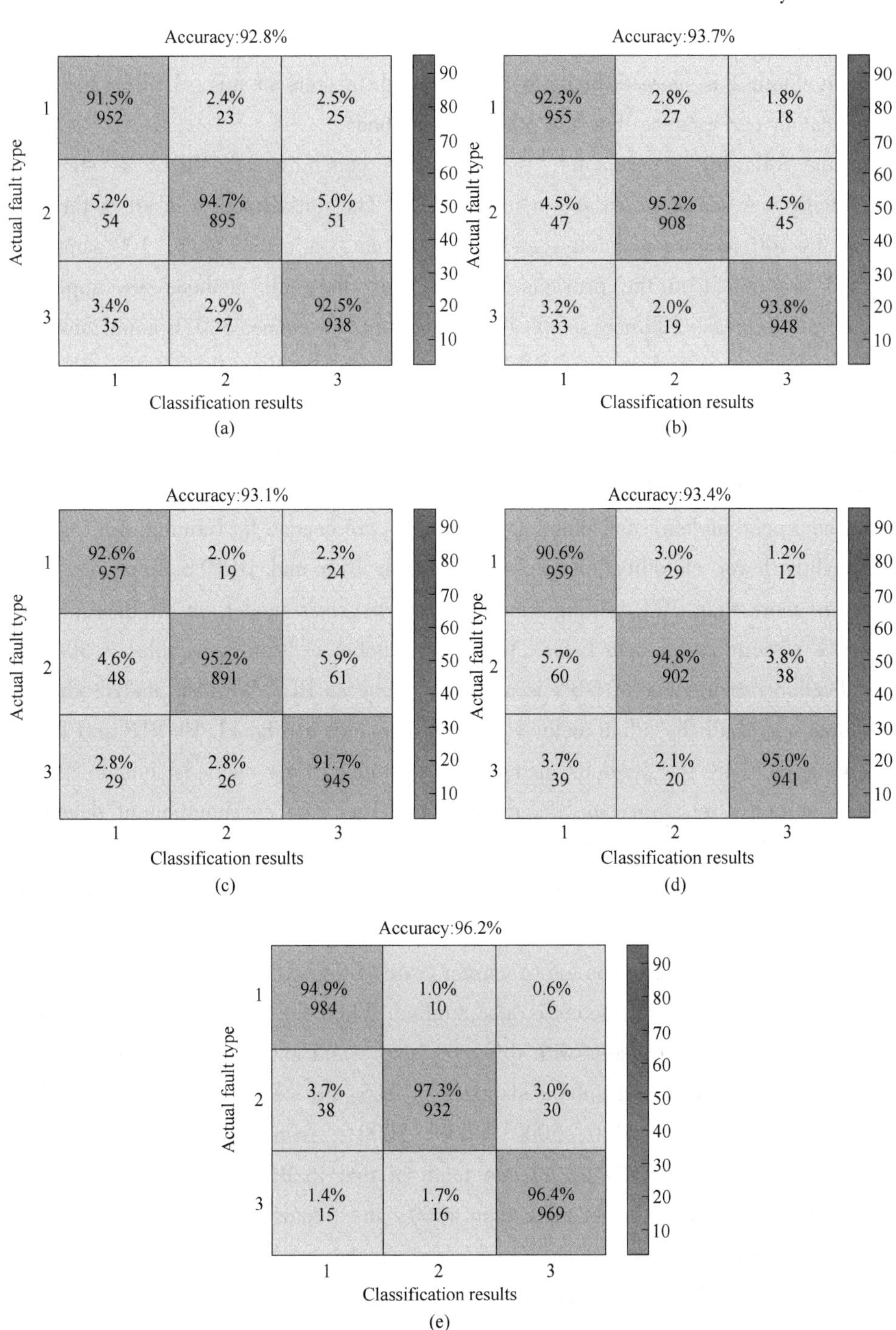

Fig. 11.8 Confusion matrices of different methods

(a) BLS; (b) XGBoost; (c) DBN; (d) SAE; (e) DBPLS

(Scan the QR code on the front of the book for color picture)

matrices of BLS, XGBoost, DBN, SAE, and DBPLS. Compared with Fault 1 and Fault 3, Fault 2 is more difficult to be identified. In spite of this, DBPLS has better classification performance than the other four methods.

In the following experiments, the extensible capability of DBPLS is shown by introducing new samples, as shown in **Fig. 11. 9**. The initial models of all methods are trained by 100 samples of Fault 1 and Fault 2. Then, as a new fault, 100 samples of Fault 3 are added to the previous models. After that, all methods are applied to dynamically increase 100 new samples each time until it reaches 500. It is obviously that the classification performance of DBPLS is always better than that of the other four methods. When there are a small number of training samples, such as 100 or 200, BLS and DBPLS perform much better than DBN and SAE. It is due to the linear optimization problems in BLS and DBPLS, which are much simpler than those in DBN and SAE based nonlinear models, and thus, fewer samples are needed for training. And XGBoost has slightly lower classification performance than BLS and DBPLS. Moreover, when there are more than 300 training samples, the diagnosis models of XGBoost, DBN, and SAE perform better than before. When the number of training samples is 500, the classification performance of DBN is about the same as BLS. We also analyze the time consumptions of all these five methods, which is shown in **Fig. 11. 10**. BLS and DBPLS can be updated by the previous model, but the other three methods include the new fault samples by retraining the diagnosis model. The time consumption of developing DBPLS model is higher than BLS and XGBoost. This is mainly because of the estimations of quality variables and the calculations of fusion nodes. However, the DBPLS model can be developed with less time than DBN and SAE.

Furthermore, the influence on Accuracy with different number of feature nodes, enhanced nodes and fusion nodes is also studied. When the number of fusion nodes is fixed, the experimental results with different feature and enhanced nodes are shown in **Fig. 11. 11**. The feature and enhanced nodes in BLS_P are set to be {100, 200, 300, 400, 500} and {1000, 2000, 3000, 4000, 5000}, respectively. The corresponding nodes in BLS_Q are normally set to one tenth of that in BLS_P. When the number of enhanced nodes in BLS_P is not more than 4000, the Accuracy is more than 90%. The influence of feature nodes and enhanced nodes on Accuracy is not significant, which shows that DBPLS has good robustness. The Accuracy reaches the maximum value when the feature and the enhanced nodes in BLS_P are 300 and 2000, respectively.

The experimental results with different fusion nodes are shown in **Fig. 11. 12**. The number of fusion nodes in DBPLS is initially set to be 500. Then, we add 500 fusion

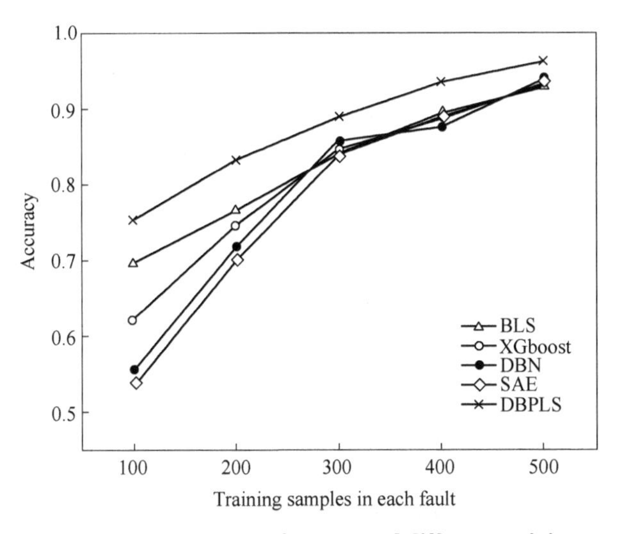

Fig. 11.9 The classification performance of different training samples

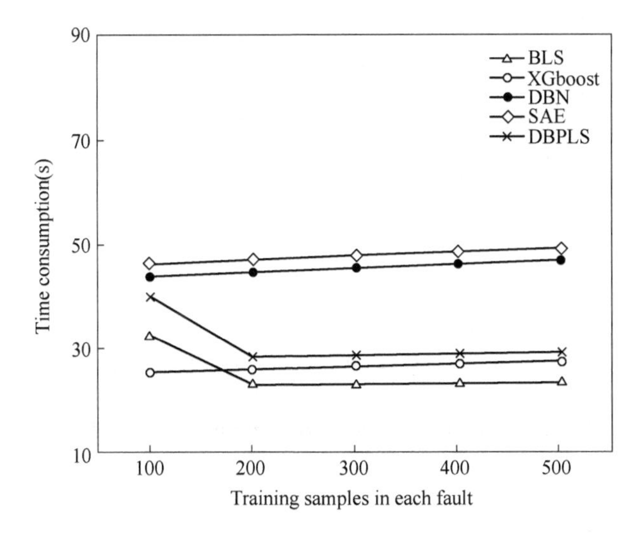

Fig. 11.10 The time consumptions of different methods

nodes to the model until it reaches 5000. The Accuracy increases at first and then decreases with an increase in the number of fusion nodes. When the number of fusion nodes is 2500, the Accuracy reaches the maximum value.

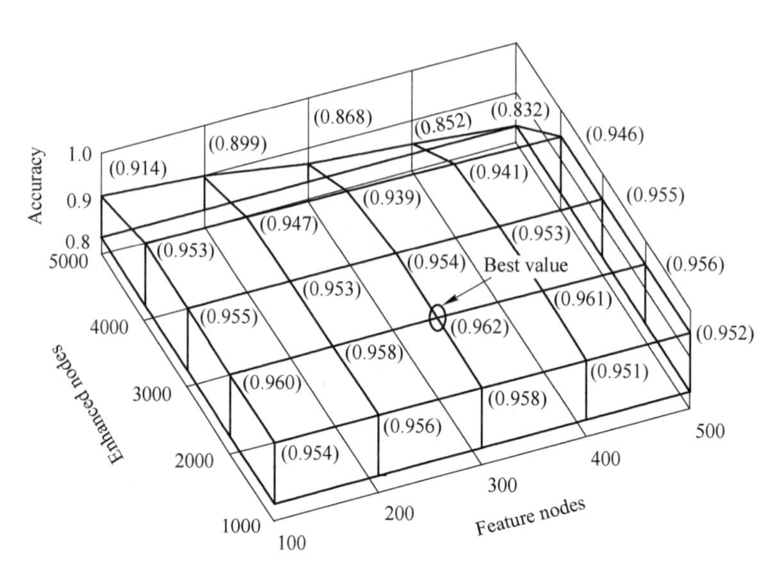

Fig. 11. 11 The Accuracy with different feature and enhanced nodes

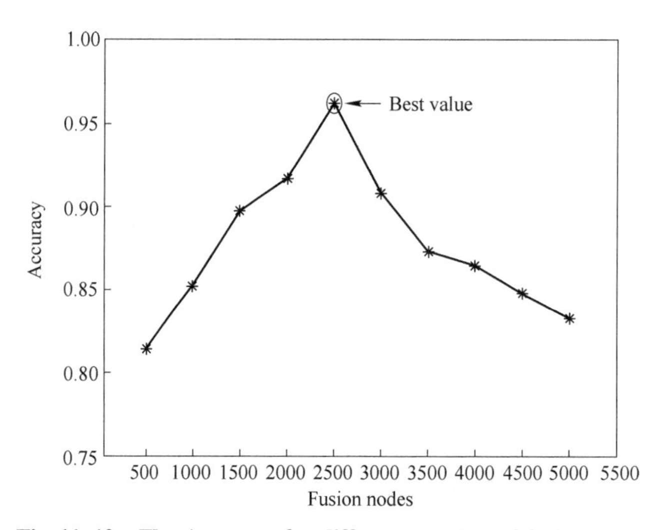

Fig. 11. 12 The Accuracy for different number of fusion nodes

11. 4 Conclusions

In this chapter, an extensible fault isolation framework based on DBPLS has been proposed focusing on the quality-related process monitoring in the hot rolling process. The proposed DBPLS jointly incorporated the well-known BLS and PLS methods. Firstly, process data and quality data were used as the input to different BLSs to obtain feature and enhanced nodes, then PLS was used to generate the fusion nodes. Finally, all these nodes were combined as the expanded input matrix to connect the output layer for quality-related fault isolation. Using JITL method, another important role of PLS in DBPLS is the soft sensor modeling of quality data to realize online fault isolation. Furthermore, owing to its extensible capability, DBPLS could be quickly remodeled to improve the classification performance using new fault samples without a retraining process. The effectiveness of the proposed framework was validated using data from a real hot rolling process. Compared with BLS, XGBoost, DBN, and SAE, DBPLS had more accurate fault classification results. Furthermore, to improve the classification performance, how to choose the numbers of different nodes are also studied.

References

[1] Askarian M, Benitez R, Graells M, et al. Data-based fault detection in chemical processes: Managing records with operator intervention and uncertain labels [J]. Expert Systems with Applications, 2016, 63: 35-48.

[2] Ge Z Q, Song Z H, Ding S X, et al. Data mining and analytics in the process industry: The role of machine learning [J]. IEEE Access, 2017, 5 (99): 20590-20616.

[3] Jiang Y C, Yin S. Recursive total principle component regression based fault detection and its application to vehicular cyber-physical systems [J]. IEEE Transactions on Industrial Informatics, 2018, 14 (4): 1415-1423.

[4] Peng K X, Zhang K, You B, et al. A quality-based nonlinear fault diagnosis framechapter focusing on industrial multimode batch processes [J]. IEEE Transactions on Industrial Electronics, 2016, 63 (3): 2615-2624.

[5] Ding S X, Yin S, Peng K X, et al. A novel scheme for key performance indicator prediction and diagnosis with application to an industrial hot strip mill [J]. IEEE Transactions on Industrial Informatics, 2013, 9 (4): 2239-2247.

[6] Ding S X. Model-based fault diagnosis techniques [M]. 2nd ed. Berlin, Heidelberg: Springer, 2013.

[7] Li L L, Ding S X, Yang Y, et al. A fault detection approach for nonlinear systems based on data-driven realizations of fuzzy kernel representations [J]. IEEE Transactions on Fuzzy Systems, 2018, 26 (4): 1800-1812.

[8] Qin S J. Survey on data-driven industrial process monitoring and diagnosis [J]. Annual Reviews in Control, 2012, 36 (2): 220-234.

[9] Yin S, Li X W, Gao H J, et al. Data-based techniques focused on modern industry: An overview [J]. IEEE Transactions on Industrial Electronics, 2015, 62 (2): 657-667.

[10] MacGregor J F, Jaeckle C, Kiparissides C, et al. Process monitoring and diagnosis by

muliblock PLS methods [J]. AIChE Journal, 1994, 40 (5): 826-838.

[11] Zhou D H, Li G, Qin S J. Total projection to latent structures for process monitoring [J]. AIChE Journal, 2010, 56 (1): 168-178.

[12] Qin S J, Zheng Y Y. Quality-relevant and process-relevant fault monitoring with concurrent projection to latent structures [J]. AIChE Journal, 2013, 59 (2): 496-504.

[13] Yin S, Zhu X, Kaynak O. Improved PLS focused on key-Performance-indicator-related fault diagnosis [J]. IEEE Transactions on Industrial Electronics, 2015, 62 (3): 1651-1658.

[14] Yu M, Wang D W, Luo M, et al. Fault detection, isolation and identification for hybrid systems with unknown mode changes and fault patterns [J]. Expert Systems with Applications, 2012, 39: 9955-9965.

[15] Chiang L H, Russell E L, Braatz R D. Fault diagnosis in chemical processes using Fisher discriminant analysis, discriminant partial least squares, and principal component analysis [J]. Chemometrics and Intelligent Laboratory Systems, 2000, 50 (2): 243-252.

[16] Zhu Z B, Song Z H. A novel fault diagnosis system using pattern classification on kernel FDA subspace [J]. Expert Systems with Applications, 2011, 38: 6895-6905.

[17] Hu Q, He Z, Zhang Z, et al. Fault diagnosis of rotating machinery based on improved wavelet package transform and SVMs ensemble [J]. Mechanical Systems and Signal Processing, 2007, 21 (2): 688-705.

[18] Chiang L H, Kotanchek M E, Kordon A K. Fault diagnosis based on Fisher discriminant analysis and support vector machines [J]. Computers and Chemical Engineering, 2004, 28 (8): 1389-1401.

[19] Perez L G, Flechsig A J, Meador J L, et al. Training an artificial neural netchapter to discriminate between magnetizing inrush and internal faults [J]. IEEE Transactions on Power Delivery, 1994, 9 (1): 431-441.

[20] Shatnawi Y, Al-Khassaweneh M. Fault diagnosis in internal combustion engines using extension neural netchapter [J]. IEEE Transactions on Industrial Electronics, 2014, 61 (3): 1434-1443.

[21] Hinton G E, Osindero S, Teh Y W. A fast learning algorithm for deep belief nets [J]. Neural Computation, 2006, 18 (7): 1527-1554.

[22] Vincent P, Larochelle H, Lajoier I, et al. Stacked denoising autoencoders: learning useful representations in a deep netchapter with a local denoising criterion [J]. Journal of Machine Learning Research, 2010, 11 (12): 3371-3408.

[23] Krizhevsky A, Sutskever I, Hinton G E. Imagenet classification with deep convolutional neural netchapters [J]. Advances in Neural Information Processing Systems, 2012, 25: 1097-1105.

[24] Zhang Z, Zhao J. A deep belief netchapter based fault diagnosis model for complex chemical processes [J]. Computers and Chemical Engineering, 2017, 107: 395-407.

[25] Shao H, Jiang H, Zhang X, et al. Rolling bearing fault diagnosis using an optimization deep belief netchapter [J]. Measurement Science and Technology, 2015, 26 (11): 115002.

[26] Wen L, Gao L, Li X Y. A new deep transfer learning based on sparse auto-encoder for fault diagnosis [J]. IEEE Transactions on Systems, Man, and Cybernetics: Systems, 2019, 46 (1): 136-144.

[27] Yu W K, Zhao C H. Broad convolutional neural netchapter based industrial process fault diagnosis with incremental learning capability [J]. IEEE Transactions on Industrial Electronics, 2020, 67 (6): 5081-5091.

[28] Chen C L P, Liu Z. Broad learning system: An effective and efficient incremental learning system without the need for deep architecture [J]. IEEE Transactions on Neural Netchapters and Learning Systems, 2018, 29 (1): 10-24.

[29] Zhang K, Dong J, Peng K X. A novel dynamic non-Gaussian approach for quality-related fault diagnosis with application to the hot strip mill process [J]. Journal of the Franklin Institute, 2017, 354 (2): 702-721.

[30] Fujiwara K, Kano M, Hasebe S, et al. Soft-sensor development using correlation-based just-in-time modeling [J]. AIChE Journal, 2009, 55 (7): 1754-1765.

[31] Wise B M, Gallagher N B. The process chemometrics approach to process monitoring and fault detection [J]. Journal of Process Control, 1996, 6 (6): 329-348.

[32] Dayal B S, MacGregor J F. Improved PLS algorithms [J]. Journal of Chemometrics, 1997, 11 (1): 73-85.

[33] Jong S D. SIMPLS: An alternative approach to partial least squares regression [J]. Chemometrics and Intelligent Laboratory Systems, 1993, 18 (3): 251-263.

[34] Ma L, Dong J, Peng K X, et al. Hierarchical monitoring and root-cause diagnosis framechapter for key performance indicator-related multiple faults in process industries [J]. IEEE Transactions on Industrial Informatics, 2019, 15 (4): 2091-2100.

[35] Ma L, Dong J, Peng K X. A novel robust semi-supervised classification framechapter for quality-related coupling faults in manufacturing industries [J]. IEEE Transactions on Industrial Informatics, 2020, 16 (5): 2946-2955.

Chapter 12 A Novel Fault Detection Method Based on the Extraction of Slow Features for Dynamic Nonstationary Processes

Process monitoring for large-scale and integrated industrial systems has become a research hotspot, in order to ensure the stable operation of modern industrial processes and the continuous growth of production benefits. With the rapid development of Industrial Internet of Things (IIoT), a large amount of data reflecting the operating status can be collected and stored. Therefore, data-driven process monitoring methods develop rapidly and have successfully applied to the process industry[1-3].

With the urgent demands of social development for multi-variety, multi-specification and high-quality products, modern process industries increasingly rely on complex industrial processes that can produce many varieties and types of products. For continuous industrial processes, the uncertainty of raw materials and external environment, the slow time-varying behavior of the equipment or the process will lead to the nonstationary characteristic. It often shows that the statistical indexes of some process variables, such as mean, variance and covariance, change with time. In addition, the batch processes also have nonstationary characteristic, because they contain multiple operation stages. For example, in the actual hot rolling process (HRP), the rolling system produces different grades of steel under different operation modes, and the strip steels are produced coil by coil. Each coil represents a batch, and this process will change significantly. For process monitoring, complex nonstationary characteristics lead to different states of the process, so it is impossible to effectively monitor the whole process with a fixed model.

As one of the most popular data-driven methods, multivariate statistical process monitoring (MSPM) is widely used in the actual industrial process. MSPM requires a little prior knowledge to establish a monitoring model under normal condition. For intercoupling and large-scale modern industries, the challenges brought by nonlinearity and dynamics are fully considered. Kernel based learning approach[4-5] and neural network[6] are mainly proposed for nonlinear problems. The expansion matrix[7] and the time series correlation analysis[8,32] are utilized to solve the problem in dynamic processes. Zhang et al. [33] proposed a modified canonical variate analysis based on dynamic kernel decomposition for dynamic nonlinear process quality monitoring. Tao et al. [34] proposed a parallel quality-related dynamic principal component regression method for chemical process monitoring. Although the process characteristics of nonlinearity and dynamics have been taken into account in fault detection, most MSPM methods assume that the operating conditions are stable and nonstationarity is not fully considered. The dynamic bias in the normal operating state is confused with a fault signal, which will lead to high false alarm rate (FAR). Meanwhile, the fault signals may be submerged by the nonstationary signals. As a result, it is difficult to extract fault features correctly, which will cause low fault detection rate (FDR).

For the fault detection of nonstationary processes, lots of work have been done in time-varying parameter models and the extraction of stationary components. Considering the time-varying characteristic of actual industrial processes, the adaptive variable time window[9-10] was used to update the detecting model. Moving window solved the problem that model parameters cannot reflect slow time-varying process characteristics and was used in nonlinear dynamic processes successfully. In addition, He et al. [11] adopted dynamic partial least squares (DPLS) to model each moving window, and distinguished each stable subsegment according to the similarity matrix. The subsegments were modeled dynamically to identify and monitor the nonstationary transition process. Time-varying parameter methods can be adopted to early failure conditions and slow time-varying conditions, but are not suitable for long-term nonstationary trends. It is another common way to deal with nonstationarity by different models and stationary subspace analysis. Berthouex et al. [12] established autoregressive integrated moving average (ARIMA) model to predict the operation state of wastewater treatment process and Sharma et al. [13] applied ARIMA to develop a forecasting model for nonstationary time series. Meanwhile, stationary subspace analysis[28] was applied to extract the global stationary features and establish a global monitoring model for the time-invariant information. However, the extraction of stationary components leads to the loss of dynamic information and fault characteristics.

In 1978, Engle and Granger[14] first proposed cointegration analysis (CA): cointegrated variables are generally unstable in their orders, but exhibit mean-reverting cointegrating relationship that force the variables to move around common stochastic trends. For several nonstationary time series of same order in a dynamic system, the linear combination of them reduce the order of the single integer and obtain a stable residual sequence. As an effective method to describe the relationship among nonstationary variables, CA combines the advantages of short-term and long-term models of time series analysis, and has been widely used in the fields of economy[15] and environmental science[16] for many years.

In the industrial processes, due to the constraints of physical and chemical conditions within the system, or the intercoupling and correlation among subsystems, the variables present the common trend of certainty and maintain the long-term dynamic equilibrium relationship. These provide the basis for the existence of cointegration relationship, and mean that the cointegration model of a residual sequence can be established to satisfy the stochastic stationary relationship. At the same time, the residual sequence represents the dynamic equilibrium error of nonstationary process. Chen[17] and Li[18] et al. proved that CA was effective in revealing the long-term equilibrium relationship among nonstationary variables in industrial processes. When the equilibrium state is broken, it means that some faults occur. Zhong et al. [19] considered the nonstationary, non-Gaussian and dynamic characteristics of processes at the same time, and adopted Bayesian inference to integrate multiple feature subspaces to obtain comprehensive fault detection results. The detecting strategy extracts multi-fault feature information effectively by the partition of variables in large-scale distributed dynamic processes. Recently, a dynamic distributed monitoring strategy[29] was proposed to separate the dynamic variations from the steady states, and it was applied to large-scale nonstationary processes monitoring under closed-loop control. Meanwhile, considering the inexact disturbance information in practice, Xue[30] and Wan[31] et al. proposed distributionally robust optimization approach, which providing a new route for nonstationarity fault detection. Lin et al. [20] applied common trend analysis based on cointegration method to extract stationary and nonstationary factors. Meanwhile they defined two statistics to respectively monitor the two factors to well describe the sequence correlation among process variables. In addition, Wu et al. [27] proposed the output-relevant common trend analysis (OCTA) method to model the input-output relationship, and achieved key performance indicators (KPIs)-related nonstationary process monitoring. To overcome the difficulty of high dimensionality and the residuals not obeying normal distribution, a modified general trend framework based on Chigiria procedure was proposed[21], and it effectively reduced FAR by the actual industrial dataset verification. Zhao et al. [22] used sparse cointegration analysis to fully decompose the different cointegration relationships of nonstationary variables. They established a two-level detecting model to comprehensively consider the interpretability of the detecting results.

Slow feature analysis (SFA)[35] has been proved its effectiveness in dynamic process monitoring[36-37] by extracting invariant features which change slowly with time. Besides, the improved SFA methods[38-39] have been studied to achieve solid feature extraction. Considering the long-term equilibrium and dynamic relationship, Zhao et al.[23] proposed a full condition monitoring strategy based on CA and SFA, and the physical meaning of the four monitoring spaces was more clear. Both CA and SFA are linear algorithms, which have poor ability to deal with nonlinear processes. Zhang et al.[24] used t-distributed stochastic neighbor embedding (t-SNE) to extract nonlinear principal components (NPC). Meanwhile, a dissimilarity index was defined for condition identification, and developed a novel nonlinear full condition process monitoring model.

Multiblock and distributed nonstationary processes monitoring strategies considering the different characteristics of variables, such as nonstationary, non-Gaussian, nonlinear, have gained significant interest. However, some methods of subspace monitoring further integrate the monitoring results, which belong to the decision level fusion. The methods of separately modeling stationary and nonstationary variables ignore the inherent causal relationship among variables. Therefore, considering feature level fusion and variable coupling, more potential feature information can be discovered. Although the corresponding algorithms designed for different characteristics improve FDR effectively, they are not satisfactory in terms of online running speed and calculation cost. In addition, in actual industrial process, the cost of false alarm cannot be ignored. Simply considering the detection rate without paying attention to FAR cannot fully reflect the effectiveness of the algorithm.

The main contributions in this work are as follows. In order to solve above problems and promote energy conservation and emission reduction, a novel exergy-related process monitoring method based on improved support tensor data description (ISTDD) is proposed for HSMP with analytics of energy flow and spatial information. The main objectives of this chapter are summarized as follows:

(1) Considering the nonstationary characteristic of the process, the variables are divided according to the unit root test of time series, and the stationary residual sequence components of nonstationary variables are extracted. And a novel fault detection framework based on feature fusion is proposed.

(2) For the extracted stationary components, SFA algorithm is used to obtain slow features. Feature level fusion is much helpful to eliminate redundant information and improve the accuracy and real-time performance of the algorithm.

(3) In view of the high FAR caused by the single training set, the online testing is

regarded as an independent process, and the K-nearest neighbor (KNN) algorithm based on Mahalanobis distance is used to set the dynamic control limits. It can improve FDR and reduce FAR effectively.

12.1 Preliminaries

12.1.1 The stationarity of univariate time series

For a time series. For a third-order tensor $\boldsymbol{X}_t$, $t \in T$, it is said to be strictly stationary, if the joint distribution $F_{t_1, t_2, \cdots, t_m}(\boldsymbol{X}_1, \boldsymbol{X}_2, \cdots, \boldsymbol{X}_m)$ is identical to that of $F_{t_1+\tau, t_2+\tau, \cdots, t_m+\tau}(\boldsymbol{X}_1, \boldsymbol{X}_2, \cdots, \boldsymbol{X}_m)$ for all τ, where m is an arbitrary positive integer and $(t_1, t_2, \cdots, t_m)$ is a collection of m positive integers.

However, it is a very strict condition to obtain the joint distribution of a time series. Therefore, strictly stationary time series is only applied in theory. In practice, a weaker version of stationary, called weakly stationary, is used.

We say $\boldsymbol{X}_t$ is weakly stationary if all the second-order characteristic statistics are time invariant. In the condition of weak stationary, we assume that the characteristics of time series are mainly determined by low order statistics. A weakly stationary time series X_t should satisfy the following conditions.

(1) For arbitrary integers t, $j \in T$ the variance is a constant, $\mathrm{Var}(\boldsymbol{X}_t) = \mathrm{Var}(\boldsymbol{X}_{t-j}) = \sigma^2$.

(2) For arbitrary integers t, $j \in T$, the variance is a constant, $E(\boldsymbol{X}_t) = E(\boldsymbol{X}_{t-j}) = \mu$.

(3) For arbitrary integerst, s, $j \in T$, the autocovariance only depends on the lag s, $\gamma(t, t-s) = \gamma(t-j, t-s-j) = \gamma(s)$.

In the actual process, due to the condition switching, disturbance, etc., the observed variables include stationary and nonstationary ones. They should be modeled by different methods to describe the process characteristics. Therefore, before modeling and monitoring, it is necessary to test the stationarity of variables.

12.1.2 Augmented Dickey-Fuller test

Unit root test is a standardized approach to test the stationarity of a time series. The common methods include ADF test, PP test, KPSS test, etc. ADF is used as an example to introduce the principle of unit root test in this paper, as it outperforms other methods in industrial process[22].

For the following p order autoregression model

$$X_t = a_1 X_{t-1} + a_2 X_{t-2} + \cdots + a_p X_{t-p} + \varepsilon_t \tag{12.1}$$

where, ε_t is the independent identical distribution innovation process with zero mean.

Replace the parameter a_i, $i = 1, \cdots, p$ and rewrite equation (12.1) as

$$\boldsymbol{X}_t = \rho \boldsymbol{X}_{t-1} + \theta_1 \Delta \boldsymbol{X}_{t-1} + \theta_2 \Delta \boldsymbol{X}_{t-2} + \cdots + \theta_{p-1} \Delta \boldsymbol{X}_{t-p+1} + \varepsilon_t \tag{12.2}$$

where, $\rho = \theta_0 = \sum_{j=1}^{F} a_j$, $\theta_i = -\sum_{j=i+1}^{p} a_j$, $i = 1, 2, 3, \cdots, p-1$, the lag order p are generally determined by Akaike Information Criterion (AIC), Δ is the difference operator.

A difference operation $\Delta X_t = X_t - X_{t-1}$, is performed on both sides of equation (12.2), and the ADF equation can be expressed as follows:

$$\Delta X_t = \alpha + \beta t + (\rho - 1)\boldsymbol{X}_{t-1} + \sum_{i=1}^{p} \theta_i \Delta \boldsymbol{X}_{t-i} + \varepsilon_t \tag{12.3}$$

where, α is the mean of time series; βt is the linear trend; ρ and θ_i are constant coefficients.

In general, the least square method is used to estimate the above parameters. If the characteristic root $\rho < 1$, the time series is stationary. Conversely, if $\rho \geqslant 1$, then there are more than one unit root. The detailed steps about ADF can be obtained in reference [18].

12.1.3 Johansen cointegration analysis

If the observed series X_t is stationary, the population variance of series is a constant, denoted $X_t \sim I(0)$. After d differencing, a series with no deterministic component which has a stationary is said to be integrated of order d, denoted $X_t \sim I(0)$.

For the multivariate nonstationary time variables with same order, the Johansen cointegration analysis can be applied to determine the cointegration relationship and estimate the cointegration vector. Cointegration formulated the phenomenon that nonstationary processes can have linear combinations that are stationary. More precisely, assuming an N-dimensional time series $\boldsymbol{X} = [\boldsymbol{x}_1, \boldsymbol{x}_2, \cdots, \boldsymbol{x}_N]$, $\boldsymbol{X} \in R^{M \times N}$ and $x_t = (\boldsymbol{x}_1, x_2, \cdots, x_N)$, $t = 1, \cdots, M$ is the time series vector at t, where M is the number of samples. If some linear combination with a vector $\boldsymbol{\beta} = (\beta_1, \beta_2, \cdots, \beta_N)^{\mathrm{T}}$, is stationary,

$$\xi_t = \beta_1 x_1 + \beta_2 x_2 + \cdots + \beta_N x_N = \boldsymbol{\beta}^{\mathrm{T}} \boldsymbol{x}_t \tag{12.4}$$

The time series $\boldsymbol{X}$ is cointegrated and ξ_t is a stationary residual sequence. The cointegrating vector $\boldsymbol{\beta}$ can be solved by following procedure.

Johansen proposed the method based on vector autoregressive (VAR) to obtain the cointegrating matrix[25]. Assuming a vector $\boldsymbol{x}_t \sim I(1)$, obtain the VAR model for decentralized data:

$$\boldsymbol{x}_t = \boldsymbol{\Pi}_1 \boldsymbol{x}_{t-1} + \cdots + \boldsymbol{\Pi}_p \boldsymbol{x}_{t-p} + \boldsymbol{\varepsilon}_t \tag{12.5}$$

where, $\boldsymbol{\Pi}_1, \cdots, \boldsymbol{\Pi}_p$ are $N \times N$ autoregressive coefficient matrices; the lag order p can be determined by AIC.

Subtract $\boldsymbol{x}_{t-1}$ from both sides of equation (12.5), and the vector error correction model (VECM) can be obtained:

$$\Delta \boldsymbol{x}_t = \boldsymbol{\Pi} \boldsymbol{x}_{t-1} + \sum_{i=1}^{q} \boldsymbol{\Gamma}_i \Delta \boldsymbol{x}_{t-i} + \boldsymbol{\varepsilon}_t \tag{12.6}$$

where, $q = p - 1$, $\boldsymbol{I} = \sum_{i=1}^{p} \boldsymbol{\Pi}_i - \boldsymbol{I}$, $\boldsymbol{\Gamma}_i = -\sum_{j=i+1}^{p} \boldsymbol{\Pi}_j$; $\boldsymbol{x}_t$ is a nonstationary variable; $\Delta \boldsymbol{x}_t$, $\boldsymbol{\varepsilon}_t$ are both stationary.

To keep the balance of stationarity, $\boldsymbol{\Pi x}_t$ should be a stationary variable, and $\boldsymbol{\Pi}$ should be a singular matrix. Decompose $\boldsymbol{\Pi}$ into the product of column full rank matrix $\boldsymbol{B} = [\boldsymbol{\beta}_1, \cdots, \boldsymbol{\beta}_r]$ and the adjustment coefficients matrix $\boldsymbol{A}$, that is $\boldsymbol{\Pi} = \boldsymbol{A}\boldsymbol{B}^{\mathrm{T}}$, where $\boldsymbol{A}, \boldsymbol{B} \in R^{N \times r}$, $0 < r < N$ and parameter r is the number of cointegrating relationship.

$$\xi_{t-1} = \boldsymbol{B}^{\mathrm{T}} \boldsymbol{x}_{t-1} = (\boldsymbol{A}^{\mathrm{T}} \boldsymbol{A})^{-1} \boldsymbol{A}^{\mathrm{T}} \left(\Delta \boldsymbol{x}_t - \sum_{i=1}^{q} \boldsymbol{\Gamma}_i \Delta \boldsymbol{x}_{t-i} - \boldsymbol{\varepsilon}_t \right) \tag{12.7}$$

The variables in the right hand side of equation (12.7) are both stationary. So is the residual sequence ξ_{t-1}. The linear combination of nonstationary variables $\boldsymbol{B}^{\mathrm{T}} \boldsymbol{x}_{t-1}$ is stationary, and each column vector of the matrix $\boldsymbol{B}$ represents a cointegration vector $\boldsymbol{\beta}_i$, $i = 1, \cdots, r$.

The maximum likelihood estimation can be used for solving $\boldsymbol{B}$. Meanwhile, it can be solved by the characteristic equation in equation (12.8)[25],

$$|\lambda \boldsymbol{S}_{11} - \boldsymbol{S}_{10} \boldsymbol{S}_{00}^{-1} \boldsymbol{S}_{01}| = 0 \tag{12.8}$$

where, $\boldsymbol{S}_{ij} = 1/M \boldsymbol{e}_i \boldsymbol{e}_j^{\mathrm{T}}$, $i, j = 0, 1$; $\boldsymbol{e}_0 = \Delta \boldsymbol{x}_t - \sum_{i=1}^{q} \boldsymbol{\Theta}_i \Delta \boldsymbol{x}_{t-i}$; $\boldsymbol{e}_1 = \boldsymbol{x}_{t-1} - \sum_{i=1}^{q} \boldsymbol{\Phi}_i \Delta \boldsymbol{x}_{t-i}$, and the coefficient $\boldsymbol{\Theta}_i$ and $\boldsymbol{\Theta}_i$ can be obtained by the least squares estimation.

The obtained eigenvalues are arranged in the order of $\lambda_1 \geqslant \lambda_2 \geqslant \cdots \geqslant \lambda_N$, and arrange the corresponding eigenvectors to form eigenvector matrix $\boldsymbol{V}$, $\boldsymbol{V} \in R^{N \times N}$. The cointegration matrix $\boldsymbol{B}$ are included in the eigenvector matrix $\boldsymbol{V}$, and the value of r is decided by the trace statistics and maximum eigenvalue statistics defined in reference [26].

Take the r-th feature vectors in matrix $\boldsymbol{V}$ as the cointegration matrix $\boldsymbol{B}$. Then, the residual sequence based on the cointegration model in equation (12.9) is stationary:

$$\xi_{ti} = \boldsymbol{\beta}_i^{\mathrm{T}} \boldsymbol{x}_t = \beta_{i1} x_1 + \beta_{i2} x_2 + \cdots + \beta_{iN} x_N, \quad i = 1, \cdots, r \tag{12.9}$$

In conclusion, for the multivariable nonstationary process with dynamic equilibrium relationship, the cointegration matrix is obtained according to the cointegration test in equation (12.8), and the stationary cointegration residual sequence can be obtained by projecting the nonstationary variables into the cointegration vector space.

12.1.4 Slow feature analysis

SFA method extracts the most slowly changing components from time-varying signals. It assumes that the local features of original sensor signal coding change rapidly, but the perceived real environment changes slowly. The components with slow changes can well reflect the characteristics of the surrounding environment.

Assuming an N-dimensional input signal $\boldsymbol{x}(t) = (x_1, x_2, \cdots, x_N)^{\mathrm{T}}$, SFA aims to find a J-dimensional nonlinear function $\boldsymbol{g}(\boldsymbol{x}) = [g_1(\boldsymbol{x}), \cdots, g_J(\boldsymbol{x})]^{\mathrm{T}}$, where each element $g_j(\boldsymbol{x})$ is the weighted sum of K nonlinear functions $h_k(\boldsymbol{x})$ and $j = 1, \cdots, J$. It can be expressed as follows:

$$g_j(\boldsymbol{x}) := \sum_{k=1}^{K} w_{jk} h_k(\boldsymbol{x}) \tag{12.10}$$

Generally speaking, $K > \max(N, J)$, with the function $\boldsymbol{h} = [h_1, \cdots, h_K]^{\mathrm{T}}$, the input signal generates a nonlinear signal $K > \max(N, J) \boldsymbol{z}_t := \boldsymbol{h}(\boldsymbol{x})$, and the nonlinear problem is turned into linear one. The weighted vector $\boldsymbol{w}_j = [w_{j1}, \cdots, w_{jK}]^{\mathrm{T}}$ is the vector to be emulated, and the output signal is:

$$y_{tj} = g_j(\boldsymbol{x}(t)) = \boldsymbol{w}_j^{\mathrm{T}} \boldsymbol{h}(\boldsymbol{x}(t)) = \boldsymbol{w}_j^{\mathrm{T}} \boldsymbol{z}_t \tag{12.11}$$

The square mean of the first derivative of a variable is used as a measurement of the rate of change to extract the slow feature variables. That is, for any j, the objective function can be rewritten as follows:

$$\min \Delta(y_j) = \langle \dot{y}_j^2 \rangle = \boldsymbol{w}_j^{\mathrm{T}} \langle \dot{\boldsymbol{z}} \dot{\boldsymbol{z}}^{\mathrm{T}} \rangle \boldsymbol{w}_j, \quad j = 1, \cdots, J \tag{12.12}$$

under the constraints

$$\langle y_j \rangle = \boldsymbol{w}_j^{\mathrm{T}} \langle \boldsymbol{z} \rangle = 0 \tag{12.13}$$

$$\langle y_j^2 \rangle = \boldsymbol{w}_j^{\mathrm{T}} \langle \boldsymbol{z} \boldsymbol{z}^{\mathrm{T}} \rangle \boldsymbol{w}_j = \boldsymbol{w}_j^{\mathrm{T}} \boldsymbol{w}_j = 1 \tag{12.14}$$

$$\forall j' < j: \langle y_{j'} y_j \rangle = \boldsymbol{w}_{j'}^{\mathrm{T}} \langle \boldsymbol{z} \boldsymbol{z}^{\mathrm{T}} \rangle \boldsymbol{w}_j = \boldsymbol{w}_{j'}^{\mathrm{T}} \boldsymbol{w}_j = 0 \tag{12.15}$$

where, $\dot{y}$ is the derivative of the feature variable, and the change rate of the j-th feature is written as $\Delta_j := \Delta(y_j) := \langle \dot{y}_j^2 \rangle$, $\langle \cdot \rangle$ denotes temporal averaging, that is $\langle y \rangle = \frac{1}{t_1 - t_0}\int_{t_0}^{t_1} y(t)\,dt$. The slow feature principal components (PCs) can be obtained by meeting the constraints above.

12.1.5 K-nearest neighbor fault detection based on Mahalanobis Distance

KNN algorithm is based on the assumption that similar samples are "close", and dissimilar samples are "far" away from each other in the feature space. It measures the similarity between the unlabeled samples and the training set labeled samples to achieve classification.

The traditional KNN algorithm is based on Euclidean distance, while Mahalanobis Distance (MD) is the Euclidean distance in the normalized principal component space. MD is not affected by the dimension. The distance between two points is unconcerned with the original unit of measurement. Considering the distribution characteristics of the data set, MD can improve the sensitivity to the differences in some dimensions, and eliminate the correlation interference between variables after coordinate transformation. Therefore, Mahalanobis Distance is employed to measure the similarity between samples in this paper.

Given a training sample $x_i \in R^N$, $i = 1, 2, \cdots, M$, the Mahalanobis distance between x_i and x_j is defined as

$$\mathrm{MD}(x_i, x_j) = \sqrt{(x_i - x_j) S^{-1} (x_i - x_j)^{\mathrm{T}}} \tag{12.16}$$

where, $i \neq j$, and S is the covariance matrix of multidimensional variables.

In the training phase, the threshold can be determined by calculating the KNN distance of each offline sample and distribution of the whole sample. In the online detecting phase, whether each online state is abnormal or not is determined by the relationship between the threshold and the samples distance of KNN.

The Mahalanobis distance of KNN is defined as:

$$D_j^2 = \sum_{i=1}^{K} \mathrm{MD}_i^2 \tag{12.17}$$

where, MD_i is the Mahalanobis distance between sample x_j and the i-th nearest neighbor. The corresponding control limit of D_j^2 is estimated by kernel density estimation (KDE) with given confidence level, generally 95%, and set the value of the point as T_α.

12.2 Nonstationary process monitoring framework based on SFA-MDKNN

In this section, the nonstationary fault detection framework based on SFA and MDKNN is proposed, and the detailed steps of designing dynamic control limits are presented.

Training dataset $\boldsymbol{X} \in R^{M \times N}$ is consisted of M samples, and each sample is an N-dimensional vector. Firstly, standardize $\boldsymbol{X}$ by Z-Score in equation (12.18), and discriminate nonstationary variables set $\boldsymbol{X}_{\mathrm{NS}} = [\boldsymbol{x}_{\mathrm{NS},1}, \boldsymbol{x}_{\mathrm{NS},2}, \cdots, \boldsymbol{x}_{\mathrm{NS},N_{\mathrm{NS}}}]$, $\boldsymbol{X}_{\mathrm{NS}} \in R^{M \times N_{\mathrm{NS}}}$, and stationary variables set $\boldsymbol{X}_{\mathrm{S}} = [\boldsymbol{x}_{\mathrm{S},1}, \boldsymbol{x}_{\mathrm{S},2}, \cdots, \boldsymbol{x}_{\mathrm{S},N_{\mathrm{S}}}]$, $\boldsymbol{X}_{\mathrm{S}} \in R^{M \times N_{\mathrm{S}}}$ by employing the ADF test, where $\boldsymbol{x}_{\mathrm{S}}$ and $\boldsymbol{x}_{\mathrm{NS}}$ are stationary/nonstationary variable.

$$x^* = \frac{x - \bar{x}}{\sigma} \tag{12.18}$$

where, x is the raw data; $\bar{x}$ is the mean of training set; σ is the standard deviation of the training set.

Secondly, modeling nonstationary variables by cointegration analysis, the residual sequence obtained by cointegration model is stationary. When a fault occurs in the process, some variables are affected and appear abnormal. At this time, the original long-term equilibrium relationship is broken. The cointegration relationship between variables no longer exists, and the residual sequence becomes unstable. The following cointegration model is established for nonstationary variables at time t:

$$\xi_t = \boldsymbol{B}^{\mathrm{T}} \boldsymbol{X}_{\mathrm{NS}}(t), \quad t = 1, \cdots, M \tag{12.19}$$

where, $\boldsymbol{X}_{\mathrm{NS}}(t)$ is the sample at time t.

Cointegration matrix $\boldsymbol{B} = (\boldsymbol{\beta}_1, \cdots, \boldsymbol{\beta}_r)$, $\boldsymbol{B} \in R^{N_{\mathrm{NS}} \times r}$ is determined by equation (12.19), and the number of cointegration vectors reserved r is given by Johansen test. The stationary residuals extracted from nonstationary variables are

$$\xi = \begin{bmatrix} \xi_{11} & \xi_{12} & \cdots & \xi_{1r} \\ \xi_{21} & \xi_{22} & \cdots & \xi_{2r} \\ \vdots & \vdots & \ddots & \vdots \\ \xi_{M1} & \xi_{M2} & \cdots & \xi_{Mr} \end{bmatrix}, \quad \xi \in R^{M \times r} \tag{12.20}$$

$$\xi_{ti} = \boldsymbol{\beta}_i^{\mathrm{T}} \boldsymbol{X}_{\mathrm{NS}}(t) = \beta_{i1} x_{t1} + \beta_{i2} x_{t2} + \cdots + \beta_{i\mathrm{NS}} x_{t\mathrm{NS}}, \quad i = 1, \cdots, r \tag{12.21}$$

The combination matrix $V_S = [x_{S,1}, x_{S,2}, \cdots, x_{S,N_S}, \xi_1, \xi_2, \cdots, \xi_r] \in R^{M\times(N_S+r)}$ is obtained by composing the stationary residual sequence and the stationary variables.

Thirdly, extract slow feature PCs from the combination matrix V_S, which is fusing the stationarity data at the feature level. Then, the front l PCs are chosen as the feature for next stage, and the parameter l is determined by the cross validation method. After extracting the slow feature, the training dataset is $Y = (y_1, y_2, \cdots, y_M)^T$, $Y \in R^{M\times l}$, and calculate the sum of Mahalanobis distances of K nearest neighbors of each training point, according to equation (12.20). In this paper $K = 7$, and the corresponding control limit can be calculated.

Fault detection based on KNN (FD-KNN) algorithm has the characteristics of nonlinearity, and its classification only depends on similar samples, which makes it be widely used in nonlinear and multimodal fault detection. However, for FD-KNN method, based on the measurement of similar samples, the fault detection index is closely related to the distribution and sparsity of data. In addition, the random change and large fluctuation of operation state can cause false detection easily. In order to improve the effect of process monitoring, we propose a dynamic threshold setting method:

(1) Step 1: Compute the sum of KNN Mahalanobis distance D_j^2 of training sample x_n in the feature space according to equation (12.20), and estimate the control limit by KDE, which is set as fixed threshold T_α.

(2) Step 2: For online sample x_i in testing period, the K nearest neighbor samples $x_{n_j} (j = 1, 2, \cdots, K)$ in the feature space of training set are determined, and compute $D^2 = \sum_{n_j=1}^{K} \mathrm{MD}^2(x_{n_j}, x_i)$, where $\{x_{n_j}, j = 1, 2, \cdots, K\} \subset \{x_n, n = 1, 2, \cdots, M\}$, x_n is the training sample, and x_i is the testing sample.

(3) Step 3: For each nearest neighbor sample x_{n_j}, search and calculate the distance of K nearest neighbors $D_{n_j}^2 = \sum_{i=1}^{K} \mathrm{MD}_i^2$ in training feature space, and the mean is calculated by $\overline{D_{n_j}^2} = \frac{1}{K}\sum_{i=1}^{K} \mathrm{MD}_i^2 = T_{\beta_t}$.

Take $K=3$ as an example, and **Fig. 12.1** is used to describe the steps above.

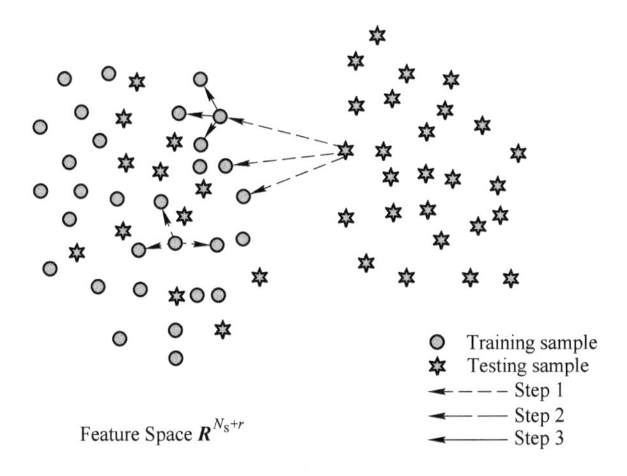

Fig. 12.1 The procedure of dynamic control limit setting (K=3)

(4) Step 4: Combine the fixed threshold T_α and dynamic threshold T_{β_t}, and the integrated control limit is $T_D^t = (T_\alpha + T_{\beta_t})/2$. Meanwhile, the fault detection logic is:

$$\begin{cases} D^2 < T_D^t & \text{normal} \\ D^2 > T_D^t & \text{fault} \end{cases} \tag{12.22}$$

For basic FD-KNN, the common threshold setting method is the coverage ratio of training samples under a given confidence. The threshold is usually in the form of the fixed value, which can not adapt to dynamic characteristics. However, whether each online sample is abnormal depends on its distance from the K nearest neighbors in the offline sample. Therefore, the detection of each online sample can be regarded as an independent process, which provides the possibility and rationality for the setting of dynamic threshold. The dynamic threshold setting method proposed above comprehensively calculates the threshold under a given confidence and the mean value of K nearest neighbors distance. The dynamic threshold of fault detection can reduce false alarm and missing alarm caused by the state changing. In addition, the entire nonstationary process monitoring framework is shown in **Fig. 12.2**.

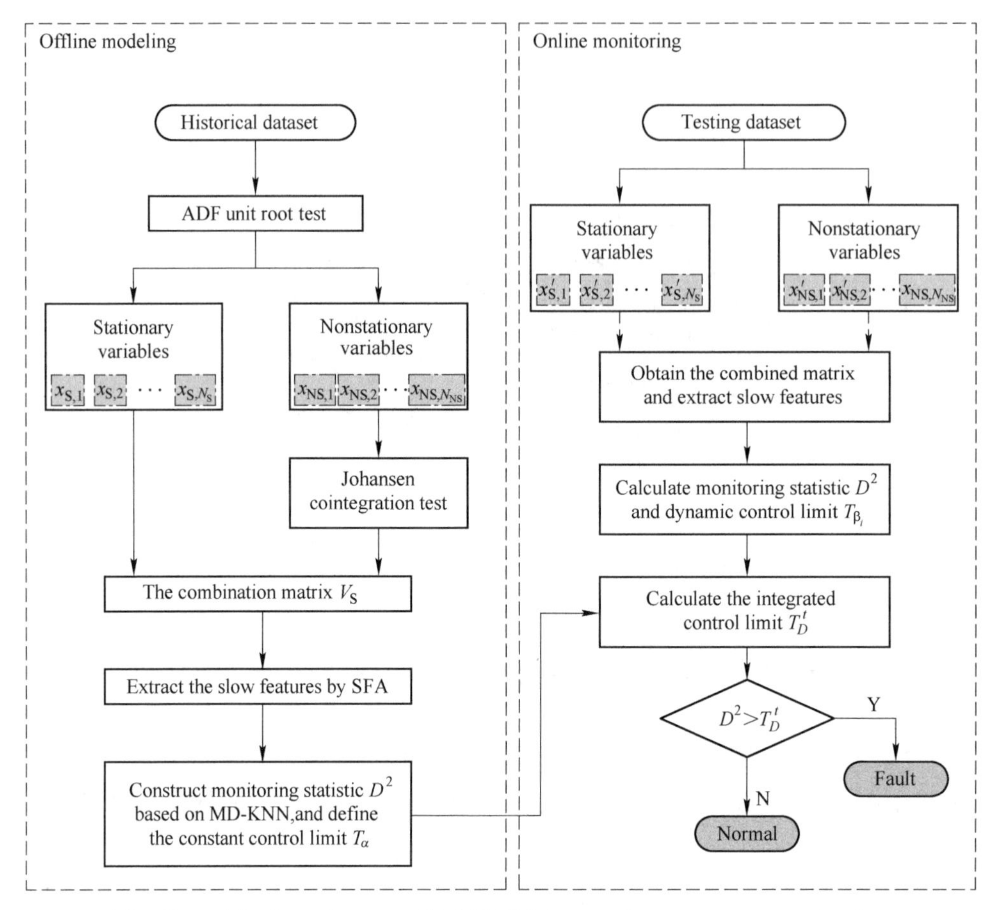

Fig. 12. 2 The flow chart of SFA-MDKNN nonstationary process monitoring

12. 3 Case study

In this section, Tennessee Eastman process, as a continuous process, and hot rolling process, as a batch one, are employed to illustrate the effectiveness of the proposed framework. Meanwhile, some nonstationary algorithms are compared with the proposed method. All experiments are completed on MATLAB r2020a platform, and the hardware.

12. 3. 1 Introduction of TE process

Tennessee Eastman process is a simulation benchmark based on an actual chemical industrial process, and is widely used to verify the proposed algorithm concerning fault diagnosis and process control. **Fig. 12. 3** is the flow chart of TE process, and there are

12 manipulation variables and 41 measurement variables (including 22 process variables and 19 component variables) in the process. In this paper, 22 process variables and 12 manipulation variables are selected. 480 samples of normal operation data are used as the training set, and 960 samples of abnormal operation data are used as the test set. Among them, the 161_{st} and above samples in test dataset are the fault samples.

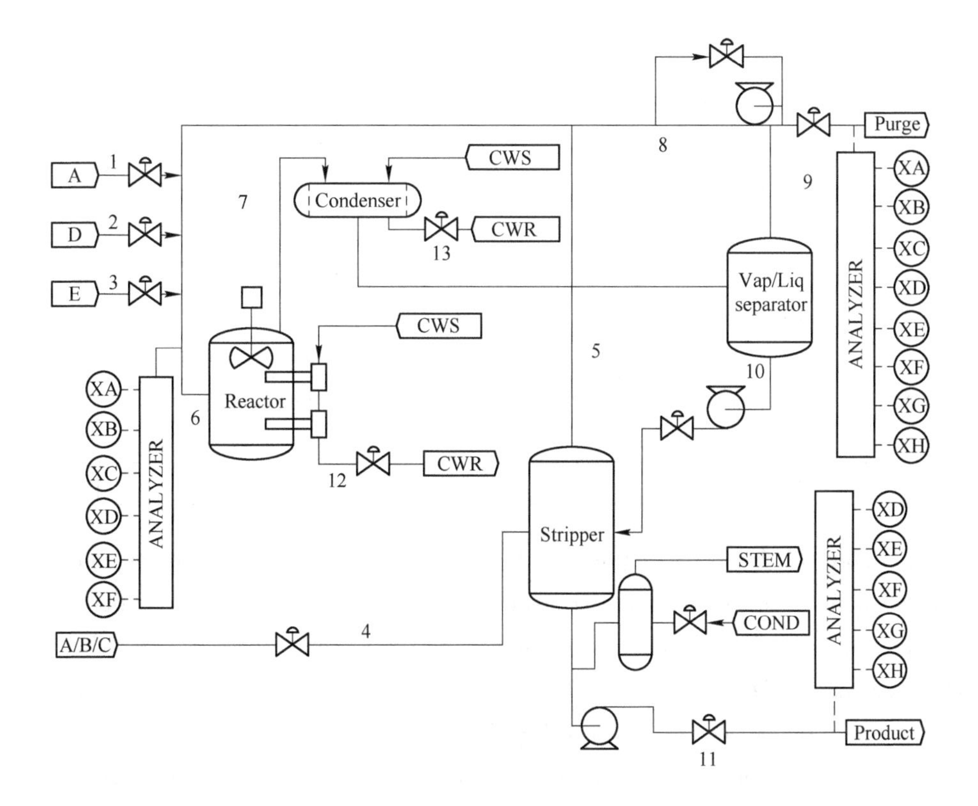

Fig. 12.3 The flowchart of TE benchmark process

12.3.2 Results and analysis of the proposed method in TE

480 original normal samples are for the unit root nonstationary test. The lag order is determined by AIC. Meanwhile the constant and linear time trend are taken into account to identify nonstationary series in TE process. The results of the stationarity of 33 variables by ADF test are shown in **Table 12.1**.

Table 12.1 Nonstationary tests for 22 process and 11 manipulation variables

No.	Description	ADF	No.	Description	ADF
1	A feed (stream 1)	1	18	stripper temperature	0
2	D feed (stream 2)	1	19	stripper steam flow	0
3	E feed (stream 3)	1	20	compressor work	0
4	Total feed (stream 4)	1	21	reactor cooling water outlet temp	1
5	Recycle flow (stream 8)	1	22	condenser cooling water outlet temp	1
6	Reactor feed rate (stream 6)	1	23	D feed flow (stream 2)	1
7	Reactor pressure	1	24	E feed flow (stream 3)	1
8	Reactor level	1	25	A feed flow (stream 1)	1
9	Reactor temperature	1	26	total feed flow (stream 4)	1
10	Purge rate (stream 9)	1	27	compressor recycle valve	1
11	separator temperature	0	28	purge valve (stream 9)	1
12	separator level	1	29	separator pot liquid flow	1
13	separator pressure	1	30	stripper liquid product flow	1
14	separator underflow	1	31	tripper steam valve	0
15	stripper level	1	32	reactor cooling water flow	1
16	stripper pressure	1	33	condenser cooling water flow	1
17	stripper underflow	1			

In **Table 12.1**, "0" represents the rejection to the hypothesis of unit root nonstationarity, and the variable is stationary, "1" represents the acception of the hypothesis, and the variable is nonstationary. From Table 12.1, it is obviously observed that variables 11, 18, 19, 20 and 31 are nonstationary, and the five nonstationary sequences are shown in **Fig. 12.4**.

Take Johansen cointegration analysis for nonstationary variables, and we find there are 2 groups of cointegration relationship, which means that $r = 2$. Normalizing the cointegration vector matrix $\boldsymbol{B} = (\boldsymbol{\beta}_1, \boldsymbol{\beta}_2) \in R^{5\times 2}$, where $\boldsymbol{\beta}_1 = [17.99\ \ -6.67\ \ 2.08\ \ 1.00\ \ -7.42]^{\mathrm{T}}$, $\boldsymbol{\beta}_2 = [-27.84\ \ 43.77\ \ 1.00\ \ -1.21\ \ -9.88]^{\mathrm{T}}$, there are 2 stationary residual sequences, as shown in **Fig. 12.5**.

The parameter of KNN is chosen as $K = 7$ according to the cross validation method, and the number of slow feature principal components is selected as $l = 16$. The confidence of kernel density estimation is 95%. FDR and FAR are important indexes to evaluate the performance of detection algorithms. In this paper, correct monitoring rate (CMR) is used, which is more effectively accurate measure of the algorithms.

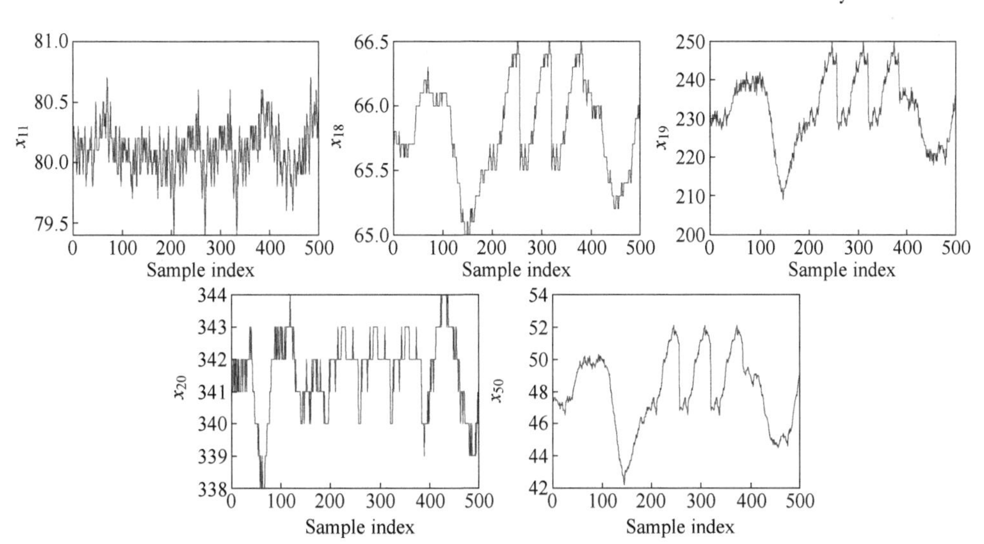

Fig. 12.4 Five nonstationary sequences tested by ADF in TE process

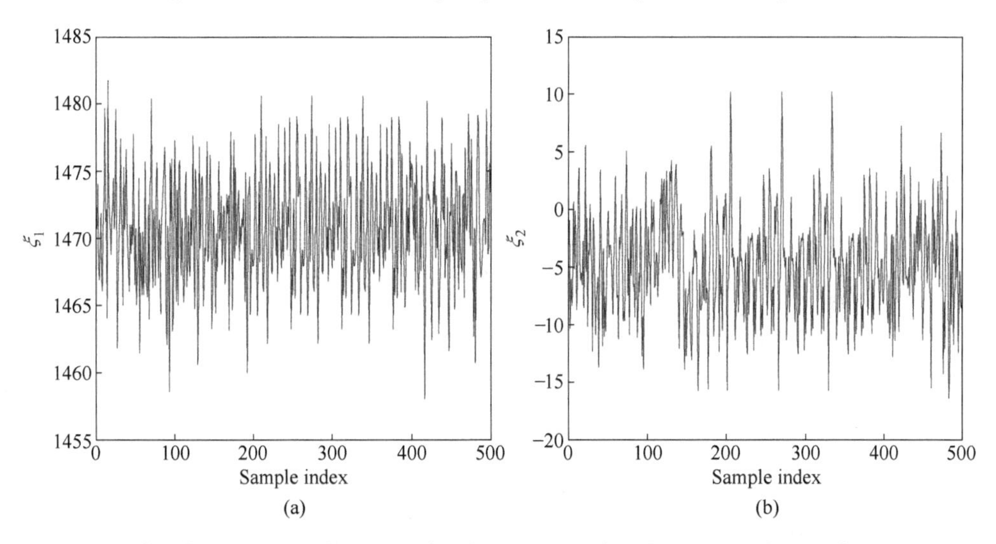

Fig. 12.5 Two stationary residual sequences after Johansen cointegration

$$\text{CMR} = \frac{\text{FDR} + (1 - \text{FAR})}{2} \tag{12.23}$$

where

$$\text{FDR} = \frac{\text{no. of samples } (D^2 > T_D^t \mid \text{faulty})}{\text{no. of samples(faulty)}} \tag{12.24}$$

$$\text{FAR} = \frac{\text{no. of samples } (D^2 > T_D^t \mid \text{normal})}{\text{no. of samples(normal)}} \tag{12.25}$$

The proposed method is compared with SFA and PCA-MDKNN methods. The monitoring results of the 18 types of faults are listed in **Table 12. 2** after removing the Faults 3,9 and 15 of TE process. Among them, SFA uses the original method based on S^2 statistics, and the number of slow feature PCs is $l_1 = 16$. PCA-MDKNN is similar to SFA-MDKNN, but PCA is used in combination matrix feature fusion, and the number of PCs is selected as $l_2 = 17$ according to cumulative variance criterion. Taking Fault 1 and Fault 4 as examples, the monitoring results are shown in **Fig. 12. 6** and **Fig. 12. 7**, respectively.

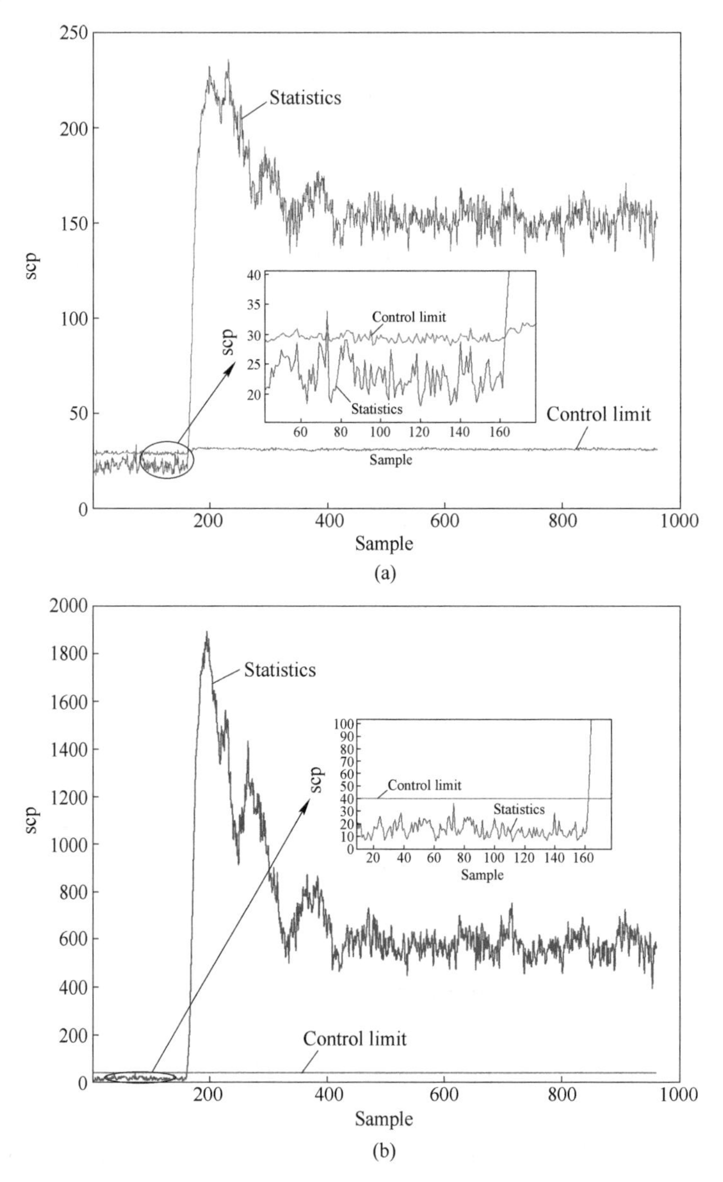

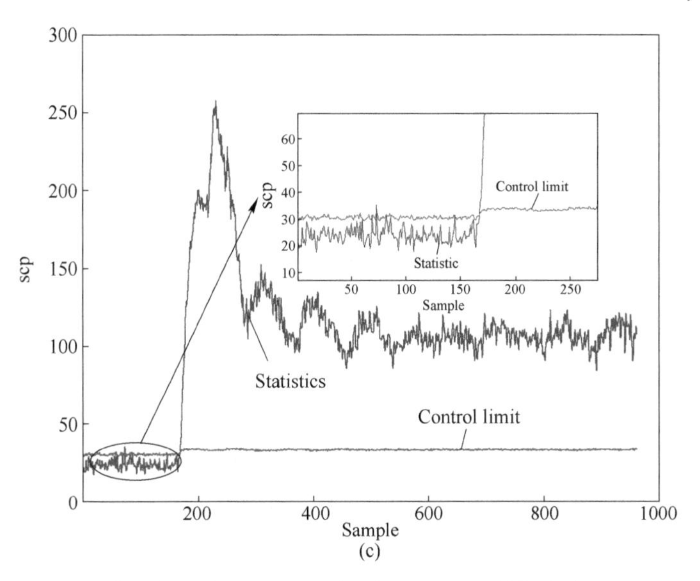

Fig. 12.6 Process monitoring results of Fault 1

(a) SFA-MDKNN; (b) SFA-S^2; (c) PCA-MDKNN

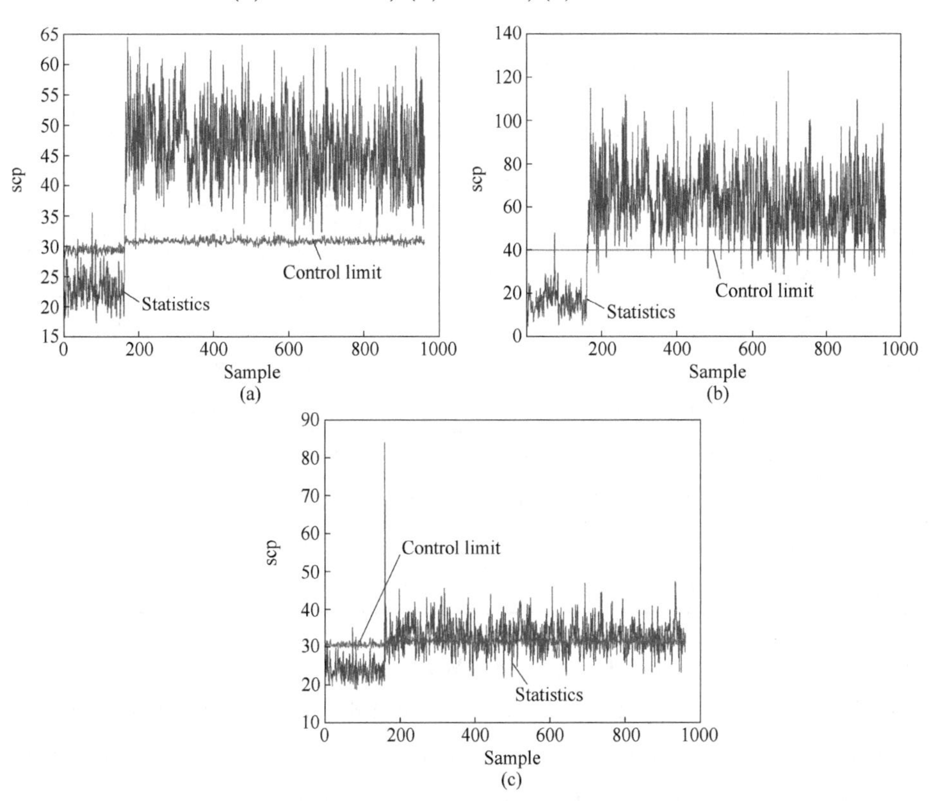

Fig. 12.7 Process monitoring results of Fault 4

(a) SFA-MDKNN; (b) SFA-S^2; (c) PCA-MDKNN

The experiments above are completed on MATLAB r2020a platform, and the hardware configuration is as follows: i5@ 1. 00GHz. For each fault training and testing, the average running time of the program is 7. 931s, which can basically meet the requirements of real-time detection.

Table 12. 2 FDRs and FARs for TE process

Fault type	SFA-MDKNN			SFA- S^2			PCA-MDKNN		
	FDR	FAR	Time (s)	FDR	FAR	Time (s)	FDR	FAR	Time (s)
Fault 1	0. 997	0. 006	8. 229	**0. 998**	**0**	0. 284	0. 991	0. 013	8. 249
Fault 2	**0. 985**	0. 006	8. 025	0. 984	**0**	0. 253	0. 984	0. 013	8. 098
Fault 4	**0. 996**	0. 019	8. 031	0. 936	**0. 006**	0. 251	0. 644	0. 013	8. 052
Fault 5	**1**	0. 019	7. 938	**1**	**0. 006**	0. 252	0. 243	0. 013	7. 895
Fault 6	**1**	**0**	8. 095	**1**	**0**	0. 269	0. 989	0. 025	7. 980
Fault 7	**1**	0. 006	8. 015	**1**	**0**	0. 248	**1**	0. 006	8. 066
Fault 8	**0. 981**	0. 006	8. 099	0. 979	**0**	0. 249	0. 968	0. 013	9. 682
Fault 10	0. 764	0. 019	7. 961	**0. 873**	**0**	0. 278	0. 561	0. 013	7. 884
Fault 11	**0. 593**	0. 013	7. 856	0. 454	**0**	0. 254	0. 580	0. 006	7. 956
Fault 12	**0. 999**	0. 031	7. 854	**0. 999**	**0**	0. 265	0. 983	0. 019	8. 061
Fault 13	**0. 951**	**0**	7. 737	0. 948	**0**	0. 254	0. 944	**0**	8. 331
Fault 14	0. 758	0. 013	7. 897	0. 504	**0**	0. 256	**1**	0. 013	8. 198
Fault 16	**0. 891**	0. 088	7. 859	0. 881	**0**	0. 268	0. 861	0. 050	8. 023
Fault 17	**0. 914**	0. 031	7. 796	0. 838	**0**	0. 255	0. 818	0. 025	7. 991
Fault 18	**0. 903**	0. 013	7. 915	0. 899	**0**	0. 249	0. 896	0. 031	8. 188
Fault 19	**0. 790**	**0**	7. 890	0. 529	**0**	0. 277	0. 394	0. 006	8. 005
Fault 20	0. 871	0. 006	7. 825	**0. 879**	**0**	0. 270	0. 373	0. 019	8. 029
Fault 21	0. 500	0. 088	7. 731	**0. 525**	**0. 006**	0. 263	0. 401	0. 025	8. 061
Average	**0. 883**	0. 020	7. 931	0. 846	**0. 001**	0. 261	0. 757	0. 017	8. 153
CMR		**0. 932**			0. 923			0. 870	

Compared with SFA, the FDR of SFA-MDKNN has generally better performance, especially in Fault 4, 11, 13, 14, 17 and 19. This indicates that the treatment of nonstationary variables and the design of threshold dynamic control limit can significantly improve the detection rate. For some faults, such as Fault 10 and 21, the detection rate of the proposed method is worse than that of the traditional SFA method. It may be caused by the disadvantage that KNN method depends on data distribution. Compared with PCA-MDKNN, the PCA method is based on static feature modeling, not describing the dynamic features like SFA method. As a result, the detection performance is generally poor. It is found that SFA-MDKNN is more accurate and valid than the other two methods from the mean of CMR for 18 types of faults. In addition, the time cost of SFA-MDKNN is lower than that of PCA-MDKNN.

12.3.3 Introduction of hot rolling process

Hot rolling is a key process of iron and steel production process. Modern hot rolling process mainly includes heating furnace, roughing mill, transfer table and shear, finishing mill, laminar cooling device, coiler and other devices. The finishing mill area (FMA) is one of the most crucial sections in HRP, which guarantees the production safety and quality stability, as shown in **Fig. 12.8**. FMA consists of 7 stands. In each stand, rolling and bending forces are offered by a hydraulic system, and the rolls are driven by an electromechanical system.

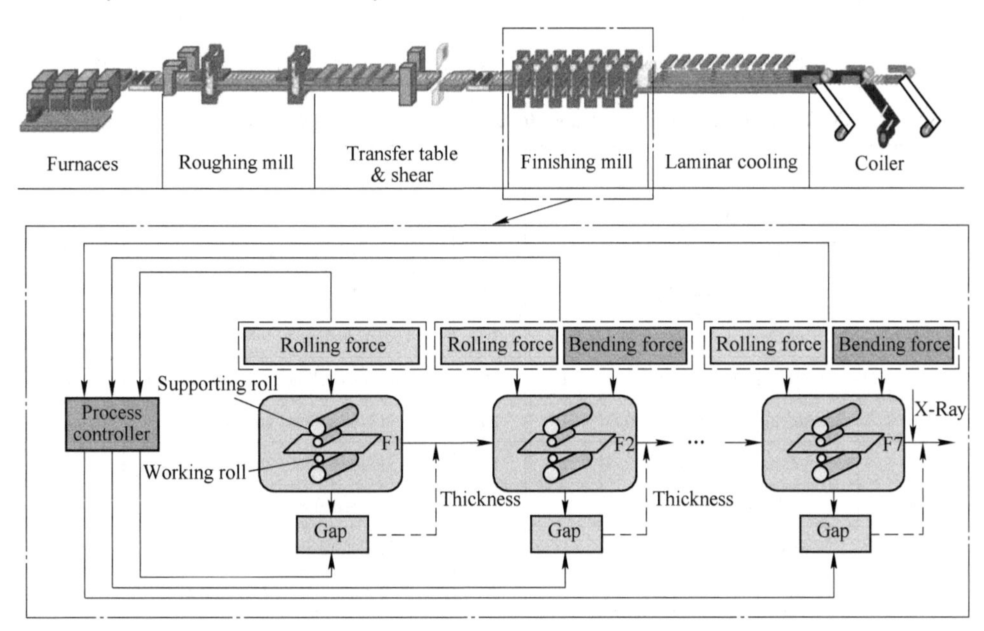

Fig. 12.8 Schematic layout of HRP

In this paper, the real data collected from HRP (1700 mm HRP production line of Anshan Iron and Steel Group Company Limited.) are used for the experimental study. 20 process variables in FMA are selected, and a typical fault is selected as the simulation verification. The sensor for measuring roll bending force of the 5_{th} stand fails, which is manifested in step jump of the measured value. 2000 samples of normal operation data are used as the training set, and 2000 samples of abnormal operation data are used as the testing set. Among them, the 1001_{st} and above samples in the test dataset are fault samples.

12.3.4 Results and analysis of the proposed method in HRP

2000 original normal samples are for the unit root nonstationary test. The results of the stationarity of 20 variables by ADF test are shown in **Table 12.3**. Obviously observed that variables 5, 6, 7, 8, 10, 14, 17 and 20 are nonstationary. Take Johansen cointegration analysis for nonstationary variables, and we find there are 3 groups of cointegration relationship, which means that $r = 3$. Normalizing the cointegration vector matrix $\boldsymbol{B} = (\boldsymbol{\beta}_1, \boldsymbol{\beta}_2, \boldsymbol{\beta}_3) \in R^{8\times 3}$,

where

$$\boldsymbol{\beta}_1 = [62.41 \quad -131.85 \quad -2.19 \quad -2.55 \quad 4.78 \quad -1.99 \quad -1.00 \quad 32.58]^{\mathrm{T}}$$

$$\boldsymbol{\beta}_2 = [-131.81 \quad 10.79 \quad -55.31 \quad 1.52 \quad -1.00 \quad -10.21 \quad -7.78 \quad -1.97]^{\mathrm{T}}$$

$$\boldsymbol{\beta}_3 = [-676.32 \quad 87.23 \quad 283.71 \quad 9.35 \quad 7.29 \quad 55.55 \quad -56.58 \quad -1.00]^{\mathrm{T}}$$

Table 12.3 Nonstationary tests for 20 process variables of HRP

No.	Description	ADF	No.	Description	ADF
1	Gap of the 1st stand	1	11	Stripper temperature	0
2	D feed (stream 2)	1	12	Stripper steam flow	0
3	E feed (stream 3)	1	13	Compressor work	0
4	Total feed (stream 4)	1	14	Reactor cooling water outlet temp	1
5	Recycle flow (stream 8)	1	15	Condenser cooling water outlet temp	1
6	Reactor feedrate (stream 6)	1	16	D feed flow (stream 2)	1
7	Reactor pressure	1	17	E feed flow (stream 3)	1
8	Reactor level	1	18	A feed flow (stream 1)	1
9	Reactor temperature	1	19	Total feed flow (stream 4)	1
10	Purge rate (stream 9)	1	20	Compressor recycle valve	1

The parameter of KNN is chosen as $K = 5$ according to the cross validation method, and the number of slow feature principal components is selected as $l = 6$. The monitoring results of sensor fault of bending force in the 5_{th} stand are listed in **Table 12.4**. Among them, SFA uses the original method based on S^2 statistics, and the number of slow feature PCs is $l_1 = 11$. PCA-MDKNN is similar to SFA-MDKNN, but PCA is used in combination matrix feature fusion, and the number of PCs is selected as $l_2 = 6$ according to cumulative variance criterion. The monitoring results of sensor fault of bending force in the 5_{th} are shown in **Fig. 12.9**, respectively.

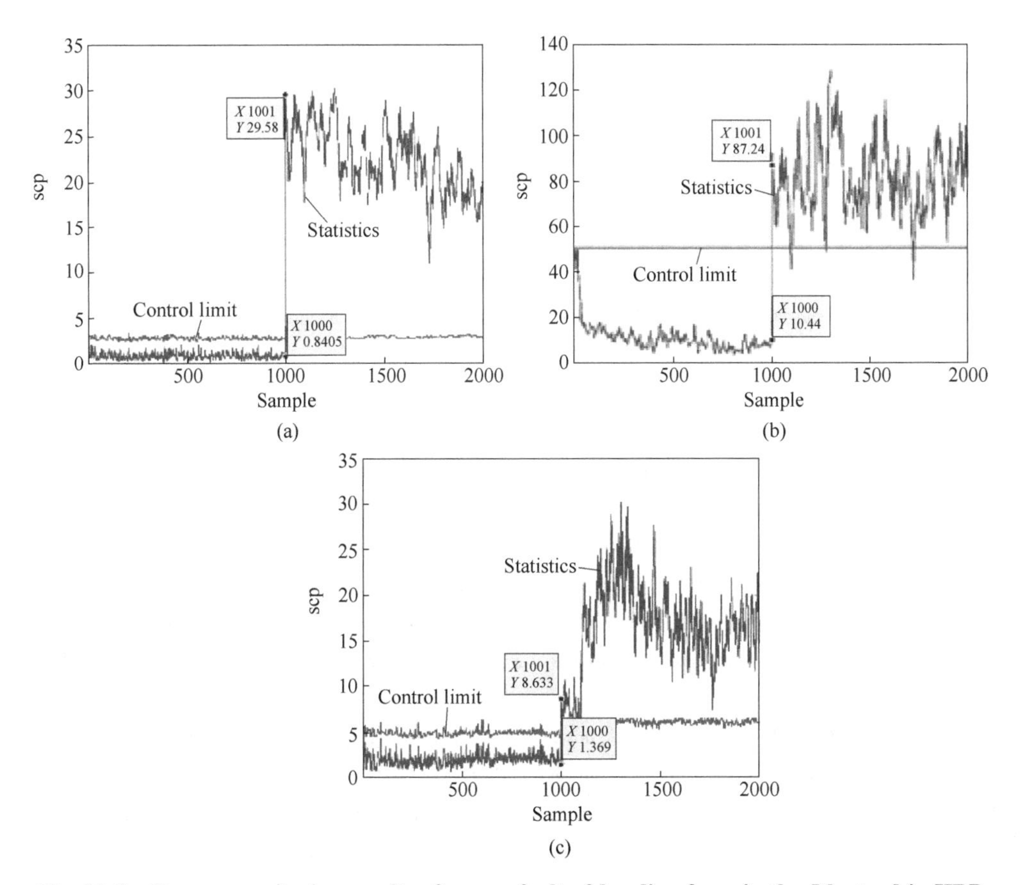

Fig. 12.9 Process monitoring results of sensor fault of bending force in the 5th stand in HRP

(a) SFA-MDKNN; (b) SFA-S^2; (c) PCA-MDKNN

As summarized in **Table 12.4**, the SFA-MDKNN method can effectively detect all abnormal samples in HRP, comparing with the other two methods. At the first 50 points when the equipment is just started, the status is nonstationary which can be seen in **Fig. 12.9** (b). The proposed method extracts the slow features of stationary residuals and stationary variables, which greatly reduce the influence of nonstationary characteristic on the detection results. Meanwhile, the dynamic control limit can adapt to the operation state, and improve CMRs.

Table 12.4 FDRs and FARs for HRP

Fault type	SFA-MDKNN			SFA- S^2			PCA-MDKNN		
	FDR	FAR	Time (s)	FDR	FAR	Time (s)	FDR	FAR	Time (s)
Fault 1	**1**	**0**	14.331	0.983	0.001	0.259	0.995	**0**	15.075
CMR	**1**			0.991			0.998		

References

[1] Qin S J. Survey on data-driven industrial process monitoring and diagnosis [J]. Annual Reviews in Control, 2012, 36 (2): 220-234.

[2] Yin S, Ding S X, Xie X C, et al. A Review on Basic Data-Driven Approaches for Industrial Process Monitoring [J]. IEEE Transactions on Industrial Electronics, 2014, 61 (11): 6418-6428.

[3] Ge Z Q, Chen J H. Plant-Wide Industrial Process Monitoring: A Distributed Modeling Framework [J]. IEEE Transactions on Industrial Informatics, 2016, 12 (1): 310-321.

[4] Choi S W, Lee C, Lee J M, et al. Fault detection and identification of nonlinear processes based on kernel PCA [J]. Chemometrics and Intelligent Laboratory Systems, 2005, 75 (1): 55-67.

[5] Rosipal R, Trejo L J. Kernel Partial Least Squares Regression in Reproducing Kernel Hilbert Space [J]. Journal of Machine Learning Research, 2002, 2: 97-123.

[6] Kämpjärvi P, Sourander M, Komulainen T, et al. Fault detection and isolation of an on-line analyzer for an ethylene cracking process [J]. Control Engineering Practice, 2008, 16 (1): 1-13.

[7] Hsu C C, Chen M C, Chen L S. A novel process monitoring approach with dynamic independent component analysis [J]. Control Engineering Practice, 2010, 18 (3): 242-253.

[8] Shang C, Huang B, Yang F, et al. Probabilistic slow feature analysis-based representation learning from massive process data for soft sensor modeling [J]. Aiche Journal, 2015, 61 (12): 4126-4139.

[9] Jaffel I, Taouali O, Harkat M F, et al. Moving window KPCA with reduced complexity for nonlinear dynamic process monitoring [J]. ISA transactions, 2016, 64: 184-192.

[10] Yao L, Ge Z Q. Online Updating Soft Sensor Modeling and Industrial Application Based on Selectively Integrated Moving Window Approach [J]. IEEE Transactions on Instrumentation and Measurement, 2017, 66 (8): 1985-1993.

[11] He Y C, Zhou L, Ge Z Q, et al. Dynamic mutual information similarity based transient process identification and fault detection [J]. Canadian Journal of Chemical Engineering, 2018, 96 (7): 1541-1558.

[12] Berthouex P M, Box G E P. Time series models for forecasting wastewater treatment plant performance [J]. Water Research, 1996, 30: 1865-1875.

[13] Sharma R R, Kumar M, Maheshwari S, et al. EVDHM-ARIMA-Based Time Series Forecasting

Model and Its Application for COVID-19 Cases [J]. IEEE Transactions on Instrumentation and Measurement, 2021, 70: 1-10.

[14] Engle R F, Granger C W J. Co-integration and error correction: representation, estimation and testing [J]. Econometrica, 1987, 55: 251-276.

[15] Arčabič V, Gelo T, Sonora R J, et al. Cointegration of electricity consumption and GDP in the presence of smooth structural changes [J]. Energy Economics, 2021, 97: 105196.

[16] Rafaqet A, Bukhsh K, Yasin M A. Impact of urbanization on CO_2 emissions in emerging economy: Evidence from Pakistan [J]. Sustainable Cities and Society, 2019, 48: 101553.

[17] Chen Q, Kruger U, Leung A Y T, et al. Cointegration Testing Method for Monitoring Nonstationary Processes [J]. Industrial & Engineering Chemistry Research, 2009, 48 (7): 3533-3543.

[18] Li G, Qin S J, Yuan T. Nonstationarity and cointegration tests for fault detection of dynamic processes [J]. IFAC Proceedings Volumes, 2014, 47 (3): 10616-10621.

[19] Zhong K, Sun X, Han M. Bayesian inference based reorganized multiple characteristics subspaces fusion strategy for dynamic process monitoring [J]. Control Engineering Practice, 2021, 112: 104816.

[20] Lin Y L, Kruger U, Chen Q. Monitoring Nonstationary Dynamic Systems Using Cointegration and Common-Trends Analysis [J]. Industrial & Engineering Chemistry Research, 2017, 56: 8895-8905.

[21] Lin Y L, Kruger U, Gu F S, et al. Monitoring nonstationary and dynamic trends for practical process fault diagnosis [J]. Control Engineering Practice, 2019, 84: 139-158.

[22] Zhao C H, Sun H, Tian F, et al. Total Variable Decomposition Based on Sparse Cointegration Analysis for Distributed Monitoring of Nonstationary Industrial Processes [J]. IEEE Transactions on Control Systems Technology, 2020, 28 (4): 1542-1549.

[23] Zhao C H, Huang B. A full-condition monitoring method for nonstationary dynamic chemical processes with cointegration and slow feature analysis [J]. Aiche Journal, 2018, 64: 1662-1681.

[24] Zhang C F, Peng K X, Dong J. A nonlinear full condition process monitoring method for hot rolling process with dynamic characteristic [J]. ISA Transactions, 2020, 112: 363-372.

[25] Johansen S. Estimation and Hypothesis Testing of Cointegration Vectors in Gaussian Vector Autoregressive Models [J]. Econometrica, 1991, 59: 1551-1580.

[26] Lin Y L, Qian C. Online non-stationary process monitoring by common trends model [J]. CIESC Journal, 2017, 68: 178-187.

[27] Wu D H, Zhou D H, Chen M Y, et al. Output-Relevant Common Trend Analysis for KPI-Related Nonstationary Process Monitoring with Applications to Thermal Power Plants [J]. IEEE Transactions on Industrial Informatics, 2021, 17 (10): 6664-6675.

[28] Yu W K, Zhao C H, Huang B, et al. Stationary Subspace Analysis-Based Hierarchical Model for Batch Processes Monitoring [J]. IEEE Transactions on Control Systems Technology, 2021,

29 (1): 444-453.

[29] Zhao C H, Sun H. Dynamic Distributed Monitoring Strategy for Large-Scale Nonstationary Processes Subject to Frequently Varying Conditions Under Closed-Loop Control [J]. IEEE Transactions on Industrial Electronics, 2019, 66: 4749-4758.

[30] Xue T, Ding S X, Zhong M Y, et al. A distribution independent data-driven design scheme of optimal dynamic fault detection systems [J]. Journal of Process Control, 2020, 95: 1-9.

[31] Wan Y M, Ma Y J, Zhong M Y. Distributionally robust trade-off design of parity relation-based fault detection systems [J]. International Journal of Robust and Nonlinear Control, 2021, 31 (18): 9149-9174.

[32] Zhang H W, Shang J, Zhang J X, et al. Nonstationary Process Monitoring for Blast Furnaces Based on Consistent Trend Feature Analysis [J]. IEEE Transactions on Control Systems Technology, 2021, 30: 1257-1267.

[33] Zhang M Q, Luo X L. Modified canonical variate analysis based on dynamic kernel decomposition for dynamic nonlinear process quality monitoring [J]. ISA transactions, 2020, 45: 106-120.

[34] Tao Y, Shi H, Tan S. Parallel quality-related dynamic principal component regression method for chemical process monitoring [J]. Journal of Process Control, 2019, 54: 33-45.

[35] Wiskott L, Terrence J S. Slow Feature Analysis: Unsupervised Learning of Invariances [J]. Neural Computation, 2002, 14: 715-770.

[36] Shang C, Yang F, Gao X Q, et al. Concurrent monitoring of operating condition deviations and process dynamics anomalies with slow feature analysis [J]. AIChE Journal, 2015, 61: 3666-3682.

[37] Shang C, Yang F, Huang B, et al. Recursive Slow Feature Analysis for Adaptive Monitoring of Industrial Processes [J]. IEEE Transactions on Industrial Electronics, 2018, 65: 8895-8905.

[38] Yu W K, Zhao C H. Recursive Exponential Slow Feature Analysis for Fine-Scale Adaptive Processes Monitoring with Comprehensive Operation Status Identification [J]. IEEE Transactions on Industrial Informatics, 2019, 15 (6): 3311-3323.

[39] Ranjith C, Huang B. Siamese Neural Network-Based Supervised Slow Feature Extraction for Soft Sensor Application [J]. IEEE Transactions on Industrial Electronics, 2021, 68 (9): 8953-8962.

Chapter 13 Distributed Quality-Related Process Monitoring Framework Using Parallel DVIB-VAE-mRMR for Large-Scale Processes

The large-scale processes usually refer to continuous manufacturing processes with multiple operation units and many variables[1]. There are many complex characteristics such as serious coupling among operation units and strong correlation between process variables[2]. To meet the growing demands on product quality, modeling and monitoring for large-scale processes are facing more challenges than those of the traditional industrial processes. Process monitoring, as an effective method to distinguish the operating status of the process system, has recently been widely concerned from industry and academy[3-4]. It is generally categorized as model-based and data-driven methods[5]. The model-based methods need precise mathematical models and are primarily built for small-scale processes. However, the large-scale processes are difficult or even impossible to conduct mechanism analysis and gain kinematic equations, which severely limits their application. Therefore, data-driven process monitoring methods are considered to be the mainstream ones in largescale processes, as they can extract meaningful information from process data and do not require expert experiences[6].

In the past few decades, multivariate statistical process monitoring (MSPM), represented by partial least squares (PLS), principal component analysis (PCA), fisher discriminant analysis (FDA) and independent component analysis (ICA), have been extensively implemented as major data-driven techniques in industrial applications[7-8]. However, the direct application of these MSPM methods for process monitoring have inherent limitations[9]. Firstly, the construction of a single monitoring model for large-scale processes causes high complexity and severely affects the monitoring performance. Secondly, a single monitoring model tends to ignore important local information and is also not conducive to the local fault location. To overcome the drawback in the traditional MSPM methods, multi-block or distributed statistical MSPM methods were proposed and applied. The large-scale processes are decomposed into different sub-blocks, each of which includes one local monitoring model[10]. Ge et al. developed the distributed PCA (DPCA). DPCA was a beginning of data-driven process monitoring methods as well[11]. Subsequently, multi-block PCA (MBPCA) and kernel PCA (KPCA) were introduced and widely applied to monitor the large-scale processes[12]. Choi and Lee introduced multi-block PLS (MBPLS) for monitoring and fault diagnosis[13]. Zhu et al. proposed a distributed parallel PCA method for process monitoring in large-scale processes[14]. Jiang and Yan designed a hierarchical monitoring model based on canonical correlation analysis method for multi-unit chemical processes[15].

These traditional distributed MSPM methods of process monitoring or fault diagnosis are used in a way that requires process data with obvious statistical rules. However, in large-scale processes, process data often coexist with complex features such as nonlinearity and coupled correlations among variables, which may affect the accuracy of monitoring and fault diagnosis. In addition, the main purpose of the mentioned above methods is to detect potential anomalies or faults. Due to the control loops, some anomalies or faults can be compensated and corrected, which do not lead to quality anomalies or faults in the large-scale industrial systems. Furthermore, enterprise managers and engineers are more concerned with anomalies or faults that cause obviously changes of product quality and productivity[16]. Therefore, quality-related process monitoring methods are more in line with the actual characteristics and production requirements of industrial scene. Early quality-related modeling methods focused on shallow learning, such as partial least squares (PLS), support vector regression (SVR), principal component regression (PCR), and Gaussian mixture model (GMM)[17]. PLS was also the first MSPM methods applied to monitor quality-related faults. Because operational data in real large-scale processes exhibit complex characteristics, effective feature extraction of operational data is one of the important challenges for quality-related monitoring and fault diagnosis.

However, shallow learning shows certain limitations as they have weak learning ability and can only extract primary features. The neural network of deep learning contains multiple hidden layers, which can extract deep features in large-scale processes[18]. Hence, quality-related modeling methods based on deep learning have become a research hotspot. Among them, autoencoder (AE), variational autoencoder (VAE), deep belief networks (DBN), restricted Boltzmann machines and convolutional neural networks (CNN) are the most commonly used methods[19]. Zhao et al. combined graph CNN with long short-term memory (LSTM) to extract the quality-related spatio-temporal features[20]. Yuan et al. used LSTM to carry out supervised quality modeling[21]. Tang et al. proposed a nonlinear quality-related fault detection method using combined deep variational information bottleneck (DVIB) and VAE[22]. Deep learning methods can extract the implied deep features by combining shallow features. Given that deep learning methods are capable of mining potentially important information by extracting deep features, the combinations of distributed frameworks and deep learning methods are more helpful to increase the performance of quality-related process monitoring in large-scale processes. Yao et al. proposed distributed parallel deep learning methods for multimodal quality prediction with big process data[23]. Rong et al. utilized the idea of community partitioning in complex networks to design large-scale supervised process monitoring based on quality indicators[24]. However, most of the quality-related deep learning methods do not take into account distributed monitoring models.

The large-scale processes consist of multiple processes in series, which are characterized by coupling among operation units and strong correlation among process variables. Thus, the quality fluctuations of the upstream operating units will be inherited to the downstream ones with serious consequences on production safety. In summary, the complex propagation mechanism of abnormal quality information flowing among sub-blocks brings a great challenge to the rational sub-block division and local modeling of the large-scale processes. Meanwhile, there are great limitations that the redundant correlations generated during the sequential transfer of variable information among sub-blocks in process monitoring and fault location. To address the above issues, a distributed process monitoring framework for large-scale processes is proposed. The main contributions in this paper are summarized as follows.

(1) A novel distributed process monitoring framework based on parallel DVIB-VAE-mRMR is proposed, where the DVIB-VAE is utilized to precisely extract the quality-related and quality-unrelated features, and DVIB-mRMR method is used to eliminate the effects of redundant information and solve the quality inheritance of sub-blocks in series.

(2) Considering that different process variables in each sub-block can cause different effects on product quality, the correlation assessment indices between process variables and quality variables are constructed using mutual information (MI) which can classify the process variables as highly, moderately and weakly correlated with quality variables.

(3) Combining the parallel DVIB-VAE-mRMR and Bayesian inference, a multi-block synergetic strategy is constructed to effectively monitor large-scale processes.

13.1 Review of VAE, mRMR, and DVIB

13.1.1 Variational autoencoder

Variational autoencoder (VAE) was first proposed in 2013 as a probabilistic model incorporating deep neural network, which can deeply extract potentially important information embedded in the process data[25]. The predictive model generates the variational probability distribution of the hidden variables by variational inference of the input data, while the generative model regains the approximate probability distribution of the input data based on the variational probability distribution of the hidden variables. Given a training dataset $\boldsymbol{L} = \{\boldsymbol{l}_i\}_{i=1}^{N}$ consisting of N input data of continuous random process variables $\boldsymbol{l}$. The input data are assumed to be independent and identically distributed (IID). The dataset of latent variables is $\boldsymbol{Z} = \{z_j\}_{j=1}^{M}$, where z denotes one of the latent variables. The outputs are the mean and variance associated to the conditional probability distribution $p_\theta(\boldsymbol{l}|z)$. The relationship between $\boldsymbol{l}$ and z can be described as multidimensional Gaussian distribution as following form,

$$p_\theta(\boldsymbol{l}|z) = N(\boldsymbol{\mu}(\boldsymbol{l}|z;\ \theta),\ \mathrm{diag}(\boldsymbol{\sigma}^2(\boldsymbol{l}|z;\ \theta))) \tag{13.1}$$

where, θ represents the parameter of the generated model; $\boldsymbol{\mu}(\boldsymbol{l}|z;\ \theta)$ and $\mathrm{diag}(\boldsymbol{\sigma}^2(\boldsymbol{l}|z;\ \theta))$ are the mean vector and covariance matrix of the Gaussian distribution, respectively.

If $p_\theta(\boldsymbol{l}|z)$ and $p(z)$ are known, the process variables $\boldsymbol{l}$ with complex distribution are calculated as

$$p(\boldsymbol{l}) = \int_z p_\theta(\boldsymbol{l}|z)p(z)\,\mathrm{d}z \tag{13.2}$$

where, the prior probability distribution $p(z)$ of z is usually set to the normal distribution, that means $p(z) = N(\boldsymbol{0},\ \boldsymbol{I})$.

Since the mapping of latent variables to process data is nonlinear, the calculation of $p_\theta(\boldsymbol{l}|z)$ in the generative model is irreversible[25], which means the corresponding $p_\theta(z|\boldsymbol{l})$ is difficult to be directly measured and computed backwards through $p_\theta(\boldsymbol{l}|z)$. Hence, the predictive model $q_\phi(z|\boldsymbol{l})$ is introduced to approximate the unavailable $p_\theta(\boldsymbol{l}|z)$ and ϕ is the parameter of the predictive model. Similar to the generative model $p_\theta(\boldsymbol{l}|z)$, the predictive model $q_\phi(z|\boldsymbol{l})$ also satisfies the multivariate Gaussian distribution. To describe the approximation between $q_\phi(z|\boldsymbol{l})$ and the actual distribution $p_\theta(z|\boldsymbol{l})$, Kullback-Leibler divergence (KL divergence, D_{KL}) is utilized for determining the dissimilarity of them. The greater the difference of $q_\phi(z|\boldsymbol{l})$ and $p_\theta(z|\boldsymbol{l})$

is, the greater the value of $D_{KL}(q_\phi(z|\boldsymbol{l}) \| p_\theta(z|\boldsymbol{l}))$ is[26]. Therefore, the objective of VAE can be obtained

$$\mathcal{X}(\theta, \phi; \boldsymbol{l}) = E_{q_\phi(z|\boldsymbol{l})}[\lg p_\theta(\boldsymbol{l}|z)] - D_{KL}[q_\phi(z|\boldsymbol{l}) \| p(z)] \tag{13.3}$$

where, $D_{KL}[q_\phi(z|\boldsymbol{l}) \| p(z)] = \frac{1}{2}\left(\sum_{j=1}^{M} \mu_j^2(\boldsymbol{l}|z; \theta) + \sigma_j^2(\boldsymbol{l}|z; \theta) - \lg\sigma_j^2(\boldsymbol{l}|z; \theta) - 1\right)$.

$\mathcal{X}(\theta, \phi; \boldsymbol{l})$ contains the expected error and KL divergence. The training objective of VAE is to minimize D_{KL} term while maximizing the expected error. After several iterations, the value of $\mathcal{X}(\theta, \phi; \boldsymbol{l})$ gradually converges.

13.1.2 Minimal redundancy maximal relevance

Due to the possible redundancy between features, the most relevant m features may not get better results than the other m features. When selecting features, we should consider the influence of two factors: redundancy among feature variables and the correlation between feature variables and target value. Considering the adverse effects of the above problems on the feature selection, mRMR algorithm was first proposed by Peng et al., which can maximize the relevance between feature variables and the target value while minimizing the redundancy between features[27]. MI is used as a criterion for measuring the relevance and redundancy in mRMR.

Both correlation and MI are indicators used to measure the relationship between random variables. In linear regression models, the correlation allows us to well analyze which factors have the greatest impact on the target variable. However, the correlation can only capture the linear relationship between variables but cannot obtain their nonlinear relationship. Therefore, MI that can effectively reflect the nonlinear relationship between variables is adopted under nonlinear conditions.

Given two random variables $\boldsymbol{l}_i$ and $\boldsymbol{l}_j$ with the probability distribution functions $p(\boldsymbol{l}_i)$ and $p(\boldsymbol{l}_j)$ respectively, the joint probability distribution function is $p(\boldsymbol{l}_i, \boldsymbol{l}_j)$. The relationship between the process variables based on MI is expressed as

$$\boldsymbol{I}(\boldsymbol{l}_i, \boldsymbol{l}_j) = \iint_{\boldsymbol{l}_i, \boldsymbol{l}_j} p(\boldsymbol{l}_i, \boldsymbol{l}_j) \lg \frac{p(\boldsymbol{l}_i, \boldsymbol{l}_j)}{p(\boldsymbol{l}_i)p(\boldsymbol{l}_j)} \mathrm{d}\boldsymbol{l}_i \mathrm{d}\boldsymbol{l}_j \tag{13.4}$$

We want to obtain the m most relevant feature variables so that the feature subset $\boldsymbol{C}$ satisfies the following equation

$$\max_{l_i, v} D(\boldsymbol{C}, v); \ D = \frac{1}{|\boldsymbol{C}|} \sum_{l_i \in C} \boldsymbol{I}(\boldsymbol{l}_i; v) \tag{13.5}$$

where, v is the target value, $\boldsymbol{I}(\boldsymbol{l}_i; v)$ is the MI of v and the best m feature variables.

$\boldsymbol{C}$ likely contains some feature variables that are highly correlated. Thus, these feature variables can cause redundancy. The redundancy is expressed as follows

$$\min_{l_i,l_j} R(\boldsymbol{C})\ ;\ R = \frac{1}{|\boldsymbol{C}|^2} \sum_{l_i,l_j \in C} \boldsymbol{I}(\boldsymbol{l}_i,\ \boldsymbol{l}_j) \tag{13.6}$$

The final goal of mRMR is to obtain the set of maximum correlation-minimum redundancy, so the optimal objective function is presented as below

$$\max G(D,\ R)\ ;\ G = D - R \tag{13.7}$$

13.1.3 Deep variational information bottleneck

DVIB is capable of obtain effective information that are relevant to the target output and remove irrelevant information from them. The principle of DVIB is to use deep neural network and variational theory to parameterize the information bottleneck (IB) model. The IB criterion aims to learn an optimal encoder $\boldsymbol{U}$ to extract feature information of the input dataset $\boldsymbol{L} = \{\boldsymbol{l}_i\}_{i=1}^{N}$, which in turn maximizes the information associated with the output dataset $\boldsymbol{T} = \{\boldsymbol{t}_i\}_{i=1}^{N}$. Its objective function is expressed as

$$\chi_{\mathrm{IB}}(\delta) = I(\delta;\ \boldsymbol{U},\ \boldsymbol{T}) - \beta I(\delta;\ \boldsymbol{L},\ \boldsymbol{U}) \tag{13.8}$$

where, $I(\boldsymbol{U},\ \boldsymbol{T})$ is the MI of $\boldsymbol{U}$ and $\boldsymbol{T}$, and $I(\boldsymbol{L},\ \boldsymbol{U})$ denotes the MI of $\boldsymbol{L}$ and $\boldsymbol{U}$, δ is the parameter to be optimized, and β is the Lagrange multiplier $I(\boldsymbol{U},\ \boldsymbol{T})$ that regulates the weights of $I(\delta;\ \boldsymbol{U},\ \boldsymbol{T})$ and $I(\delta;\ \boldsymbol{L},\ \boldsymbol{U})$.

To achieve the deep extraction of the features associated with $\boldsymbol{L}$ and $\boldsymbol{T}$, DVIB designs the encoder $q_{\phi\mathrm{D}}(\boldsymbol{u}\,|\,\boldsymbol{t})$ and decoder $p_{\theta\mathrm{D}}(\boldsymbol{t}\,|\,\boldsymbol{u})$, where $p_{\theta\mathrm{D}}(\boldsymbol{t}\,|\,\boldsymbol{u})$ is used to approximate the true probability distribution $p_{\mathrm{D}}(\boldsymbol{t}\,|\,\boldsymbol{u})$. Then, the variational low bound of $I(\boldsymbol{U},\ \boldsymbol{T})$ is obtained

$$\boldsymbol{I}(\boldsymbol{U},\boldsymbol{T}) \geqslant \iint_{\boldsymbol{T},\boldsymbol{U}} p_{\mathrm{D}}(\boldsymbol{l},\ \boldsymbol{t}) E_{q_{\phi\mathrm{D}}(\boldsymbol{u}|\boldsymbol{t})}[\lg p_{\theta\mathrm{D}}(\boldsymbol{t}\,|\,\boldsymbol{u})]\,\mathrm{d}\boldsymbol{l}\mathrm{d}\boldsymbol{t} \tag{13.9}$$

Accordingly, the priori hypothetical distribution $r(\boldsymbol{u})$ is used to estimate the true distribution $p_{\mathrm{D}}(\boldsymbol{u})$, then the variational upper bound on $\boldsymbol{I}(\boldsymbol{L},\ \boldsymbol{U})$ can be given as below,

$$\boldsymbol{I}(\boldsymbol{L},\ \boldsymbol{T}) \geqslant \int_{l} p_{\mathrm{D}}(\boldsymbol{l}) D_{\mathrm{KL}}[q_{\phi\mathrm{D}}(\boldsymbol{u} \mid \boldsymbol{l}) \parallel r(\boldsymbol{u})]\,\mathrm{d}\boldsymbol{l} \tag{13.10}$$

The variational forms of $I(\boldsymbol{U},\ \boldsymbol{T})$ and $I(\boldsymbol{L},\ \boldsymbol{U})$ are combined and the distribution is approximated in practice. Then, the optimization function of DVIB can be approximated as

$$\chi_{\mathrm{DVIB}}(\theta,\ \phi;\ \boldsymbol{L},\ \boldsymbol{T}) = I(\theta,\ \phi;\ \boldsymbol{U},\ \boldsymbol{T}) - \beta I(\theta,\ \phi;\ \boldsymbol{L},\ \boldsymbol{U}) \tag{13.11}$$

More details about DVIB is presented in reference [28] .

13.2 Processes monitoring based on the parallel DVIB-VAE-mRMR

13.2.1 Process decomposition and variables selection

The large-scale processes usually consist of multiple interacting operation units with complex process characteristics. To improve the local monitoring effect and decrease the computational complexity, the large-scale processes are decomposed into different sub-blocks. An effective process decomposition should fully consider the physical constraints and the topology of operation units, as well as the strong coupling and quality inheritance in different operation units. The large-scale processes can be decomposed into B sub-blocks by the product processing order and spatial location. For example, if there are 7 operation units $A \to B \to C \to D \to E \to F \to G$ in a process, they can be divided into upstream, midstream and downstream, i.e. $A \to B \to C$ (sub-block 1), $D \to E$ (sub-block 2), $F \to G$ (sub-block 3) . On this basis, the coupling between the sub-blocks is reflected by the sequential transfer of quality information.

In addition to process decomposition, variables selection and segmentation within each sub-block are crucial for potential feature extraction and local model construction. Firstly, the variables and quality indicators are standardized for eliminating the effect of different magnitudes. Assume that the process variables are represented as $\boldsymbol{X} = [\boldsymbol{X}_1, \boldsymbol{X}_2, \cdots, \boldsymbol{X}_B] \in \mathrm{R}^{N \times M}$, where N is the number of samples, and M is the dimension of process variables, $\boldsymbol{X}_b \in \mathrm{R}^{N \times M_b} (b = 1, 2, \cdots, B)$ is the variable groups in the sub-block b and M_b represents the number of variables in the sub-block b and $\sum_{b=1}^{B} M_b = M$. Let $\boldsymbol{X}_{bj} (j = 1, 2, \cdots, M_b)$ be an arbitrary one-dimensional process variables in the sub-block b , $\boldsymbol{Y} \in \mathrm{R}^{N \times r}$ denote the variable groups of quality indicators, where r is the dimensions of the quality variables, and $\boldsymbol{y}_i (i = 1, 2, \cdots, r)$ are an arbitrary one-dimensional quality variables.

The MI with the random variable $\boldsymbol{x}_{bj}$ in sub-block b and quality variables $\boldsymbol{y}_i$ is expressed as

$$MI_b(\boldsymbol{x}_{bj}, \boldsymbol{y}_i) = \iint_{\boldsymbol{x}_{bj}, \boldsymbol{y}_i} p(\boldsymbol{x}_{bj}, \boldsymbol{y}_i) \lg \frac{p(\boldsymbol{x}_{bj}, \boldsymbol{y}_i)}{p(\boldsymbol{x}_{bj}) p(\boldsymbol{y}_i)} \mathrm{d}\boldsymbol{x}_{bj} \mathrm{d}\boldsymbol{y}_i \tag{13.12}$$

In each sub-block, different variables hold different effects on the quality variables. The thresholds α_b and $\beta_b (0 < \alpha_b < \beta_b < 1)$ are set to evaluate these relationships of process variables and quality ones as follows,

Rank 1: if $\beta_b \leqslant MI_b(\boldsymbol{x}_{bj}, \boldsymbol{y}_i) < 1$, then $\boldsymbol{x}_{bj}$ is highly correlated with quality variables;

Rank 2: if $\alpha_b < MI_b(\boldsymbol{x}_{bj}, \boldsymbol{y}_i) < \beta_b$, then $\boldsymbol{x}_{bj}$ is moderately correlated with quality variables;

Rank 3: if $0 < MI_b(\boldsymbol{x}_{bj}, \boldsymbol{y}_i) \leqslant \alpha_b$, then $\boldsymbol{x}_{bj}$ is weakly correlated with quality variables;

The variables meet Rank 1 and Rank 2 are combined to build a quality-related dataset $\boldsymbol{x}_{b\mathrm{h}}(b = 1, 2, \cdots, B)$. The variables meet Rank 2 and Rank 3 are combined to build a quality-unrelated dataset $\boldsymbol{x}_{b\mathrm{w}}(b = 1, 2, \cdots, B)$.

13.2.2 Parallel DVIB-VAE-mRMR model based on quality inheritance

The complex correlations between process variables in large-scale processes require deep feature extraction for process variables in various sub-blocks under consideration of quality inheritance. In addition, faults occurring in the system have different effects on product quality and it is necessary to pay attention to the correlation between the faults and the product quality. To improve the monitoring performance of large-scale processes, we design a monitoring framework which can characterize the quality inheritance among sub-blocks. A parallel DVIB-VAE-mRMR model is proposed as well. A local parallel DVIB-VAE model under deep learning network is utilized to extract quality-related latent variables and quality-unrelated features. Meanwhile, the quality-related latent variables extracted from the previous sub-block are sequentially passed to the next sub-block. To eliminate the redundant information generated between correlated sub-blocks, the quality-related latent variables are selected using mRMR method. The selected quality-related latent variables from the previous sub-block are used as a constraint in the next sub-block, which expresses the quality information transfer across all sub-blocks. The specific modeling steps are as follows.

Firstly, the quality-related latent variables and quality-unrelated features are extracted hierarchically. $\boldsymbol{x}_{b\mathrm{h}}$ and $\boldsymbol{x}_{b\mathrm{w}}$ contain rich quality-related and quality-unrelated information, whose corresponding features should be effectively extracted, respectively. Specifically, each sub-block contains two DVIB networks and one VAE network. The design of the corresponding parallel DVIB-VAE stems from the requirement for deep extraction of quality-related and quality-unrelated features in each sub-block. Due to the limitations in the network structure of VAE, the features extracted by

VAE cannot characterize a real physical meaning[29]. Meanwhile, DVIB can extract quality-related features and remove quality-unrelated information. We use one DVIB to extract a small amount of quality information contained in $\boldsymbol{x}_{bw}$, where the quality-unrelated information can be used as a constraint to assist VAE in extracting quality-unrelated features. We use another DVIB to further extract the quality-related information in $\boldsymbol{x}_{bh}$ after receiving the quality information extracted by the previous DVIB. Therefore, two DVIB networks have different roles in feature extraction. The sub-block is divided into the upper layer $\boldsymbol{x}_{bh}$ and the lower layer $\boldsymbol{x}_{bw}$ based on the input dataset. In the upper layer of sub-block 1, $\boldsymbol{x}_{1h}$ is inputted into $DVIB_1$ for training until $DVIB_1$ converges. The corresponding encoder $q_{\phi_{\boldsymbol{v}_1}}(\boldsymbol{v}_1 \mid \boldsymbol{x}_{1h})$ and decoder $p_{\boldsymbol{\theta}_{y_{1h}}}(\boldsymbol{y} \mid \boldsymbol{v}_1)$ of $DVIB_1$ can be obtained, where $\phi_{\boldsymbol{v}_1}$ and $\theta_{y_{1h}}$ are the network parameters. Hence, the quality-related latent variables can be fully extracted in the upper layer of sub-block 1.

It is worth noting that $\boldsymbol{x}_{1w}$ mainly contains a large amount of quality-unrelated information, but it still has a small amount of quality-relevant information, which may have an impact on the whole production process. Therefore, both quality-related latent variables and quality-unrelated features of $\boldsymbol{x}_{1w}$ are extracted in the lower layer of sub-block 1. $\boldsymbol{x}_{1h}$ is inputted into $DVIB_2$ for training. When the network parameters $q_{\phi_{\boldsymbol{v}_{1w}}}(\boldsymbol{v}_{1w} \mid \boldsymbol{x}_{1w})$, $p_{\theta_{y_{1w}}}(\boldsymbol{y} \mid \boldsymbol{v}_{1w})$, $\phi_{\boldsymbol{v}_{1w}}$ and $\theta_{y_{1w}}$ of the $DVIB_2$ converge, the corresponding encoder $q_{\phi_{\boldsymbol{v}_{1w}}}(\boldsymbol{v}_{1w} \mid \boldsymbol{x}_{1w})$ is fixed as one auxiliary part of the VAE to extract the quality-unrelated features. The extracted information using VAE is divided into two parts: the quality-related potential variables $\boldsymbol{v}_{1w}$ and the quality-unrelated features $\boldsymbol{u}_1$ in the lower layer. Since the prior assumption meets the multidimensional Gaussian distribution in VAE, $\boldsymbol{v}_{1w}$ and $\boldsymbol{u}_1$ are independent of each other. The generative model established by VAE is expressed as

$$p_\theta(\boldsymbol{x}_{1w}, \boldsymbol{v}_{1w}, \boldsymbol{u}_1) = p_{\theta_x}(\boldsymbol{x}_{1w} \mid \boldsymbol{v}_{1w}, \boldsymbol{u}_1) p_\theta(\boldsymbol{v}_{1w}) p_\theta(\boldsymbol{u}_1) \tag{13.13}$$

Correspondingly, the predictive model by VAE is written as

$$q_\phi(\boldsymbol{v}_{1w}, \boldsymbol{u}_1 \mid \boldsymbol{x}_{1w}) = q_{\phi_{\boldsymbol{v}}}(\boldsymbol{v}_{1w} \mid \boldsymbol{x}_{1w}, \boldsymbol{u}_1) q_{\phi_{\boldsymbol{u}}}(\boldsymbol{u}_1 \mid \boldsymbol{x}_{1w}) \tag{13.14}$$

When the corresponding encoder $q_{\phi v_{1w}}(\boldsymbol{v}_{1w} \mid \boldsymbol{x}_{1w})$ is fixed, according to equation (13.11), equation (13.12) and equation (13.13), the optimization objective function of VAE is expanded as follows,

$$\begin{aligned} \chi(\theta, \phi; \boldsymbol{x}_i) = {} & E_{q_{\phi_{\boldsymbol{v}}}(q_{\phi_{\boldsymbol{v}}}(\boldsymbol{v}_{1w} \mid x) \mid x) q_{\phi_{\boldsymbol{u}}}(\boldsymbol{u}_1 \mid x)}[\lg p_{\theta_x}(\boldsymbol{x}_i \mid \boldsymbol{u}_1, \boldsymbol{v}_{1w})] - \\ & D_{KL}(q_{\phi_{\boldsymbol{v}}}(\boldsymbol{v}_{1w} \mid \boldsymbol{x}_i) \| p_\theta(\boldsymbol{v}_{1w})) - D_{KL}(q_{\phi_{\boldsymbol{u}}}(\boldsymbol{u}_1 \mid \boldsymbol{x}_i) \| p_\theta(\boldsymbol{u}_1)) \end{aligned} \tag{13.15}$$

Therefore, it can be concluded that minimizing $D_{KL}(q_{\phi_{\boldsymbol{v}}}(\boldsymbol{v}_{1w} \mid \boldsymbol{x}_i) \| p_\theta(\boldsymbol{v}_{1w}))$

and $D_{KL}(q_{\phi_u}(\boldsymbol{u}_1 \mid \boldsymbol{x}_i) \| p_\theta(\boldsymbol{u}_1))$ is equivalent to minimizing the MI of $I(\boldsymbol{V}, \boldsymbol{X})$ and $I(\boldsymbol{U}, \boldsymbol{X})$. Obviously, the weaker the correlation between $\boldsymbol{v}_{1w}$ and $\boldsymbol{u}_1$ is, the better the reconstruction performance of VAE is. When the network parameters of $q_{\phi_{v_{1w}}}(\boldsymbol{v}_{1w} \mid \boldsymbol{x}_{1w})$ are fixed, more quality-unrelated features $\boldsymbol{u}_1$ can be extracted.

The quality-related latent variables extracted from sub-block 1 are written as

$$\boldsymbol{v}_{key,1} = \boldsymbol{v}_1 \cup \boldsymbol{v}_{1w} \tag{13.16}$$

Secondly, the redundant information of quality-related latent variables is eliminated. The presence of redundancy among variables increases the computational complexity and affects the accuracy of monitoring performance. The redundancy among variables in $\boldsymbol{v}_{key,1}$ originates from two factors. On the one hand, the datasets $\boldsymbol{x}_{1h}$ and $\boldsymbol{x}_{1w}$ used in the extraction of quality-related latent variables $\boldsymbol{v}_{key,1}$ have partial intersection. On the other hand, only the correlation between the latent variables and the quality variables should be considered and the redundancy among the latent variables can be ignored. Thus, mRMR method is introduced to screen the redundant variables of $\boldsymbol{v}_{key,1}$ and eliminates redundancy.

Finally, the quality information of the series sub-blocks is transmitted sequentially. Quality inheritance is applied throughout the large-scale processes. When modeling, the quality inheritance is considered to better describe the operational characteristics of the actual industrial system, which in turn improves the modeling accuracy and reliability. Therefore, the extracted quality-related latent variables from the previous sub-block can be considered the constraint for the next sub-block to supplement the original input data from sub-block 2 to sub-block B. The input data of each sub-block after merging are expressed as

$$\begin{aligned} \boldsymbol{S}_{sub1} &= \boldsymbol{x}_{1h} \cup \boldsymbol{x}_{1w} \\ \boldsymbol{S}_{sub2} &= \boldsymbol{v}_{1s} \cup \boldsymbol{x}_{2h} \cup \boldsymbol{x}_{2w} \\ &\vdots \\ \boldsymbol{S}_{subB} &= \boldsymbol{v}_{(B-1)s} \cup \boldsymbol{x}_{Bh} \cup \boldsymbol{x}_{Bw} \end{aligned} \tag{13.17}$$

The correlations among sub-blocks are established on the basis of data layer fusion, and the quality-related latent variables and quality-unrelated features are extracted using the constructed parallel DVIB-VAE-mRMR model. In summary, the structure of the parallel DVIB-VAE-mRMR model based on the quality inheritance is shown in **Fig. 13. 1**.

13. 2. 3 Local-global synergetic process monitoring

In our large-scale process monitoring framework which considers the quality inheritance,

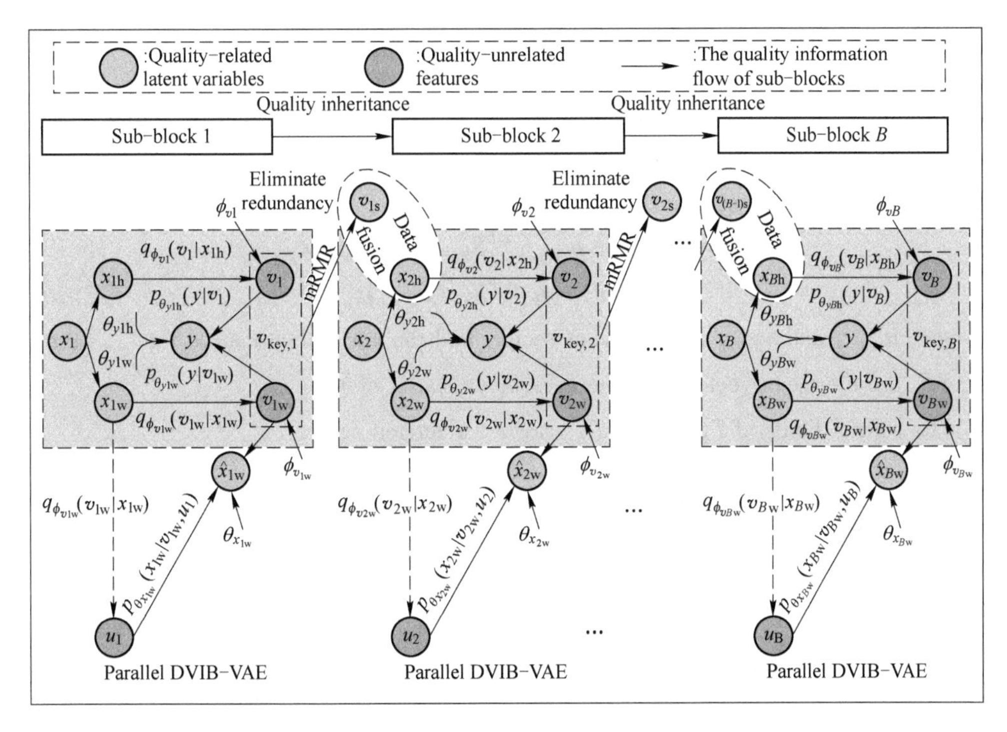

Fig. 13.1 The structure of the parallel DVIB-VAE-mRMR model based on the quality inheritance

we need monitor the faulty of each sub-block, and justify whether the fault effects the final product quality by the information fusion of features extracted from multiple sub-blocks. Local process monitoring is used to detect the fault which has been described in section 1.2.3. A more comprehensive and accurate process monitoring result can be obtained by the global monitoring which combines the information of features extracted from different sub-blocks. $\boldsymbol{v}_{\text{key},b}$ and $\boldsymbol{u}_b$ denote the quality-related latent variables and quality-unrelated features of the sub-block b, respectively. The monitoring statistics corresponding to each sub-block are designed using $\boldsymbol{v}_{\text{key},b}$ and $\boldsymbol{u}_b$. The output mean and covariance of $q_{\phi_{vb}}(\boldsymbol{v}_b \mid \boldsymbol{x}_{b\text{h}})$, $q_{\phi_{vb\text{w}}}(\boldsymbol{v}_{b\text{w}} \mid \boldsymbol{x}_{b\text{w}})$ and $q_{\phi_{u1}}(\boldsymbol{u}_{b\text{w}} \mid \boldsymbol{x}_{b\text{w}})$ are obtained according to the parallel DVIB-VAE model. Given $\boldsymbol{v}_{\text{key},b} = \boldsymbol{v}_b \cup \boldsymbol{v}_{b\text{w}}$, the mean of $\boldsymbol{v}_{\text{key},b}$ is expressed as

$$\bar{\boldsymbol{v}}_{\text{key},b} = \frac{n_b \times d_b}{n_b \times d_b + n_{b\text{w}} \times d_{b\text{w}}} \bar{\boldsymbol{v}}_b + \frac{n_{b\text{w}} \times d_{b\text{w}}}{n_b \times d_b + n_{b\text{w}} \times d_{b\text{w}}} \bar{\boldsymbol{v}}_{b\text{w}} \tag{13.18}$$

where, $\bar{\boldsymbol{v}}_b$ is the output mean of samples $\boldsymbol{v}_b$; n_b is the number of samples $\boldsymbol{v}_b$; d_b is the dimension of samples $\boldsymbol{v}_b$; $\bar{\boldsymbol{v}}_{b\text{w}}$ is the output mean of samples $\boldsymbol{v}_{b\text{w}}$; $n_{b\text{w}}$ is the number of samples $\boldsymbol{v}_{bw}$ and d_{bw} is the dimension of samples $\boldsymbol{v}_{bw}$.

Because of $\boldsymbol{v}_{\text{key},b} = \boldsymbol{v}_b \cup \boldsymbol{v}_{bw}$, we can obtain $n_b = n_{bw}$. Equation (13.18) can be simplified.

According to the above description of $\boldsymbol{v}_{\text{key},b}$ and $\boldsymbol{u}_b$, the designed quality-related monitoring statistic $T^2_{v_{\text{key},b}}$ and the quality-unrelated monitoring statistic $T^2_{\boldsymbol{u}_b}$ are expressed as

$$T^2_{v_{\text{key},b}} = (\bar{\boldsymbol{v}}_{\text{key},b} - \boldsymbol{\mu}_{\text{key},b})^{\mathrm{T}} \boldsymbol{S}^{-1}_{\text{key},b} (\bar{\boldsymbol{v}}_{\text{key},b} - \boldsymbol{\mu}_{\text{key},b}) \tag{13.19}$$

$$T^2_{\boldsymbol{u}_b} = (\bar{\boldsymbol{u}}_b - \boldsymbol{\mu}_b)^{\mathrm{T}} \boldsymbol{S}_b^{-1} (\bar{\boldsymbol{u}}_b - \boldsymbol{\mu}_b) \tag{13.20}$$

where, $\boldsymbol{\mu}_{\text{key},b}$ and $\boldsymbol{S}_{\text{key},b}$ are the mean vectors and the covariance matrix of $\boldsymbol{v}_{\text{key},b}$; $\boldsymbol{\mu}_b$ and $\boldsymbol{S}_b$ are the mean vectors and the covariance matrix of $\bar{\boldsymbol{u}}_b$.

The control limits of $T^2_{\boldsymbol{v}_{\text{key},b}}$ and $T^2_{\boldsymbol{u}_b}$ can be calculated as

$$T^2_{\boldsymbol{v}_{\text{key},b}(\text{th})} = \frac{(d_b + d_{bw})(n_b^2 - 1)}{n_b(n_b - (d_b + d_{bw}))} \times F(\alpha,\ d_b + d_{bw},\ n_b - (d_b + d_{bw})) \tag{13.21}$$

$$T^2_{\boldsymbol{u}_b(\text{th})} = \frac{k_{\boldsymbol{u}_b}(n^2_{\boldsymbol{u}_b} - 1)}{n_{\boldsymbol{u}_b}(n_{\boldsymbol{u}_b} - k_{\boldsymbol{u}_b})} F(\alpha,\ k_{\boldsymbol{u}_b},\ n_{\boldsymbol{u}_b} - k_{\boldsymbol{u}_b}) \tag{13.22}$$

where, $d_b + d_{bw}$ and $k_{\boldsymbol{u}_b}$ are the dimensions of $\boldsymbol{v}_{\text{key},b}$ and $\boldsymbol{u}_b$, respectively; n_b and $n_{\boldsymbol{u}_b}$ are the numbers of $\boldsymbol{v}_{\text{key},b}$ and $\boldsymbol{u}_b$, and $n_b = n_{\boldsymbol{u}_b}$; α is the given confidence level.

Thus, $F(\alpha,\ d_b + d_{bw},\ n_b - (d_b + d_{bw}))$ and $F(\alpha,\ k_{\boldsymbol{u}_b},\ n_{\boldsymbol{u}_b} - k_{\boldsymbol{u}_b})$ denote the F distributions of $T^2_{\boldsymbol{v}_{\text{key},b}}$ and $T^2_{\boldsymbol{u}_b}$.

The local monitoring strategy enables effective monitoring of abnormal fluctuations in each sub-block. However, it is difficult for the local monitoring model to achieve a comprehensive evaluation of the large-scale processes. To solve the above problems, Bayesian inference constructs a global monitoring index to monitor the large-scale processes.

The conditional probabilities $P^b_{T^2_{\boldsymbol{v}_{\text{key}}}}(\boldsymbol{v}_{\text{key},b} \mid F)$ and $P^b_{T^2_{\boldsymbol{v}_{\text{key}}}}(\boldsymbol{v}_{\text{key},b} \mid N)$ in faulty and normal conditions of $\boldsymbol{v}_{\text{key},b}$ can be obtained as follow

$$P^b_{T^2_{\boldsymbol{v}_{\text{key}}}}(\boldsymbol{v}_{\text{key},b} \mid F) = \exp\left(-\frac{T^2_{\boldsymbol{v}_{\text{key},b}(\text{th})}}{T^2_{\boldsymbol{v}_{\text{key},b}}}\right) \tag{13.23}$$

$$P^b_{T^2_{\boldsymbol{v}_{\text{key}}}}(\boldsymbol{v}_{\text{key},b} \mid N) = \exp\left(-\frac{T^2_{\boldsymbol{v}_{\text{key},b}}}{T^2_{\boldsymbol{v}_{\text{key},b}(\text{th})}}\right) \tag{13.24}$$

Correspondingly, the quality-related global monitoring statistic BIC_{re} is expressed as

$$BIC_{re} = \frac{\sum_{b=1}^{B} (P_{T^2_{\boldsymbol{v}_{key}}}^{b}(\boldsymbol{v}_{key,b} \mid F) P_{T^2_{\boldsymbol{v}_{key}}}^{b}(F \mid \boldsymbol{v}_{key,b}))}{\sum_{b=1}^{B} P_{T^2_{\boldsymbol{v}_{key}}}^{b}(\boldsymbol{v}_{key,b} \mid F)} \tag{13.25}$$

Similar to BIC_{re}, the quality-unrelated global monitoring statistic BIC_{un} is given as below

$$BIC_{un} = \frac{\sum_{b=1}^{B} (P_{T^2_{\boldsymbol{u}}}^{b}(\boldsymbol{u}_b \mid F) P_{T^2_{\boldsymbol{u}}}^{b}(F \mid \boldsymbol{u}_b))}{\sum_{b=1}^{B} P_{T^2_{\boldsymbol{u}}}^{b}(\boldsymbol{u}_b \mid F)} \tag{13.26}$$

where, F is the faulty conditions; $P_{T^2_{\boldsymbol{v}_{key}}}^{b}(F)$ can be defined as the prior probabilities of $\boldsymbol{v}_{key,b}$ in faulty conditions; $P_{T^2_{\boldsymbol{v}_{key}}}^{b}(F \mid \boldsymbol{v}_{key,b})$ is the fault probability of $\boldsymbol{v}_{key,b}$; $P_{T^2_{\boldsymbol{u}}}^{b}(F)$ is the prior probability of $\boldsymbol{u}_b$ in faulty conditions; $P_{T^2_{\boldsymbol{u}}}^{b}(\boldsymbol{u}_b \mid F)$ denotes the conditional probabilities of $\boldsymbol{u}_b$ in faulty condition.

During the online testing phase, the global process monitoring can be performed by

$$\left.\begin{array}{l} BIC_{re} < \alpha_{re} \\ BIC_{un} < \alpha_{un} \end{array}\right\} \Rightarrow \text{normal}$$

$$BIC_{re} > \alpha_{re} \Rightarrow \text{quality-related fault} \tag{13.27}$$

$$\left.\begin{array}{l} BIC_{re} < \alpha_{re} \\ BIC_{un} > \alpha_{un} \end{array}\right\} \Rightarrow \text{quality-unrelated fault}$$

where, α_{re} and α_{un} are the control limits of the global monitoring statistics BIC_{re} and BIC_{un}, respectively. Both of which are equivalent to $1 - \alpha$.

The detailed discription of the proposed monitoring framework using the parallel DVIB-VAE-mRMR for large-scale processes is provided. The flowchart can be presented in **Fig. 13.2**.

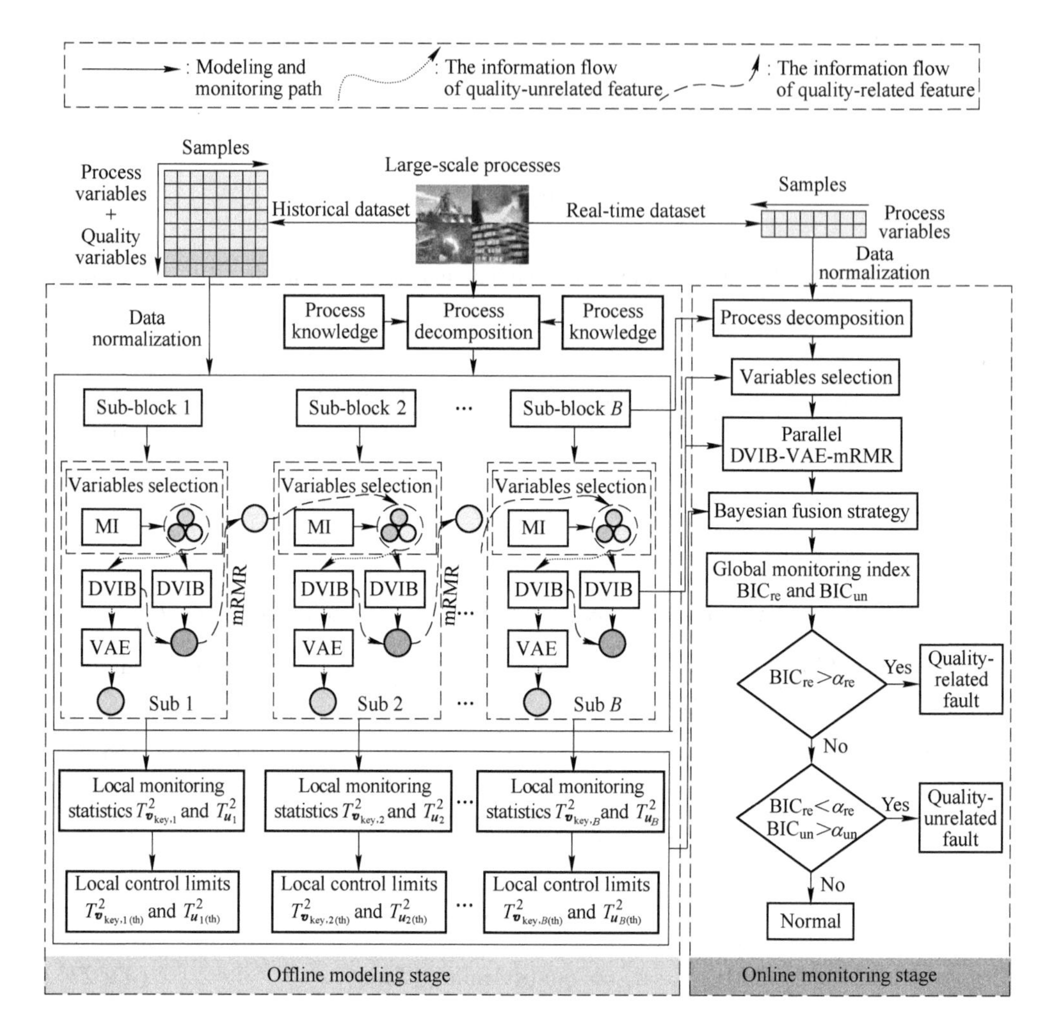

Fig. 13. 2 The monitoring framework based on the parallel DVIB-VAE-mRMR for large-scale processes

13. 3 Case study

The performances of the proposed monitoring methods under quality inheritance are verified to the real hot strip mill process (HSMP). Parallel DVIB-VAE-mRMR method is also compared with DPCA and Distributed DVIB in this paper.

13. 3. 1 Descriptions of the hot strip mill process

As the continuous and serial large-scale processes, HSMP has complex correlations in different control loops and process variables, and its quality abnormities are the result of multi-process coupling and accumulation. The whole rolling process involves complex physical and chemical changes. Once the fault occurs, it is difficult for HSMP to guarantee quality stability through automatic control systems. Therefore, a real HSMP dataset, from 1700-mm production line, Ansteel Corporation in Liaoning Province of P. R. China, is used to realize the process monitoring. The acquired data duration is 30 s and the sampling period is 0.01 s.

As the most critical area of HSMP, the finishing mill process (FMP) can ensure process safety and the final product quality of large-scale processes, and its schematic is presented in **Fig. 13.3**. FMP is made up of seven stands in series. In each stand, there are four rolls including a pair of working rolls and a pair of supporting rolls. The looper device between the rolling mills controls the strip tension to ensure a flat strip shape. The electromechanical systems are used to rotate the rolls to move the strip forward smoothly. Meanwhile, a large number of sensors and rolling force monitoring devices are deployed in the finishing rolling units.

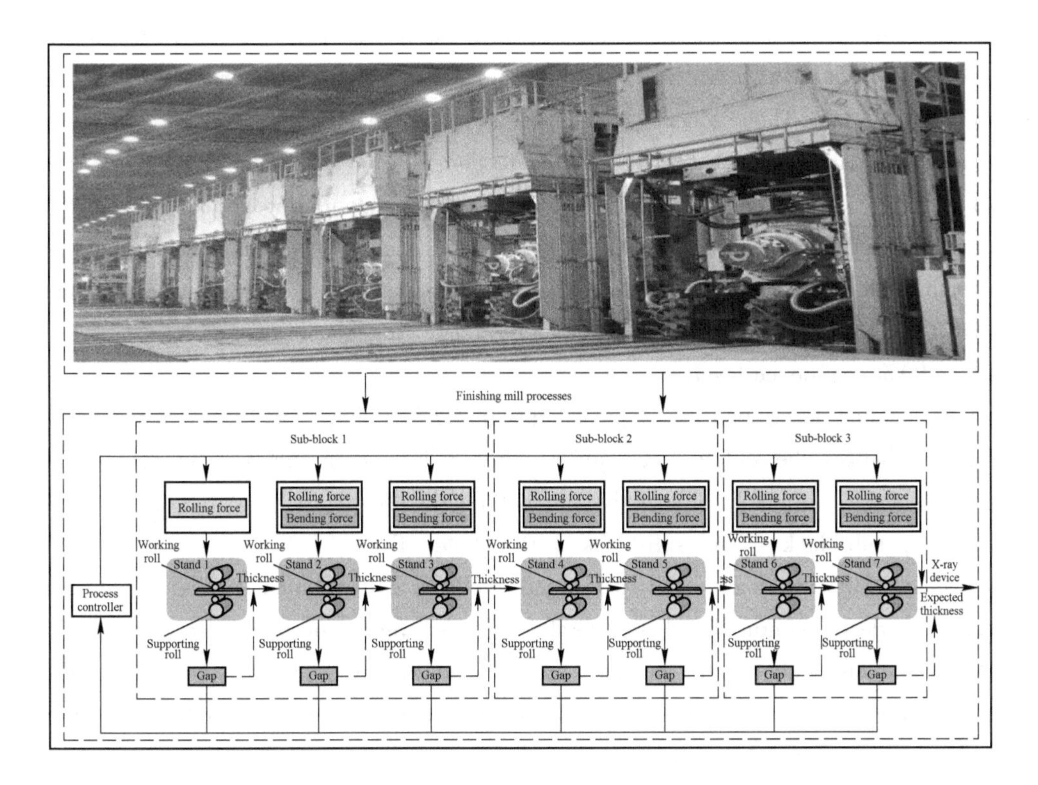

Fig. 13.3 Schematic of the finishing mill process in the HSMP

13.3.2 Experiment results and analysis

The strip is continuously rolled by 7 rolling mills to ensure that the final quality reaches the set value. On the one hand, the finishing mill stands are connected sequentially. The failure of the previous finishing mill stand will affect the operation of the latter one. Therefore, the spatial distribution of different finishing mill stands is worthy of attention. On the other hand, FMP is a large-scale process, and its sub-process decomposition enables effective feature extraction and process monitoring. Based on the above process knowledge, the spatial distribution of the FMP should be retained when performing sub-blocks decomposition. Thus, 7 finishing mill stands are divided into 3 sub-blocks, with stands 1-3 as sub-block 1, stands 4-5 as sub-block 2 and stands 6-7 as sub-block 3. The process variables contained multiple stands are then decomposed into different sub-blocks, respectively. For evaluating the effectiveness of proposed method, a group of historical data containing 3000 samples are firstly trained in the offline modeling stage to determine the confidence limits. The sampling rate is 0.01 s. There are 20 process variables and 1 quality variable in each sample. Secondly, a group of test data containing 3000 samples are used for process monitoring in the online monitoring stage. Then, local monitoring results are combined using Bayesian inference. Finally, the local-global fault detection rate (FDR) and false alarm rate (FAR) for different faults are calculated to inspect the monitoring performance. In the proposed method, the parallel DVIB-VAE-mRMR model is set as 3-layer network structure, the activation function is Tanh, the learning rate is 10^{-3}, the mini-batch size is 64, the model uses the Adam gradient update strategy and the control limit is set to the 99% confidence level.

The detailed description of 3 typical faults in FMP is shown in **Table 13.1**. Fault 1 directly changes the rolling force and bending roll force of the 3_{rd} and subsequent stands, which ultimately affects the product quality. Therefore, faults occur in sub-block 1, sub-block 2 and sub-block 3. Distributed PCA (DPCA) is a distributed statistical MSPM methods, and distributed DVIB is a distributed process monitoring based on deep learning. For better verify the effectiveness of the proposed method, DPCA and distributed DVIB methods are selected as the comparison ones. **Fig. 13.4** (a), (b) and (c) represent these experimental results of three methods in Fault 1. As shown in **Fig. 13.4**, all the above three methods for global monitoring can get good detection results. However, few faults can be detected in the sub-block 2 by DPCA which indicates the limitations of DPCA for local monitoring in Fault 1. The distributed

DVIB can detect faults in all sub-blocks, but it cannot determine whether the fault affects product quality. The occurrence of quality-related faults can affect quality-unrelated information, whereas quality-unrelated faults are not. Compared with DPCA and distributed DVIB, the proposed method can better determine local-global monitoring results and also determine Fault 1 as a quality-related fault.

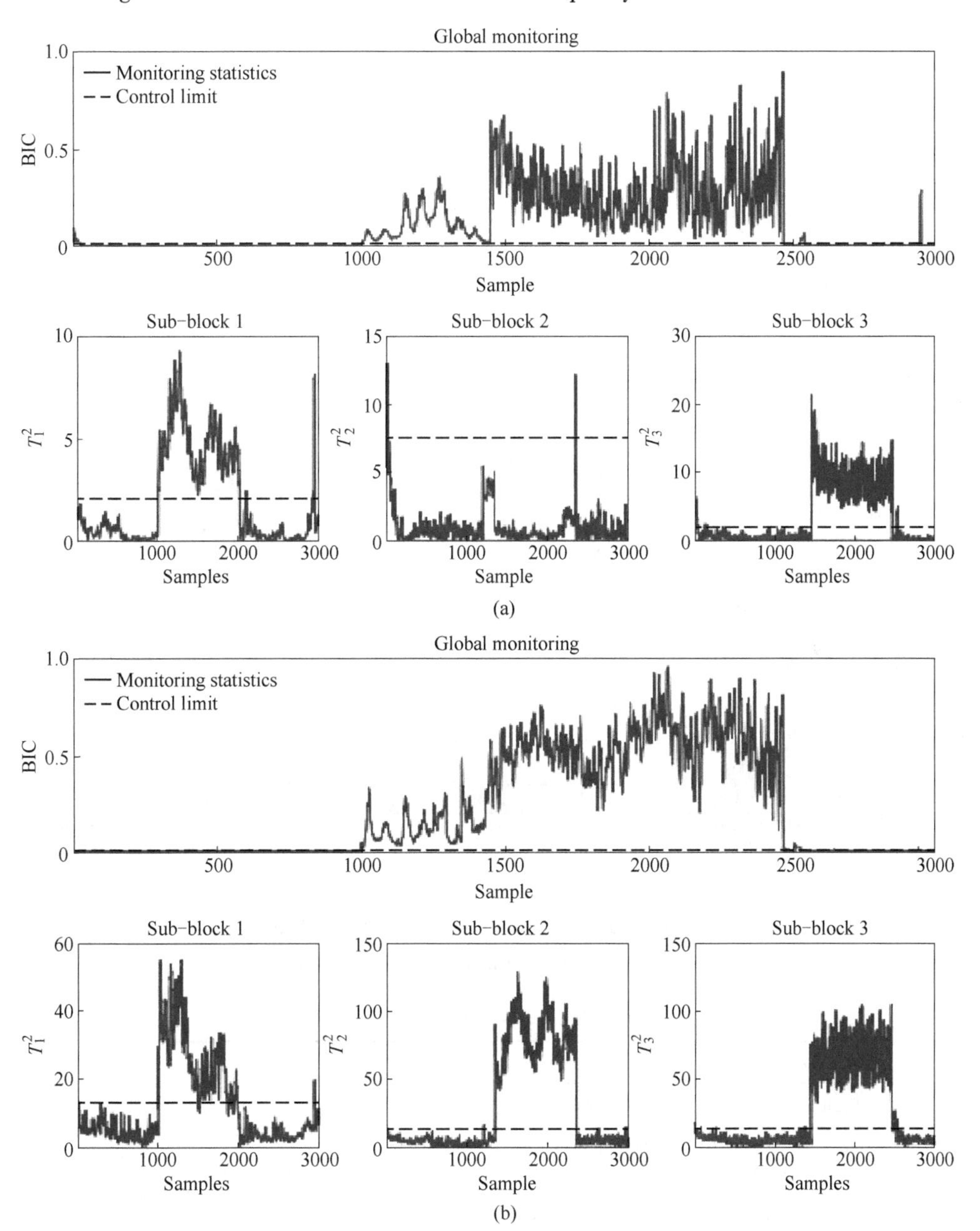

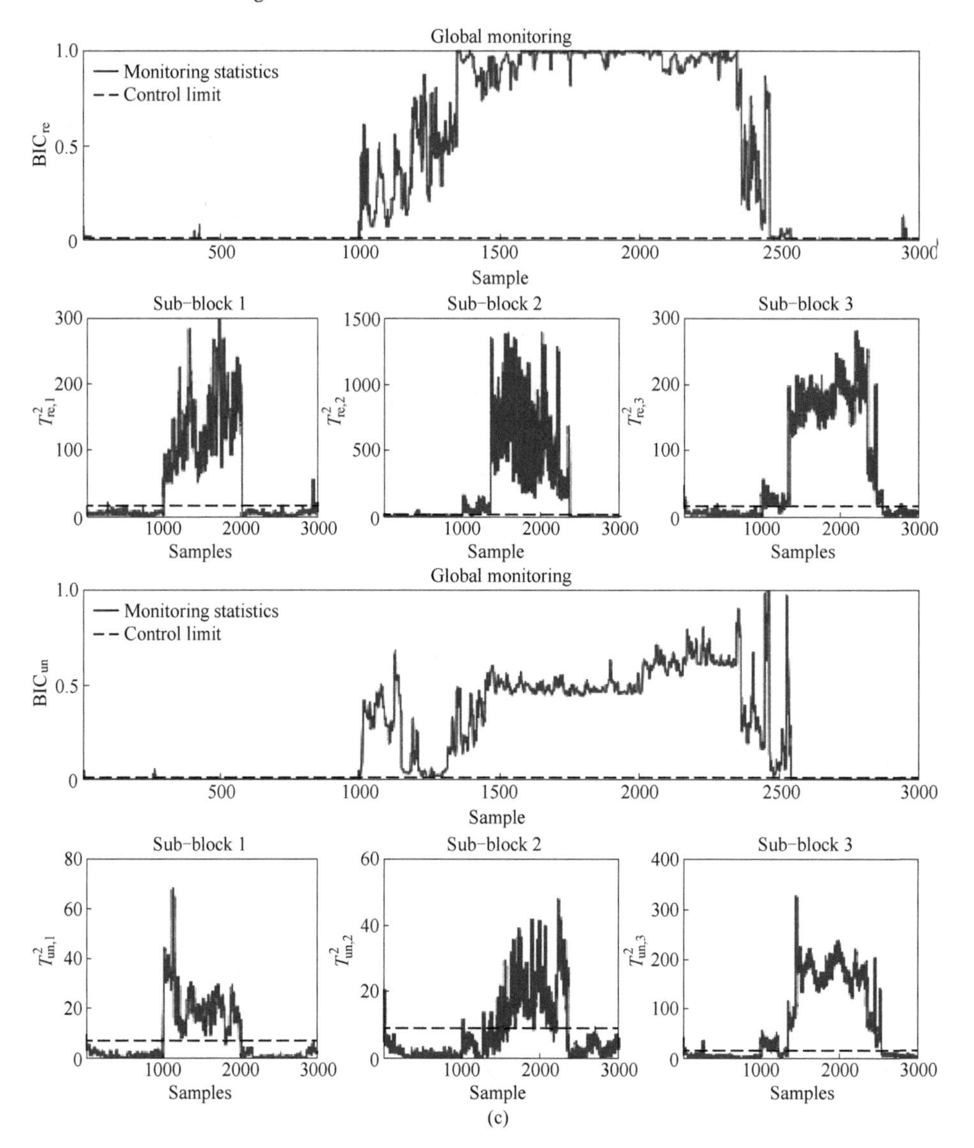

Fig. 13.4 Process monitoring results for Fault 1

(a) DPCA; (b) Distributed DVIB; (c) The proposed method

Table 13.1 3 typical faults in FMP

Fault No.	Occurrence and duration	Type
Fault 1	Occurs in the 10_{th} s and lasts 15 s	Quality-related
Fault 2	Occurs in the 20_{th} s and lasts forever	Quality-related
Fault 3	Occurs in the 10_{th} s and lasts 10 s	Quality-unrelated

After Fault 2 occurs, it can directly affect the sampling values of the roll gap and total rolling force of the 4th stands, and those of all subsequent stands will also change, thus affecting the product quality. Based on the above description, Fault 2 appears in sub-block 2 and affects sub-block 3. The monitoring results by DPCA are presented in **Fig. 13. 5** (a). As can be seen in **Fig. 13. 5** (a), DPCA only monitors the fault in sub-block 3 and does not detect the location where the fault occurs in time. The

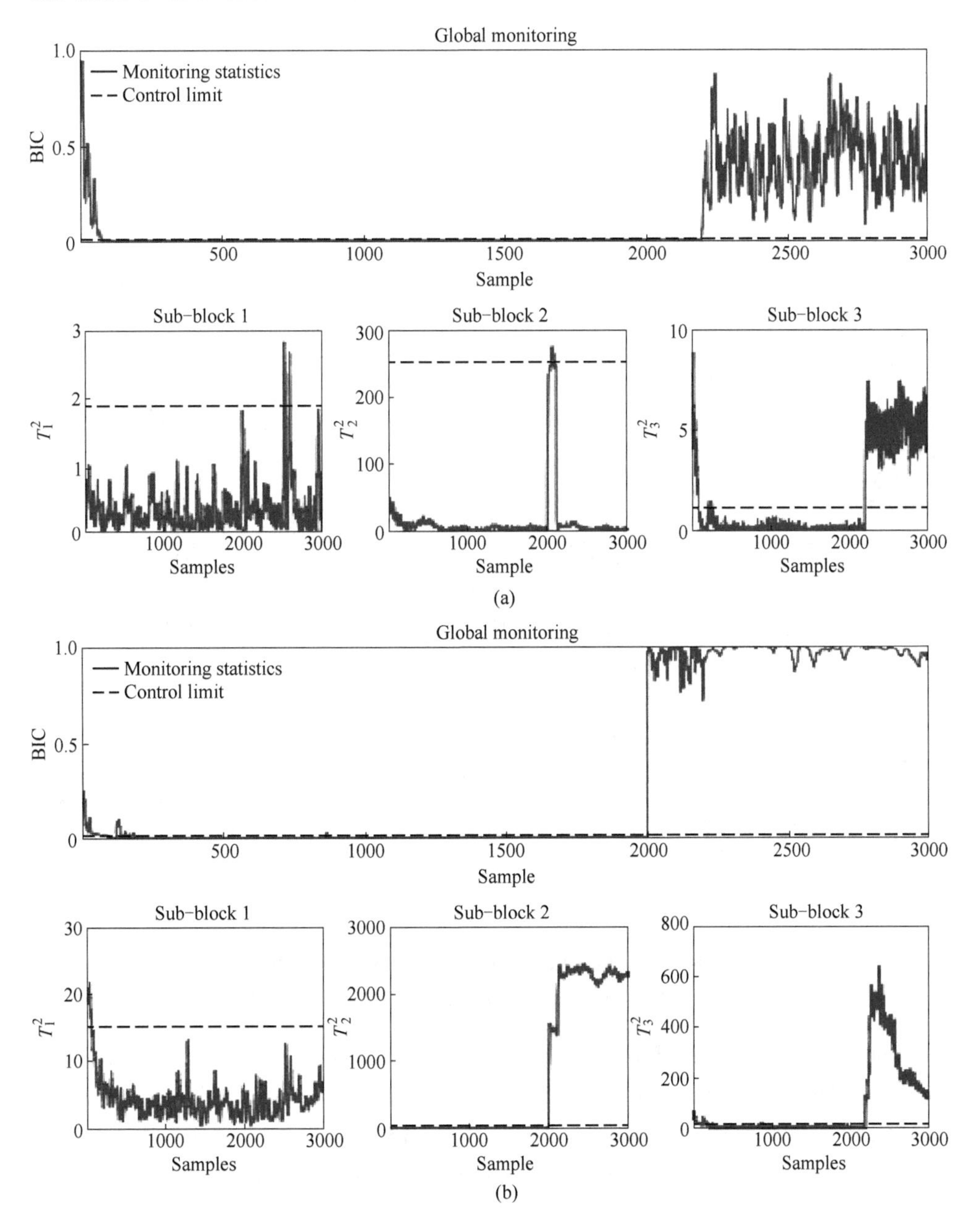

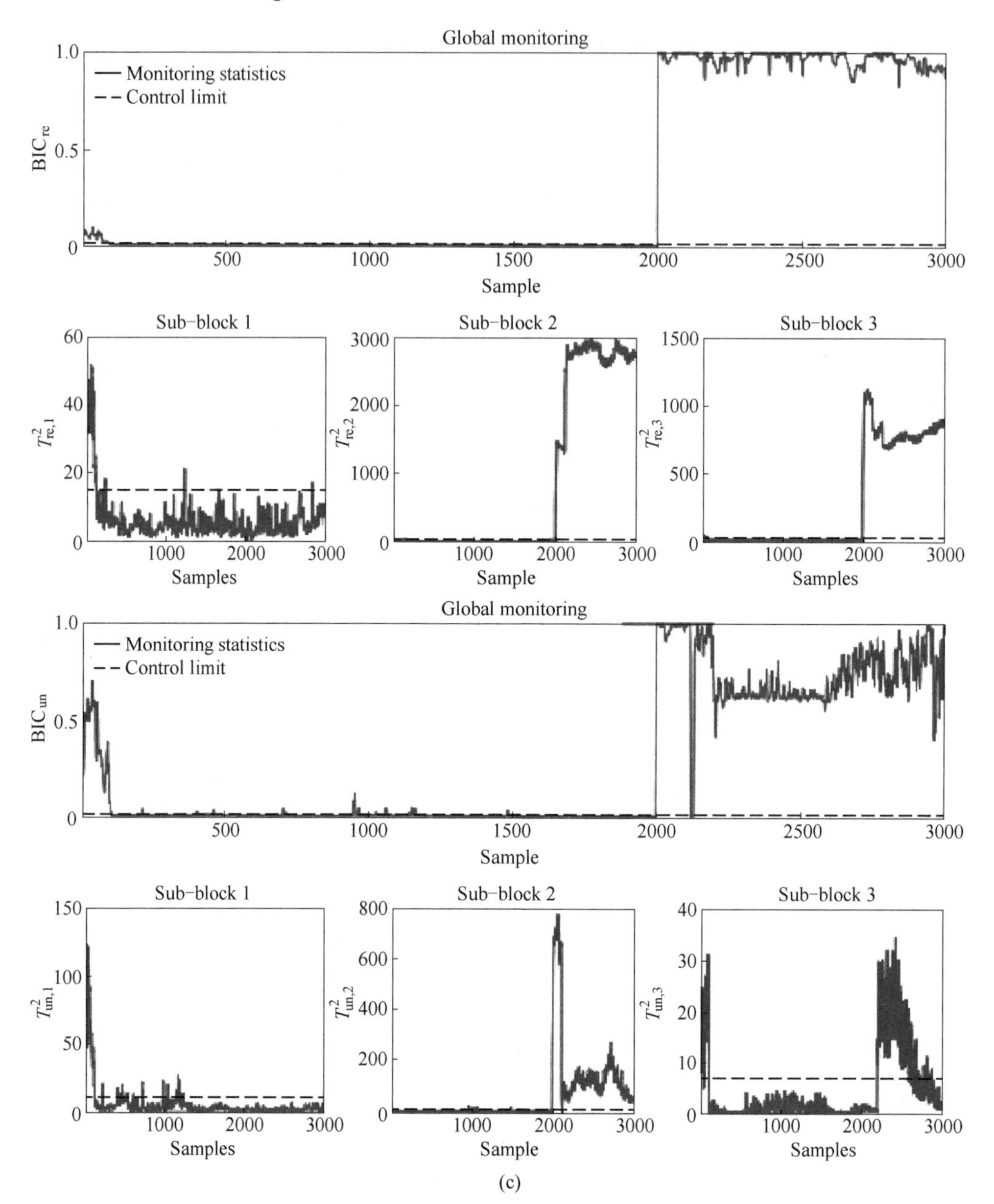

(c)

Fig. 13. 5 Process monitoring results for Fault 2

(a) DPCA; (b) Distributed DVIB; (c) The proposed method

monitoring results in **Fig. 13. 5** (b) show that the distributed DVIB can locate the fault very well. However, it is difficult to determine whether the fault affects the product quality. From the monitoring results shown in **Fig. 13. 5** (c), the proposed method is more effective in monitoring the fault in time and can determine Fault 2 is quality-related fault.

Fault 3 does not affect the thickness and is considered as a quality-unrelated fault. **Fig. 13.6** (a), (b) and (c) show the monitoring results of three methods for Fault 3. On the one hand, it can be seen that the proposed method can distinguish that Fault 3 has no effect on the final product quality. On the other hand, the effectiveness of the proposed method is significantly better than those of DPCA and distributed DVIB.

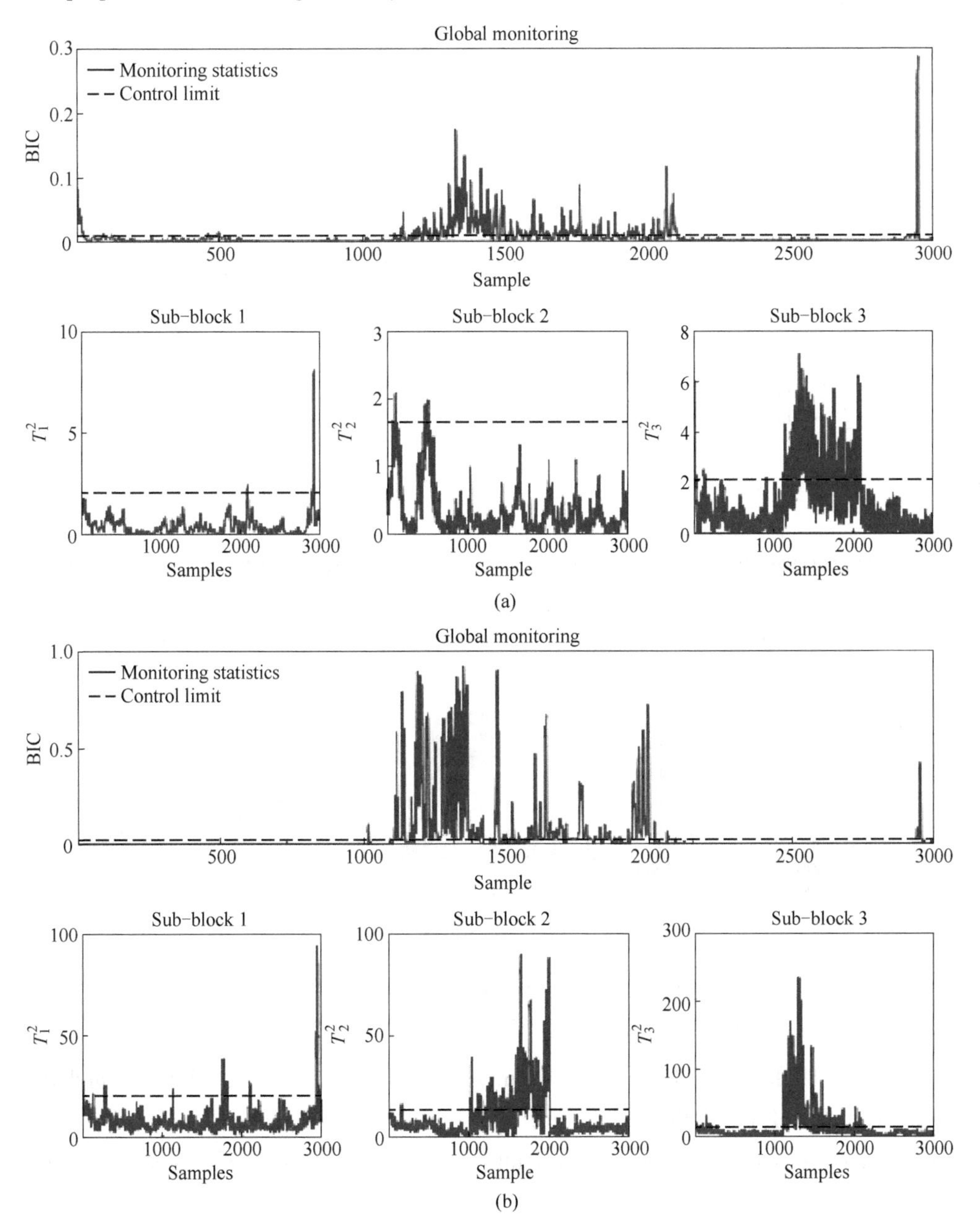

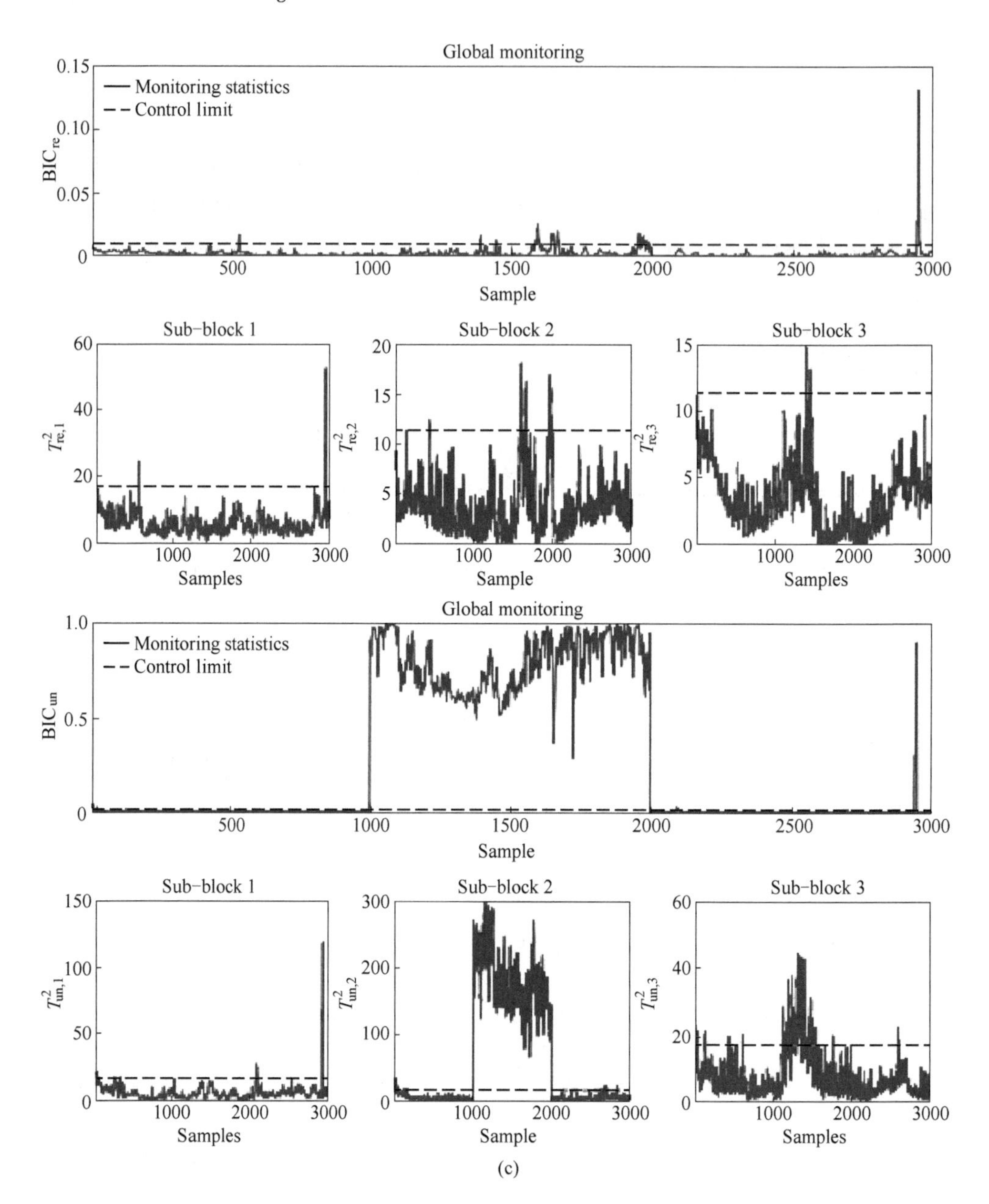

Fig. 13. 6 Process monitoring results for Fault 3

(a) DPCA; (b) Distributed DVIB; (c) The proposed method

Fault 1, Fault 2 and Fault 3 are monitored by DPCA, distributed DVIB and the proposed method. From the monitoring results, we can conclude that the proposed method has good performance for the faults, and can also effectively identify quality-related and quality-unrelated faults in time.

The FDR and FAR results of DPCA, distributed DVIB and the proposed method for the three faults are listed in **Table 13. 2. Table 13. 2** demonstrates the global monitoring results. It can be seen that the proposed method has the best monitoring performance. The proposed method has the highest FDR for all three faults. Among them, the FDR of Fault 1 is 99. 93%, Fault 2 and Fault 3 reaches 100%, and there is a lowest FAR for Fault 3.

Table 13. 2 Comparison of global FDR and FAR results for 3 faults (%)

Fault No.	DPCA		Distributed DVIB		The proposed method	
	FAR	FDR	FAR	FDR	FAR	FDR
Fault 1	3. 80	97. 80	**2. 13**	98. 40	6. 80	**99. 93**
Fault 2	**4. 65**	81. 80	8. 15	**100**	5. 25	**100**
Fault 3	5. 55	46. 40	3. 90	73. 90	**3. 50**	**100**

13. 4 Conclusions

In this study, a novel process monitoring framework based on quality inheritance for large-scale processes is proposed. Firstly, a priori knowledge is used to divide sub-blocks and evaluate the correlation between process variables and quality indicators to construct quality-related and quality-unrelated datasets. Then, a parallel DVIB-VAE-mRMR model is designed to achieve deep extraction and redundancy analysis of quality-related and quality-unrelated features. Finally, the results of each sub-block are fused by Bayesian inference for comprehensive large-scale processes monitoring. The effectiveness of the proposed framework is demonstrated by a real HSMP. Future research will be devoted to constructing the propagation networks and fault root cause diagnosis scheme under quality inheritance.

References

[1] Zhang H J, Zhang C, Dong J, et al. A new key performance indicator oriented industrial process monitoring and operating performance assessment method based on improved Hessian locally linear embedding [J]. International Journal of Systems Science, 2022, 53 (16): 3538-3555.

[2] Li W Q, Zhao C H. Hybrid fault characteristics decomposition based probabilistic distributed fault diagnosis for large-scale industrial processes [J]. Control Engineering Practice, 2019, 84: 377-388.

[3] Fan H, Lai X, Du S, et al, Distributed monitoring with integrated probability PCA and mRMR

for drilling processes [J]. IEEE Transactions on Instrument and Measurement, 2022, 71: 1-13.

[4] Kong X, Yang Z, Luo J, et al. Extraction of reduced fault subspace based on KDICA and its application in fault diagnosis [J]. IEEE Transactions on Instrument and Measurement, 2022, 71: 1-12.

[5] Cao F, Zhang Z, He X. Active fault isolation of over-actuated systems based on a control allocation approach [J]. IEEE Transactions on Instrument and Measurement, 2022, 71: 1-10.

[6] Ji H Q, He X, Shang J, et al. Incipient fault detection with smoothing techniques in statistical process monitoring [J]. Control Engineering Practice,, 2017, 62: 11-21.

[7] Yao L, Shao W M, Ge Z Q. Hierarchical quality monitoring for large-scale industrial plants with big process data [J]. IEEE Transactions on Neural Networks and Learning Systems, 2021, 32 (8): 3330-3341.

[8] Ge Z, Chen J. Plant-wide industrial process monitoring: A distributed modeling framework [J]. IEEE Transactions on Industrial Informatics, 2016, 12 (1): 310-321.

[9] Yuan X, Feng L, Wang K, et al. Deep learning for data modeling of multirate quality variables in industrial processes [J]. IEEE Transactions on Instrument and Measurement, 2021, 70: 1-11.

[10] Tong C, Song Y, Yan X. Distributed statistical process monitoring based on four-subspace construction and Bayesian inference [J]. Industrial & Engineering Chemistry Research, 2013, 52 (29): 9897-9907.

[11] Ge Z, Song Z. Distributed PCA model for plant-wide process monitoring [J]. Industrial & Engineering Chemistry Research, 2013, 52 (5): 1947-1957.

[12] Tong C D, Yan X F. A Novel Decentralized Process Monitoring Scheme Using a Modified Multiblock PCA Algorithm [J]. IEEE Transactions on Automation Science and Engineering, 2017, 14 (2): 1129-1138.

[13] Choi S W, Lee I B. Multiblock PLS-based localized process diagnosis [J]. Journal of Process Control, 2005, 15 (3): 295-306.

[14] Zhu J, Ge Z, Song Z. Distributed parallel PCA for modeling and monitoring of large-scale plant-wide processes with big data [J]. IEEE Transactions on Industrial Informatics, 2017, 13 (4): 1877-1885.

[15] Jiang Q C, Yan X F. Hierarchical monitoring for multi-unit chemical processes based on local-global correlation features [J]. Acta Automatica Sinica, 2020, 46 (9): 1770-1782.

[16] Miao M, Yu J. A deep domain adaptative network for remaining useful life prediction of machines under different working conditions and fault modes [J]. IEEE Transactions on Instrument and Measurement, 2021, 70: 1-14.

[17] Peng K X, Zhang K, You B, et al. Quality-related prediction and monitoring of multi-mode processes using multiple PLS with application to an industrial hot strip mill [J]. Neurocomputing, 2015, 168 (30): 1094-1103.

[18] Bao Y, Wang B, Guo P D, et al. Chemical process fault diagnosis based on a combined deep learning method [J]. The Canadian Journal of Chemical Engineering, 2022, 100 (1): 54-66.

[19] Zhao R, Yan R Q, Chen Z H, et al. Deep learning and its applications to machine health monitoring [J]. Mechanical Systems and Signal Processing, 2019, 115: 213-237.

[20] Chang S C, Zhao C H. A spatio-temporal synergistic graph convolution long short-term memory network and its application for industrial soft sensors [J]. Control and Decision, 2022, 37 (1): 77-86.

[21] Yuan X F, Li L, Wang Y L. Nonlinear dynamic soft sensor modeling with supervised long short-term memory network [J]. IEEE Transactions on Industrial Informatics, 2020, 16 (5): 3168-3176.

[22] Tang P, Peng K X, Dong J. Nonlinear quality-related fault detection using combined deep variational information bottleneck and variational autoencoder [J]. ISA Transactions, 2021, 114: 444-454.

[23] Yao L, Ge Z Q. Distributed parallel deep learning of hierarchical extreme learning machine for multimode quality prediction with big process data [J]. Engineering Applications of Artificial Intelligence, 2019, 81: 450-465.

[24] Rong M Y, Shi H B, Tan S. Large-scale supervised process monitoring based on distributed modified principal component regression [J]. Industrial & Engineering Chemistry Research, 2019, 53 (89): 18223-18240.

[25] Kingma D P, Welling M. Auto-encoding variational bayes [J]. arXiv preprint, arXiv: 2013, 131 2. 6114.

[26] Michal R, Dominik Z, Georg M. Variational autoencoders pursue PCA directions (by accident) [C]// Proceedings of the IEEE/CVF Conference on Computer Vision and Pattern Recognition (CVPR), CVF Publishing, 2019: 12406-12415.

[27] Peng H C, Long F H, Ding C. Feature selection based on mutual information criteria of max-dependency, max-relevance, and min-redundancy [J]. IEEE Transactions on Pattern Analysis & Machine Intelligence, 2005, 27 (8): 1226-1238.

[28] Alemi A A, Fischer I, Dillon J V, et al. Deep variational information bottleneck [J]. arXiv preprint, arXiv: 2016, 1612. 00410.

[29] Qian D, Cheung W K. Enhancing variational autoencoders with mutual information neural estimation for text generation [C]// Proceedings of the 2019 Conference on Empirical Methods in Natural Language Processing and the 9th International Joint Conference on Natural Language Processing, 2013: 4047-4057.

Chapter 14 Lifecycle Operating Performance Assessment Framework Based on RKCVA

HSMP is an essential part of the modern iron and steel industry. In face of increasing competition, improving product quality and maximizing economic benefits are the goals of steel enterprises. However, due to the existence of interference and uncertainty, it is generally known that the operating condition may deviate from the initial optimal operating point as time goes on[1-3]. Nonoptimal operating conditions may lead to the decline of product yields and economic losses. Moreover, once faults occur, they may result in disastrous consequences and hazards for personnel, plant, and the environment. Automation engineers have the responsibility to know the process conditions in time to avoid the occurrences of the above situations. Therefore, it is of great theoretical and practical significance to assess operating performance of the production process, so as to facilitate production managements and process improvements[4-6].

In the last ten years, some studies have been published on the operating performance assessment for different industrial processes. As the first researchers in this field, Ye et al. proposed an online probabilistic assessment framework for multimode industrial processes[7]. Based on the Gaussian mixture model (GMM), assessment indices were designed for each mode and a classification method was developed to divide them into different operating performance grades. Feital et al. proposed a multimodal performance monitoring approach based on maximum likelihood principal component analysis (MLPCA) and a component-wise identification of operating modes[8]. Sedghi and Huang proposed an optimality assessment framework for multimode and multiregion processes[9]. The model-based clustering discriminant analysis (MclustDA) was used for mode detection and sequential forward floating search (SFFS) was introduced for tracing the nonoptimal behavior. Liu et al. proposed assessment strategies for both steady and transient modes. Based on a comprehensive economic index, partial least squares (PLS) was used to predicted the assessment index for different modes[10]. Then, Liu et al. proposed an online assessment method based on performance similarity and total projection to latent structures (T-PLS) for non-Gaussian multimode processes[11].

PLS-based methods focus on maximizing the covariances between process variables and quality variables. It may extract latent information of process variables that have a large magnitude of variations but not necessarily highly correlated to the quality variables. However, canonical variable analysis (CVA) can overcome this drawback as it focuses on maximizing correlations between process variables and quality variables[12]. Moreover, several singular value decompositions (SVDs) are involved behind PLS while processing large-scale industrial data. CVA is more efficient than PLS, because it only needs one-step SVD. Therefore, CVA-based model and it extensions have been deeply studied in recent years, which have largely improved monitoring performance for processes characterized by strong auto-correlated and cross correlated variables[13-14].

Recently, operating performance assessment for plantwide process has been a new research hotpot[15-16]. Zou et al. proposed a hierarchical multiblock assessment method for a gold hydrometallurgy process[17]. Based on fuzzy probabilistic rough sets (FPRS), the operating performance grades were properly defined at the subblock level. The operating performance grade of the global level was directly determined by the assessment results at the subblock level. Chang et al. proposed a two-level multiblock assessment framework for multimode plant-wide process[18]. Quantitative subblocks and qualitative subblocks in each mode were modeled respectively using quantitative and qualitative methods. Based on the two-level hybrid model, the global performance grade was determined by the subblock performance grades. In order to achieve the online assessment for hydrocracking process, Li et al. proposed a two-layer fuzzy synthetic operating performance assessment framework. Based on the correlations with the product quality, the process variables were classified into different subblocks[19]. Then, an assessment system was built based on the concept of positive deviation of the quality variables.

Since the process is not always in normal operating mode, and sometimes faults may occur, it is not enough to only evaluate the normal operating conditions, and abnormal operating conditions should also be evaluated[20]. Therefore, it is necessary to study the lifecycle operating performance assessment for the process. The concept of lifecycle refers to the lifecycle of the process, which consists of fault grade assessment part and normal operating grade assessment part, which is shown in **Fig. 14. 1**.

The lifecycle operating performance assessment for the HSMP has been so far little studied. In this work, a new lifecycle operating performance assessment framework is proposed for the HSMP. As a typical nonlinear plant-wide process, the HSMP consists of many different parts and operation units. From raw materials to the final products, the

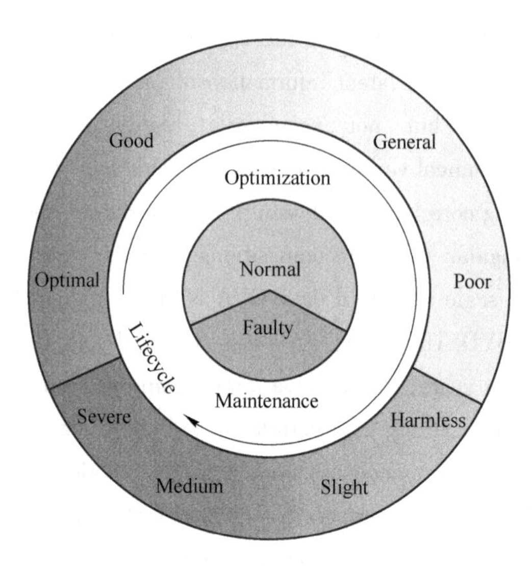

Fig. 14. 1 Lifecycle operating performance assessment

HSMP has obvious automation hierarchy, mainly including equipment level, real-time control level, process control level, manufacturing execution level, business management level, and business strategy level. They work together to maintain a highly efficient operation and ensure stable product quality[21]. In addition, the CVA-based method assumes that the process data is well collected and outliers do not exist. However, for a real HSMP, such assumption is no longer tenable and outliers can be hardly avoided. According to the causes, outliers appear in two main aspects. On one hand, outliers can be occasionally introduced by errors in data collection from measurement devices, data transmission and management; on the other hand, due to the complex operating conditions, HSMP usually has large process noise and disturbances, which generate heavy-tailed data[22]. The existence of outliers has a serious adverse effect on the modeling, which may skew the parameter estimation and lead to the misspecification of the model. To reduce the impacts of outliers, partial robust M-regression (PRM) is proposed[23]. By selecting appropriate weighting strategy, the bad leverage points are weighted down. Compared with other robust methods, PRM is resistant to bad leverage points[24-25]. Therefore, kernel canonical variable analysis (KCVA) is integrated with PRM to overcome both the nonlinearity of the HSMP and the effects of outliers. Two system levels are included in this framework, namely, the real-time control level (L1) and the process control level (L2). In the L1 level, the HSMP is further divided into upstream, midstream and downstream (three-stream) according to the production process. The lifecycle operating performance assessment includes two steps. First, assess whether the process is under normal or faulty operating condition. Then, based on different predesigned rules, fault grades and normal operating grades can be evaluated in the corresponding conditions. The main contributions of this chapter are:

(1) A robust kernel canonical variable analysis (RKCVA) model is developed, which considers nonlinearities and outliers simultaneously.

(2) Based on RKCVA, a fault grade assessment method is proposed to evaluate the faults that occur in the HSMP.

(3) Based on RKCVA, a normal operating grade assessment method is proposed to evaluate the normal operating condition.

14.1 Review of CVA and PRM

14.1.1 Canonical variable analysis

CVA was first proposed by H. Hotelling in the 1930s[26], which is a method for exploring the relationships between two sets of multidimensional variables. Suppose that two datasets are collected from real process and given as:

$$
\begin{aligned}
\boldsymbol{X} &= [\boldsymbol{x}_1, \boldsymbol{x}_2, \cdots, \boldsymbol{x}_n] \in R^{m\times n} \\
\boldsymbol{Y} &= [\boldsymbol{y}_1, \boldsymbol{y}_2, \cdots, \boldsymbol{y}_n] \in R^{q\times n}
\end{aligned}
\tag{14.1}
$$

where, $\boldsymbol{X}$ and $\boldsymbol{Y}$ are process variables and quality variables, m and q are the numbers of corresponding variables, n is the number of samples. The core concept of CCA is to find the optimal projection vectors $\boldsymbol{J}$ and $\boldsymbol{L}$ to maximize the correlation between $\boldsymbol{J}^{\mathrm{T}}\boldsymbol{X}$ and $\boldsymbol{L}^{\mathrm{T}}\boldsymbol{Y}$:

$$
\mathrm{cor}(\boldsymbol{J}^{\mathrm{T}}\boldsymbol{X}, \boldsymbol{L}^{\mathrm{T}}\boldsymbol{Y}) = \frac{\langle \boldsymbol{J}^{\mathrm{T}}\boldsymbol{X}, \boldsymbol{L}^{\mathrm{T}}\boldsymbol{Y}\rangle}{\|\boldsymbol{J}^{\mathrm{T}}\boldsymbol{X}\| \, \|\boldsymbol{L}^{\mathrm{T}}\boldsymbol{Y}\|}
\tag{14.2}
$$

Usually, this is formulated as a constraint optimization problem:

$$
\begin{aligned}
(\boldsymbol{J}, \boldsymbol{L}) &= \mathrm{argmax}\boldsymbol{J}^{\mathrm{T}} \sum\nolimits_{XY} \boldsymbol{L} \\
\text{s. t.} \quad \boldsymbol{J}^{\mathrm{T}} \Sigma_{XX}\boldsymbol{J} &= \boldsymbol{L}^{\mathrm{T}} \Sigma_{YY}\boldsymbol{L} = 1
\end{aligned}
\tag{14.3}
$$

where, $\Sigma_{XX} = E(\boldsymbol{X}\boldsymbol{X}^{\mathrm{T}})$ and $\Sigma_{YY} = E(\boldsymbol{Y}\boldsymbol{Y}^{\mathrm{T}})$ are the autocovariance matrices of $\boldsymbol{X}$ and $\boldsymbol{Y}$, and $\Sigma_{XY} = E(\boldsymbol{X}\boldsymbol{Y}^{\mathrm{T}})$ is the cross-covariance matrix between $\boldsymbol{X}$ and $\boldsymbol{Y}$. Kuss and Graepel showed how this optimization problem can be transformed to an eigenproblem[27].

14.1.2 Partial robust M-regression

Partial robust M-regression (PRM) is a robust form of weighted iterative PLS algorithm with single output variable. The impacts of outliers on the model is overcome by assigning different weights to the samples (the weights of outliers are close to 0, the weights of normal data points are close to 1). Two types of outliers are considered in the PRM model: one is bad leverage points (i. e. outliers in the space of predictor variables), and the other is high residual points. The bad leverage point is located far away from the majority of the data in the input space. The high residual point is the object with large absolute difference between the observed and predicted value of their output. The high residual weight of the i_{th} sample can be calculated:

$$w_i^r = f\left(\frac{r_i}{\tilde{r}},\ c\right) \tag{14.4}$$

$$f(z,\ c) = \frac{1}{(1 + |z/c|)^2} \tag{14.5}$$

where, $f(\cdot)$ is the weight function; r_i is the difference between the actual and predicted value of the i_{th} sample; $\tilde{r} = \underset{i}{\mathrm{median}} \left| r_i - \underset{j}{\mathrm{median}}\ r_j \right|$ is a robust median estimate of residual scale; c is a tuning constant.

The bad leverage weight of the i_{th} sample can be calculated as:

$$w_i^l = f\left(\frac{\| \boldsymbol{t}_i - \mathrm{med}_{\mathrm{L}_1}(\boldsymbol{T}) \|}{\underset{i}{\mathrm{median}} \| \boldsymbol{t}_i - \mathrm{med}_{\mathrm{L}_1}(\boldsymbol{T}) \|},\ c\right) \tag{14.6}$$

where, $\| \cdot \|$ is the Euclidean norm, $\boldsymbol{t}_i$ is the score vector of the i_{th} sample, and $\mathrm{med}_{\mathrm{L}_1}(\boldsymbol{T})$ is the L1-median of score vectors.

Based on the high residual weight w_i^r and bad leverage weight w_i^l, the general weight of the i_{th} sample can be calculated as:

$$w_i = \sqrt{w_i^r w_i^l} \tag{14.7}$$

PRM algorithm updates the general weights iteratively until the convergence is achieved.

14.2 Lifecycle OPA framework based on RKCVA

The operating performance assessment consists of three steps: First, a key performance indicator (KPI) is defined. Then, based on the historical training data set, the prediction model between KPI and process variables is established. Finally, the performance grade can be predicted by the trained models. The KPI is usually based on product quality[28-29]. Unfortunately, product quality is a process specific index, which is difficult to directly apply to other processes. Meanwhile, there are also some drawbacks in the existing methods for plant-wide process. First, the concept of hierarchy does not fit with the concept of automation hierarchy. The so-called hierarchy only considers the concept of local and global models. The automation hierarchy in the HSMP are shown in **Fig. 14.2**. Obviously, previous works only used L1 data. Second, traditional multiblock method has no difference in the importance of each subblock by default, which is obviously a simplification and cannot meet real industrial processes. For the HSMP, the upstream block is far less important than the other three blocks (midstream, downstream, and L2). Third, most of the existing methods only evaluate the operating performance grade of normal operating conditions. Considering the logic and integrity of operating performance assessment, whether the process is in normal or faulty condition should be assessed first. Although the probability of fault occurrence is low, it should not be ignored. Moreover, different strategies should be adopted for different fault grades.

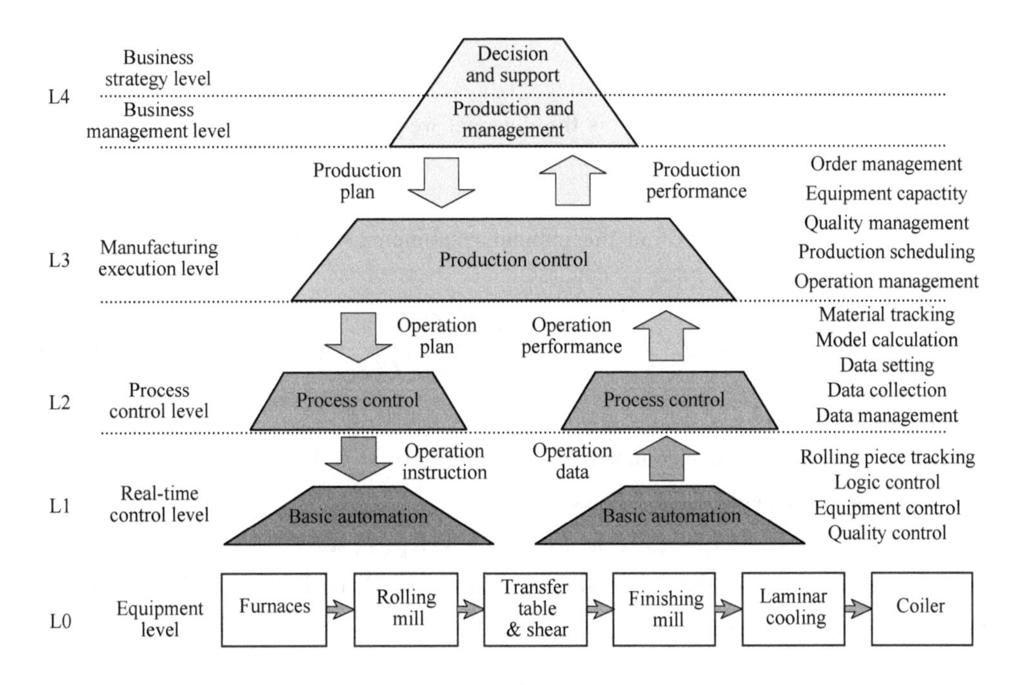

Fig. 14.2 Automation hierarchy in the HSMP

14.2.1 RKCVA model

In order to deal with both the nonlinear characteristic and outliers in the HSMP, kernel technique and PRM is incorporated into the CVA model. Define the normalized input process data as $\boldsymbol{X} = [x_1, x_2, \cdots, x_n]^{\mathrm{T}} \in R^{n\times m}$, where n and m are the number of samples and variables, respectively. Let the normalized output data be $\boldsymbol{Y} = [y_1, y_2, \cdots, y_n]^{\mathrm{T}} \in R^{n\times 1}$. $\boldsymbol{X}$ is first mapped from the original space into a high dimensional feature space by $\boldsymbol{\Phi}$: $\boldsymbol{X} \rightarrow \boldsymbol{\Phi}(\boldsymbol{X})$. The kernel matrix $\boldsymbol{K}$ can be represented as $\boldsymbol{\Phi}(\boldsymbol{X})\boldsymbol{\Phi}^{\mathrm{T}}(\boldsymbol{X})$. The Gaussian kernel function is used to calculate each entry of $\boldsymbol{K}$ as follows:

$$\boldsymbol{K}(i, j) = \langle \boldsymbol{\Phi}(x_i), \boldsymbol{\Phi}(x_j) \rangle = \exp\left(-\frac{\| x_i - x_j \|^2}{\sigma^2}\right) \tag{14.8}$$

where, $\langle \cdot \rangle$ is an inner product operator, and σ is the standard deviation parameter, $i, j = 1, 2, \cdots, n$. Then $\boldsymbol{K}$ and $\boldsymbol{Y}$ are processed by PRM algorithm to eliminate outliers:

$$\begin{aligned} \boldsymbol{K}_W &= (\boldsymbol{W\Phi}(\boldsymbol{X}))(\boldsymbol{W\Phi}(\boldsymbol{X}))^{\mathrm{T}} = \boldsymbol{WKW} \\ \boldsymbol{Y}_W &= \boldsymbol{WY} \end{aligned} \tag{14.9}$$

where, $\boldsymbol{W} = \begin{bmatrix} w_1 & 0 & \cdots & 0 \\ 0 & w_2 & \ddots & \vdots \\ \vdots & \ddots & \ddots & 0 \\ 0 & \cdots & 0 & w_n \end{bmatrix}$ is the diagonal weight matrix.

The goal of RKCVA is to find the optimal combinations of $\boldsymbol{K}_W$ and $\boldsymbol{Y}_W$, where the canonical correlation is defined as follows:

$$\rho(\boldsymbol{J}, \boldsymbol{L}) = \frac{\boldsymbol{J}^{\mathrm{T}}\boldsymbol{K}_W\boldsymbol{C}_Y\boldsymbol{L}}{(\boldsymbol{J}^{\mathrm{T}}\boldsymbol{K}_W^2\boldsymbol{J})^{1/2}(\boldsymbol{L}^{\mathrm{T}}\boldsymbol{C}_Y^2\boldsymbol{L})^{1/2}} \tag{14.10}$$

where, $\boldsymbol{C}_Y = E(\boldsymbol{Y}_W^{\mathrm{T}}\boldsymbol{Y}_W)$ is the autocovariance matrix of $\boldsymbol{Y}_W$, and the coefficient matrices $\boldsymbol{J}$ and $\boldsymbol{L}$ can be calculated by equation (14.3). For any x_i, the reduced projection can be calculated as follows:

$$\boldsymbol{t}_i = \langle \boldsymbol{J}_p, w_i\boldsymbol{\Phi}(x_i) \rangle = \boldsymbol{J}_p^{\mathrm{T}}\boldsymbol{K}_W(:, i) \tag{14.11}$$

where, p is the order of dimensionality reduction, which is determined by minimizing the Akaike information criterion (AIC)[30].

14.2.2 Two-level and three-stream strategy

From raw materials to final products, the HSMP has plant-wide characteristic, which consists of many production units. Meanwhile, there are different levels in the automation system of the HSMP. The material flow, energy flow, and information flow make each production unit and each system level interconnected. In order to simplify the scale of operating performance assessment and make its model have strong physical meaning, the HSMP can be divided into upstream, midstream, and downstream. From the system level perspective, L1 and L2 level are used for lifecycle operating performance assessment. **Fig. 14.3** shows the two-level and three-stream strategy for the HSMP.

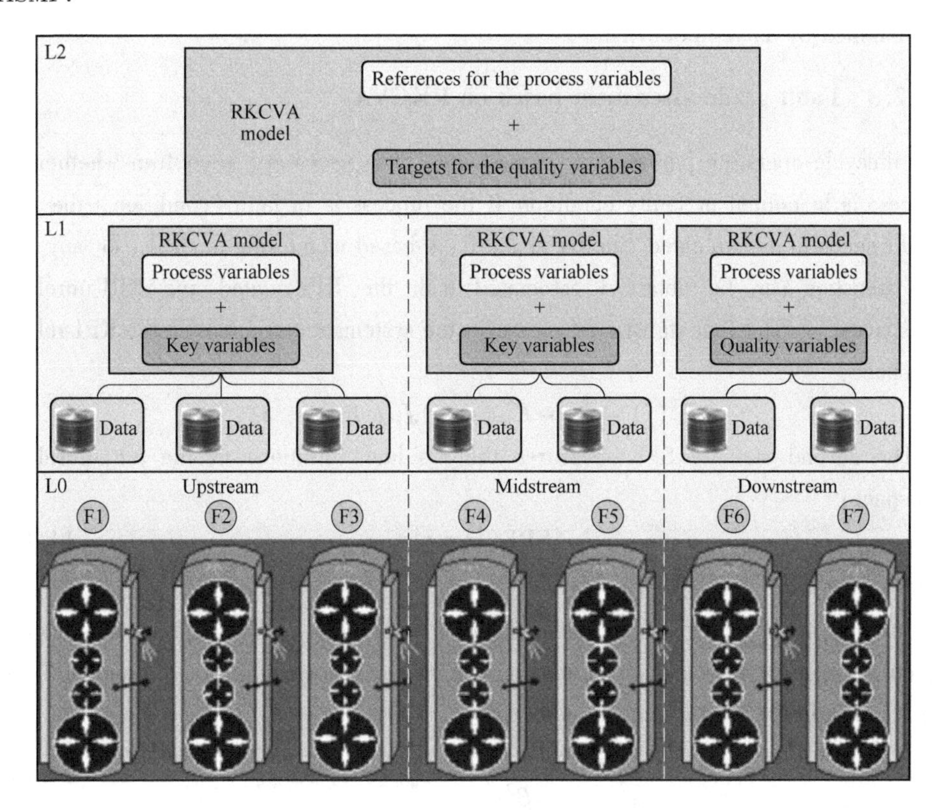

Fig. 14.3 The illustration of a two-level, three-stream strategy for the HSMP

The HSMP is usually made up of seven stands in series. Once one of them is in abnormal operating condition, it will affect the subsequent stands, even the final product quality, which may bring economic losses to the enterprise. Owing to the automatic control systems, the anomalies of upstream stands can be compensated by the adjustment of the subsequent stands. However, when anomalies occur in the midstream or downstream stands, they are difficult to compensate. Thus, the fault levels are diverse in different streams. Based on the process knowledge in the equipment level (L0), the upstream, midstream, and downstream RKCVA models are trained using the real-time data of the first three, middle two, and last two stands respectively. The references of process variables and targets of quality variables are used to train the RKCVA model of L2 level. Restricted by high temperature, the outlet thickness of each stand is not measurable, but it can be approximately expressed by the gap value of supporting rolls. Therefore, the roll gap values of the 4_{th} and 6_{th} stands are taken as the key variables of the upstream and midstream RKCVA models, respectively. The exit thickness is taken as the quality variable of the downstream RKCVA model, which can be obtained by an X-ray device.

14.2.3 Fault grade assessment based on RKCVA

For lifecycle operating performance assessment, it is necessary to confirm whether the process is in normal or faulty condition. If the process is in faulty condition, the fault grade needs to be evaluated. Similar to the CVA-based monitoring method, for any x_i, two statistics can be obtained associated with the KPI-related and KPI-unrelated subspaces[31]. The first statistic T^2 measures the systematic variations in the KPI-related subspace:

$$T^2(i) = \boldsymbol{t}_i^{\mathrm{T}}\boldsymbol{t}_i = \boldsymbol{K}_W^{\mathrm{T}}(:,\ i)\boldsymbol{J}_P\boldsymbol{J}_P^{\mathrm{T}}\boldsymbol{K}_W(:,\ i) \tag{14.12}$$

The second statistic SPE measures the residual variations in the KPI-unrelated subspace:

$$\mathrm{SPE}(i) = \boldsymbol{r}_i^{\mathrm{T}}\boldsymbol{r}_i \tag{14.13}$$

where, $\boldsymbol{r}_i = (\boldsymbol{I} - \boldsymbol{J}_P\boldsymbol{J}_P^{\mathrm{T}})\boldsymbol{K}_W(:,\ i)$.

The thresholds for T^2 and SPE are defined as α^2 and β^2, respectively, which can be calculated using kernel density estimation (KDE)[32]. For simplicity, T^2 and SPE can be combined as a mixed statistic as follows:

$$Y^2(i) = \frac{T^2(i)}{\alpha^2} + \frac{\mathrm{SPE}(i)}{\beta^2} = \boldsymbol{K}_W^{\mathrm{T}}(:,\ i)\boldsymbol{\Psi}\boldsymbol{K}_W(:,\ i) \tag{14.14}$$

$$\boldsymbol{\Psi} = \frac{\boldsymbol{J}_P\boldsymbol{J}_P^{\mathrm{T}}}{\alpha^2} + \frac{(\boldsymbol{I} - \boldsymbol{J}_P\boldsymbol{J}_P^{\mathrm{T}})^{\mathrm{T}}(\boldsymbol{I} - \boldsymbol{J}_P\boldsymbol{J}_P^{\mathrm{T}})}{\beta^2} \tag{14.15}$$

where, the threshold of Y^2 is defined as γ^2, which can be calculated by KDE.

As shown in **Fig. 14.3**, there are four mixed statistics with different thresholds, which is difficult to fuse directly. In order to form a concise decision, the statistics can be transformed to fault probabilities based on Bayesian inference. Thus, the fault probability of x_i in the upstream RKCVA model is calculated as:

$$P_{\mathrm{up}}(F \mid x_{i,\mathrm{up}}) = \frac{P_{\mathrm{up}}(x_{i,\mathrm{up}} \mid F)P_{\mathrm{up}}(F)}{P_{\mathrm{up}}(x_{i,\mathrm{up}})} \tag{14.16}$$

$$P_{\mathrm{up}}(x_{i,\mathrm{up}}) = P_{\mathrm{up}}(x_{i,\mathrm{up}} \mid F)P_{\mathrm{up}}(F) + P_{\mathrm{up}}(x_{i,\mathrm{up}} \mid N)P_{\mathrm{up}}(N) \tag{14.17}$$

where, $P_{\mathrm{up}}(F)$ and $P_{\mathrm{up}}(N)$ are the prior probabilities of faulty condition F and normal condition N, respectively.

When $P_{up}(F)$ is selected as the significance level l, $P_{up}(N)$ can be defined as $1-l$. As defined in reference [33], the conditional probabilities $P_{up}(x_{i,up} \mid F)$ and $P_{up}(x_{i,up} \mid N)$ can be calculated as:

$$
\begin{aligned}
P_{up}(x_{i,up} \mid F) &= \exp\left[-\frac{Y_{up}^2(i)}{\gamma_{up}^2}\right] \\
P_{up}(x_{i,up} \mid N) &= \exp\left[-\frac{\gamma_{up}^2}{Y_{up}^2(i)}\right]
\end{aligned}
\tag{14.18}
$$

where, γ_{up}^2 is the threshold of Y_{up}^2 in the upstream. Similarly, the fault probabilities of x_i in the midstream, downstream, and L2 RKCVA model can be calculated, which are defined as $P_{mid}(F \mid x_{i,mid})$, $P_{down}(F \mid x_{i,down})$ and $P_{L2}(F \mid x_{i,L2})$, respectively. Based on the above four fault probabilities, a concise weighted statistic can be calculated as:

$$
P(F \mid x_i) = \sum_s \frac{P_s(x_{i,s} \mid F) P_s(F \mid x_{i,s})}{\sum_s P_s(x_{i,s} \mid F)} \tag{14.19}
$$

where, $s = \{up, mid, down, L2\}$.

For a new sample x_{new}, if the concise statistic $P(F \mid x_{new}) > l$, it means that a fault occurs in the HSMP. Then four fault grades should be evaluated according to the following procedures:

(1) Severe faults ($P_{L2}(F \mid x_{new,L2}) > l$): Faults occur at the L2 level and endanger process safety. Due to human factors, system parameter setting errors make the process difficult to meet normal production requirements. In order to ensure process safety and avoid serious economic losses, parameter settings need to be modified immediately.

(2) Medium fault ($P_{down}(F \mid x_{new,down}) > l$): Faults occur in the downstream stands and degrade key process operating performance. In this case, the product quality is seriously affected. Moreover, potential safety hazards may exist in the process.

(3) Slight faults ($P_{mid}(F \mid x_{new,mid}) > l$): Faults occur in the midstream stands and degrade non-key process operating performance. The fault probability of the downstream stands is increased, and thus such faults may affect key process operating performance.

(4) Harmless faults ($P_{up}(F \mid x_{new,up}) > l$): Faults occur in the upstream stands

and have no harmful influence on process operating performance. Owing to intrinsic control effect of the HSMP, such faults can be compensated or even eliminated.

14.2.4 Normal operating grade assessment based on RKCVA

If the concise statistic $P(F \mid x_{\text{new}})$ is lower than its threshold, the HSMP is under normal operating condition. In such condition, the goal of monitoring changes from production safety to CEB maximization. Therefore, it is necessary to assess normal operating grades of the HSMP for further adjusting equipment and gaining higher profits. There are many factors affecting the KPI of the HSMP, such as material, energy cost, and product quality, etc. However, considering the complexity of the simulation experiment and the difficulty of data collection, product quality is used to represent the KPI in this chapter. Among many quality indicators, the strip thickness is a widely used key performance indicator in the HSMP. Compared with upstream stands, midstream and downstream stands have much more influences on the strip thickness. Therefore, only the operating grades of midstream and downstream stands are evaluated in this chapter, and the assessment results are weighted to obtain an overall operating grade of the HSMP.

Based on the national standard GB/T 709—2006 and expert experience, the division criterions of normal operating grades for different products are listed in **Table 14.1**. For simplicity, the four operating grades in **Table 14.1**, i.e. optimal, good, general and poor are numbered as 1, 2, 3, and 4. Assuming that the training datasets of midstream and downstream stands are $(\boldsymbol{X}_{\text{mid}}^c, \boldsymbol{Y}_{\text{mid}}^c)$ and $(\boldsymbol{X}_{\text{down}}^c, \boldsymbol{Y}_{\text{down}}^c)$, $c = 1, 2, 3, 4$. Based on RKCVA model, the midstream and downstream canonical correlation features of grade c can be obtained as follows:

$$
\begin{aligned}
\boldsymbol{T}_{\text{mid}}^c &= \langle \boldsymbol{J}_{P_{\text{mid}}^c}, \boldsymbol{W}_{\text{mid}}^c \boldsymbol{\Phi}(\boldsymbol{X}_{\text{mid}}^c) \rangle = \boldsymbol{J}_{P_{\text{mid}}^c}^{\text{T}} \boldsymbol{K}_{W_{\text{mid}}}^c \\
\boldsymbol{T}_{\text{down}}^c &= \langle \boldsymbol{J}_{P_{\text{down}}^c}, \boldsymbol{W}_{\text{down}}^c \boldsymbol{\Phi}(\boldsymbol{X}_{\text{down}}^c) \rangle = \boldsymbol{J}_{P_{\text{down}}^c}^{\text{T}} \boldsymbol{K}_{W_{\text{down}}}^c
\end{aligned}
\tag{14.20}
$$

where, $\boldsymbol{J}_{P_{\text{mid}}^c}$ and $\boldsymbol{J}_{P_{\text{down}}^c}$ are the midstream and downstream coefficient matrices of grade c, respectively; $\boldsymbol{K}_{W_{\text{mid}}}^c$ and $\boldsymbol{K}_{W_{\text{down}}}^c$ are the midstream and downstream weighted kernel matrices of grade c, respectively.

Table 14.1 Normal operating grade division based on thickness

Thickness (mm)	Optimal	Good	General	Poor
0.80 to 1.50	±0.04	±0.08	±0.12	±0.15
1.50 to 2.00	±0.05	±0.09	±0.13	±0.17
2.00 to 2.50	±0.05	±0.10	±0.14	±0.18

Continued Table 14.1

Thickness (mm)	Optimal	Good	General	Poor
2.50 to 3.00	±0.05	±0.10	±0.15	±0.20
3.00 to 4.00	±0.05	±0.11	±0.17	±0.22

For online operating grade assessment, only one sample is not effective to characterize the process performance and is easily disturbed by outliers. Therefore, a window of the dataset with width h is defined as the basic unit, i.e. $\boldsymbol{X}_{\mathrm{mid}}(t) = [\boldsymbol{x}_{\mathrm{mid},\, t-h+1}, \boldsymbol{x}_{\mathrm{mid},\, t-h+2}, \cdots, \boldsymbol{x}_{\mathrm{mid},\, t}]^{\mathrm{T}}$ and $\boldsymbol{X}_{\mathrm{down}}(t) = [\boldsymbol{x}_{\mathrm{down},\, t-h+1}, \boldsymbol{x}_{\mathrm{down},\, t-h+2}, \cdots, \boldsymbol{x}_{\mathrm{down},\, t}]^{\mathrm{T}}$, where t is the time index.

Similarly, the canonical correlation features of grade c for $\boldsymbol{X}_{\mathrm{mid}}(t)$ and $\boldsymbol{X}_{\mathrm{down}}(t)$ can be obtained as:

$$
\begin{aligned}
\hat{\boldsymbol{T}}^{c}_{\mathrm{mid}}(t) &= \boldsymbol{J}^{\mathrm{T}}_{P^{c}_{\mathrm{mid}}} \hat{\boldsymbol{K}}^{c}_{W_{\mathrm{mid}}}(t) \\
\hat{\boldsymbol{T}}^{c}_{\mathrm{down}}(t) &= \boldsymbol{J}^{\mathrm{T}}_{P^{c}_{\mathrm{down}}} \hat{\boldsymbol{K}}^{c}_{W_{\mathrm{down}}}(t)
\end{aligned}
\tag{14.21}
$$

where, $\hat{\boldsymbol{K}}^{c}_{W_v}(t) = \boldsymbol{W}^{c}_{v}\boldsymbol{\Phi}(\boldsymbol{X}^{c}_{v})\boldsymbol{\Phi}^{\mathrm{T}}(\boldsymbol{X}_{v}(t))$; $v = \{\mathrm{mid},\ \mathrm{down}\}$.

Inspired by the successful use of PCA similarity factor for measuring the similarity of two datasets[34-36], for grade c, a joint difference factor D^{c}_{joint} is defined as:

$$
\begin{aligned}
D^{c}_{\mathrm{mid}} &= \frac{1}{P^{c}_{\mathrm{mid}}} \left| \mathrm{trace}\left(\boldsymbol{T}^{c}_{\mathrm{mid}} (\boldsymbol{T}^{c}_{\mathrm{mid}})^{\mathrm{T}} - \hat{\boldsymbol{T}}^{c}_{\mathrm{mid}}(t) (\hat{\boldsymbol{T}}^{c}_{\mathrm{mid}}(t))^{\mathrm{T}}\right) \right| \\
D^{c}_{\mathrm{down}} &= \frac{1}{P^{c}_{\mathrm{down}}} \left| \mathrm{trace}\left(\boldsymbol{T}^{c}_{\mathrm{down}} (\boldsymbol{T}^{c}_{\mathrm{down}})^{\mathrm{T}} - \hat{\boldsymbol{T}}^{c}_{\mathrm{down}}(t) (\hat{\boldsymbol{T}}^{c}_{\mathrm{down}}(t))^{\mathrm{T}}\right) \right| \\
D^{c}_{\mathrm{joint}} &= \eta D^{c}_{\mathrm{mid}} + (1-\eta) D^{c}_{\mathrm{down}}
\end{aligned}
\tag{14.22}
$$

where, P^{c}_{mid} and P^{c}_{down} are the orders of midstream and downstream dimensionality reduction, respectively; $0 \leqslant \eta \leqslant 1$ is a joint coefficient, which avoids the instability and unreliability of D^{c}_{joint} and makes the assessment result more reasonable.

For the convenience of the assessment, a weighted assessment index[37] based on D^{c}_{joint} is defined for grade c:

$$
D^{c}_{\mathrm{RKCVA}} = \begin{cases} \dfrac{1/D^{c}_{\mathrm{joint}}}{\sum_{c=1}^{4} 1/D^{c}_{\mathrm{joint}}}, & \text{if } D^{c}_{\mathrm{joint}} \neq 0 \\ D^{c}_{\mathrm{RKCVA}} = 1 \text{ and } D^{c'}_{\mathrm{RKCVA}} = 0 (c' \neq c), & \text{if } D^{c}_{\mathrm{joint}} = 0 \end{cases}
\tag{14.23}
$$

where, $\sum_{c=1}^{4} D^{c}_{\mathrm{RKCVA}} = 1$, $0 \leqslant D^{c}_{\mathrm{RKCVA}} \leqslant 1$.

Obviously, the closer D^c_{RKCVA} is to 1, the closer the process is operating on grade c. At time t, the normal operating grade of the HSMP Grade(t) can be represented as follows:

$$\mathrm{Grade}(t) = \mathrm{argmax}\{D^c_{\mathrm{RKCVA}},\ c = 1,\ 2,\ 3,\ 4\} \tag{14.24}$$

In summary, the entire procedures of the proposed lifecycle operating performance assessment framework are shown in **Fig. 14. 4**.

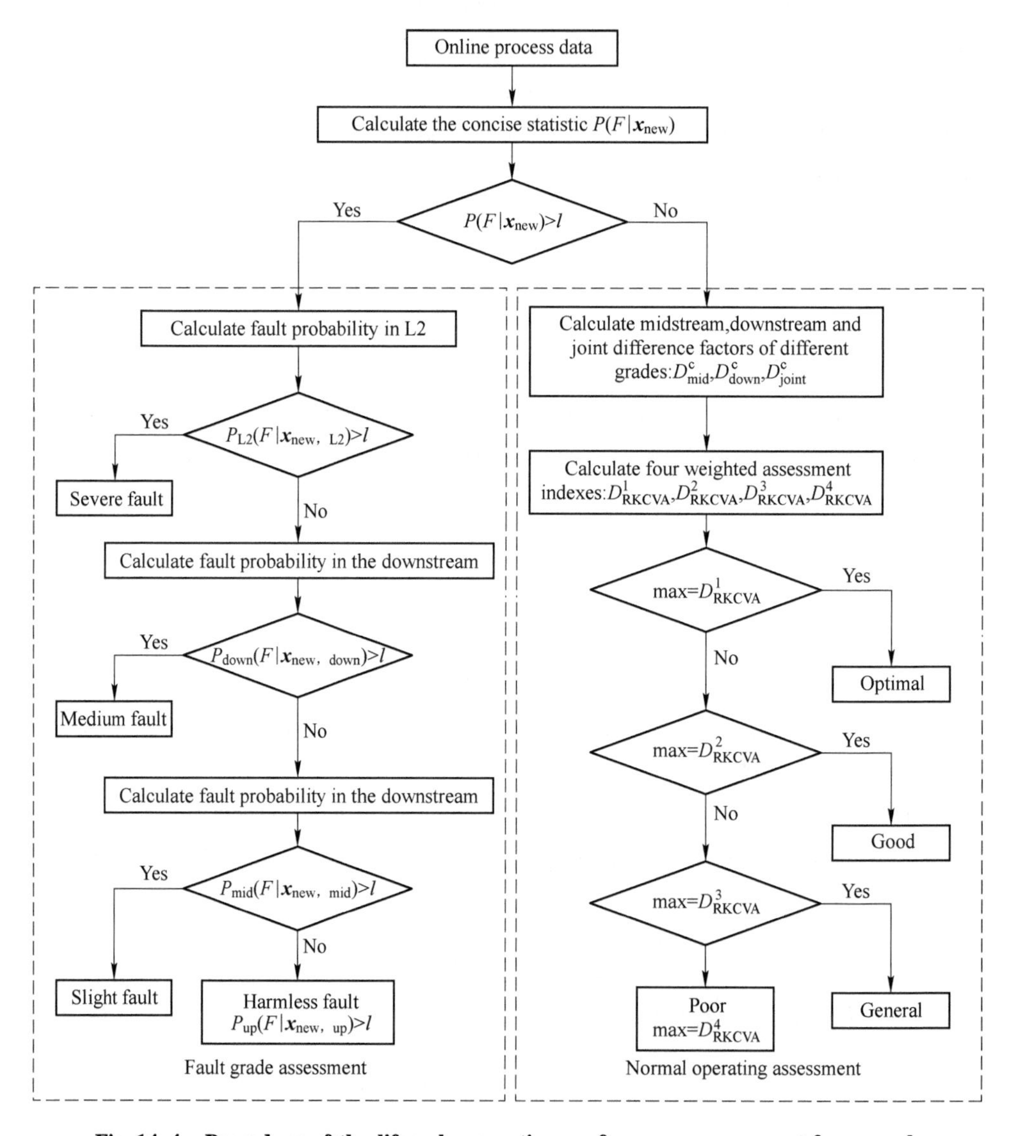

Fig. 14. 4 Procedure of the lifecycle operating performance assessment framework

14.3 Case study

In this section, the proposed lifecycle operating performance assessment framework is applied to a real HSMP. The assessment results based on RKCVA model are compared with some existing methods to illustrate the advantages and validity of the proposed method. In this chapter, the thickness is selected as the KPI, and 40 variables closely related to this index at the L1 and L2 levels are used for lifecycle operating performance assessment, which are shown in **Table 14.2**. Four hundred samples in each normal operating grade (1600 normal operating samples in total) are used to train the two-level, three-stream RKCVA model, and 5% of them are randomly selected to be introduced with disturbances. Furthermore, the significance level l is defined as 5%.

Table 14.2 Process variables in the HSMP

Variable	Description	Unit
G_i (L1)	Roll gap at the i_{th} stand, $i=1, \cdots, 7$	mm
F_i (L1)	Rolling force at the i_{th} stand, $i=1, \cdots, 7$	MN
B_i (L1)	Bending force at the i_{th} stand, $i=2, \cdots, 7$	MN
S_i (L1)	Rolling speed at the i_{th} stand, $i=1, \cdots, 7$	mm/s
Gs_i (L2)	Roll gap reference at the i_{th} stand, $i=1, \cdots, 7$	mm
Bs_i (L2)	Bending force reference at the i_{th} stand, $i=2, \cdots, 7$	MN

14.3.1 Fault grade assessment resultsand analysis

In order to validate the fault grade assessment performance of the proposed model, it is compared with a KCVA-based model. Two typical faults in the HSMP are used in this subsection. For each fault, there are 500 testing samples.

14.3.1.1 Case 1 (medium fault)

Fault 1 is caused by a bending force sensor failure in the 5_{th} stand, which can significantly affect the measurements of the corresponding bending force. It occurs at the 150_{th} testing sample and ends at the 300_{th} testing sample. **Fig. 14.5** presents the detection results of KCVA-based method. It can be found that there are some false alarm samples in the three streams (L1) and L2 level. Compared with the KCVA-based method, the detection results of the proposed method are shown in **Fig. 14.6**. Obviously, the RKCVA-based method has higher fault detection rates (FDRs) and

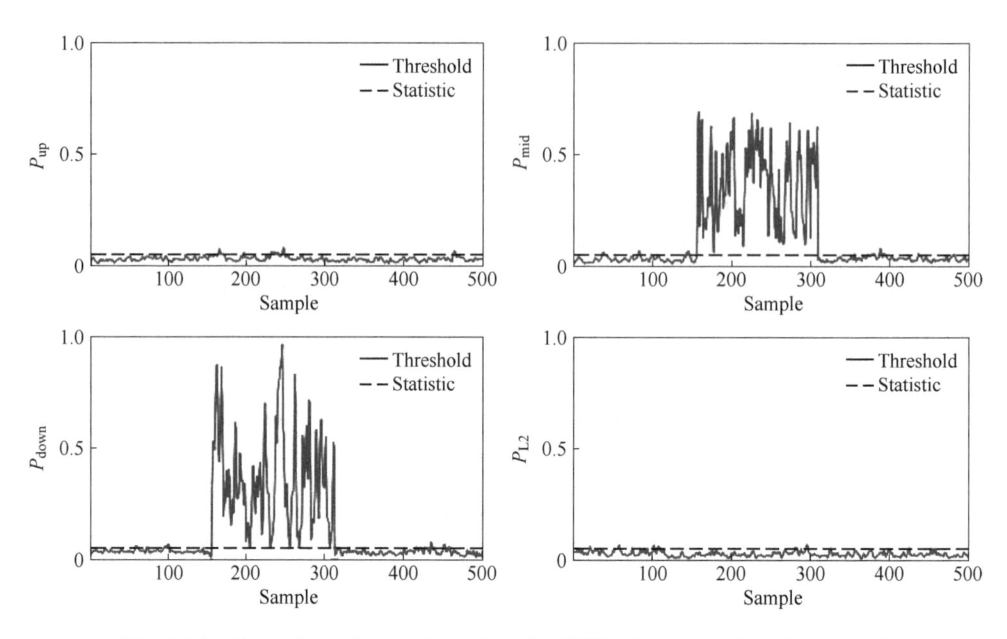

Fig. 14. 5 Fault detection results using the KCVA-based method for Fault 1

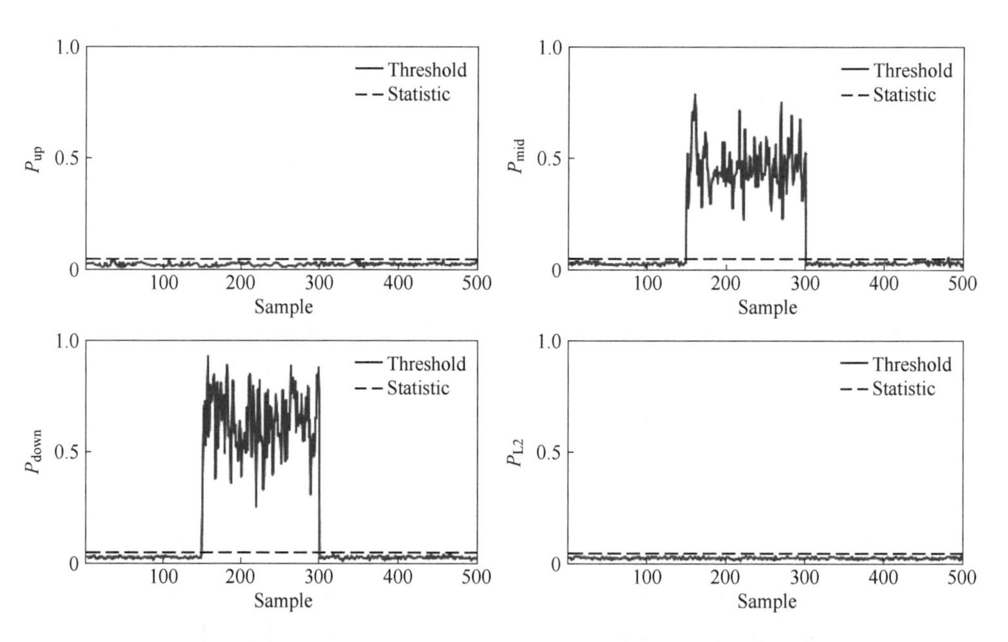

Fig. 14. 6 Fault detection results using the proposed method for Fault 1

lower false alarm rates (FARs) in the three streams and L2 level. Moreover, it can be clearly observed that Fault 1 originates in the midstream stands and propagates to the downstream. According to the fault grade assessment procedures, this fault belongs to medium grade. The assessment result is consistent with the actual situation in the 5_{th} stand.

14.3.1.2 Case 2 (severe fault)

Fault 2 is caused by the malfunction of gap control loop in the 4_{th} stand, which can leads to the abnormal setting of the gap and affect the final rolling thickness. It occurs at the 300_{th} testing sample and lasts to the 500_{th} testing sample. **Fig. 14.7** and **Fig. 14.8** show the detection results given by the above two methods. In each subgraph of **Fig. 14.8**, it can be seen that the KCVA-based method has more false alarm rates in the first 300 samples. In addition, the proposed method can promptly detect Fault 2 at the 300_{th} testing sample, while the KCVA-based one has detection delay.

Unlike Fault 1, the L2 level detects this fault, which indicates Fault 2 belongs to the severe grade. In order to ensure the safety of production and the economic benefits of enterprises, field engineers need to modify the parameter setup of the gap control loop.

The FDRs and FARs of the KCVA and proposed method are summarized in **Table 14.3** and **Table 14.4**. Based on the above analysis, the proposed framework can, not only detect the faults accurately, but also evaluate fault grades effectively.

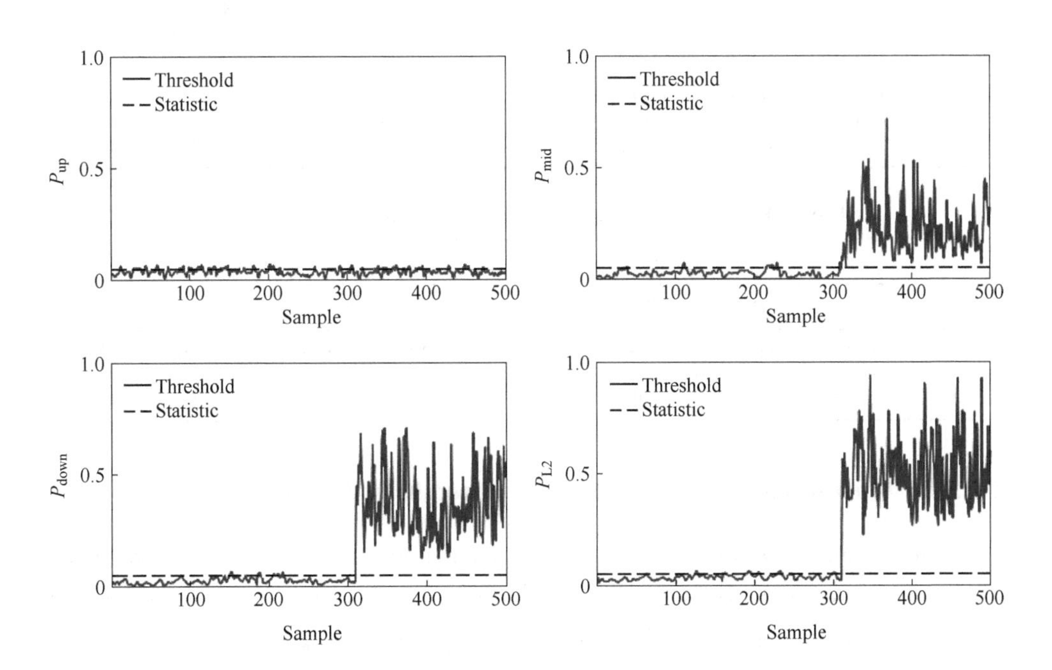

Fig. 14.7 Fault detection results using the KCVA-based method for Fault 2

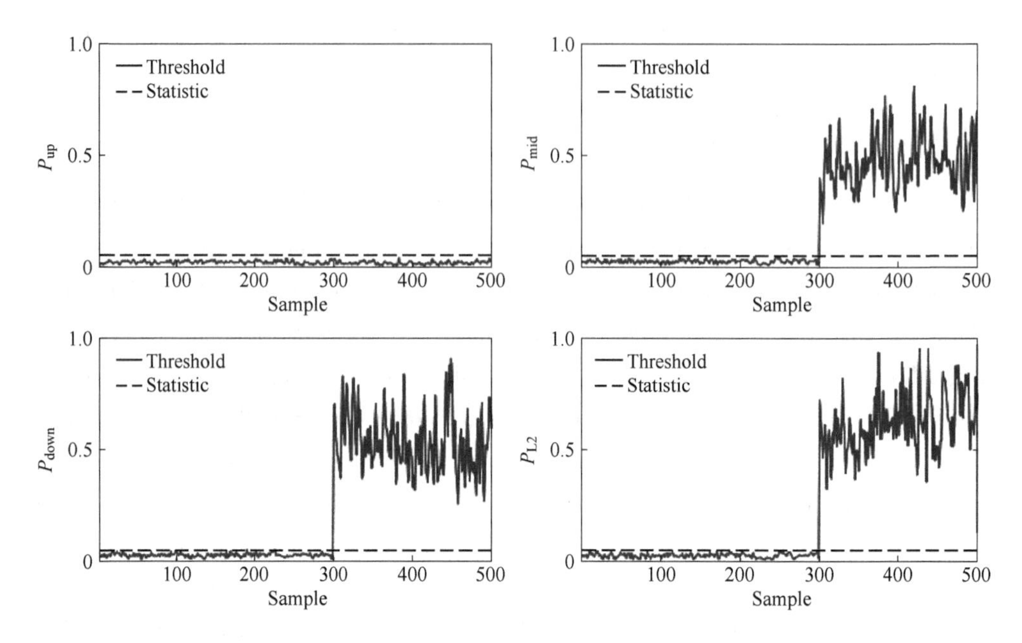

Fig. 14. 8 Fault detection results using the proposed method for Fault 2

Table 14. 3 FDRs of the KCVA-based and the proposed method

Fault No.	KCVA-based				Proposed			
	P_{up}	P_{mid}	P_{down}	P_{L2}	P_{up}	P_{mid}	P_{down}	P_{L2}
1	14. 0%	**95. 3%**	**96. 0%**	4. 7%	0%	**100%**	**100%**	0%
2	13. 5%	**95. 5%**	**95. 5%**	**96. 5%**	0%	**100%**	**100%**	**100%**

Table 14. 4 FARs of the KCVA-based and the proposed method

Fault No.	KCVA-based				Proposed			
	P_{up}	P_{mid}	P_{down}	P_{L2}	P_{up}	P_{mid}	P_{down}	P_{L2}
1	0. 9%	**11. 7%**	**12. 3%**	6. 6%	0%	**0. 6%**	**0. 3%**	0%
2	15. 7%	**5. 7%**	**7. 0%**	**10. 7%**	0%	**0. 3%**	**1. 7%**	**0. 3%**

14. 3. 2 Normal operating grade assessment results and analysis

In order to prove the validity of normal operating grade assessment, the proposed method is applied to the HSMP under normal conditions. A total of 500 samples are used to test the RKCVA-based model, including four performance grades: optimal, good, general, and poor, corresponding to 125, 125, 125 and 125 samples, respectively. The parameters used in normal operating grade assessment are set as follows: the width of dataset $h = 7$, the joint coefficient $\eta = 0.4$.

For comparison, the evaluation model is also established using KCVA and total kernel partial M-regression (T-KPRM). In this work, an assessment accuracy rate (AAR) of each normal operating grade is used to evaluate the assessment performance, which is defined by:

$$\mathrm{AAR}_s = \frac{\text{samples(correctly identified)}}{\text{total samples}}, \quad s = 1, 2, 3, 4 \tag{14.25}$$

Table 14.5 presents the AAR_s of the KCVA-based method, the T-KPRM-based method, and the proposed method. For each normal operating grade, the RKCVA-based model has higher accuracy than the KCVA-based model and the T-KPRM-based model. In addition, the average accuracy of the proposed model is nearly 13% higher than that of the KCVA-based model and 5% higher than that of the T-KPRM-based model. **Fig. 14.9**, **Fig. 14.10**, and **Fig. 14.11** show the assessment results given by the above three methods, respectively.

Table 14.5 AARs of the KCVA-based, the T-KPRM-based, and the proposed method

Method	Optimal	Good	General	Poor	Average
KCVA-based	85.6%	84.0%	83.2%	86.4%	84.8%
T-KPRM-based	92.8%	93.6%	92.0%	92.8%	92.8%
Proposed	97.6%	98.4%	96.8%	96.8%	97.4%

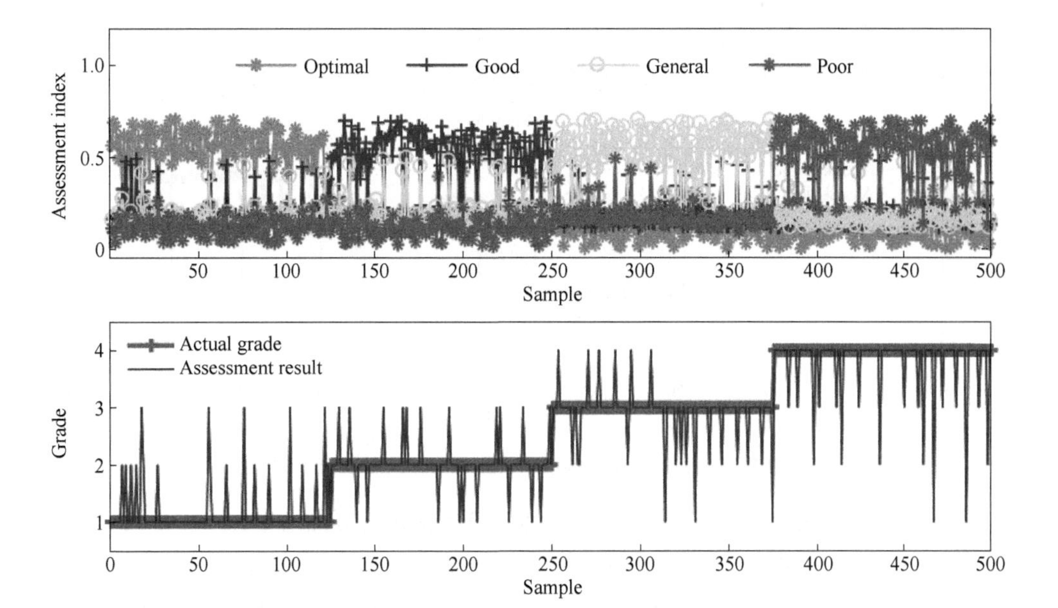

Fig. 14.9 The assessment result of the KCVA-based method

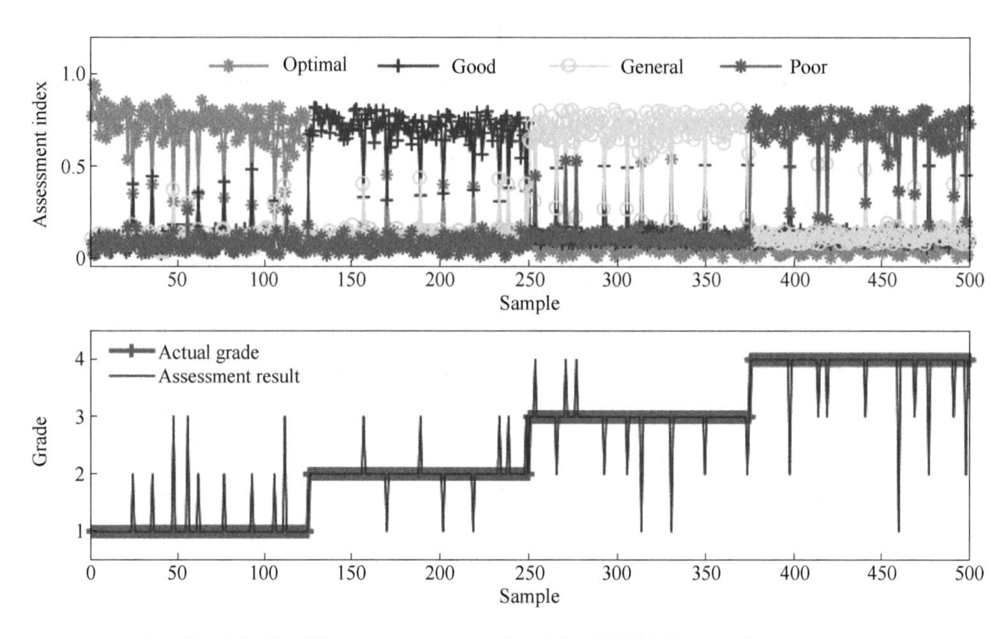

Fig. 14. 10 The assessment result of the T-KPRM-based method

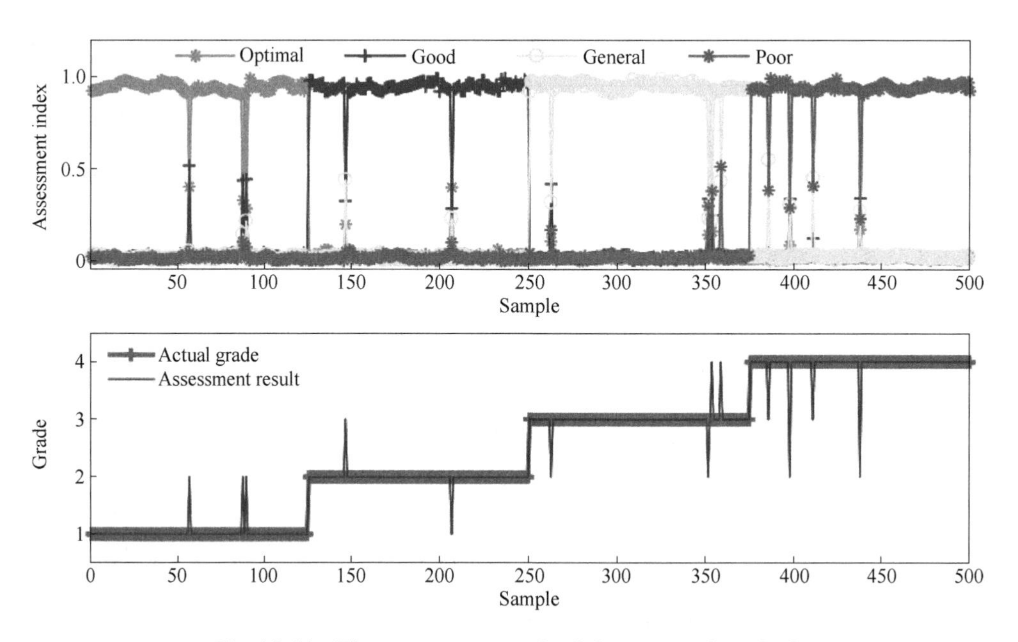

Fig. 14. 11 The assessment result of the proposed method

In **Fig. 14. 9**, it can be found that the KCVA-based method makes the most wrong assessments. The T-KPRM-based method has fewer wrong assessments as shown in **Fig. 14. 10**. Compared with these two methods, the proposed method has the fewest wrong assessments as shown in **Fig. 14.11**. In addition, the operating grade of the testing samples can be clearly confirmed according to the assessment index of the proposed method. However, the assessment indexes of KCVA-based method and T-KPRM-based method may be difficult for field engineers to make decisions. This is because that RKCVA can, not only deal with outliers, but also identify the correlations between process variables and quality variables. The joint coefficient η is a key parameter for normal operating grade assessment. Thus, the average AARs with different values of η are presented in **Fig. 14.12**. The value of η in RKCVA model is initially set to be 0. Then, η is increased by 0.1 each time until it reaches to 1. From **Fig. 14.12**, it can be observed that the average AAR increases at first and then decreases with the increasing of η . When the value of η is 0.4, the average ASR reaches its optimal value.

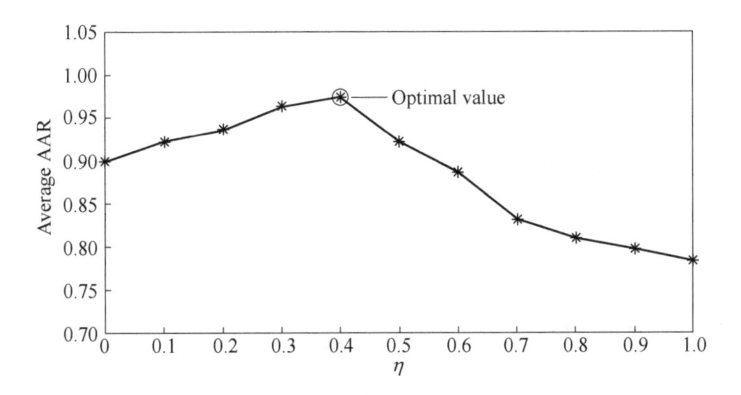

Fig. 14.12 The ARRs for different values of η

Based on the above simulation results and analysis, it can be concluded that the proposed framework not only works well in fault grade assessment, but also achieves good results in normal operating assessment, which will provide theoretical basis and technical support for the safety monitoring and quality assurance of the HSMP.

14.4 Conclusions

In this chapter, a new lifecycle operating performance assessment framework based on RKCVA is proposed for the HSMP. The RKCVA model can deal with nonlinearities and outliers simultaneously by integrating the advantages of KCVA and PRM. In online assessment, the proposed framework includes not only fault grade assessment unit, but also normal operating grade assessment unit, which is called the lifecycle operating performance assessment. First, a concise statistic based on Bayesian inference is calculated to determine whether the process is in faulty or normal operating condition. If the process is in faulty operating condition, the fault grade can be evaluated by the fault probability. Then if the process is in normal operating condition, the normal operating grade can be assessed by the weighted assessment index. Finally, the proposed framework is applied to the process data from the HSMP. Compared with the KCVA-based and T-KPRM-based methods, RKCVA can not only deal with outliers, but also identify the correlations between process variables and quality variables. Thus, the assessments results of the proposed framework show the highest AARs among the three methods.

References

[1] Ding S X, Yin S, Peng K X, et al. A novel scheme for key performance indicator prediction and diagnosis with application to an industrial hot strip mill [J]. IEEE Transactions on Industrial Informatics, 2013, 9 (4): 2239-2247.

[2] Peng K X, Zhang K, Dong J, et al. Quality-relevant fault detection and diagnosis for hot strip mill process with multi-specification and multi-batch measurements [J]. Journal of the Franklin Institute, 2015, 352 (3): 987-1006.

[3] Shardt Y A W, Mehrkanoon S, Zhang K, et al. Modelling the strip thickness in hot steel rolling mills using least-squares support vector machines [J]. The Canadian Journal of Chemical Engineering, 2018, 96: 171-178.

[4] Huang B, Kadali R. Dynamic modeling, predictive control and performance monitoring [J]. Landon: Springer-Verlag, 2008.

[5] Yin S, Lou H, Ding S X. Real-time implementation of fault-tolerant control systems with performance optimization [J]. IEEE Transactions on Industrial Electronics, 2014, 61 (5): 2402-2411.

[6] Wang Y Q, Wei S L, Zhou D H, et al. Control performance assessment for ILC-controlled batch processes in a 2-D system framework [J]. IEEE Transactions on Systems, Man, and Cybernetics: Systems, 2018, 48 (9): 1493-1504.

[7] Ye L, Liu Y, Fei Z, et al. Online probabilistic assessment of operating performance based on safety and optimality indices for multimode industrial processes [J]. Industrial & Engineering Chemistry Research, 2009, 48 (24): 10912-10923.

[8] Feital T, Kruger U, Dutra J, et al. Modeling and performance monitoring of multivariate multimodal processes [J]. AIChE Journal, 2012, 59: 1557-1569.

[9] Sedghi S, Huang B. Real-time assessment and diagnosis of process operating performance [J]. Engineering, 2017, 3 (2): 214-219.

[10] Liu Y, Wang F L, Chang Y Q, et al. Comprehensive economic index prediction based operating optimality assessment and nonoptimal cause identification for multimode processes [J].

Chemical Engineering Research & Design, 2015, 97: 77-90.

[11] Liu Y, Wang F L, Chang Y Q, et al. Operating optimality assessment and nonoptimal cause identification for non-Gaussian multimode processes with transitions [J]. Chemical Engineering Science, 2015, 137: 106-118.

[12] Liu Q, Zhu Q Q, Qin S Q, et al. Dynamic concurrent kernel {CCA} for strip-thickness relevant fault diagnosis of continuous annealing processes [J]. Journal of Process Control, 2018, 67: 12-22.

[13] Chen Z W, Ding S X, Peng T, et al. Fault detection for non-Gaussian processes using generalized canonical correlation analysis and randomized algorithms [J]. IEEE Transactions on Industrial Electronics, 2018, 65 (2): 1559-1567.

[14] Wang Y, Jiang Q C, Yan X F, et al. Joint-individual monitoring of large-scale chemical processes with multiple interconnected operation units incorporating multiset CCA [J]. Chemometrics & Intelligent Laboratory Systems, 2017, 166: 14-22.

[15] Ge Z Q. Review on data-driven modeling and monitoring for plant-wide industrial processes [J]. Chemometrics & Intelligent Laboratory Systems, 2017, 171: 16-25.

[16] Zhao C H, Sun H. Dynamic distributed monitoring strategy for large-scale nonstationary processes subject to frequently varying conditions under closed-loop control [J]. IEEE Transactions on Industrial Electronics, 2019, 66 (6): 4749-4758.

[17] Zou X Y, Wang F L, Chang Y Q, et al. Two-level multi-block operating performance optimality assessment for plant-wide processes [J]. The Canadian Journal of Chemical Engineering, 2018, 96: 2395-2407.

[18] Chang Y Q, Zou X Y, Wang F L, et al. Multi-mode plant-wide process operating performance assessment based on a novel two-level multi-block hybrid model [J]. Chemical Engineering Research & Design, 2018, 136: 721-733.

[19] Li L, Yan X F, Wang Y L, et al. A two-layer fuzzy synthetic strategy for operational performance assessment of an industrial hydrocracking process [J]. Control Engineering Practice, 2019, 93 (9): 104187.

[20] Luo L J, Lovelett R J, Ogunnaike B Q. Hierarchical monitoring of industrial processes for fault detection, fault grade evaluation, and fault diagnosis [J]. AIChE Journal, 2017, 63 (7): 2781-2795.

[21] Ma L, Dong J, Peng K X. A novel hierarchical detection and isolation framework for quality-related multiple faults in large-scale processes [J]. IEEE Transactions on Industrial Electronics, 2020, 67 (2): 1316-1327.

[22] Zhu J L, Ge Z Q, Song Z H, et al. Review and big data perspectives on robust data mining approaches for industrial process modeling with outliers and missing data [J]. Annual Reviews in Control, 2018, 46: 107-133.

[23] Serneels S, Croux C, Filzmoser P, et al. Partial robust M-regression [J]. Chemometrics & Intelligent Laboratory Systems, 2000, 79: 55-64.

[24] Jia R D, Mao Z Z, Chang Y Q, et al. Kernel partial robust M-regression as a flexible robust nonlinear modeling technique [J]. Chemometrics & Intelligent Laboratory Systems, 2010, 100: 91-98.

[25] Chu F, Dai W, Shen J, et al. Online complex nonlinear industrial process operating optimality assessment using modified robust total kernel partial {M-regression} [J]. Chinese Journal of Chemical Engineering, 2018, 26: 775-785.

[26] Hotelling H. Relations between two sets of variates [J]. Biometrika, 1936, 28: 312-377.

[27] Kuss M, Graepel T. The geometry of kernel canonical correlation analysis [J]. Technical Report 108, Max-Planck Institute for Biol. Cybernetics, 2003.

[28] Shardt Y A W, Hao H Y, Ding S X. A new soft sensor-based process monitoring scheme incorporating infrequent KPI measurements [J]. IEEE Transactions on Industrial Electronics, 2015, 62: 3843-3851.

[29] Ma L, Dong J, Peng K X, et al. Hierarchical monitoring and root cause diagnosis framework for key performance indicator-related multiple faults in process industries [J]. IEEE Transactions on Industrial Informatics, 2019, 15 (4): 2091-2100.

[30] Chiang L H, Russell E L, Braatz R D. Fault detection and diagnosis in industrial systems [J]. New York: Springer, 2001.

[31] Zhu Q Q, Liu Q, Qin S J. Concurrent quality and process monitoring with canonical correlation analysis [J]. Journal of Process Control, 2017, 60: 95-103.

[32] Odiowei P E P, Cao Y. Nonlinear dynamic process monitoring using canonical variate analysis and kernel density estimations [J]. Computer Aided Chemical Engineering, 2009, 27: 1557-1562.

[33] Ge Z Q, Chen J H. Plant-wide industrial process monitoring: A distributed modeling framework [J]. IEEE Transactions on Industrial Informatics, 2016, 12 (1): 310-321.

[34] Krzanowski W J. Between-groups comparison of principal components [J]. Publications of the American Statistical Association, 1979, 74 (367): 310-321.

[35] Johannesmeyer M C, Singhal A, Seborg D E. Pattern matching in historical data [J]. AIChE Journal, 2002, 48 (9): 2022-2038.

[36] Deng X G, Tian X M, Chen S, et al. Nonlinear process fault diagnosis based on serial principal component analysis [J]. IEEE Transactions on Neural Networks & Learning Systems, 2018, 29 (3): 560-572.

[37] Liu Y, Wang F L, Chang Y Q, et al. Performance-relevant kernel independent component analysis based operating performance assessment for nonlinear and {non-Gaussian} industrial processes [J]. Chemical Engineering Science, 2019, 209: 115167.

Chapter 15 Comprehensive Operating Performance Assessment Based on DSGRU

With the development of automation and information technology, modern industrial processes have become more and more integrated and complicated. Due to the uncertainty factors such as raw material change, equipment aging, operation environment change, industrial processes may gradually deviate from the optimal work point and enter the non-optimal operating conditions, which will affect the economic benefits of enterprises[1-3]. Therefore, developing effective assessment methods for process operating performance is the necessary trend and urgent need for the improvement of process monitoring. However, taking the variables of the hot strip mill process (HSMP) as an example (as shown in **Fig. 15. 1**), it is difficult to analyze the inherent nonlinearity and dynamicity of the industrial process for operating performance assessment.

Operating performance assessment with the nonlinear characteristic has drawn wide attention in recent years. Strictly speaking, most industrial process variables do not have a linear relationship. Especially due to the changes of operating conditions, the nonlinearity of process variables becomes more and more significant[4]. Traditional linear methods cannot effectively extract and analyze the latent process features[5]. Therefore, nonlinear methods have gained in popularity over recent years. Based on total projection to latent structures (TPLS), Liu et al.[6] proposed a kernel version of TPLS (TKPLS) for the gold hydrometallurgical process. In order to avoiding the tedious data alignment work, Liu et al.[7] proposed a novel assessment method based on nonlinear optimality related variation information (NORVI). Considering linear and nonlinear relationships coexisting in some complex processes, Li et al.[8] proposed a linearity evaluation and variable subset partition based hierarchical modeling and monitoring method. With the development of deep learning, some methods based on deep neural network have also been proposed[9]. Yan et al.[10] first used the autoencoder (AE) to extract the nonlinear characteristics, and established a corresponding monitoring model for nonlinear process. Yu et al.[11] further combined denoising autoencoder (DAE)

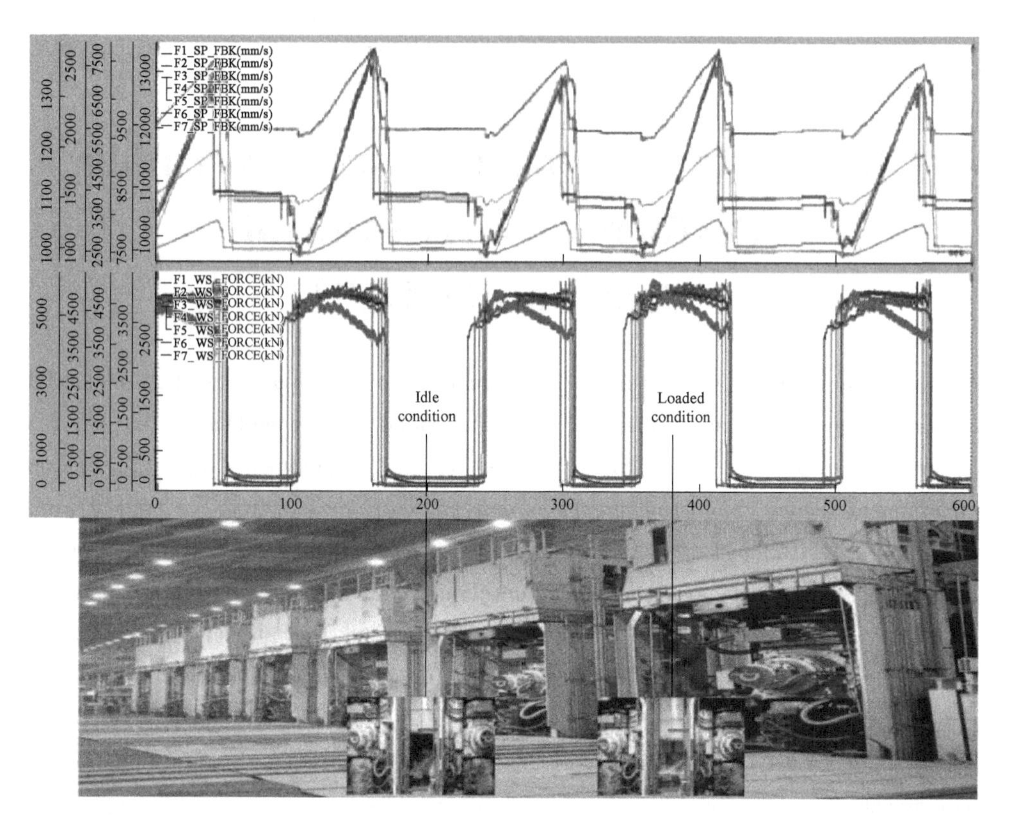

Fig. 15. 1 The illustration of a two-level, three-stream strategy for the HSMP

(Scan the QR code on the front of the book for color picture)

and elastic network (EN) to deal with process nonlinearity and noise interference simultaneously.

Traditional assessment methods often assume that the process data is static and ignore its temporal correlations. However, modern industrial processes usually have dynamic characteristic, which refers to the time-related characteristics of industrial process data[12]. It is closely related to the internal mechanisms of the process, including the operating stage of the process, the mechanism of physical and chemical reactions, noise and disturbances[13]. To address this issue, dynamic models have been proposed in recent years. Zou and Zhao employed cointegration analysis to analyze the long-term equilibrium relations of dynamic process for performance assessment[14]. Dong and Qin proposed a novel dynamic principal component analysis (DiPCA) model for dynamic data modeling and process monitoring[15]. Zhang et al.[16] combined canonical variate analysis (CVA) and slow feature analysis (SFA) for fine-scale identification of dynamic process operation statuses. Zou and Zhao proposed a concurrent operating performance assessment method based on CVA and SFA to analyze the process static and dynamic features[17].

In order to deal with nonlinearity and dynamicity simultaneously, recurrent neural network (RNN) network has been widely used[18]. Gugulothu et al. combined RNN with autoencoder (AE) for remaining useful life (RUL) prediction[19]. Compared with traditional RNN, long short-term memory (LSTM) can solve the gradient vanishing and explosion problems[20]. Ye and Yu proposed a LSTM convolutional autoencoder method for health condition monitoring of machines[21]. However, LSTM model has complex structure and long training time[22]. As a variant of LSTM, gated recurrent unit (GRU) removes cell state and uses hidden state to transmit information. Its performance is similar to LSTM, but its computational complexity is much lower[23]. Wang et al. combined hidden Markov model (HMM) and improved GRU for degradation evaluation of slewing bearing[24]. Yuan et al. integrated vanilla RNN, LSTM, and GRU to evaluate machine health condition[25]. From the perspective of classification, operation performance assessment is a multi-classification issue. Siamese neural network (SNN) is a measure of feature similarity proposed and widely used in the analysis of comparatively similar samples[26-28]. It is a symmetric network composed of two identical networks in parallel, in which the weights and thresholds are shared. The main purpose of SNN is to minimize the distance between samples from the same class and to maximize the distance between samples from different classes. Therefore, this chapter presents a Siamese gated recurrent unit network (SGRU) by combining the advantages of GRU and SNN.

Since abnormal conditions caused by instrumentation failure, process executor failure and improper operation may lead to the destruction of the whole system, or even catastrophic accidents[29-31], it is not enough to merely assess operating performance under normal condition. The assessments of fault level under abnormal condition is also important. Therefore, it is necessary to investigate the comprehensive assessment for process operating performance. The concept of comprehensive operating performance assessment includes two aspects: on the one hand, it can deal with nonlinear and dynamic characteristics of the process simultaneously; on the other hand, it can effectively assess both operating performance level under normal conditions and fault level under abnormal conditions, which is illustrated in **Fig. 15. 2**. It is worth noting that traditional distributed assessment methods only use the information of each subsystem to establish the corresponding independent local model, ignoring the internal relationship between different subsystems, which may lead to the loss of important process information provided by other subsystems. In practical industrial processes, some subsystems cannot complete the information interaction. It is also unreasonable to assume that all subsystems can communicate and interact with each other.

In order to solve above issues, a distributed Siamese gated recurrent unit network (DSGRU) framework with partial communication is proposed for the hot strip mill process (HSMP). Based on the DSGRU framework, comprehensive operating performance assessment is carried out for both normal and abnormal conditions. The main contributions are summarized as follows:

(1) Based on SNN and Encoder-Decoder network (EDN) with GRUs, a Siamese gated recurrent unit network is developed for dealing with the nonlinear and dynamic characteristics in the HSMP at the same time.

(2) Considering the actual communication interaction between different subsystems, the partial communication fusion data is updated by principal component analysis (PCA) and the structure matrix, and the DSGRU model with partial communication is developed.

(3) A DSGRU-based comprehensive assessment framework is proposed for operating performance assessment under normal condition and fault level assessment under abnormal condition. In addition, when the HSMP is in non-optimal condition, a nonoptimal cause identification strategy is developed to trace the non-optimal variables.

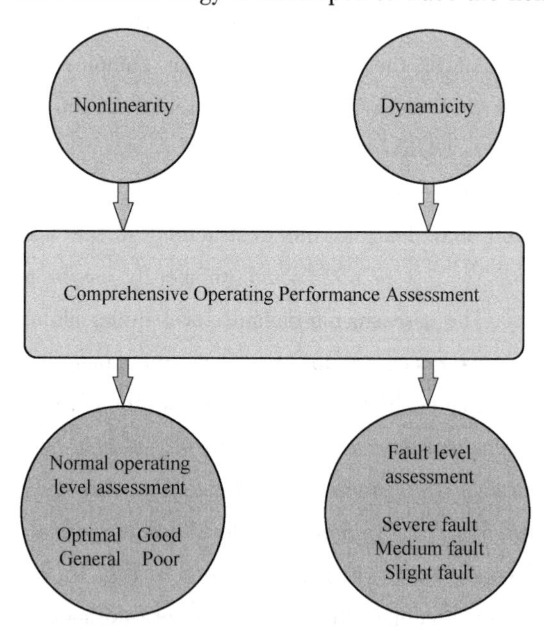

Fig. 15. 2 Comprehensive operating performance assessment

15.1 Review of SNN and GRU

15.1.1 Siamese neural network

SNN is a feature similarity measurement method proposed by Bromley et al.[26], which is widely used in the analysis of similar samples. SNN is a cooperative network, which consists of two identical neural networks in parallel. The weights of the two networks are shared. The main purpose of SNN is to reduce the value of loss function between similar samples and enlarge the that between dissimilar samples. The structure of SNN is presented in **Fig. 15.3**. $\boldsymbol{x}_1$ and $\boldsymbol{x}_2$ are two different input samples of SNN. They can be encoded into the hidden feature $\boldsymbol{H}(\boldsymbol{x}_1)$ and $\boldsymbol{H}(\boldsymbol{x}_2)$ by symmetric Network 1 and Network 2. The output of SNN is the distance metric between $\boldsymbol{H}(\boldsymbol{x}_1)$ and $\boldsymbol{H}(\boldsymbol{x}_2)$. The loss function for SNN can be calculated as:

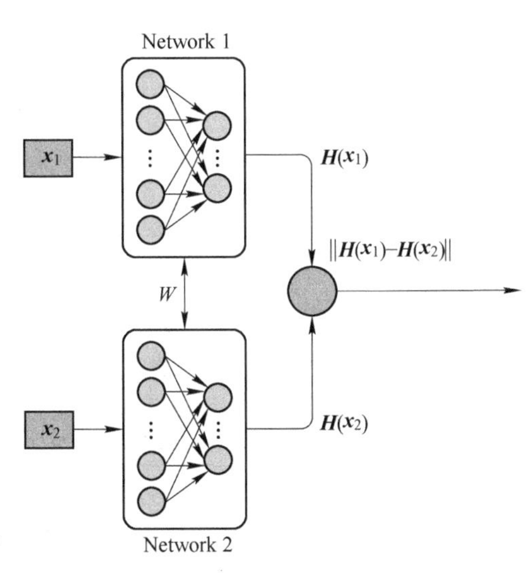

Fig. 15.3 Structure of Siamese neural network

$$L_{\mathrm{SNN}} = \begin{cases} d\{\boldsymbol{H}(\boldsymbol{x}_1),\ \boldsymbol{H}(\boldsymbol{x}_2)\},\ \boldsymbol{x}_1 \text{ is similar to } \boldsymbol{x}_2 \\ \max(\eta - d\{\boldsymbol{H}(\boldsymbol{x}_1),\ \boldsymbol{H}(\boldsymbol{x}_2)\},\ 0),\ \text{otherwise} \end{cases} \tag{15.1}$$

where, $d\{\boldsymbol{H}(\boldsymbol{x}_1),\ \boldsymbol{H}(\boldsymbol{x}_2)\}$ is a distance metric between hidden features; the hyper-parameter η is the threshold.

In this chapter, Euclidean distance is taken as the distance metric and $d\{\boldsymbol{H}(\boldsymbol{x}_1),\ \boldsymbol{H}(\boldsymbol{x}_2)\}$ can be rewritten as $\|\boldsymbol{H}(\boldsymbol{x}_1) - \boldsymbol{H}(\boldsymbol{x}_2)\|_2^2$. While training SNN, the weights of Network 1 and Network 2 are updated in the same way, which makes the networks show the same nonlinear function.

15.1.2 Encoder-Decoder network with GRU

Encoder-Decoder network (EDN), also known as sequence-to-sequence model, has been widely used in nonlinear time-series modeling tasks[32]. It maps an input sequence to a fixed length output, and then extracts a dynamic feature vector to represent the input sequence. The mapping operation is performed by GRUs in this chapter. The structure of EDN with GRUs is depicted in **Fig. 15.4**. It consists of input sequence,

Encoder GRU cells, nonlinear dynamic features, Decoder GRU cells and output Sequence, which are equivalent to a five-layer structure. Encoder GRU maps the m-dimensional input sequence data $\{x_t \in R^{1\times m}, t=1, 2, \cdots, T\}$ to the d-dimensional hidden nodes $\{h_t \in R^{1\times d}, t=1, 2, \cdots, T\}$, where h_T is considered as the nonlinear dynamic feature, T is the sequence length. Then h_T is transferred to the decoder GRU, which generates an reconstructed sequence data $\{\hat{x}_t \in R^{1\times m}, t=1, 2, \cdots, T\}$. The loss function of EDN is to minimize the reconstruction error $\sum_{t=1}^{T} \| x_t - \hat{x}_t \|$.

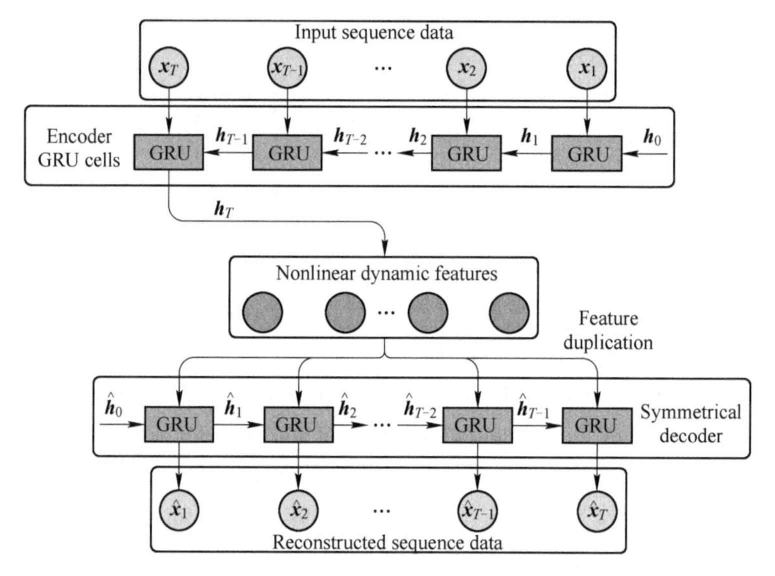

Fig. 15.4 Structure of EDN with GRUs

As a popular and simpler variant of LSTM, GRU uses the same gate control mechanism as LSTM[33]. While overcoming the problem of gradient disappearance, GRU also synthesizes the forget gate and input gate of the LSTM into an update gate, which means the cell state and hidden state are mixed. GRU consists of only two gates, namely the update gate z_t and reset gate r_t. The incorporation of new input with the previous memory is adjusted by r_t. The preservation of the previous memory is controlled by z_t. **Fig. 15.5** shows the structure of GRU and its updating equations are given as:

$$z_t = \sigma(W_z x_t + V_z h_{t-1} + b_z) \tag{15.2}$$

$$r_t = \sigma(W_r x_t + V_r h_{t-1} + b_r) \tag{15.3}$$

$$\tilde{h}_t = \tanh(W_{\tilde{h}} x_t + V_{\tilde{h}}(r_t \odot h_{t-1}) + b_{\tilde{h}}) \tag{15.4}$$

$$h_t = (1 - z_t) \odot h_{t-1} + z_t \odot \tilde{h}_t \tag{15.5}$$

where, $\boldsymbol{x}_t$, $\boldsymbol{h}_t$ and $\tilde{\boldsymbol{h}}_t$ are input vector, output hidden states and candidate hidden states at time t; $\{\boldsymbol{W}_z, \boldsymbol{V}_z\}$, $\{\boldsymbol{W}_r, \boldsymbol{V}_r\}$ and $\{\boldsymbol{W}_{\tilde{h}}, \boldsymbol{V}_{\tilde{h}}\}$ are weight matrices of the update gate, reset gate and candidate hidden state; $\boldsymbol{b}_z$, $\boldsymbol{b}_r$ and $\boldsymbol{b}_{\tilde{h}}$ are the corresponding biases; $\odot$ denotes the Hadamard product; σ is the Sigmoid function; tanh is the hyperbolic tangent function.

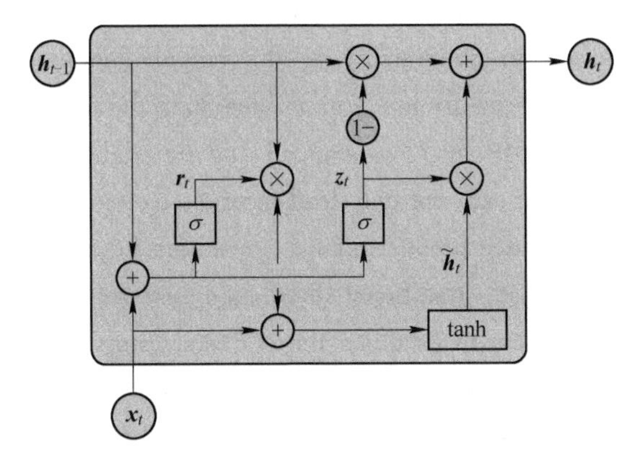

Fig. 15.5 Graph of GRU

15.2 Comprehensive OPA framework based on DSGRU

HSMP has high degree of automation, the main equipment are closely connected in a streamlined layout. Due to frequent changes of raw materials and violent fluctuations of operating modes, HSMP has inherent nonlinear and dynamic characteristics. In the face of small and medium-sized systems, traditional centralized operating performance assessment (OPA) methods can more or less achieve satisfactory results. However, for such a large-scale complex system as HSMP, the centralized OPA methods are not effective because of the large number of process variables, the distance between operation units, the physical connection restrictions, and so on. The relationship between process variables is too complex to identify, and even if centralized models are obtained, high dimensions and complexity inevitably lead to computational problems in integrated decision-making and control. Therefore, it is necessary to investigate decomposition methods to integrate large processes into a certain number of interconnected subsystems and obtain distributed sub-models. A typical HSMP consists of seven rolling stands, through deep research on the process knowledge, it can be divided into three subsystems: upstream (1_{st} - 3_{rd} stands), midstream (4_{th} - 5_{th} stands), and downstream (6_{th} - 7_{th} stands). In the normal production process, the non-optimal operating condition of any work area may affect the whole production line and the final product quality, so it is necessary to evaluate the operating performance in real time. In addition, once a fault occurs, it is also necessary to effectively identify the type and severity of the fault, and develop different maintenance strategies for different faults. For illustration, the thickness anomalies caused by upstream stand faults can be compensated to a certain extent by the rolling action of the next stand and this harmless fault does not need to be alarmed. However, when the fault is difficult to be compensated and causes an abnormality in the output thickness of midstream and downstream stands, it is necessary to alarm and maintain the fault. Once maintenance is carried out, the entire HSMP must be stopped and the slab has to be heated in the furnace for a long time. This will not only lead to excessive oxidation of slab and reduce product yield, but also reduce product quality, resulting in production waste and huge economic losses. Besides that, traditional distributed assessment methods only use the information of each subsystem to establish the corresponding independent local model, ignoring the internal relationship between different subsystems, which may lead to the loss of important process information provided by other subsystems. In practical HSMP, not all of the three subsystems can realize the information interaction because of the actual physical location of different rolling stands. It is also unreasonable to assume that all subsystems can communicate and interact with each other. In order to effectively monitor the normal operating conditions of HSMP and avoid production interruption caused by unimportant faults, it is necessary to study a novel comprehensive operating performance assessment method with partial communication.

15.2.1 Siamese gated recurrent unit network

In this subsection, a Siamese GRU network is proposed to form the base model of DSGRU network. With Siamese GRU, nonlinear dynamic features can be extracted from time-sequence data to build a classification model for fault and operating performance identification. In this chapter, sliding window technique is used to construct the initial data into time-sequence data. For illustration, the time sequence length is selected as $T = 4$. The data processing procedure is given in **Fig. 15.6**. $\boldsymbol{X}$, $\boldsymbol{Y}$ and $\boldsymbol{S}$ denote the initial data, data label and time sequence data, respectively.

The configuration of Siamese GRU network is shown in **Fig. 15.7**. In order to enlarge the gap between the optimal working condition samples and the other samples (non-optimal samples and fault samples) and improve the classification accuracy of similar samples, Siamese GRU has two inputs: s^{in} (input of $\mathbf{GRU}_{\text{in}}$) are samples from the training set, s^{re} (input of $\mathbf{GRU}_{\text{re}}$) are reference samples which are randomly selected from the training set for calculating similarity. The EDN with GRU is used to extract nonlinear dynamic features of $\boldsymbol{s}^{\text{in}}$ and $\boldsymbol{s}^{\text{re}}$. Assuming that $\boldsymbol{h}_T^{\text{in}}$ and $\boldsymbol{h}_T^{\text{re}}$ are the nonlinear dynamic features, which can be expressed as:

$$\begin{cases} \boldsymbol{h}_T^{\text{in}} = (1 - z_T^{\text{in}}) \odot \boldsymbol{h}_{T-1}^{\text{in}} + z_T^{\text{in}} \odot \tilde{\boldsymbol{h}}_T^{\text{in}} \\ \boldsymbol{h}_T^{\text{re}} = (1 - z_T^{\text{re}}) \odot \boldsymbol{h}_{T-1}^{\text{re}} + z_T^{\text{re}} \odot \tilde{\boldsymbol{h}}_T^{\text{re}} \end{cases} \tag{15.6}$$

where, z_T^{in} and z_T^{re} denote the update gates, $\tilde{\boldsymbol{h}}_T^{\text{in}}$ and $\tilde{\boldsymbol{h}}_T^{\text{re}}$ denote the candidate states at time T . Furthermore, $\boldsymbol{h}_T^{\text{in}}$ and $\boldsymbol{h}_T^{\text{re}}$ are transferred to decoder GRU, where the outputs $\hat{\boldsymbol{s}}^{\text{in}}$ and $\hat{\boldsymbol{s}}^{\text{re}}$ can be regarded as the reconstructed sequence data. Then Siamese GRU is separated into similarity comparison part and classification part (fault and operating performance identification).

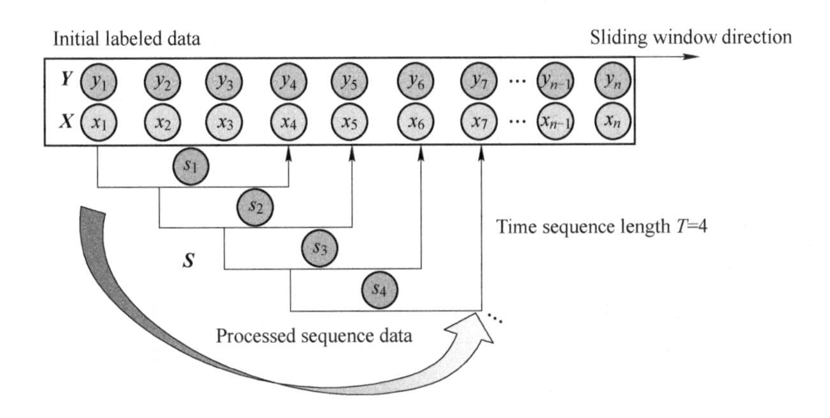

Fig. 15.6 Illustration of serialization for initial data

In the similarity comparison part, the similarity between nonlinear dynamic features $\boldsymbol{h}_T^{\text{in}}$ and $\boldsymbol{h}_T^{\text{re}}$ is compared. The Euclidean distance between $\boldsymbol{h}_T^{\text{in}}$ and $\boldsymbol{h}_T^{\text{re}}$ is calculated as follows:

$$S = \| \boldsymbol{h}_T^{\text{in}} - \boldsymbol{h}_T^{\text{re}} \|_2^2 \tag{15.7}$$

In the classification part, the fully connected layer uses softmax function to performs fault and operating performance identification. Assuming that $p(y^{\text{in}} = k \mid \boldsymbol{h}_T^{\text{in}};\ \theta)$ represents the probability that input $\boldsymbol{h}_T^{\text{in}}$ will be identified as Class k, for a K-class classifier, the output is given as:

$$h_{\theta}(\boldsymbol{h}_T^{\text{in}}) = \begin{bmatrix} p(y^{\text{in}} = 1 \mid \boldsymbol{h}_T^{\text{in}};\ \theta) \\ p(y^{\text{in}} = 2 \mid \boldsymbol{h}_T^{\text{in}};\ \theta) \\ \vdots \\ p(y^{\text{in}} = K \mid \boldsymbol{h}_T^{\text{in}};\ \theta) \end{bmatrix} = \frac{1}{\sum_{j=1}^{K} \exp(\theta_j^{\text{T}} h_T^{\text{in}})} \begin{bmatrix} \exp(\theta_1^{\text{T}} \boldsymbol{h}_T^{\text{in}}) \\ \exp(\theta_2^{\text{T}} \boldsymbol{h}_T^{\text{in}}) \\ \vdots \\ \exp(\theta_K^{\text{T}} \boldsymbol{h}_T^{\text{in}}) \end{bmatrix} \tag{15.8}$$

where, $\boldsymbol{\theta} = [\theta_1 \quad \theta_2 \quad \cdots \quad \theta_K]^{\text{T}}$ is the weight matrix, $\sum_{j=1}^{K} \exp(\theta_j^{\text{T}} \boldsymbol{h}_T^{\text{in}})$ is the normalized function.

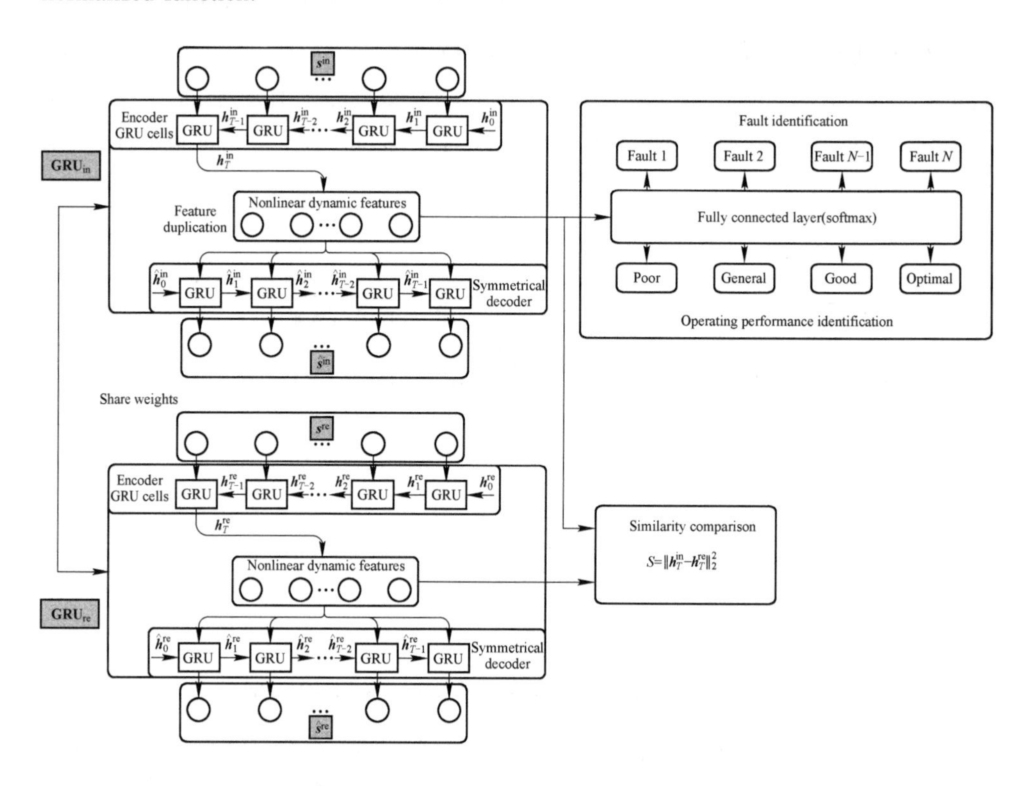

Fig. 15.7 The structure of Siamese GRU

The purpose of network training is to find the optimal parameters of the network and minimize the error of loss function. In this chapter, the joint loss function is formed through combining the loss of EDN with GRU, SNN and the softmax part, which can be calculated as:

$$L_{\text{SGRU}} = \underbrace{\frac{\alpha}{n - T + 1} \sum_{i=1}^{n-T+1} \sum_{t=1}^{T} \left((\boldsymbol{s}_t^{\text{in},i} - \hat{\boldsymbol{s}}_t^{\text{in},i})^2 + (\boldsymbol{s}_t^{\text{re},i} - \hat{\boldsymbol{s}}_t^{\text{re},i})^2 \right)}_{\text{loss of EDN with GRU}} +$$

$$\underbrace{\frac{\beta}{2(n-T+1)}\sum_{i=1}^{n-T+1}\left(1\{y_i^{\mathrm{in}}=y_i^{\mathrm{re}}\}S^2+(1-1\{y_i^{\mathrm{in}}=y_i^{\mathrm{re}}\})\max(\eta-S,\ 0)^2\right)}_{\text{loss of SNN}}+$$

$$\underbrace{\frac{\gamma}{n-T+1}\sum_{i=1}^{n-T+1}\sum_{k=1}^{K}\left(1\{y_i^{\mathrm{in}}=k\}\ln p(y_i^{\mathrm{in}}=k\mid \boldsymbol{h}_T^{\mathrm{in},i};\ \boldsymbol{\theta})\right)}_{\text{loss of softmax}} \tag{15.9}$$

where, α, β and γ are regularization parameters; $n-T+1$ is the length of serialized training samples; y_i^{in} and y_i^{re} are fault type or operating performance grade of the input sequence sample $\boldsymbol{s}_t^{\mathrm{in},i}$ and $\boldsymbol{s}_t^{\mathrm{re},i}$, respectively.

$\hat{\boldsymbol{s}}_t^{\mathrm{in},i}$ and $\hat{\boldsymbol{s}}_t^{\mathrm{re},i}$ represent the reconstructed sequence sample. $1\{\cdot\}$ denotes an indicative function. If it is true, it is equal to 1, otherwise it is equal to 0. In this chapter, back propagation through time (BPTT) algorithm is used to calculate the gradient of each parameter[35], and then Adam optimizer is used to optimize the network. Adam optimizer can adjust the learning rate more flexibly by calculating the first moment estimation and the second moment estimation of gradient, which is suitable for parameter optimization of complex networks[36].

15.2.2 DSGRU framework with partial communication

HSMP consists of several production equipment, which cooperate and associate with each other. In order to realize the effective operating performance assessment, the process variables are divided into different sub-blocks according to the process knowledge. In each sub-block, an assessment model will be established. Assuming that the initial process variables $\boldsymbol{X}\in R^{n\times m}$ are divided into several sub-blocks as:

$$\boldsymbol{X}=[\boldsymbol{X}_1,\ \boldsymbol{X}_2,\ \cdots,\ \boldsymbol{X}_b,\ \cdots,\ \boldsymbol{X}_B] \tag{15.10}$$

where, $b=1,\ 2,\ \cdots,\ B$, B is the number of sub-blocks, sub-block $\boldsymbol{X}_b\in R^{n\times m_b}$ consists of m_b variables.

The sampling frequencies of different variables are often are diverse. In this chapter, down-sampling technique is used to synchronize different variables.

Then, the process data in each sub-block is preliminarily compressed. As a simple and effective feature extraction method, PCA has been widely used. In this chapter, PCA is applied to project $\boldsymbol{X}_b$ into a low-dimensional space as:

$$\boldsymbol{X}_b=\boldsymbol{T}_b\boldsymbol{P}_b^{\mathrm{T}}+E_b \tag{15.11}$$

where, $\boldsymbol{T}_b\in R^{n\times A_b}$ is the score matrix; $\boldsymbol{P}_b\in R^{m_b\times A_b}$ is the loading matrix; A_b is the number of principle components (PCs), which can be determined by many methods; $\boldsymbol{E}_b\in R^{n\times m_b}$ is the residual matrix.

Cumulative percent variance (CPV) is used to determine the number of PCs:

$$\mathrm{CPV}_b(l_b) = 100\left[\frac{\sum_{j=1}^{l_b}\lambda_j}{\sum_{j=1}^{m_b}\lambda_j}\right]\% \tag{15.12}$$

where, l_b is the retained first l PCs of $\boldsymbol{X}_b$.

Obviously, the more PCs are retained, the greater the CPV. The decision becomes a balance between the amount of parsimony and comprehensiveness of the data. In this chapter, the CPV is set as 90%[37].

Afterwards, a fusion data set $\boldsymbol{I}_b \in R^{n\times(m_b+\sum_{i=1(i\neq b)}^{B} A_i)}$ is constructed, which contains the information of the other sub-blocks and can be calculated as follows:

$$\boldsymbol{I}_b = [\boldsymbol{X}_1\boldsymbol{P}_1, \cdots, \boldsymbol{X}_b, \boldsymbol{X}_{b+1}\boldsymbol{P}_{b+1}, \cdots, \boldsymbol{X}_B\boldsymbol{P}_B] \tag{15.13}$$

Considering the communication between sub-blocks, a structure matrix $\boldsymbol{C} \in R^{B\times B}$ is constructed in this chapter. The element c_{ij} in the matrix indicates whether the sub-block $\boldsymbol{X}_i$ and $\boldsymbol{X}_j$ are connected. If they are connected, c_{ij} is equal to 1, otherwise, c_{ij} is equal to 0. It should be noted that the connectivity between sub-blocks has directionality, sometimes $c_{ij} \neq c_{ji}$ may exist. For HSMP, the process is divided into three sub-blocks (upstream, midstream and downstream), and the corresponding structure matrix is given as:

$$\boldsymbol{C} = \begin{bmatrix} 1 & 1 & 0 \\ 0 & 1 & 1 \\ 0 & 0 & 1 \end{bmatrix} \tag{15.14}$$

where, $c_{12} = 1$ denotes upstream and midstream are directly related, $c_{13} = 0$ denotes upstream is not connected with downstream. Based on the structure matrix $\boldsymbol{C}$, the fusion data set $\boldsymbol{I}_b$ is according to the following rules:

$$\begin{cases} \text{When } c_{ib} = 0(i \neq b), \ \boldsymbol{P}_i = 0 \\ \text{When } c_{ib} = 1(i \neq b), \ \boldsymbol{P}_i = \boldsymbol{P}_i \end{cases} \tag{15.15}$$

Assume that the updated partial communication fusion data set is $\tilde{\boldsymbol{I}} = [\tilde{\boldsymbol{I}}_1, \tilde{\boldsymbol{I}}_2, \cdots, \tilde{\boldsymbol{I}}_b, \cdots, \tilde{\boldsymbol{I}}_B]$. Then a serialization operation is carried out on $\tilde{\boldsymbol{I}}$ to obtain the time-sequence data $\tilde{\boldsymbol{S}}$. Finally, we use $\tilde{\boldsymbol{S}}$ to train the Siamese GRU in each sub-block. The illustration of DSGRU framework with partial communication is shown in **Fig. 15. 8.**

15. 2. 3 DSGRU-based comprehensive OPA

The comprehensive operating performance assessment includes not only normal operating

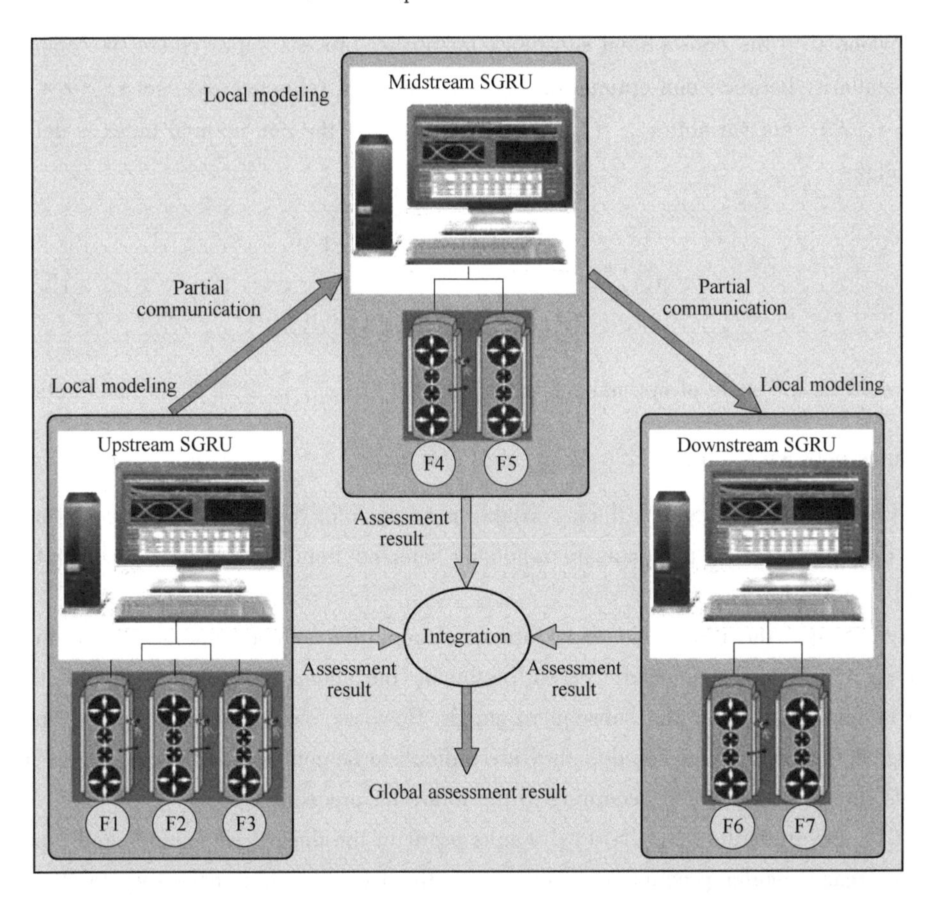

Fig. 15.8 The illustration of DSGRU framework with partial communication

grade assessment and non-optimal cause identification, but also fault identification and fault level assessment. Based on the national standard GB/T 709—2006, the normal operating grades for different products can be divided into optimal, good, general and poor. The four operating grades are numbered as 1, 2, 3, and 4. Assuming that the fault types are fault 1, fault 2, ⋯, fault F. For simplicity, they are numbered as 5, 6, ⋯, $F+4$. HSMP is divided into three sub-blocks, thus the upstream, midstream and downstream are represented as $b=1$, 2 and 3, respectively. At time t, the time-sequence data $\tilde{\boldsymbol{s}}_{t,b}$ is input to the corresponding Siamese GRU model to obtain the sub-block assessment result $G_{t,b}$, and then the global assessment result G_t can be calculated as follows:

$$G_t = \max\{G_{b,t},\ b=1,\ 2,\ 3\} \tag{15.16}$$

If $G_t = 1$, the process is in optimal condition. If $1 < G_t \leqslant 4$, the process is in non-optimal condition. For non-optimal condition, non-optimal cause identification strategy is developed in the non-optimal sub-block for further process improvement. We compare the similarity between non-optimal data $x_{t,b}$ and optimal reference data set $x_{r,b}^{\mathrm{opt}}(r = 1, 2, \cdots, R)$. For variable $x_{t,b,i}(i = 1, 2, \cdots, m_b)$, the non-optimal index is defined as follows:

$$\mathrm{nop}(i) = \frac{\sum_{r=1}^{R} d\{x_{t,b,i}, x_{r,b,i}^{\mathrm{opt}}\}/R}{\sum_{i}^{m_b}\left[\sum_{r=1}^{R} d\{x_{t,b,i}, x_{r,b,i}^{\mathrm{opt}}\}/R\right]} \tag{15.17}$$

where, R is the scale of optimal reference data set; $d\{x_{t,b,i}, x_{r,b,i}^{\mathrm{opt}}\}$ is the Euclidean distance between $x_{t,b,i}$ and $x_{r,b,i}^{\mathrm{opt}}$; $\sum_{i}^{m_b} \mathrm{nop}(i) = 1$.

If $\mathrm{nop}(i)$ is the largest, the i_{th} variable is a cause for the non-optimality. It is noted that the optimal reference data are randomly selected from the history data at optimal level.

If $G_t > 4$, the process is under abnormal condition and a fault may occur in the process. Owing to the automatic control systems, the anomalies in the upstream stands can be compensated by the subsequent stands. However, when anomalies occur in the midstream or downstream stands, they are difficult to be compensated. Thus, three fault levels should be evaluated according to the following procedures:

(1) Severe faults ($G_{3,t} > 4$): Faults occur in the downstream stands and process cannot reach regular production requirements. In this case, it is necessary to stop the entire production line and overhaul the faulty equipment to ensure production safety and product quality.

(2) Medium faults ($G_{2,t} > 4$): Faults occur in the midstream stands and degrade non-key process operating performance. Since the output slab from midstream stands would be further rolled in the downstream stands, the fault probability of the downstream stands is increased. At this time, the process will be warned. If $G_{3,t} \leqslant 4$, process maintenance should be carried out after the current batch is completed. Otherwise, the process needs to stop running and overhaul.

(3) Slight faults ($G_{1,t} > 4$): Faults occur in the upstream stands and affect the input thickness of midstream stands. Owing to intrinsic control effect of the HSMP, such faults can be compensated or even eliminated. Thus, we only record the assessment results and carry out maintenance during downtime that does not affect regular production.

In summary, the flowchart of the comprehensive operating performance assessment is shown in **Fig. 15. 9**.

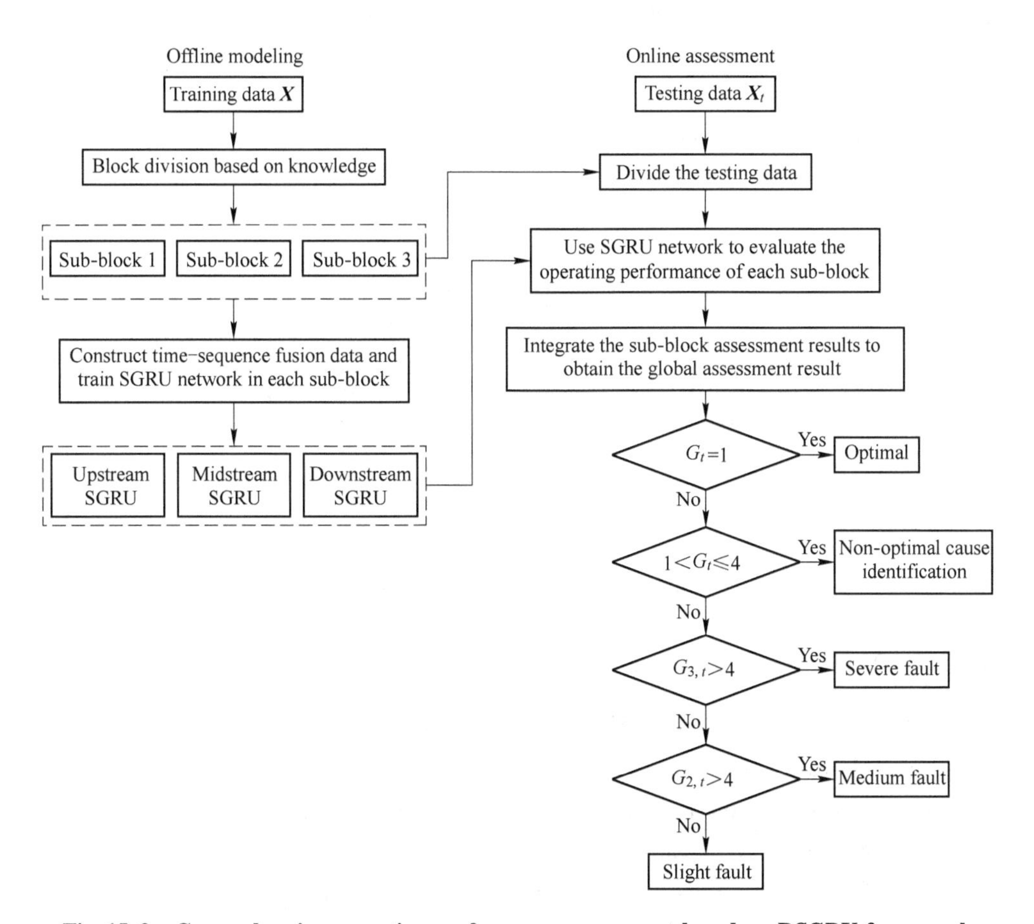

Fig. 15.9 Comprehensive operating performance assessment based on DSGRU framework

15.3 Case study

Different from other production processes in steel manufacturing, like the blast furnace and continuous casting process, the HSMP plays a key role in the steel product quality. As shown in **Fig. 15.10**, the main equipment of HSMP are seven mill stands in series. In each stand, there are two working rolls in the middle and two supporting rolls in both sides. A high-precision hydraulic system is also equipped to control the rolling and bending forces, such that a slab can be properly rolled. Before a slab enters a stand, a motor drives working rolls to the initial speed. Because of the rebounding phenomenon, the actual gap is enlarged when a slab reaches working rolls. In the HSMP, the individual stands are not independent, but are coupled with each other using hundreds of control loops. They operate cooperatively to ensure the strip quality. Rolling and bending forces can be measured by piezomagnetic and strain gauge sensors in real-time. However, due to the high temperature, the slab thickness is difficult to measure. In practice, the gap between two working rolls can be measured by the altitude difference between two supporting rolls[38-39].

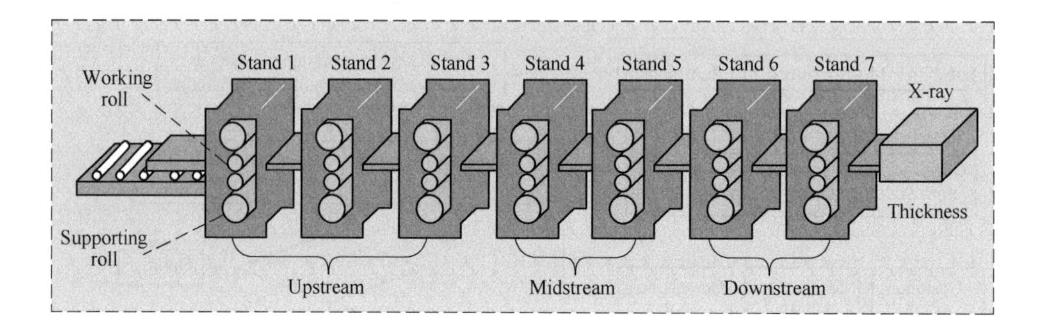

Fig. 15. 10 A schematic of the HSMP

15. 3. 1 Model training settings

The distributed sensors in the HSMP provide massive data, such as gap value, rolling speed in each stand. These informative data can be selected as the process variables that reflect the process running performance. As shown in **Table 15. 1**, 41 process variables are used for the proposed comprehensive operating performance assessment. Four normal operating levels (optimal, good, general and poor) and two typical faults (Fault 1 is the bending force sensor failure at the 5_{th} stand and Fault 2 is the malfunction of hydraulic system at the 4_{th} stand) are considered in this chapter. Moreover, in order to meet the market demands for different specifications of strip steel, the production line yields different products. Without loss of generality, three products with thickness 1. 55 mm, 2. 70 mm, and 3. 95 mm are selected for our research. In order to compare the proposed method with other existing methods, three cases are designed by the historical data from field engineering, which is shown in **Table 15. 2**. In Case 1, 200 samples of each level (optimal/good/general) are used for training the model, and 100 samples of each level (optimal/good/general) are used for testing. The data split of Case 2 and Case 3 is the same as that of Case 1. It is noted that reference data are randomly selected from training data as the input of $\mathbf{GRU}_{re}$, and the left training data are the input of $\mathbf{GRU}_{in}$.

Table 15.1 Process variables in the HSMP

Variable	Description	Unit
G_i	Roll gap at the i_{th} stand, $i=1, \cdots, 7$	mm
F_i	Rolling force at the i_{th} stand $i=1, \cdots, 7$	MN
B_i	Bending force at the i_{th} stand, $i=2, \cdots, 7$	MN
S_i	Rolling speed at the i_{th} stand, $i=1, \cdots, 7$	mm/s
C_i	Current at the i_{th} stand, $i=1, \cdots, 7$	A
P_i	Power at the i_{th} stand, $i=1, \cdots, 7$	kW

The time sequence length (T) is an important parameter, which can be determined by the training root mean square error (RMSE). In this chapter, the RMSE is calculated as:

$$\text{RMSE} = \sqrt{\sum_{i=1}^{n-T+1}\sum_{t=1}^{T}[(s_t^{\text{in},i} - \hat{s}_t^{\text{in},i})^2 + (s_t^{\text{re},i} - \hat{s}_t^{\text{re},i})^2]/(n-T+1)} \tag{15.18}$$

Fig. 15.11 shows the relationship between model performances (training RMSE) and T. Obviously, the optimal T for modeling is 13, where the training RMSE is the minimum. In addition, with the criterion CPV = 90%, the dimensions after projection (A_b) in upstream, midstream, and downstream are set as 11, 7, and 7, respectively. For the stability of the model, the training tasks are executed 50 times. With the optimal training performance, both the hidden nodes in the encoder and decoder are set as 200, and the node in the fully connected layer is set as 300.

Table 15.2 Design of the three cases for the HSMP

Case	Level No.	Description
1 (1.55 mm)	1	100 optimal level samples (1_{st}-100_{th})
	2	100 good level samples (101_{st}-200_{th})
	3	100 general level samples (201_{st}-300_{th})
2 (2.70 mm)	1	100 optimal level samples (1_{st}-100_{th})
	3	100 general level samples (101_{st}-200_{th})
	5	100 Fault 1 samples (201_{st}-300_{th})
3 (3.95 mm)	1	100 optimal level samples (1_{st}-100_{th})
	4	100 poor level samples (101_{st}-200_{th})
	6	100 Fault 2 samples (201_{st}-300_{th})

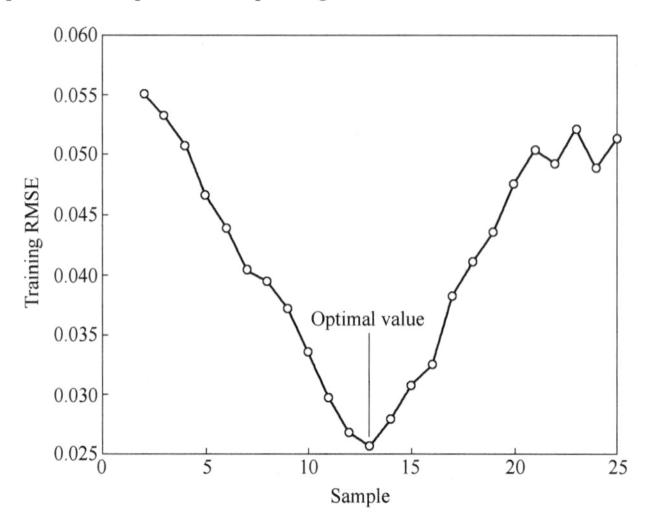

Fig. 15. 11 The relationship between model performances and *T*

15. 3. 2 Monitoring results and analysis

In order to verify the effectiveness of the proposed framework, DSGRU is compared with index driven sparse denoising autoencoder (ISDAE)[40], stacked autoencoder (SAE)[41], and variational autoencoder (VAE)[42], and Siamese recurrent neural network (SRNN)[43], respectively. An accuracy rate (the proportion of the samples that are correctly identified among all the testing samples in each case) is used to evaluate the assessment performance, which is defined as:

$$\text{Accuracy rate} = \frac{\text{samples(correctly identified)}}{\text{total samples}} \tag{15.19}$$

Table 15. 3 shows the accuracy rates and training time of the above five methods, which manifests the proposed method has higher accuracy rate than ISDAE, SAE, VAE, and SRNN method in each case. Besides, it can be seen from **Table 15. 3** that SRNN has the shortest training time, ISDAE, SAE, and VAE have moderate training time, the proposed method has the longest training time. However, the average accuracy rate of the DSGRU is nearly 2% higher than the best of the other four methods, and the increased training time is within an acceptable range.

Table 15. 3 Accuracy rates and training time of different methods

Case No.	ISDAE	SAE	VAE	SRNN	DSGRU
1	95. 00%	93. 67%	96. 00%	97. 67%	99. 00%
2	92. 67%	90. 70%	94. 00%	95. 33%	97. 33%

Continued Table 15.3

Case No.	ISDAE	SAE	VAE	SRNN	DSGRU
3	91.67%	93.33%	92.33%	94.67%	96.33%
Average	93.11%	92.57%	94.11%	95.89%	97.55%
Training time	33.1 s	34.3 s	32.9 s	28.5 s	35.2 s

15.3.2.1 Case 1 (1.55 mm)

The first 100 testing samples are running at optimal level. Then, due to increasing the roll gap at the 4_{th} stand, the operating performance level turns to good. Starting from the 201_{st} testing sample, the roll gap at the 4_{th} stand is further increased, and the operating performance is further deteriorated to general level. The assessment results of different methods are shown in **Fig. 15.12** (a)-(f). Because ISDAE, SAE, VAE, and SRNN develop a global assessment model for the whole process instead of a local assessment model for each stream of HSMP, they may not function well in capturing the relations between process variables and reflecting local behaviors. In particular, SAE has worse assessment performance than the other three methods. Owing to the similar Siamese structure, the accuracy rate of SRNN is closest to that of DSGRU. Due to the ability of the DSGRU to handle both nonlinearity and dynamics, the assessment accuracy has been improved by the proposed method. When the process works at non-optimal level, the non-optimal index is calculated at the first non-optimal sub-block. Specifically, it can be inferred from **Fig. 15.13** (a) that the roll gap at the 4_{th} stand G4 is the nonoptimal variable. Based on the process knowledge of the HSMP, it is known that the actual non-optimal cause is consistent with the result in **Fig. 15.15** (a), that is, the increase of the roll gap at the midstream stand results in the decrease of the operating performance level. Therefore, the proposed method can assess the process operating performance accurately and the non-optimal cause can be traced effectively. Then workers can make a reasonable adjustment and optimization scheme to ensure high quality and efficient operation of the HSMP.

15.3.2.2 Case 2 (2.70 mm)

The first 100 testing samples are running in optimal condition. Due to the decrease of rolling force at the 5_{th} stand, the process degenerates to general level from the 101_{st} testing sample. Then, Fault 1 is introduced to the HSMP. As a typical bending force sensor failure at the 5_{th} stand, Fault 1 occurs at the 201_{st} testing sample and lasts for

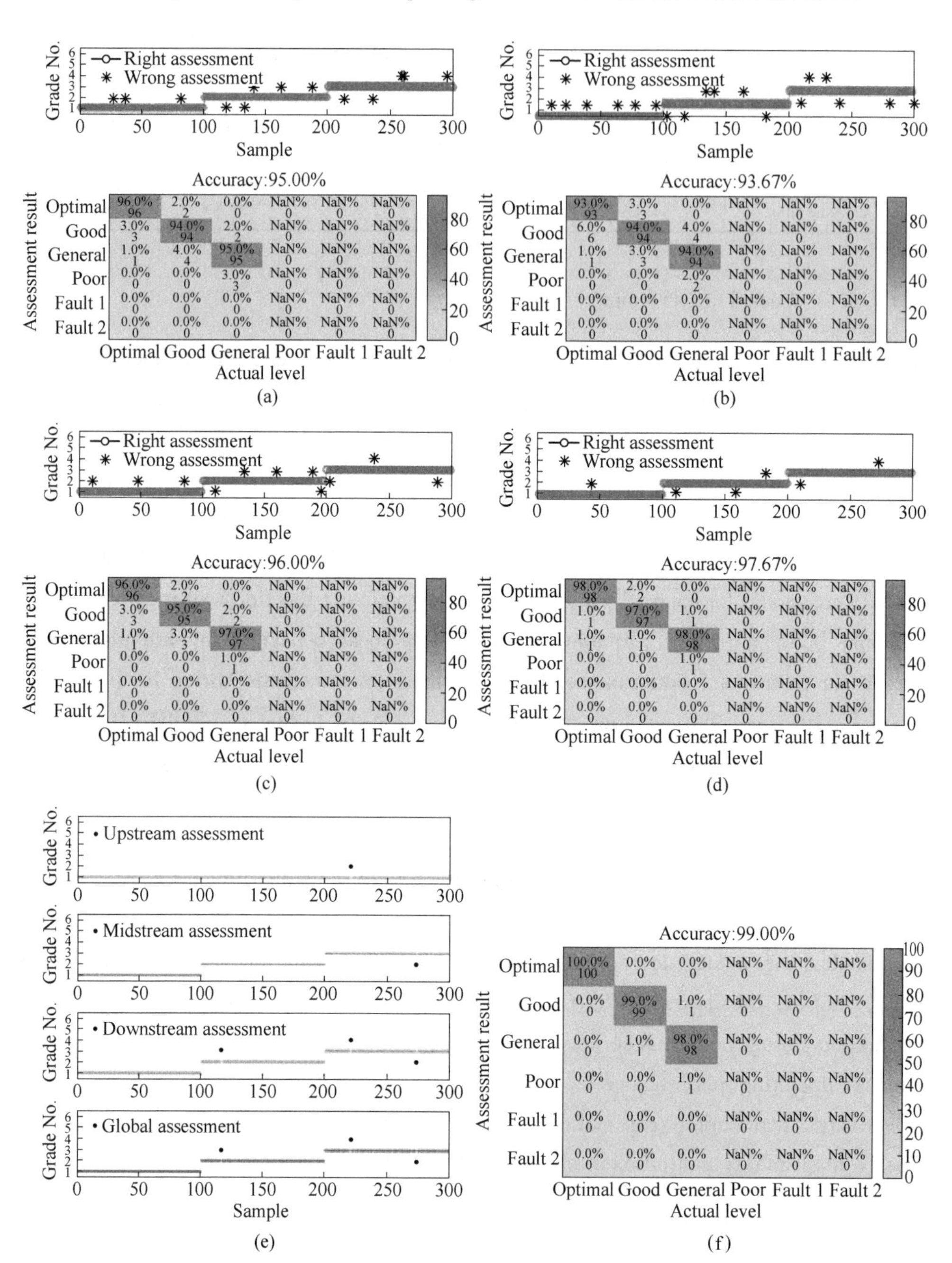

Fig. 15. 12 The simulation results of Case 1

(a) The assessment result of ISDAE; (b) The assessment result of SAE; (c) The assessment result of VAE; (d) The assessment result of SRNN; (e) The assessment result of DSGRU; (f) The confusion matrix of DSGRU

(Scan the QR code on the front of the book for color picture)

100 sampling cycles. **Fig. 15.14** (a)-(f) illustrate the assessment results and confusion matrices of ISDAE, SAE, VAE, SRNN, and the proposed method, respectively. As shown in **Fig. 15.14** (b), SAE has the most false assessment samples. From **Fig. 15.14** (c) and **Fig. 15.14** (d), it can be found that SRNN have higher accuracy rate than ISDAE, and VAE. Compared with these four methods, it is obvious that the DSGRU has the best accuracy. When the HSMP runs at general level, **Fig. 15.13** (b) shows the non-optimal index at the midstream sub-block. From **Fig. 15.13**, it can be inferred that the rolling force at the 5_{th} stand F5 is the non-optimal variable. Furthermore, when a fault occurs at the midstream, the proposed method can assess the fault level as medium fault. Hence, the assessment capability is significantly improved by the DSGRU.

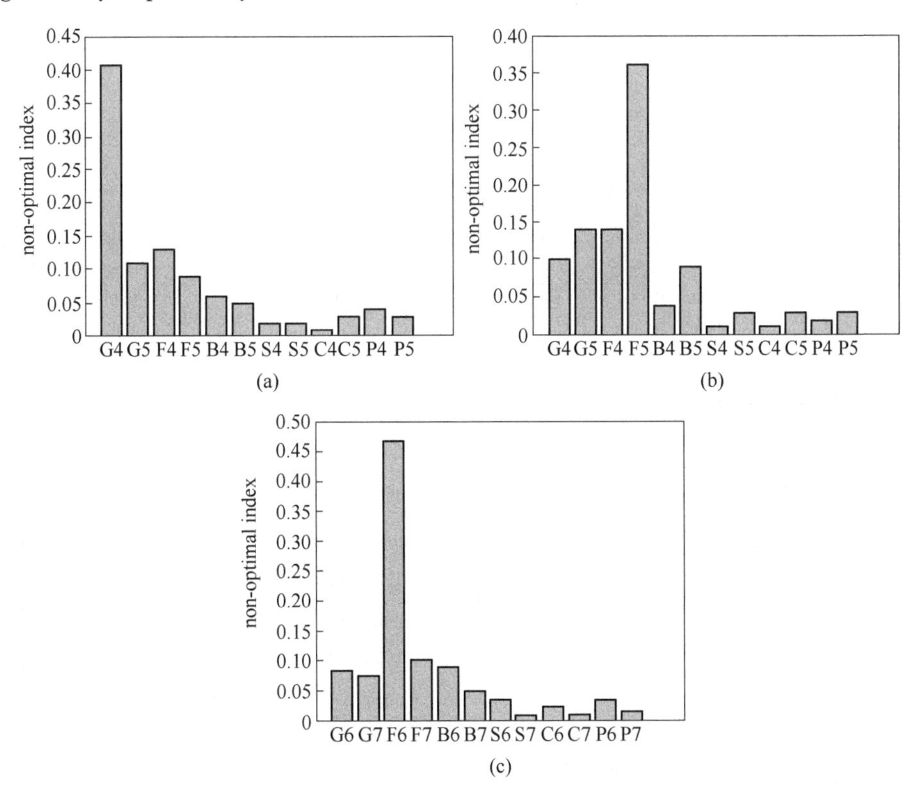

Fig. 15.13 The non-optimal cause identification results

(a) The nop-optiomal index of Case 1; (b) The nop-optiomal index of Case 2; (c) The nop-optiomal index of Case 3

15.3.2.3 Case 3 (3.95 mm)

For the first 100 testing samples, the HSMP works at optimal level. Then, due to

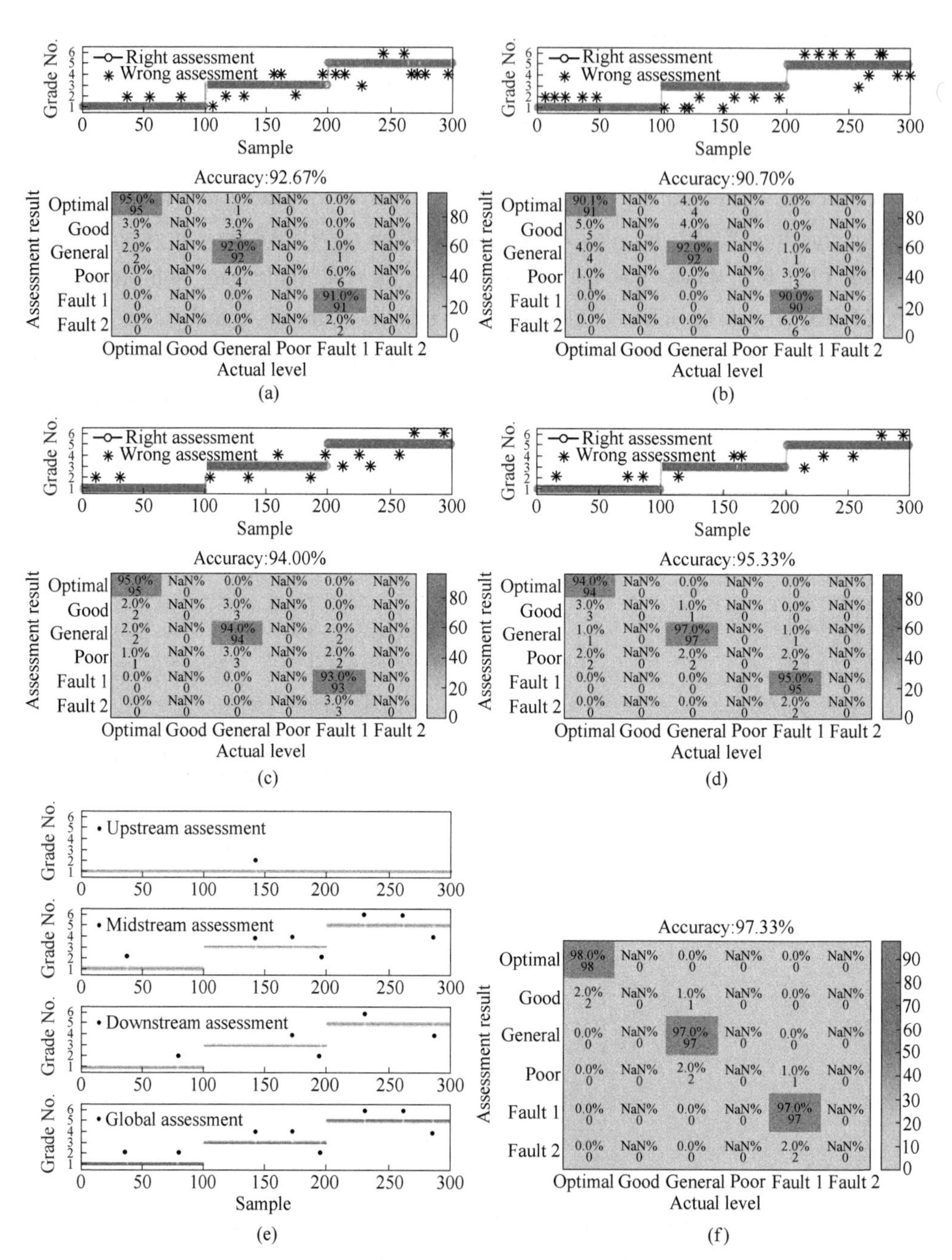

Fig. 15. 14　The simulation results of Case 2

(a) The assessment result of ISDAE; (b) The assessment result of SAE; (c) The assessment result of VAE; (d) The assessment result of SRNN; (e) The assessment result of DSGRU; (f) The confusion matrix of DSGRU

(Scan the QR code on the front of the book for color picture)

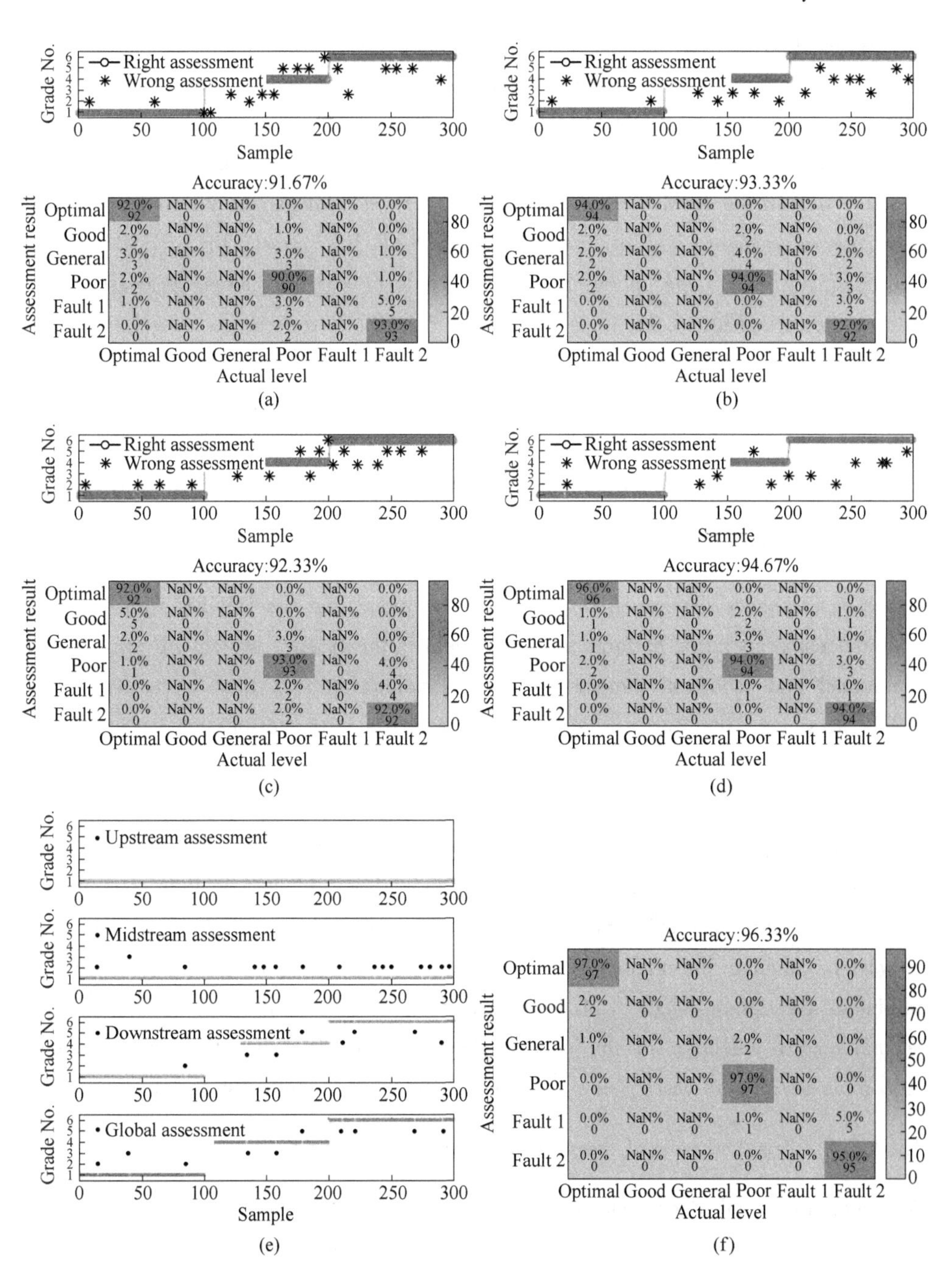

Fig. 15.15 The simulation results of Case 3

(a) The assessment result of ISDAE; (b) The assessment result of SAE; (c) The assessment result of VAE; (d) The assessment result of SRNN; (e) The assessment result of DSGRU; (f) The confusion matrix of DSGRU

(Scan the QR code on the front of the book for color picture)

decreasing the rolling force, the operating condition turns to poor level. As a representative malfunction of the hydraulic system at the 6_{th} stand, Fault 2 occurs from the 201_{st} testing sample. The assessment results and confusion matrices of ISDAE, SAE, VAE, and SRNN are shown in **Fig. 15. 15** (a)-(d). Obviously, ISDAE cannot function well in the HSMP. From **Fig. 15. 15** (b)-(d), it can be inferred that SRNN has better assessment capability than SAE, and VAE. In contrast, the proposed method can effectively assess the operating performance level with the lowest false assessment rate as shown in **Fig. 15. 15** (e)-(f). When the HSMP runs at poor level, the non-optimal index at the downstream sub-block is calculated. As shown in **Fig. 15. 13** (c), the rolling force at the 6_{th} stand F6 can be regarded as the non-optimal root cause. In addition, when a fault occurs at the downstream, the DSGRU can assess the fault level as severe fault. Therefore, the previous conclusions are also proved by these results.

15. 3. 3 Further discussions and comments

Based on the above experiments, it can be concluded that the DSGRU works well for comprehensive operating performance assessment in the HSMP, which will provide both theoretical and technical support for safety monitoring and quality control of the HSMP. However, there are still some issues worthy of further discussion.

15. 3. 3. 1 Feature selection issue

The method proposed includes twice feature selection process. The first feature selection is the preliminary compression of process data in each sub-block with PCA. The number of retained PCs in each sub-block is determined by CPV. However, there are a plethora of ways to calculate the number of PCs, such as Akaike information criterion (AIC), minimum description length (MDL), imbedded errorfunction (IEF), etc[37]. The second feature selection in the encoder-decoder with GRU has the same problem. According to different cases, how to choose an appropriate method is still an open issue.

15. 3. 3. 2 Pros and cons of the proposed framework

Although traditional deep learning methods have high classification accuracy, they needs to use large-scale data sets to train the model for a long time. In contrast, Siamese network can balance the speed and accuracy well. The fault data and the data of general level and poor level are scarce compared with the data of optimal level and good level. Thus, the proposed framework is more suitable for operating performance assessment in HSMP. In addition, the proposed framework not only considers the actual communication interaction between different subsystems, but also complete the operating performance assessment and fault grade assessment concurrently. When the HSMP is in non-optimal condition, the proposed framework can also trace the non-optimal variables. However, the proposed framework also have some drawbacks. In this chapter, the loss function is formed through combining the loss of EDN with GRU, SNN and the softmax part. Of course, softmax is a good choice, but it is not necessarily the best choice. The loss function can also choose cosine distance or other functions. Moreover, the optimization process with BPTT and Adam optimizer is relatively complex. It is necessary to study more advanced optimization methods to improve the speed of model training.

15.3.3.3 Applicability of the proposed framework

Although the proposed framework relies on some process knowledge, it is still a data-driven assessment method in essence. The advantage of data-driven method is that it can be extended to other industrial processes, such as chemical process. For different industrial processes, it may be necessary to modify the time sequence length T, the number of sub-blocks B, and the structure matrix $\boldsymbol{C}$, etc. In addition, the DSGRU model can also extended to the study of fault diagnosis, such as multi fault classification.

15.4 Conclusions

In this chapter, a new comprehensive operating performance assessment based on DSGRU is proposed for the nonlinear dynamic HSMP. Combining the advantages of SNN and EDN with GRU, the SGRU network is established to deal with the issue of nonlinear relationship and dynamic features concurrently. Then, considering the actual communications between different subsystems, the DSGRU network with partial communication is developed to assess the operating condition of different subsystems and global process. When the HSMP runs at normal condition, the proposed method can assess the normal operating level and track nonoptimal causes. Once a fault occurs in the HSMP, the proposed method can identify the fault type and assess the fault level. This is the so-called comprehensive operating performance assessment. Finally, the proposed DSGRU is applied to a practical HSMP. Compared with ISDAE, SAE, VAE, and SRNN, the proposed method shows the higher accuracy and the better interpretability.

References

[1] Qin S J, Chiang L H. Advances and opportunities in machine learning for process data analytic

[J]. Computer Chemical Engineering, 2019, 126: 465-473.

[2] Sedghi S, Huang B. Real-time assessment and diagnosis of process operating performance [J]. Engineering, 2017, 3: 214-219.

[3] Wang B, Li Z C, Dai Z W, et al. A probabilistic principal component analysis-based approach in process monitoring and fault diagnosis with application in wastewater treatment plant [J]. Applied Soft Computing, 2019, 82: 105527.

[4] Jiang Q C, Yan X F. Nonlinear plant-wide process monitoring using MI spectral clustering and Bayesian inference-based multiblock KPCA [J]. Journal of Process Control, 2015, 32: 38-50.

[5] Shen B B, Ge Z Q. Supervised nonlinear dynamic system for soft sensor application aided by variational auto-encoder [J]. IEEE Transactions on Instrument. Measurement, 2020, 69 (9): 6132-6142.

[6] Liu Y, Chang Y Q, Wang F L, et al. Complex process operating optimality assessment and nonoptimal cause identification using modified total kernel PLS [C]//The 26th Chinese Control and Decision Conference, 2014.

[7] Liu Y, Wang F L, Chang Y Q. Operating optimality assessment and cause identification for nonlinear industrial processes [J]. Chemical Engineering Research & Design, 2017, 117: 472-487.

[8] Li W Q, Zhao C H, Gao F R. Linearity evaluation and variable subset partition based hierarchical process modeling and monitoring [J]. IEEE Transactions on Industrial Electronics, 2018, 65 (3): 2683-2692.

[9] Lee S, Kwak M, Tsui K L, et al. Process monitoring using variational autoencoder for high-dimensional nonlinear processes [J]. Engineering Applications of Artificial Intelligence, 2019, 83: 13-27.

[10] Yan W W, Guo P J, Gong L, et al. Nonlinear and robust statistical process monitoring based on variant autoencoders [J]. Chemometrics & Intelligent Laboratory Systems, 2016, 158: 31-40.

[11] Yu W K, Zhao C H. Robust monitoring and fault isolation of nonlinear industrial processes using denoising autoencoder and elastic net [J]. IEEE Transactions on Control Systems Technology, 2020, 28 (3): 1083-1091.

[12] Huang B, Kadali R. Dynamic Modeling, Predictive Control and Performance Monitoring [M]. London, UK: Springer, 2008.

[13] Liu Q, Zhu Q Q, Qin S J, et al. Dynamic concurrent kernel CCA for strip-thickness relevant fault diagnosis of continuous annealing processes [J]. Journal of Process Control, 2018, 67: 12-22.

[14] Zou X Y, Zhao C H. Meticulous assessment of operating performance for processes with a hybrid of stationary and nonstationary variables [J]. Industrial & Engineering Chemistry Research, 2019, 58 (3): 1341-1351.

[15] Dong Y N, Qin S J. A novel dynamic PCA algorithm for dynamic data modeling and process

monitoring [J]. Journal of Process Control, 2008, 67: 1-11.

[16] Zhang S M, Zhao C H, Huang B. Simultaneous static and dynamic analysis for fine-scale identification of process operation statuses [J]. IEEE Transactions on Industrial Informatics, 2019, 15 (9): 5320-5329.

[17] Zou X Y, Zhao C H. Concurrent assessment of process operating performance with joint static and dynamic analysis [J]. IEEE Transactions on Industrial Informatics, 2020, 16 (4): 2776-2786.

[18] Zhao R, Yan R, Chen Z, et al. Deep learning and its applications to machine health monitoring [J]. Mechanical Systems & Signal Processing, 2019, 115: 213-237.

[19] Gugulothu N, Tv V, Malhotra P, et al. Predicting remaining useful life using time series embeddings based on recurrent neural networks [J]. 2017, Preprint arXiv: 1709.01073.

[20] Hochreiter S, Schmidhuber J. Long short-term memory [J]. Neural Computing, 1997, 9 (8): 1735-1780.

[21] Ye Z, Yu J B. Health condition monitoring of machines based on long short-term memory convolutional autoencoder [J]. Applied Soft Computing, 2021, 107: 107379.

[22] Tao Y, Wang X D, Sánchez R V, et al. Spur gear fault diagnosis using a multilayer gated recurrent unit approach with vibration signal [J]. IEEE Access, 2019, 7: 56880-56889.

[23] Jung M, Lee H, Tani J. Adaptive detrending to accelerate convolutional gated recurrent unit training for contextual video recognition [J]. Neural Networks, 2018, 105: 356-370.

[24] Wang S S, Chen J, Wang H, et al. Degradation evaluation of slewing bearing using HMM and improved GRU [J]. Measurement, 2019, 146: 385-395.

[25] Yuan M, Wu Y, Lin L. Fault diagnosis and remaining useful life estimation of aero engine using LSTM neural network [C]//2016 IEEE International Conference on Aircraft Utility Systems, 2016.

[26] Bromley J, Guyon I, LeCun Y, et al. Signature verification using a siamese time delay neural network [J]. International Journal of Pattern Recognition & Artificial Intelligence, 1993, 7 (4): 669.

[27] Melekhov I, Kannala J, Rahtu E. Siamese network features for image matching [C]//Proceedings of the 23rd International Conference on Pattern Recognition, 2016.

[28] Bertinetto L, Valmadre J, Henriques J F, et al. Fully convolutional Siamese networks for object tracking [C]//Proceedings of European Conference on Computer Vision Workshops, 2016.

[29] Qin S J. Survey on data-driven industrial process monitoring and diagnosis [J]. Annual Review of Control, 2012, 36 (2): 220-234.

[30] Yin S, Ding S X, Xie X, et al. A review on basic data-driven approaches for industrial process monitoring [J]. IEEE Transactions on Industrial Electronics, 2014, 61 (11): 6418-6428.

[31] Ge Z Q, Song Z H, Ding S X, et al. Data mining and analytics in the process industry: The role of machine learning [J]. IEEE Access, 2017, 5: 20590-20616.

[32] Yao L, Ge Z Q. Cooperative deep dynamic feature extraction and variable time-delay estimation

for industrial quality prediction [J]. IEEE Transactions on Industrial Informatics, 2021, 17 (6): 3782-3792.

[33] Zhao R, Wang D Z, Yan R Q, et al. Machine health monitoring using local feature-based gated recurrent unit networks [J]. IEEE Transactions on Industrial Electronics, 2018, 65 (2): 1539-1548.

[34] Zhang C F, Peng K X, Dong J. A nonlinear full condition process monitoring method for hot rolling process with dynamic characteristic [J]. ISA Transactions, 2021, 112: 363-372.

[35] Werbos P J. Backpropagation through time: What it does and how to do it [J]. Proceedings of the IEEE, 1990, 78 (10): 1550-1560.

[36] Khan A H, Cao X, Li S, et al. BAS-ADAM: an ADAM based approach to improve the performance of beetle antennae search optimizer [J]. IEEE/CAA Journal of Automatica Sinica, 2020, 7 (2): 461-471.

[37] Valle S, Li W H, Qin S J. Selection of the number of principal components: The variance of the reconstruction error criterion with a comparison to other methods [J]. Industrial & Engineering Chemistry Research, 1999, 38: 4389-4401.

[38] Zhang K, Dong J, Peng K X. A novel dynamic non-Gaussian approach for quality-related fault diagnosis with application to the hot strip mill process [J]. Journal of the Franklin Institute, 2017, 354 (2): 702-721.

[39] Ma L, Dong J, Peng K X. Root cause diagnosis of quality-related faults in industrial multimode processes using robust Gaussian mixture model and transfer entropy [J]. Neurocomputing, 2018, 285: 60-73.

[40] Chu F, Fu Y L, Zhao X, et al. Operating performance assessment method and application for complex industrial process based on ISDAE model [J]. Acta Automatica Sinica, 2021, 47 (4): 849-863.

[41] Lv F Y, Wen C L, Liu M Q, et al. Weighted time series fault diagnosis based on a stacked sparse autoencoder [J]. Journal of Chemometrics, 2017, 31 (9): 2912.

[42] Lee S, Kwak M, Tsui K L, et al. Process monitoring using variational autoencoder for high-dimensional nonlinear processes [J]. Engineering Applications of Artificial Intelligence, 2019, 83: 13-27.

[43] Fernández-Llaneza D, Ulander S, Gogishvili D, et al. Siamese recurrent neural network with a self-attention mechanism for bioactivity prediction [J]. ACS Omega, 2021, 6: 11086-11094.

Chapter 16 KPI-Related OPA Framework Based on Distributed ImRMR-KOCTA

In complex industrial production, process parameter drift, improper operator interventions, and other uncertain disturbances often lead to process operating performance degradation. Once the operating performance deviates from the optimum, it may be difficult to guarantee the production efficiency and product quality, or even process safety[1]. Traditional process monitoring methods usually focus on distinguishing whether the process is in normal or faulty operating condition. However, in addition to the above rough judgment, field operators are more concerned about whether the process operates at the optimum. As an extension of traditional process monitoring method, operating performance assessment can further evaluate the process operating condition and identify non-optimal causes, so as to provide guidance and suggestions for maintenance decision. Hence, operating performance assessment has attracted great attention from both industrial and academic fields[2].

In the early stage, operating performance assessment methods can be divided into two categories: quantitative information-based and qualitative information-based assessment methods. Multivariate statistical analysis is one of the most commonly used quantitative information-based methods[3-4]. For production processes with a single stable operating condition, operating performance assessment methods based on PCA, multi-set principal component analysis (MPCA), and total projection to latent structures (T-PLS) have been proposed[5-7]. According to the characteristic of multi-peak distribution of process data, the assessment method based on Gaussian mixture model regression is proposed, which realizes the optimality evaluation of operating state in multi-mode processes[8]. While the most commonly used qualitative information-based assessment methods are Bayesian network[9], fuzzy theory[10], and rough set theory[11]. In addition, methods combining quantitative and qualitative information have been gradually applied for operating performance assessment[12-13]. However, nonlinearity and dynamicity are widely existing in industrial processes, and traditional assessment methods are difficult to analyze them effectively at the same time.

For monitoring the operating conditions more effectively, the concept of key performance indicators (KPIs) are introduced, which can be divided into three categories: engineering KPIs (technical performance and product quality), maintenance KPIs (running speed, maintenance time, cost) and economic KPIs (total energy consumption or productivity of factories)[14-16]. However, those KPIs are usually not easy to measure directly and has obvious time delays, for example, the exit thickness, flatness, and crown in the hot strip mill process (HSMP)[17-18]. An intuitive solution is to establish quantitative relationships between process and KPI variables, and then monitor the process with the available process variables. Among many competitive methods, PCR, PLS, CCA and their extensions are considered as the main tools to solve the problem of KPI-related process monitoring because of their simplicity and efficiency in handling large amounts of data[19-21]. Although scholars have conducted many researches on KPI-related process monitoring, the KPI-related operating performance assessment methods have been rarely reported before.

Plant-wide characteristic of industrial processes is also a great challenge for operating performance assessment. The plant-wide process usually has a long production flow with correlated manufacturing units and coupled variables. Conventional operating performance assessment methods are directly applied to the whole process, which is difficult to obtain satisfactory accuracy. The most commonly used method to deal with plant-wide processes is multi-block or decentralized method[22-23]. They divide the process into sub-blocks based on physical characteristics or data correlation. However, there are still few studies on the operating performance assessment of plant-wide industrial processes, which are mainly for the following reasons:

(1) the large-scale process and complex mechanism make it difficult to establish an assessment model accurately;

(2) the process variables do not directly affect the KPIs, but through a series of intermediate process indexes, so it is difficult to directly extract the correlation between the process variables and the KPIs;

(3) when the process is running at non-optimal condition, it is difficult to trace the non-optimal variables quickly.

Strongly inspired by the above observations, a novel KPI-related operating performance assessment method is proposed to handle the above-mentioned problems in industrial processes. Specifically, the main contributions of this chapter are summarized as:

(1) By replacing the mutual information (MI) with the maximum information

coefficient (MIC), an improved minimum redundant maximum correlation (ImRMR) is presented to describe the relationships between process variables and KPIs, which can reduce the redundancy among variables in each subsystem.

(2) With the combination of polynomial kernel and RBF kernel, a kernel output-relevant common trend analysis (KOCTA) is attempted to analyze nonlinearity and dynamicity simultaneously.

(3) Based on ImRMR and KOCTA, a distributed ImRMR-KOCTA assessment strategy is developed to evaluate the operating performance from subsystem level and plant-wide level. In addition, when the process operates under non-optimal conditions, the non-optimal causes can be identified by a designed difference index.

16.1 Review of mRMR and OCTA

In this section, mRMR algorithm and OCTA algorithm are reviewed briefly, which will motivate the problem formulation for the proposed method.

16.1.1 Minimal redundancy maximal relevance

Mutual Information (MI) is a well-known evaluation index that describes the interdependence between two variables in information theory[24]. It considers both linear and nonlinear relationships between variables. A large value of MI indicates a close correlation between variables. Specifically, the MI of any two variables x_i and x_j is defined as:

$$
\begin{aligned}
I(x_i, x_j) &= H(x_i) + H(x_j) - H(x_i, x_j) \\
&= \sum_{x_i} p(x_i)\lg\frac{1}{p(x_i)} + \sum_{x_j} p(x_j)\lg\frac{1}{p(x_j)} - \sum_{x_i}\sum_{x_j} p(x_i, x_j)\lg\frac{1}{p(x_i, x_j)}
\end{aligned}
\tag{16.1}
$$

where, $p(x_i)$ and $p(x_j)$ are the marginal probability density functions (PDF) of x_i and x_j; $p(x_i, x_j)$ is the joint PDF.

However, for a variable set $\boldsymbol{X}$, the maximal interdependency on a target variable $\boldsymbol{c}$ is difficult to obtain because of the complex computations of the joint PDF. By considering the redundancy and relevance between variables, the mRMR algorithm can deal with this knotty issue[25].

The maximal relevance of the variables in $\boldsymbol{X}$ with variable $\boldsymbol{c}$ is denoted as:

$$
\max D(\boldsymbol{X}, \boldsymbol{c}),\ D = \frac{1}{|\boldsymbol{X}|}\sum_{x_i \in \boldsymbol{X}} I(x_i, \boldsymbol{c}) \tag{16.2}
$$

where, $|\boldsymbol{X}|$ is the scale of the variable set $\boldsymbol{X}$.

It can be seen that the selected variables may have rich redundancy due to the maximum interdependency. In order to select mutually exclusive variables, the minimal redundancy is calculated as:

$$\min R(\boldsymbol{X}),\ R = \frac{1}{|\boldsymbol{X}|^2} \sum_{x_i,\ x_j \in \boldsymbol{X}} I(x_i,\ x_j) \tag{16.3}$$

Based on D and R, a mix index $\Psi(D,\ R)$ is designed to maximize the relevance and minimize the redundancy simultaneously:

$$\max \Psi(D,\ R),\ \Psi = D - R \tag{16.4}$$

By solving the above optimization function, the variables strongly correlated with target variable $\boldsymbol{c}$ are divided into the same block.

16.1.2 Output-relevant common trend analysis

Inspired from common trend analysis[26], Wu et al. proposed the OCTA method for KPI-related nonstationary process monitoring[27]. Considering an $\boldsymbol{m}$-dimensional input data $\boldsymbol{X} \in R^{N\times m}$ and a $\boldsymbol{n}$-dimensional output data $\boldsymbol{Y} \in R^{N\times n}$, the output-relevant common trend model can be defined as:

$$\begin{cases} \boldsymbol{X} = \boldsymbol{S}\boldsymbol{A}^{\mathrm{T}} + \tilde{\boldsymbol{X}} \\ \boldsymbol{Y} = \boldsymbol{T}\boldsymbol{B}^{\mathrm{T}} + \tilde{\boldsymbol{Y}} \end{cases} \tag{16.5}$$

where, $\boldsymbol{S} \in R^{N\times a}$ and $\boldsymbol{T} \in R^{N\times a}$ are nonstationary trends; $\boldsymbol{A} \in R^{m\times a}$ and $\boldsymbol{B} \in R^{n\times a}$ are loading matrices; $\tilde{\boldsymbol{X}}$ and $\tilde{\boldsymbol{Y}}$ are stationary residuals and N is the number of the samples.

In addition, cross validation is used to determine a, which is the dimension of nonstationary trends.

From a physical point of view, the nonstationary characteristics of $\boldsymbol{X}$ and $\boldsymbol{Y}$ are derived by consistent factors. That is, $\boldsymbol{X}$ and $\boldsymbol{Y}$ share common nonstationary trends. Assuming $\boldsymbol{S} = \boldsymbol{X}\boldsymbol{C}$ and $\boldsymbol{T} = \boldsymbol{Y}\boldsymbol{D}$ are the linear combinations of $\boldsymbol{X}$ or $\boldsymbol{Y}$, the equation (16.5) can be rewritten as:

$$\begin{cases} \boldsymbol{X} = \boldsymbol{X}\boldsymbol{C}\boldsymbol{A}^{\mathrm{T}} + \tilde{\boldsymbol{X}} \\ \boldsymbol{Y} = \boldsymbol{Y}\boldsymbol{D}\boldsymbol{B}^{\mathrm{T}} + \tilde{\boldsymbol{Y}} \end{cases} \tag{16.6}$$

where, $\boldsymbol{C}\boldsymbol{A}^{\mathrm{T}}$ and $\boldsymbol{D}\boldsymbol{B}^{\mathrm{T}}$ are required to be projection matrices:

$$(\boldsymbol{C}\boldsymbol{A}^{\mathrm{T}})^2 = \boldsymbol{C}\boldsymbol{A}^{\mathrm{T}}\boldsymbol{C}\boldsymbol{A}^{\mathrm{T}} = \boldsymbol{C}\boldsymbol{A}^{\mathrm{T}} \tag{16.7}$$

$$(\boldsymbol{D}\boldsymbol{B}^{\mathrm{T}})^2 = \boldsymbol{D}\boldsymbol{B}^{\mathrm{T}}\boldsymbol{D}\boldsymbol{B}^{\mathrm{T}} = \boldsymbol{D}\boldsymbol{B}^{\mathrm{T}} \tag{16.8}$$

Then, $\boldsymbol{A}^{\mathrm{T}}\boldsymbol{C} = \boldsymbol{B}^{\mathrm{T}}\boldsymbol{D} = \boldsymbol{I}_a$ can be obtained. The estimation of $\boldsymbol{A}$, $\boldsymbol{B}$, $\boldsymbol{C}$, $\boldsymbol{D}$ can be described as an optimization problem with three objectives:

$$\min_{A,\ B,\ C,\ D} \{ \|X_0(I_m - CA^{\mathrm{T}})\|_F^2 + \|Y_0(I_n - DB^{\mathrm{T}})\|_F^2 + \|X_0C - Y_0D\|_F^2 \}$$
$$\text{s.t. } A^{\mathrm{T}}C = B^{\mathrm{T}}D = I_a \tag{16.9}$$

where, $X_0 = X - 1\mu_X^{\mathrm{T}}$ and $Y_0 = Y - 1\mu_Y^{\mathrm{T}}$ are zero-mean data, μ_X and μ_Y corresponding sample means of X and Y. Equation (16.8) can be solved by alternating direction method of multipliers (ADMM)[28].

16.2 KPI-related operating performance assessment framework

Due to the frequent changes of product specifications and drastic fluctuations of operation conditions, HSMP has inherent nonlinear and dynamic characteristics. Load variations will change statistical properties of process variables, then lead to changes of KPIs. Thus, the KPIs usually change with process variables synchronously, that is to say, they often show common trends. **Fig. 16.1** shows some historical data collected from the hot rolling plant of Ansteel Group Chaoyang Iron & Steel Company Limited, China. The first variable is a KPI, i.e., thickness. The other variables are process variables, including bending force, gap, and rolling speed. Obviously, these process variables share a positive nonstationary trend with the KPI in **Fig. 16.1**.

HSMP is a long process composed of many equipments, which can be divided into several subsystems. Once the operating performance of any subsystem degenerates, it may affect the whole production line and the final product quality, resulting in production waste and huge economic losses. Therefore, it is necessary to assess the operating performance in real time. In order to effectively assess the operating performance of HSMP, a KPI-related operating performance assessment is proposed in this chapter, which benefits from mRMR and OCTA.

16.2.1 ImRMR-KOCTA model

According to the process knowledge, the HSMP is first divided into upstream, midstream, and downstream subsystem. Assume that $X = [X_{\mathrm{up}}, X_{\mathrm{mid}}, X_{\mathrm{down}}] \in R^{N\times m}$ and $Y = [Y_{\mathrm{up}}, Y_{\mathrm{mid}}, Y_{\mathrm{down}}] \in R^{N\times n}$ are process variables and KPIs. In order to better reveal and quantify the interdependencies between X and Y in each subsystem, maximal information coefficient (MIC) is used to replace MI in the mRMR algorithm. MIC has the generality, equitability, and symmetry[29]. The range of MIC is from 0 to 1. For any variable x_i and x_j, the MIC is defined as:

$$\mathrm{MIC}(x_i, x_j) = \max_{n_{x_i}\cdot n_{x_j} < B(n)} \frac{I(x_i, x_j)}{\log_2 \min(n_{x_i}, n_{x_j})} \tag{16.10}$$

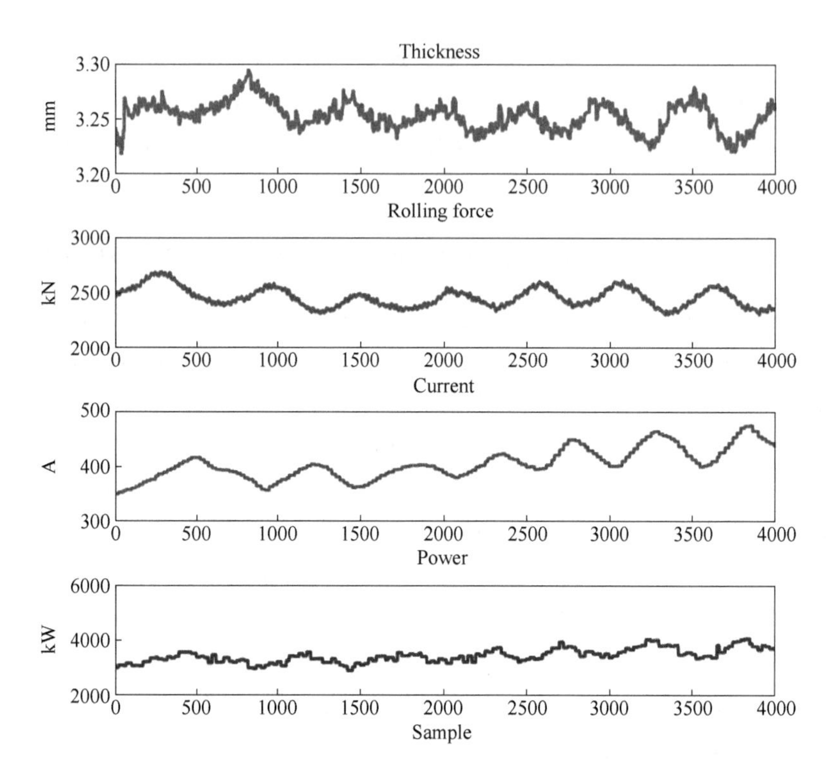

Fig. 16. 1 Common trends between process variables and the KPI

where, $n_{x_i} \cdot n_{x_j} < B(n)$, $B(n) = n^{0.6}$, n_{x_i} and n_{x_j} are the number of partitioned bins in the x_i- and x_j- axis.

The largest interdependence between process variables and KPIs in each subsystem can be calculated as:

$$\max D(\boldsymbol{X}_B, \boldsymbol{Y}_B), \ D = \frac{1}{|\boldsymbol{X}_B|} \sum_{x_i \in \boldsymbol{X}_B, \ y_i \in \boldsymbol{Y}_B} \mathrm{MIC}(x_i, y_i), \ B = \{\mathrm{up}, \mathrm{mid}, \mathrm{down}\} \tag{16.11}$$

And the minimal redundancy between process variables can be calculated as:

$$\min R(\boldsymbol{X}_B), \ R = \frac{1}{|\boldsymbol{X}_B|^2} \sum_{x_i, x_j \in \boldsymbol{X}_B} \mathrm{MIC}(x_i, x_j), \ B = \{\mathrm{up}, \mathrm{mid}, \mathrm{down}\} \tag{16.12}$$

Then, the relative MIC difference (RMICD) is designed to combine $D(\boldsymbol{X}_B, \boldsymbol{Y}_B)$ and $R(\boldsymbol{X}_B)$ in each subsystem:

$$\max Y(\boldsymbol{X}_B, \boldsymbol{Y}_B), \ Y = \frac{D - R}{R}, \ B = \{\mathrm{up}, \mathrm{mid}, \mathrm{down}\} \tag{16.13}$$

Let $Y_{tr} = \text{MIC}_{mean}/2D$ be the threshold of Y, where MIC_{mean} is the mean value of MIC between process variables and KPIs[30]. The selected process variables $\boldsymbol{X}_B^S \in R^{N \times S_B}$ are retained along with KPIs for the following KOCTA modeling, where S_B is the number of selected process variables in subsystem B.

By applying the kernel technique, $\{\boldsymbol{X}_B^S(i)\}_{i=1}^N$ is mapped into a feature space F; i.e. a nonlinear mapping $\boldsymbol{\Phi}$: $\boldsymbol{X}_B^S(i) \in R^{S_B} \rightarrow \boldsymbol{\Phi}(\boldsymbol{X}_B^S(i)) \in F$. However, the nonlinear mapping function $\boldsymbol{\Phi}$ is difficult to calculate directly, a positive semidefinite matrix $\boldsymbol{K}$ is defined to represent $\boldsymbol{\Phi}\boldsymbol{\Phi}^T$, where $\boldsymbol{K}(i, j) = \boldsymbol{K}(\boldsymbol{X}_B^S(i), \boldsymbol{X}_B^S(j)) = \langle \boldsymbol{\Phi}(\boldsymbol{X}_B^S(i)), \boldsymbol{\Phi}(\boldsymbol{X}_B^S(j)) \rangle$, $\boldsymbol{K}(\cdot, \cdot)$ is calculated using the kernel function.

In order to have good generalization, the polynomial kernel and radial basis function (RBF) kernel are integrated as follows[31]:

$$\boldsymbol{K}(\boldsymbol{X}_B^S(i), \boldsymbol{X}_B^S(j)) = \omega \underbrace{(\langle \boldsymbol{X}_B^S(i), \boldsymbol{X}_B^S(j) \rangle + 1)^b}_{\text{Polynomial kernel}} + (1 - w) \underbrace{\exp(\| \boldsymbol{X}_B^S(i) - \boldsymbol{X}_B^S(j) \|^2 / c)}_{\text{RBF kernel}} \tag{16.14}$$

where, $\omega \in [0, 1]$ is a weight coefficient.

For centralizing the mapped process variables and eliminating the mean effect in the feature space, the process variables are first preprocessed as follows:

$$\boldsymbol{K}_0 = \boldsymbol{K} - \frac{1}{N}\boldsymbol{I}_N\boldsymbol{K} - \frac{1}{N}\boldsymbol{K}\boldsymbol{I}_N + \frac{1}{N^2}\boldsymbol{I}_N\boldsymbol{K}\boldsymbol{I}_N \tag{16.15}$$

where, $\boldsymbol{I}_N = \boldsymbol{I}_N\boldsymbol{I}_N^T = \begin{bmatrix} 1 & \cdots & 1 \\ \vdots & \ddots & \vdots \\ 1 & \cdots & 1 \end{bmatrix} \in R^{N \times N}$.

After that, the linear OCTA is performed on $\boldsymbol{\Phi}(\boldsymbol{X}_B^S)$ and $\boldsymbol{Y}_B$, equation (16.6) is transformed into:

$$\begin{cases} \boldsymbol{\Phi}(\boldsymbol{X}_B^S) = \boldsymbol{\Phi}(\boldsymbol{X}_B^S)\boldsymbol{C}\boldsymbol{A}^T + \tilde{\boldsymbol{\Phi}}(\boldsymbol{X}_B^S) = \boldsymbol{S}_B\boldsymbol{A}^T + \tilde{\boldsymbol{\Phi}}(\boldsymbol{X}_B^S) \\ \boldsymbol{Y}_B = \boldsymbol{Y}_B\boldsymbol{D}\boldsymbol{B}^T + \tilde{\boldsymbol{Y}}_B = \boldsymbol{T}_B\boldsymbol{B}^T + \tilde{\boldsymbol{Y}}_B \end{cases} \tag{16.16}$$

where, $\boldsymbol{S}_B$ and $\boldsymbol{T}_B$ are nonstationary trends in subsystem $\boldsymbol{B}$. Therefore, the optimization problem of KOTCA can be described as:

$$\begin{aligned} & \min f_B(\boldsymbol{A}, \boldsymbol{B}, \boldsymbol{C}, \boldsymbol{D}) \\ = & \min_{\boldsymbol{A}, \boldsymbol{B}, \boldsymbol{C}, \boldsymbol{D}} \{ \| \boldsymbol{\Phi}(\boldsymbol{X}_B^S)(\boldsymbol{I}_{S_B} - \boldsymbol{C}\boldsymbol{A}^T) \|_F^2 + \| \boldsymbol{Y}_B(\boldsymbol{I}_{Y_B} - \boldsymbol{D}\boldsymbol{B}^T) \|_F^2 + \\ & \| \boldsymbol{\Phi}(\boldsymbol{X}_B^S)\boldsymbol{C} - \boldsymbol{Y}_B\boldsymbol{D} \|_F^2 \} \end{aligned}$$

$$\text{s. t. } \boldsymbol{A}^{\mathrm{T}}\boldsymbol{C} = \boldsymbol{B}^{\mathrm{T}}\boldsymbol{D} = \boldsymbol{I}_{a_B} \tag{16.17}$$

where, a_B and Y_B are numbers of common factors and KPIs in subsystem $\boldsymbol{B}$, respectively.

Before solving the equation (16.17) with ADMM, an augmented Lagrangian function is defined as:

$$L_\alpha = f_B(\boldsymbol{A}, \boldsymbol{B}, \boldsymbol{C}, \boldsymbol{D}) + \frac{\alpha_1}{2}\|\boldsymbol{A}^{\mathrm{T}}\boldsymbol{C} - \boldsymbol{I}_{a_B} + \boldsymbol{P}\|_F^2 - \frac{\alpha_1}{2}\|\boldsymbol{P}\|_F^2 + \frac{\alpha_2}{2}\|\boldsymbol{B}^{\mathrm{T}}\boldsymbol{D} - \boldsymbol{I}_{a_B} + \boldsymbol{Q}\|_F^2 - \frac{\alpha_2}{2}\|\boldsymbol{P}\|_F^2 \tag{16.18}$$

where, α_1 and α_2 are penalty parameters; $\boldsymbol{P}$ and $\boldsymbol{Q}$ are transformed lagrange multipliers.

The k_{th} iteration procedures of $\boldsymbol{A}$, $\boldsymbol{B}$, $\boldsymbol{C}$, $\boldsymbol{D}$, $\boldsymbol{P}$, and $\boldsymbol{Q}$ are as follows:

$$\boldsymbol{A}_{[k+1]} := \arg\min_{\boldsymbol{A}}\{\|\boldsymbol{\Phi}(\boldsymbol{X}_B^S)(\boldsymbol{I}_{S_B} - \boldsymbol{C}_{[k]}\boldsymbol{A}^{\mathrm{T}})\|_F^2 + \frac{\alpha_1}{2}\|\boldsymbol{A}^{\mathrm{T}}\boldsymbol{C}_{[k]} - \boldsymbol{I}_{a_B} + \boldsymbol{P}_{[k]}\|_F^2\} \tag{16.19}$$

$$\boldsymbol{B}_{[k+1]} := \arg\min_{\boldsymbol{B}}\{\|\boldsymbol{Y}_B(\boldsymbol{I}_{Y_B} - \boldsymbol{D}_{[k]}\boldsymbol{B}^{\mathrm{T}})\|_F^2 + \frac{\alpha_2}{2}\|\boldsymbol{B}^{\mathrm{T}}\boldsymbol{D}_{[k]} - \boldsymbol{I}_{a_B} + \boldsymbol{Q}_{[k]}\|_F^2\} \tag{16.20}$$

$$\boldsymbol{C}_{[k+1]} := \arg\min_{\boldsymbol{C}}\{\|\boldsymbol{\Phi}(\boldsymbol{X}_B^S)(\boldsymbol{I}_{S_B} - \boldsymbol{C}\boldsymbol{A}_{[k+1]}^{\mathrm{T}})\|_F^2 + \frac{\alpha_1}{2}\|\boldsymbol{A}_{[k+1]}^{\mathrm{T}}\boldsymbol{C} - \boldsymbol{I}_{a_B} + \boldsymbol{P}_{[k]}\|_F^2 + \|\boldsymbol{\Phi}(\boldsymbol{X}_B^S)\boldsymbol{C} - \boldsymbol{Y}_B\boldsymbol{D}_{[k]}\|_F^2\} \tag{16.21}$$

$$\boldsymbol{D}_{[k+1]} := \arg\min_{\boldsymbol{D}}\{\|\boldsymbol{Y}_B(\boldsymbol{I}_{Y_B} - \boldsymbol{D}\boldsymbol{B}_{[k+1]}^{\mathrm{T}})\|_F^2 + \frac{\alpha_2}{2}\|\boldsymbol{B}_{[k+1]}^{\mathrm{T}}\boldsymbol{D} - \boldsymbol{I}_{a_B} + \boldsymbol{Q}_{[k]}\|_F^2 + \|\boldsymbol{\Phi}(\boldsymbol{X}_B^S)\boldsymbol{C}_{[k+1]} - \boldsymbol{Y}_B\boldsymbol{D}\|_F^2\} \tag{16.22}$$

$$\boldsymbol{P}_{[k+1]} := \boldsymbol{P}_{[k]} + \boldsymbol{A}_{[k+1]}^{\mathrm{T}}\boldsymbol{C}_{[k+1]} - \boldsymbol{I}_{a_B} \tag{16.23}$$

$$\boldsymbol{Q}_{[k+1]} := \boldsymbol{Q}_{[k]} + \boldsymbol{B}_{[k+1]}^{\mathrm{T}}\boldsymbol{D}_{[k+1]} - \boldsymbol{I}_{a_B} \tag{16.24}$$

In order to get the updated result of $\boldsymbol{A}$, the optimal solution makes the partial derivative of equation (16.19) equal to zero:

$$-2(\boldsymbol{I}_{S_B} - \boldsymbol{A}\boldsymbol{C}_{[k]}^{\mathrm{T}})\boldsymbol{K}_0^{\mathrm{T}}\boldsymbol{C}_{[k]} + \alpha_1\boldsymbol{C}_{[k]}(\boldsymbol{C}_{[k]}^{\mathrm{T}}\boldsymbol{A} - \boldsymbol{I}_{a_B} + \boldsymbol{P}_{[k]}^{\mathrm{T}}) = 0 \tag{16.25}$$

which can be transformed to:

$$\alpha_1\boldsymbol{C}_{[k]}\boldsymbol{C}_{[k]}^{\mathrm{T}}\boldsymbol{A} + 2\boldsymbol{A}\boldsymbol{C}_{[k]}^{\mathrm{T}}\boldsymbol{K}_0^{\mathrm{T}}\boldsymbol{C}_{[k]} = 2\boldsymbol{K}_0^{\mathrm{T}}\boldsymbol{C}_{[k]} + \alpha_1\boldsymbol{C}_{[k]}(\boldsymbol{I}_{a_B} - \boldsymbol{P}_{[k]}^{\mathrm{T}}) \tag{16.26}$$

Obviously, equation (16.26) is a Sylvester equation, Hesenberg-Schur algorithm is a commonly used method to solve this problem[27].

Similarly, the updated result of $\boldsymbol{B}$ can be obtained as follows:

$$\alpha_2 \boldsymbol{D}_{[k]} \boldsymbol{D}_{[k]}^{\mathrm{T}} \boldsymbol{B} + 2\boldsymbol{B}\boldsymbol{D}_{[k]}^{\mathrm{T}} (\boldsymbol{Y}_B)^{\mathrm{T}} \boldsymbol{Y}_B \boldsymbol{D}_{[k]} = 2 (\boldsymbol{Y}_B)^{\mathrm{T}} \boldsymbol{Y}_B \boldsymbol{D}_{[k]} + \alpha_2 \boldsymbol{D}_{[k]} (\boldsymbol{I}_{a_B} - \boldsymbol{Q}_{[k]}^{\mathrm{T}}) \tag{16.27}$$

For the update of $\boldsymbol{C}$, the optimal solution makes the partial derivative of equation (16.21) equal to zero:

$$\begin{aligned} -2\boldsymbol{K}_0^{\mathrm{T}}(\boldsymbol{I}_{S_B} - \boldsymbol{C}\boldsymbol{A}_{[k+1]}^{\mathrm{T}})\boldsymbol{A}_{[k+1]} + 2\boldsymbol{K}_0^{\mathrm{T}}\boldsymbol{C} - 2\boldsymbol{\Phi}(\boldsymbol{X}_B^S)\boldsymbol{Y}_B\boldsymbol{D}_{[k]} + {} \\ \alpha_1 \boldsymbol{A}_{[k+1]}(\boldsymbol{A}_{[k+1]}^{\mathrm{T}}\boldsymbol{C} - \boldsymbol{I}_{a_B} + \boldsymbol{P}_{[k]}) = 0 \end{aligned} \tag{16.28}$$

which can be equivalent to a standard Sylvester equation:

$$\begin{aligned} &(\boldsymbol{K}_0^{\mathrm{T}})^{-1}(2\boldsymbol{K}_0^{\mathrm{T}} + \alpha_1 \boldsymbol{A}_{[k+1]}\boldsymbol{A}_{[k+1]}^{\mathrm{T}})\boldsymbol{C} + 2\boldsymbol{C}\boldsymbol{A}_{[k+1]}^{\mathrm{T}}\boldsymbol{A}_{[k+1]} \\ &= (\boldsymbol{K}_0^{\mathrm{T}})^{-1}[2\boldsymbol{K}_0^{\mathrm{T}}\boldsymbol{A}_{[k+1]} + 2\boldsymbol{\Phi}(\boldsymbol{X}_B^S)\boldsymbol{Y}_B\boldsymbol{D}_{[k]} + \alpha_1 \boldsymbol{A}_{[k+1]}(\boldsymbol{I}_{a_B} - \boldsymbol{P}_{[k]})] \end{aligned} \tag{16.29}$$

Hesenberg-Schur algorithm can be used to solve equation (16.29).

Similarly, the updated result of $\boldsymbol{D}$ can be obtained as follows:

$$\begin{aligned} &[(\boldsymbol{Y}_B)^{\mathrm{T}}\boldsymbol{Y}_B]^{-1}(2(\boldsymbol{Y}_B)^{\mathrm{T}}Y_B + \alpha_2 \boldsymbol{B}_{[k+1]}\boldsymbol{B}_{[k+1]}^{\mathrm{T}})\boldsymbol{D} + 2\boldsymbol{D}\boldsymbol{B}_{[k+1]}^{\mathrm{T}}\boldsymbol{B}_{[k+1]} \\ &= [(\boldsymbol{Y}_B)^{\mathrm{T}}\boldsymbol{Y}_B]^{-1}[2(\boldsymbol{Y}_B)^{\mathrm{T}}(\boldsymbol{Y}_B\boldsymbol{B}_{[k+1]} + \boldsymbol{\Phi}(\boldsymbol{X}_B^S)\boldsymbol{C}_{[k+1]}) + \alpha_1 \boldsymbol{B}_{[k+1]}(\boldsymbol{I}_{a_B} - \boldsymbol{Q}_{[k]})] \end{aligned} \tag{16.30}$$

When the primal and dual residuals are small, the iteration is terminated to get the optimal $\boldsymbol{A}$, $\boldsymbol{B}$, $\boldsymbol{C}$, $\boldsymbol{D}$, $\boldsymbol{P}$, and $\boldsymbol{Q}$.

For a new sample $\boldsymbol{X}_B^S(\mathrm{new})$, $\boldsymbol{K}_0^{\mathrm{new}}(i) = \boldsymbol{K}(\boldsymbol{X}_B^S(i), \boldsymbol{X}_B^S(\mathrm{new}))$ can be preprocessed as follows:

$$\boldsymbol{K}_{\mathrm{new}} = \left(\boldsymbol{I}_N - \frac{\boldsymbol{1}}{N}\boldsymbol{1}_N\boldsymbol{1}_N^{\mathrm{T}}\right)\left(\boldsymbol{K}_0^{\mathrm{new}} - \frac{1}{N}\boldsymbol{K}_0\boldsymbol{1}_N\right) \tag{16.31}$$

Similar to KPLS, the score vector of $\boldsymbol{X}_B^S(\mathrm{new})$ can be calculated as:

$$\boldsymbol{s}_B^{\mathrm{new}} = (\boldsymbol{T}_B^{\mathrm{T}}\boldsymbol{K}_0\boldsymbol{S}_B)^{-1}\boldsymbol{T}_B^{\mathrm{T}}\boldsymbol{K}_{\mathrm{new}} \tag{16.32}$$

In order to reveal the operating performance in subsystem $\boldsymbol{B}$, the KPI-related statistic S_B^2 and KPI-unrelated statistic SPE_B can be computed as:

$$S_B^2 = (\boldsymbol{s}_B^{\mathrm{new}})^{\mathrm{T}}\boldsymbol{\Lambda}^{-1}\boldsymbol{s}_B^{\mathrm{new}} \tag{16.33}$$

$$\mathrm{SPE}_B = \boldsymbol{\Phi}(\boldsymbol{X}_B^S(\mathrm{new}))\boldsymbol{\Phi}^{\mathrm{T}}(\boldsymbol{X}_B^S(\mathrm{new})) - 2\boldsymbol{s}_B^{\mathrm{new}}\boldsymbol{S}_B^{\mathrm{T}}\boldsymbol{K}_{\mathrm{new}} + \boldsymbol{s}_B^{\mathrm{new}}\boldsymbol{S}_B^{\mathrm{T}}\boldsymbol{K}_0\boldsymbol{S}_B(\boldsymbol{s}_B^{\mathrm{new}})^{\mathrm{T}} \tag{16.34}$$

where, $\boldsymbol{\Phi}(\boldsymbol{X}_B^S(\mathrm{new}))\boldsymbol{\Phi}^{\mathrm{T}}(\boldsymbol{X}_B^S(\mathrm{new})) = 1 - \frac{2}{N}\sum_{i=1}^{N}\boldsymbol{K}_0^{\mathrm{new}}(i) + \frac{1}{N^2}\sum_{i=1}^{N}\sum_{j=1}^{N}\boldsymbol{K}_0(i, j)$, S_B^2

reveals the nonstationary common trends between process variables and KPIs.

SPE_B can be used to analyze the stationary equilibrium relationships of process variables. Given a significance level α, the thresholds of S_B^2 and SPE_B are denoted as $S_{B,\alpha}^2$ and $SPE_{B,\alpha}$, which can be obtained by kernel density estimation (KDE)[32].

16.2.2 Distributed operating performance assessment

During offline modeling, the modeling data are first divided into several subsets. Each subset denotes an operating performance level. For hot rolling process, the setting of operating performance levels needs to consider the actual operation management needs. Thus, too many or too few operating performance levels are not appropriate. Based on the process knowledge and the national standard GB/T 709—2006, four operating performance grades (optimal, good, general, and poor) are numbered as 1, 2, 3, and 4. In subsystem B, the modeling data for each operating performance level are denoted as $\{\boldsymbol{X}_B^c, \boldsymbol{Y}_B^c\}$, $c = 1, 2, 3, 4$. Next, $\{\boldsymbol{X}_B^c, \boldsymbol{Y}_B^c\}$ are used to train the ImRMR-KOCTA model. $S_{B,\alpha}^{2,c}$ and $SPE_{B,\alpha}^c$ can be determined by KDE.

Let $\boldsymbol{x}_t$ be the online process data at time t. For revealing the operating performance of nonstationary and stationary part simultaneously, $S_{B,\alpha}^{2,c}$ and SPE_B^c are combined as follows:

$$Y_B^c(t) = \frac{S_B^{2,c}}{S_{B,\alpha}^{2,c}} + \frac{SPE_B^c}{SPE_{B,\alpha}^c} \tag{16.35}$$

The threshold of Y_B^c is denoted as $Y_{B,\alpha}^c$, which can be determined by KDE for $c = 1$, 2, 3, 4. Obviously, different operating performance levels have different statistics and thresholds. Bayesian inference can be used to fuse them. The probability for belonging to level c is calculated as follows:

$$P_B(L_B^c \mid Y_B^c(t)) = \frac{P_B(Y_B^c(t) \mid L_B^c)P_B(L_B^c)}{P_B(Y_B^c(t) \mid L_B^c)P_B(L_B^c) + P_B(Y_B^c(t) \mid \overline{L}_B^c)P_B(\overline{L}_B^c)} \tag{16.36}$$

where, L^c represents the operating performance belongs to level c at time t; $\overline{L}^c$ represents the operating performance does not belong to level c at time t.

The conditional probability $P_B(Y_B^c(t) \mid L^c)$ and $P_B(Y_B^c(t) \mid \overline{L}^c)$ can be calculated as:

$$P_B(Y_B^c(t) \mid L_B^c) = \exp\{-Y_B^c(t)/Y_{B,\alpha}^c\}$$
$$P_B(Y_B^c(t) \mid \overline{L}_B^c) = \exp\{-Y_{B,\alpha}^c/Y_B^c(t)\} \tag{16.37}$$

Then, the operating performance level of subsystem B at time t is assigned to according to the following rule:

$$L_B(t) = \operatorname{argmax} P_B(L_B^c \mid Y_B^c(t)) \tag{16.38}$$

After obtaining the operating performance levels of all subsystems, the operating performance level of the plant-wide process can be determined by weighting calculation:

$$L(t) = \operatorname{round}\left(\sum_B \beta_B L_B(t)\right),\quad B = \{\text{up, mid, down}\} \tag{16.39}$$

where, $\beta_B(\sum_B \beta_B = 1)$ is the weight coefficient of subsystem B; round( ) is a rounding function.

In real production process, due to the influence of uncertain factors, the operating performance level is sometimes difficult to be determined by the assessment result at a single time. For this reason, a moving window of width H is introduced, and the operating performance level with the highest frequency in the sliding window $[L(t-H+1), L(t-H+2), \cdots, L(t)]$ is taken as the assessment result, which is denoted as Level(t). If there are more than one operating performance levels with the highest frequency, the assessment result of the previous time will remain unchanged. In this chapter, H is set to 4.

When the process is under non-optimal condition, non-optimal cause needs to be identified. Similarity comparison is made between the non-optimal data $\boldsymbol{X}_B^S(t)$ and the optimal dataset $\boldsymbol{X}_B^{\text{opt}}$. For variable $x_{B,i}(i = 1, 2, \cdots, m_B)$, the difference index is defined as:

$$\operatorname{dif}(i) = 1 - \frac{\sum_{k=1}^{K} x_{B,i} \cdot x_{k,B,i}^{\text{opt}}}{Kx_{B,i}^2 + \sum_{k=1}^{K}(x_{k,B,i}^{\text{opt}})^2 - \sum_{k=1}^{K} x_{B,i} \cdot x_{k,B,i}^{\text{opt}}} \tag{16.40}$$

where, K is the number of samples in $\boldsymbol{X}_B^{\text{opt}}$.

The larger the value of dif(i) is, the more likely $x_{B,i}$ is regarded as the non-optimal variable. The flowchart of the proposed operating performance assessment method is presented in **Fig. 16.2**.

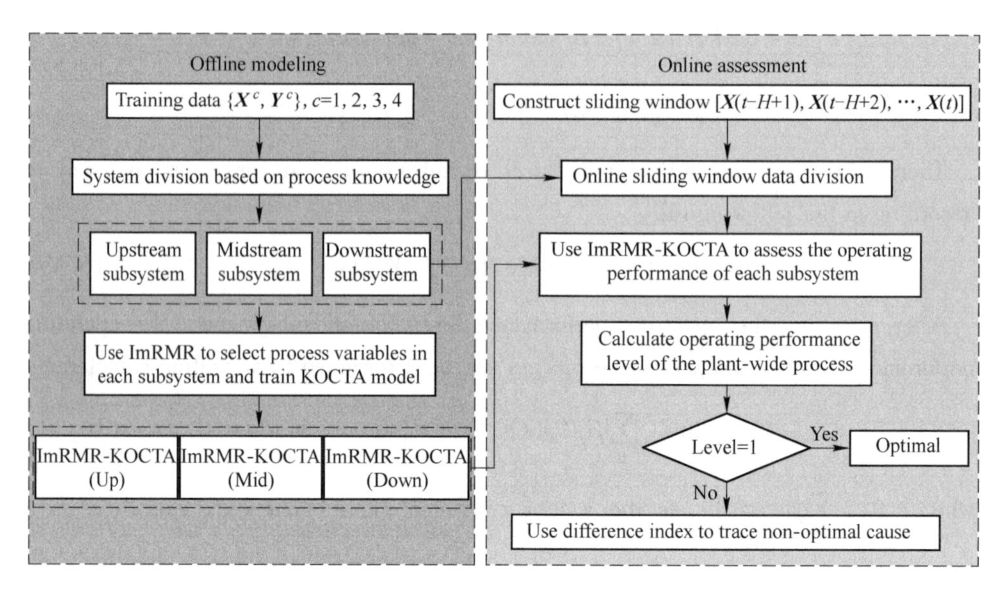

Fig. 16. 2 Flowchart of the proposed operating performance assessment method

16. 3 Case study

In this section, an industrial application to a practical HSMP is studied to illustrate the assessment performance of the proposed method.

16. 3. 1 Experimental setup

There are four rolls in each stand, including two working rolls and two supporting rolls. In order to offer rolling and bending forces, a hydraulic system is equipped to reduce the thickness of strip steel. And the strip steel can move forward smoothly through an electromechanical system. Piezomagnetic sensors and strain gauge sensors are used to measure the bending forces and rolling forces in real-time. Considering the high temperature and high speed, the unmeasurable gap between working rolls is approximately estimated by the altitude difference between supporting rolls[33]. The X-ray device is used to measure the finishing exit thickness, which is an important KPI-related variable to determine the product quality. Thus, how to extract the latent relationship among the process variables and KPI-related variable, and assess the process operating performance are opening issues. According to the thickness deviation, 41 process variables closely related to this KPI are selected for operating performance assessment of the HSMP, as shown in **Table 16. 1**. In order to meet the market demand for different specifications of strip steel, the slabs are rolled into coils with different thickness. The training and testing data used in this chapter are collected from a real HSMP, including alloy steel with 2.70 mm-thickness and carbon steel with 3.95 mm-thickness.

In order to validate the feasibility of the proposed method, it is compared with the traditional OCTA and linearity decomposition-based cointegration analysis (LDCA) method[34]. For each operating performance level, 1000 samples are used for offline modeling. As the testing set, 400 samples are generated from the HSMP, including 100 optimal samples, 100 good samples, 100 general samples and 100 poor samples. The sampling interval is 0.01 s, and significance level α is set as 0.01. Precision, recall, and accuracy are used to evaluate the performance of different assessment methods, which can be computed by:

$$\mathrm{Pre}_i = \frac{\mathrm{TP}_i}{\mathrm{TP}_i + \mathrm{FP}_i} \tag{16.41}$$

$$\mathrm{Rec}_i = \frac{\mathrm{TP}_i}{\mathrm{TP}_i + \mathrm{FN}_i} \tag{16.42}$$

$$\mathrm{Acc} = \frac{\mathrm{TP}_i + \mathrm{TN}_i}{\mathrm{TP}_i + \mathrm{FP}_i + \mathrm{TN}_i + \mathrm{FN}_i} \tag{16.43}$$

where, $i = 1, 2, 3, 4$ represents the operating performance level; TP_i, FP_i, TN_i, FN_i are the number of true positive samples, false positive samples, true negative samples, and false negative samples, respectively.

Genetic algorithm (GA) is used to tune the kernel parameters to make Acc achieve the optimal value[31].

Table 16.1 Process variables in the HSMP

Variable	Description	Unit
G_i (1-7)	Roll gap at the i_{th} stand, $i=1, \cdots, 7$	mm
F_i (8-14)	Rolling force at the i_{th} stand, $i=1, \cdots, 7$	MN
B_i (15-20)	Bending force at the i_{th} stand, $i=2, \cdots, 7$	MN
S_i (21-27)	Rolling speed at the i_{th} stand, $i=1, \cdots, 7$	mm/s
C_i (28-34)	Current at the i_{th} stand, $i=1, \cdots, 7$	A
P_i (35-41)	Power at the i_{th} stand, $i=1, \cdots, 7$	kW

16.3.2 Analysis of assessment results

(1) **Case 1**: This scenario is used to verify the effectiveness of the above

assessment methods in the production of 2. 70 mm-thickness alloy steel. During the first 100 samples, the HSMP runs at Level 1 (optimal). Then, the process operating performance degenerates to Level 2 (good) from the 101_{st} sample. For the last 200 samples, the HSMP runs at the Level 3 (general) and Level 4 (poor), respectively. In order to further analyze the above three methods, the assessment results and confusion matrices are shown in **Fig. 16. 3-Fig. 16. 5**. Obviously, compared with the traditional OCTA algorithm and the LDCA algorithm, the ImRMR-KOCTA algorithm has the fewest wrong assessing samples. The accuracy of the ImRMR-KOCTA is nearly 10% higher than that of the traditional OCTA. Although the assessment performance of the LDCA is better than the traditional OCTA, it is still inferior to the proposed method. Detailed comparison results are summarized in **Table 16. 2**.

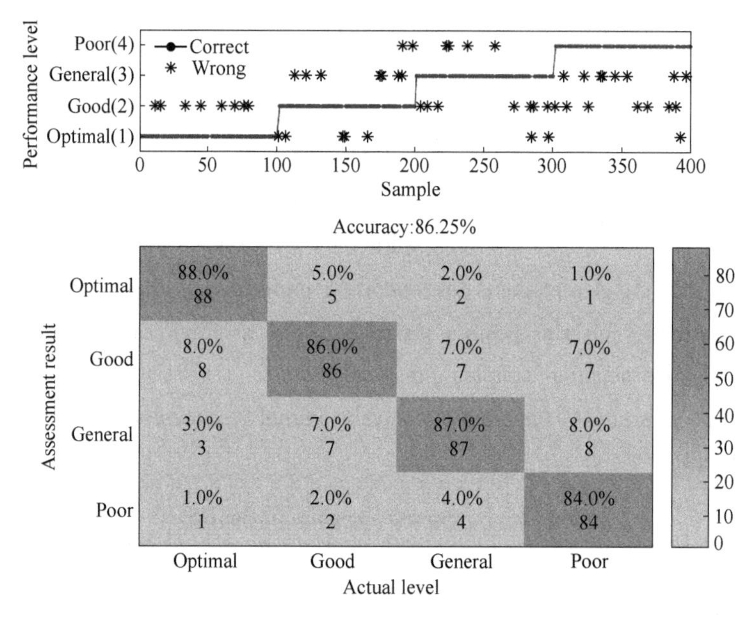

Fig. 16. 3 Assessment result of the OCTA in Case 1

(Scan the QR code on the front of the book for color picture)

Table 16. 2 Assessment performance comparison of the three methods in Case 1

Index	Pre_1	Rec_1	Pre_2	Rec_2	Pre_3	Rec_3	Pre_4	Rec_4	Acc
OCTA	91. 7%	88. 0%	79. 6%	86. 0%	82. 9%	87. 0%	92. 3%	84. 0%	86. 3%
LDCA	92. 9%	91. 0%	87. 1%	88. 0%	83. 9%	89. 0%	94. 7%	90. 0%	89. 5%
ImRMR-KOCTA	97. 9%	94. 0%	92. 3%	96. 0%	92. 2%	94. 0%	96. 9%	95. 0%	94. 8%

(2) **Case 2**: The HSMP products 3. 95 mm-thickness carbon steel. **Table 16. 3**

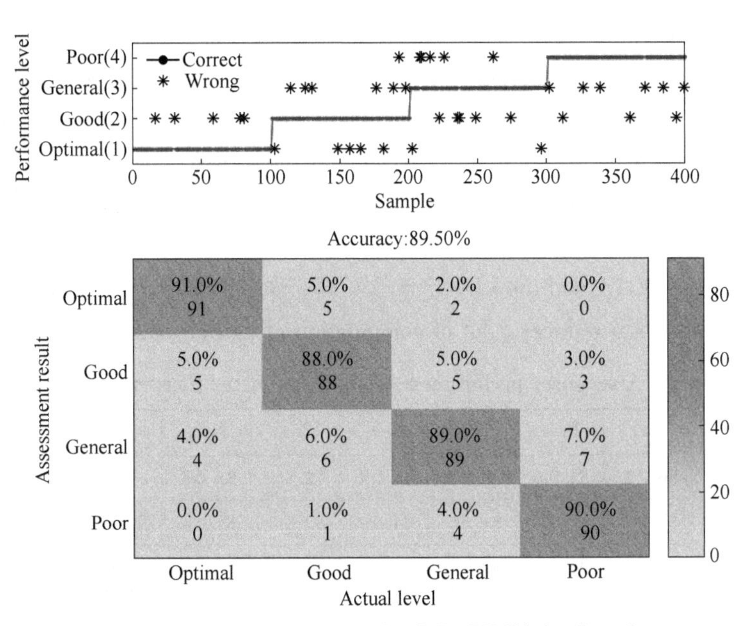

Fig. 16.4 Assessment result of the LDCA in Case 1

(Scan the QR code on the front of the book for color picture)

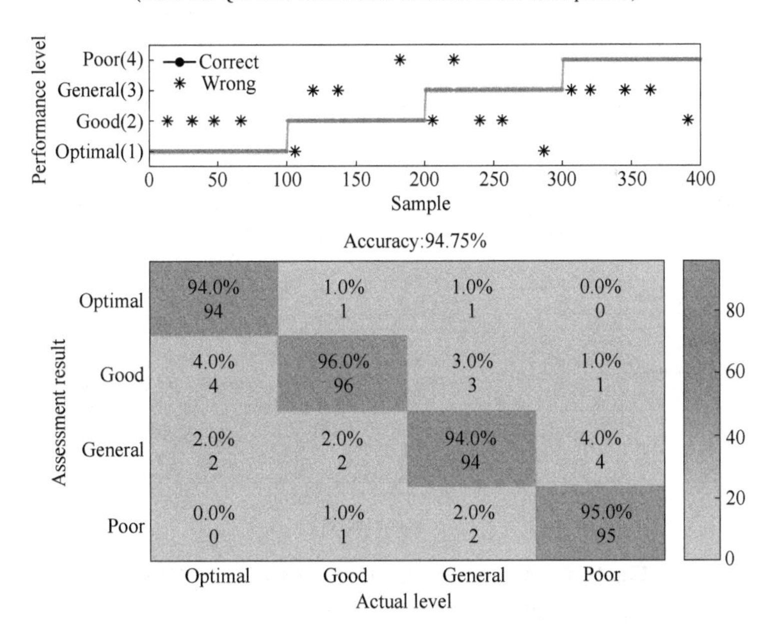

Fig. 16.5 Assessment result of the proposed method in Case 1

(Scan the QR code on the front of the book for color picture)

presents the comparison results of Case 2. The assessment accuracy of Case 2 is slightly lower than that of Case 1, which is due to the differences between different products. Similar to the assessment results of Case 1, the traditional OCTA and the LDCA have more wrong assessing samples, as shown in **Fig. 16. 6** and **Fig. 16. 7**. The confusion matrix in **Fig. 16. 8** indicates that the proposed method achieves ideal results. Compared with the traditional OCTA and the LDCA, the ImRMR-KOCTA has stronger capability to deal with dynamic nonlinearity. Moreover, the difference in time consumption is significant. The average running time of the proposed method is 14. 35 s, while that of the OCTA and the LDCA are 29. 21 s and 25. 87 s. The main reason is that the ImRMR algorithm reduces a lot of computation of kernel matrices.

Table 16. 3　Assessment performance comparison of the three methods in Case 2

Index	Pre_1	Rec_1	Pre_2	Rec_2	Pre_3	Rec_3	Pre_4	Rec_4	Acc
OCTA	86. 7%	85. 0%	78. 9%	82. 0%	79. 8%	83. 0%	87. 2%	82. 0%	83. 0%
LDCA	87. 9%	87. 0%	81. 6%	84. 0%	82. 5%	85. 0%	90. 5%	86. 0%	85. 5%
ImRMR-KOCTA	96. 8%	91. 0%	89. 7%	96. 0%	90. 4%	94. 0%	96. 8%	92. 0%	93. 3%

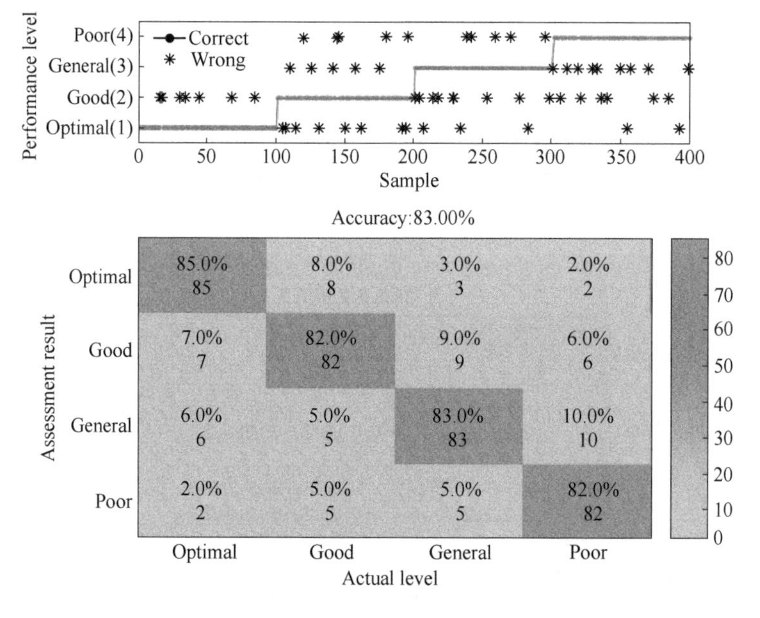

Fig. 16. 6　Assessment result of the OCTA in Case 2

(Scan the QR code on the front of the book for color picture)

When the HSMP is running at non-optimal condition, it is necessary to trace the non-optimal cause. In this chapter, two cases at Level 3 and Level 4 are taken as examples to illustrate the non-optimal cause identification. First, the non-optimal subsystems are preliminarily determined according to the assessment results of each subsystem, and

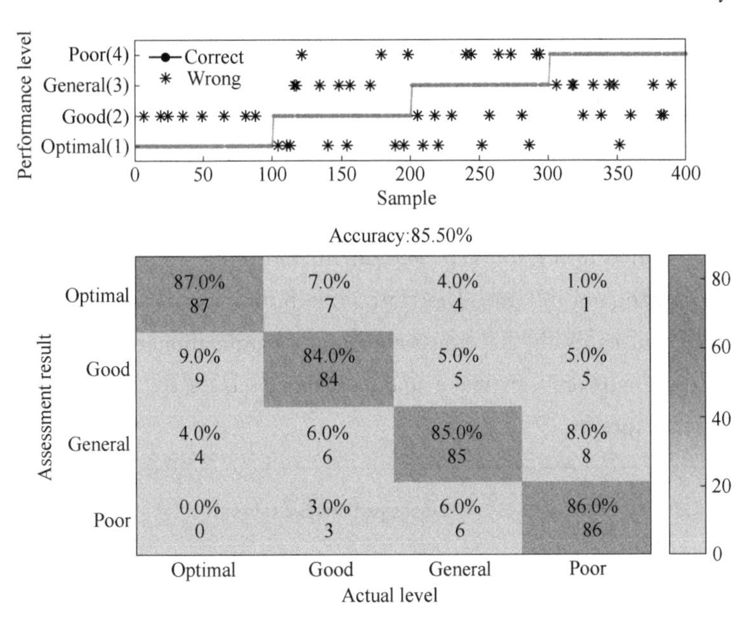

Fig. 16.7 Assessment result of the LDCA in Case 2

(Scan the QR code on the front of the book for color picture)

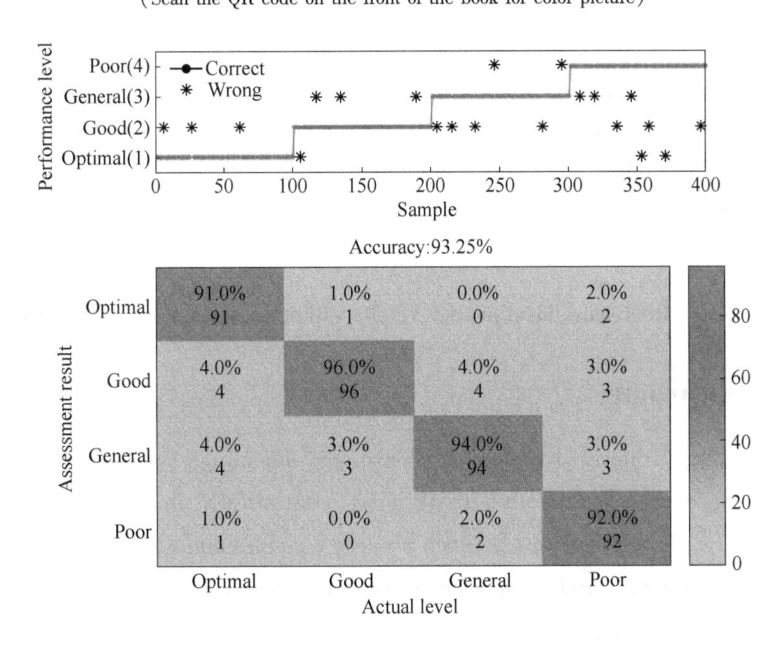

Fig. 16.8 Assessment result of the proposed method in Case 2

(Scan the QR code on the front of the book for color picture)

then the non-optimal variables are determined by the difference index. The identification results are shown in **Fig. 16.9**. Variable 4 in midstream, Variables 6, and Variable 13 in downstream have relatively large difference index, so they are regarded as the main factors leading to non-optimal condition. Considering the process knowledge of the HSMP, it can be seen that the actual reason for the non-optimal condition is consistent with the identification results in **Fig. 16. 9**. The influence of the downstream stands on the finishing exit thickness is higher than that of the midstream stands, and the rolling force is closely related to the roll gap. The fluctuation of the process operating condition will lead to the change of roll gap. Therefore, the proposed method can effectively trace the non-optimal causes. With the assessment and identification results of the proposed method, process engineers can make maintenance decision in time to ensure process safety and product quality.

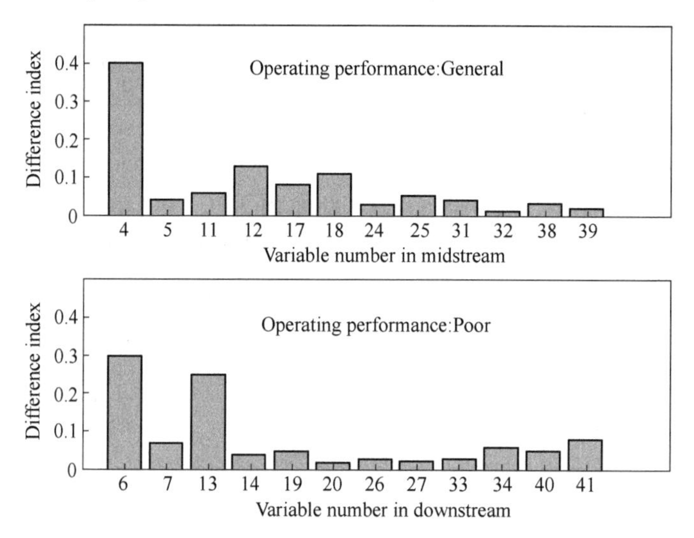

Fig. 16. 9 Root cause identification results within midstream and downstream

16. 4 Conclusions

In this chapter, a novel KPI-related distributed operating performance assessment method has been proposed for the HSMP. First, the mRMR algorithm is improved by MIC to describe the relationships between process variables and KPIs as well as reducing the redundancy among variables in each subsystem. Second, a kernel version of OCTA is developed for analyzing the dynamic and nonlinear characteristics. Then, the process operating performance is assessed accurately by the predefined statistics. Moreover, when the HMSP runs at nonoptimal operating condition, the root cause can be identified by the difference index. Finally, the proposed assessment method is illustrated with real HSMP data, and the simulation results show the superiority of the developed method comparing with the traditional OCTA and the LDCA. Future work will be concentrated on quality anomaly monitoring and performance degradation prediction for plant-wide process industries with spatio-temporal coordination.

References

[1] Qin S J, Chiang L H. Advances and opportunities in machine learning for process data analytics [J]. Computers and Chemical Engineering, 2019, 126: 465-473.

[2] Zou X Y, Wang F L, Chang Y Q. Assessment of operating performance using cross-domain feature transfer learning [J]. Control Engineering Practice, 2019, 89: 143-153.

[3] Zhao C H, Gao F R. Critical-to-fault-degradation variable analysis and direction extraction for online fault prognostic [J]. IEEE Transactions on Control Systems Technology, 2017, 25 (3): 842-854.

[4] Li W Q, Zhao C H, Gao F R. Linearity evaluation and variable subset partition based hierarchical process modeling and monitoring [J]. IEEE Transactions on Industrial Electronics, 2018, 65 (3): 2683-2692.

[5] Liu Y, Wang F L, Chang Y Q. Online fuzzy assessment of operating performance and cause identification of nonoptimal grades for industrial processes [J]. Industrial & Engineering Chemistry Research, 2013, 52 (50): 18022-18030.

[6] Liu Y, Wang F L, Chang Y Q. Operating optimality assessment based on optimality related variations and nonoptimal cause identification for industrial processes [J]. Journal of Process Control, 2016, 39: 11-20.

[7] Liu Y, Chang Y Q, Wang F L. Online process operating performance assessment and nonoptimal cause identification for industrial processes [J]. Journal of Process Control, 2014, 24 (10): 1548-1555.

[8] Liu Y, Wang F L, Chang Y Q, et al. Operating optimality assessment and nonoptimal cause identification for non-Gaussian multimode processes with transitions [J]. Chemical Engineering Science, 2015, 37: 106-118.

[9] Hosack G R, Hayes K R, Dambacher J M. Assessing model structure uncertainty through an analysis of system feedback and Bayesian networks [J]. Ecological Applications, 2008, 18 (4): 1070-1082.

[10] Biglarfadafan M, Danehkar A, Pourebrahim S, et al. Application of strategic fuzzy assessment for environmental planning; case of bird watch zoning in wetlands [J]. Open Journal of Geology, 2016, 6 (11): 1380-1400.

[11] Kusiak A. Rough set theory: A data mining tool for semiconductor manufacturing [J]. IEEE Transactions on Electronics Packaging Manufacturing, 2001, 24 (1): 44-50.

[12] Zou X Y, Wang F L, Chang Y Q, et al. Process operating performance optimality assessment and non-optimal cause identification under uncertainties [J]. Chemical Engineering Research & Design, 2017, 120: 348-359.

[13] Zou X Y, Chang Y Q, Wang F L, et al. Process operating performance optimality assessment with coexistence of quantitative and qualitative information [J]. Canadian Journal of Chemical Engineering, 2018, 96 (1): 179-188.

[14] Ding S X, Yin S, Peng K X, et al. A novel scheme for key performance indicator prediction and diagnosis with application to an industrial hot strip mill [J]. IEEE Transactions on Industrial Informatics, 2013, 9 (4): 2239-2247.

[15] Ge Z Q. Distributed predictive modeling framework for prediction and diagnosis of key performance index in plant-wide processes [J]. Journal of Process Control, 2018, 65: 107-117.

[16] Ma L, Dong J, Peng K X. A novel key performance indicator oriented hierarchical monitoring and propagation path identification framework for complex industrial processes [J]. ISA Transactions, 2020, 96: 1-13.

[17] Peng K X, Zhang K, Dong J, et al. A new data-driven process monitoring scheme for key performance indictors with application to hot strip mill process [J]. Journal of the Franklin Institute, 2014, 351 (9): 2239-2247.

[18] Shardt Y A W, Hao H Y, Ding S X. A new soft-sensor-based process monitoring scheme incorporating infrequent KPI measurements [J]. IEEE Transactions on Industrial Electronics, 2015, 62 (6): 3843-3851.

[19] Peng K X, Zhang K, Dong J, et al. Quality-relevant fault detection and diagnosis for hot strip mill process with multi-specification and multi-batch measurements [J]. Journal of the Franklin Institute, 2015, 352 (3): 987-1006.

[20] Li G, Qin S J, Zhou D H. Geometric properties of partial least squares for process monitoring [J]. Automatica, 2010, 46 (1): 204-210.

[21] Zhu Q Q, Qin S J. Supervised diagnosis of quality and process faults with canonical correlation analysis [J]. Industrial & Engineering Chemistry Research, 2019, 58 (26): 11213-11223.

[22] Jiang Q C, Yan X F, Huang B. Review and perspectives of data-driven distributed monitoring for industrial plant-wide processes [J]. Industrial & Engineering Chemistry Research, 2009, 58 (29): 12899-12912.

[23] Zhong L, Chang Y Q, Wang F L, et al. Distributed operating performance assessment of the plant-wide process based on data-driven hybrid characteristics decomposition [J]. Industrial & Engineering Chemistry Research, 2000, 59 (35): 15682-15696.

[24] Kraskov A, Stogbauer H, Andrzejak R G, et al. Hierarchical clustering using mutual information [J]. Europhysics Letters, 2005, 70 (2): 278-284.

[25] Peng H C, Long F H, Ding C. Feature selection based on mutual information criteria of max-dependency, max-relevance, and min-redundancy [J]. IEEE Transactions on Pattern Analysis and Machine Intelligence, 2005, 27 (8): 1226-1238.

[26] Gonzalo J, Granger C. Estimation of common long-memory components in cointegrated systems [J]. Journal of Business & Economic Statistics, 1995, 13 (1): 27-35.

[27] Wu D H, Zhou D H, Chen M Y, et al. Output-relevant common trend analysis for KPI-related

nonstationary process monitoring with applications to thermal power plants [J]. IEEE Transactions on Industrial Informatics, 2021, 17 (10): 6664-6675.

[28] Wang S N, Shen C G, Liang H. Distributed graph hashing [J]. IEEE Transactions on Cybernetics, 2019, 49 (5): 1896-1908.

[29] Reshef D, Reshef Y, Finucane H, et al. Detecting novel associations in large data sets [J]. Science, 2011, 334: 1518-1524.

[30] Xu C, Zhao S Y, Liu F. Distributed plant-wide process monitoring based on PCA with minimal redundancy maximal relevance [J]. Chemometrics & Intelligent Laboratory Systems, 2021, 169: 53-63.

[31] Ma L, Dong J, Hu C J, et al. A novel decentralized detection framework for quality-related faults in manufacturing industrial processes [J]. Neurocomputing, 2021, 48: 30-41.

[32] Odiowei P E P, Cao Y. Nonlinear dynamic process monitoring using canonical variate analysis and kernel density estimations [J]. IEEE Transactions on Industrial Informatics, 2009, 6 (1): 36-45.

[33] Zhang C F, Peng K X, Dong J. An extensible quality-related fault isolation framework based on dual broad partial least squares with application to the hot rolling process [J]. Expert Systems with Applications, 2021, 167: 114166.

[34] Zou X Y, Zhao C H, Gao F R. Linearity decomposition-based cointegration analysis for nonlinear and nonstationary process performance assessment [J]. Industrial & Engineering Chemistry Research, 2020, 59: 3052-3063.

Chapter 17 Spatiotemporal Synergetic OPA Framework Based on RCM-DISSIM

Process monitoring has been successfully applied in process industries such as metallurgy, petrochemical, and chemical engineering, and achieved certain economic benefits[1-4]. However, traditional process monitoring methods only focus on the occurrence of abnormal conditions. When no significant abnormal conditions occur, disturbances and uncertainties can lead to non-optimal operating conditions frequently[5-8]. Especially in the steel industry, there are frequent changes in raw materials, harsh operating environments, severe fluctuations in operating conditions. Product quality and process parameters cannot be comprehensively monitored in real-time, and the control effect is difficult to meet actual production requirements. In order to ensure high-quality and efficient production, OPA has gradually attracted the attention of researchers[9-11]. OPA can further distinguishes and identifies the operating condition of a process under normal production condition, helping process engineers obtain real-time information of the process[12]. Then, process engineers can use such information to adjust and optimize the process. Therefore, OPA has important theoretical value and practical significance for improving the economic benefits of enterprises.

In recent years, many OPA methods have been widely studied and applied to process industries. Fan et al.[13] proposed a decentralized OPA scheme based on multi-block total projection to latent structures (MB-T-PLS) and Bayesian inference for geological drilling process. Based on improved Hessian locally linear embedding (IHLLE), Zhang et al.[14] developed a new key performance indicator (KPI) assessment method for hot strip mill process. Chu et al.[15] applied the assessment approach based on supervised probabilistic slow feature analysis (SPSFA) to an actual dense medium coal preparation process. In order to deal with nonlinearity and outliers in the plant-wide hot strip mill process (PHSMP), Zhang et al.[16] proposed a lifecycle OPA method based on robust kernel canonical variable analysis (RKCVA). With the continuous development of artificial intelligence, some work attempts to use deep learning methods for OPA. Lu et al.[17] developed an OPA method based on semi-supervised cluster generative adversarial network (SSClusterGAN) for gold flotation process. Liu et al.[18] provided a new layer attention-based stacked performance-relevant denoising autoencoder (LA-SPDAE) to assess the performance of gold cyanide leaching process. Although the above assessment methods have achieved satisfying assessment results in related fields, most of them focus on modeling tasks with aligned data. In other words, the time-delay problem has seldom been considered for OPA.

Time-delay characteristic is a common attribute of most process industries, which is mainly caused by the difference of production equipment in spatial and temporal distribution[19]. Taking PHSMP as an example, seven continuous hot strip mills are distributed in different spatial positions in the finishing mill area. The sampling data of different variables exported from ibaPDA are illustrated in **Fig. 17. 1**. Obviously, there are time-delays between different variables. The existence of time-delay characteristic not only increases the difficulty of OPA modeling, but also increases the difficulty of product quality control[20]. When the change of raw materials or external interference causes the change of stability working point, process engineers generally need to rely on

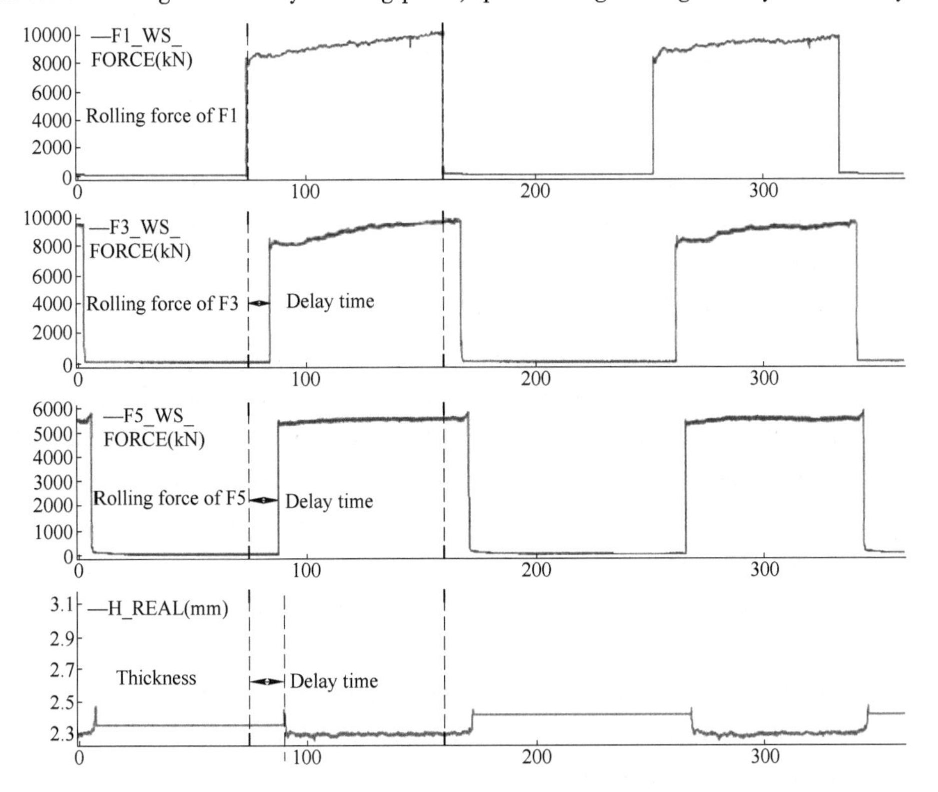

Fig. 17. 1 Real process data in PHSMP

production experience to adjust and gradually transition to the new stability working point[21]. Such dynamic adjustment is extensive and subjective. The low yield of strip products caused by the long-term working condition instability will bring a lot of energy and resource waste to iron and steel enterprises. Therefore, it is of great significance to analyze and accurately estimate the time-delay characteristic and parameters of PHSMP.

In fact, the qualitative evaluation of process condition can also be reduced to some KPIs containing deep-level running information, which is used to represent the important indicators that decision makers are most concerned about in PHSMP, which together with process variables such as roll gap, rolling force and rolling speed constitute multi-rate data. Strip thickness is a typical KPI, so this chapter mainly focuses on estimating time-delay parameters of process variables relative to strip thickness. The time-delay characteristic between process variables and the strip thickness have obvious multiplicity and correlation, which is named multi-correlated spatiotemporal-delay (MCSD) in this chapter. The first delay is caused by the difference in spatial and temporal distribution of the rolling equipment, and the second delay is caused by the uncertain delay in the field manual sampling, cooling retesting process and testing information input and review. In addition, the physical and chemical reactions in the rolling process make the process variables affect each other, which leads to MCSD between process variables and KPI.

The existing time-delay analysis methods are based on expert experience combined with correlation coefficient to obtain the maximum correlation of a certain lag time. The step corresponding to the maximum correlation coefficient of process variables and quality variables under different delay steps is the delay time[22]. Although this method is easy to implement, the influence of other auxiliary process variables on strip thickness and the mutual influence between process variables are ignored during calculation. With the continuous improvement of industrial information, the database has stored a large number of data that can reflect the MCSD characteristics of the process. Therefore, from a data-driven perspective, a spatiotemporal synergetic OPA (SSOPA) method based on reconstructed correlation matrix (RCM) and dissimilarity analytics is proposed in this chapter. And the main contributions are summarized as follows:

(1) The RCM is constructed and a new mixed grey relational coefficient (MGRC) is introduced to gray relation analysis (GRA). Based on them, the MCSD parameters between process variables and the KPI are estimated.

(2) According to the MCSD parameters, the process data are aligned. Then the

distributed dissimilarity analysis (DISSIM) model is developed in each subsystem of PHSMP for assessing the operating condition.

17.1 Review of GRA and DISSIM

17.1.1 Gray relation analysis

Initially proposed by Deng, GRA has been widely used to analyze the dynamic development of a system and examine whether the relationships between various variables are tight[23]. More specifically, the procedures of GRA are summarized as follows[24]:

(1) Step 1: Generate the reference sequence $\boldsymbol{x}_0 = [x_0^1, \cdots, x_0^j, \cdots, x_0^N]^{\mathrm{T}}$, where $j = 1, 2, \cdots, N$ is the sequence number. And $\boldsymbol{x}_i = [x_i^1, \cdots, x_i^j, \cdots, x_i^N]^{\mathrm{T}}$ is the compared sequence, where $i = 1, 2, \cdots, M$. The compared sequences $\boldsymbol{x}_i$ are represented in a matrix:

$$\boldsymbol{X} = \begin{bmatrix} x_1^1 & \cdots & x_i^1 & \cdots & x_M^1 \\ \vdots & \ddots & \vdots & \ddots & \vdots \\ x_1^j & \cdots & x_i^j & \cdots & x_M^j \\ \vdots & \ddots & \vdots & \ddots & \vdots \\ x_1^N & \cdots & x_i^N & \cdots & x_M^N \end{bmatrix}_{N \times M} \tag{17.1}$$

(2) Step 2: Normalize the data set according to one of three types of criteria as follows:

Larger-is-better:

$$\tilde{x}_i^j = \frac{x_i^j - \min\limits_j \{x_i^j\}}{\max\limits_j \{x_i^j\} - \min\limits_j \{x_i^j\}} \tag{17.2}$$

Smaller-is-better:

$$\tilde{x}_i^j = \frac{\max\limits_j \{x_i^j\} - x_i^j}{\max\limits_j \{x_i^j\} - \min\limits_j \{x_i^j\}} \tag{17.3}$$

Nominal-is-best:

$$\tilde{x}_i^j = 1 - \frac{|x_i^j - x_i^{\mathrm{obj}}|}{\max\{\max\limits_j \{x_i^j\} - x_i^{\mathrm{obj}},\ x_i^{\mathrm{obj}} - \min\limits_j \{x_i^j\}\}} \tag{17.4}$$

where, $x_i^{\mathrm{obj}} \in [\min\limits_j \{x_i^j\}, \max\limits_j \{x_i^j\}]$ is the target value of variable $\boldsymbol{x}_i$.

Based on equation (17.2) to equation (17.4), the normalized reference sequence and compared sequences becomes $\tilde{\boldsymbol{x}}_0 = [\tilde{x}_0^1, \cdots, \tilde{x}_0^j, \cdots, \tilde{x}_0^N]$ and $\tilde{\boldsymbol{X}}$.

(3) Step 3: Compute the difference between $\tilde{x}_0^j$ and $\tilde{x}_i^j$, and construct the difference matrix as follows:

$$\Delta_{0i}^j = |\tilde{x}_0^j - \tilde{x}_i^j| \tag{17.5}$$

$$\boldsymbol{\Delta} = \begin{bmatrix} \Delta_{01}^1 & \cdots & \Delta_{0i}^1 & \cdots & \Delta_{0M}^1 \\ \vdots & \ddots & \vdots & \ddots & \vdots \\ \Delta_{01}^j & \cdots & \Delta_{0i}^j & \cdots & \Delta_{0M}^j \\ \vdots & \ddots & \vdots & \ddots & \vdots \\ \Delta_{01}^N & \cdots & \Delta_{0i}^N & \cdots & \Delta_{0M}^N \end{bmatrix}_{N\times M} \tag{17.6}$$

(4) Step 4: Calculate the grey relational coefficient (GRC) γ_{0i}^j with the following equation:

$$\gamma_{0i}^j = \frac{\min_i \min_j \Delta + \zeta \max_i \max_j \Delta}{\Delta_{0i}^j + \zeta \max_i \max_j \Delta} \tag{17.7}$$

where, $\zeta (0 \leqslant \zeta \leqslant 1)$ is the distinguishing coefficient.

The larger ζ is, the lower is its distinguishability. According to existing references, when ζ is equal to 0.5, the grey relational coefficient matrix can better reflect the correlation relationship between different sequences[26].

(5) Step 5: Compute the degree of grey coefficient γ_{0i}, which is a weighted sum of the grey relational coefficients:

$$\gamma_{0i} = \sum_{j=1}^{N} [w_i^j \gamma_{0i}^j], \quad \sum_{j=1}^{N} w_i^j = 1 \tag{17.8}$$

where, w_i^j is the weight of γ_{0i}^t, which is commonly assigned by the user based on the context of specific applications.

17.1.2 Dissimilarity analytics

DISSIM was a statistical process monitoring proposed by method Kano et al[25]. It utilizes the dissimilarity between the distribution of different data sets to detect changes in operating conditions. Given two data sets $\boldsymbol{X}_1 \in R^{N_1\times M}$ and $\boldsymbol{X}_2 \in R^{N_2\times M}$, where $N_i (i=1, 2)$ is the number of samples, M is the number of variables. Normalize each variable in $\boldsymbol{X}_1$ and $\boldsymbol{X}_2$ to zero mean and unit variance, and then calculate the corresponding covariance matrices as:

$$\boldsymbol{R}_i = \frac{\boldsymbol{X}_i^{\mathrm{T}} \boldsymbol{X}_i}{N_i - 1}, \quad i = 1, 2 \tag{17.9}$$

Then, the covariance matrix of the mixture data is given by R :

$$\boldsymbol{R} = \frac{1}{N-1}\begin{bmatrix}\boldsymbol{X}_1\\\boldsymbol{X}_2\end{bmatrix}^{\mathrm{T}}\begin{bmatrix}\boldsymbol{X}_1\\\boldsymbol{X}_2\end{bmatrix} = \frac{N_1-1}{N-1}\boldsymbol{R}_1 + \frac{N_2-1}{N-1}\boldsymbol{R}_2 \tag{17.10}$$

where, $N = N_1 + N_2$ is the sum of samples in $\boldsymbol{X}_1$ and $\boldsymbol{X}_2$. Performing eigenvalue decomposition, the orthogonal matrix $\boldsymbol{P}$ can diagonalize $\boldsymbol{R}$ as follows:

$$\boldsymbol{P}^{\mathrm{T}}\boldsymbol{R}\boldsymbol{P} = \boldsymbol{\Lambda} \tag{17.11}$$

The original data set $\boldsymbol{X}_i$ can be transformed to be into $\boldsymbol{T}_i$

$$\boldsymbol{T}_i = \sqrt{\frac{N_i-1}{N-1}}\boldsymbol{X}_i\boldsymbol{P}\boldsymbol{\Lambda}^{-1/2} = \sqrt{\frac{N_i-1}{N-1}}\boldsymbol{X}_i\boldsymbol{M},\ i = 1,\ 2 \tag{17.12}$$

where, $\boldsymbol{M} = \boldsymbol{P}\boldsymbol{\Lambda}^{-1/2}$ is a transform matrix.

Thus, the covariance matrix of $\boldsymbol{T}_i$ can be obtained:

$$\boldsymbol{S}_i = \frac{1}{N_i-1}\boldsymbol{T}_i^{\mathrm{T}}\boldsymbol{T}_i = \frac{N_i-1}{N-1}\boldsymbol{M}^{\mathrm{T}}\boldsymbol{R}_i\boldsymbol{M},\ i = 1,\ 2 \tag{17.13}$$

Based on the equation (17.10) and equation (17.13), we have $\boldsymbol{S}_1 + \boldsymbol{S}_2 = \boldsymbol{I}$. It is easily verified that the eigenvalues of $\boldsymbol{S}_1$ and $\boldsymbol{S}_2$ satisfy $\lambda_m^1 + \lambda_m^2 = 1$, where λ_m^i is the m_{th} eigenvalue of $\boldsymbol{S}_i$.

And accordingly, the dissimilarity index D is designed based on the eigenvalues to evaluate the dissimilarity of $\boldsymbol{X}_1$ and $\boldsymbol{X}_2$:

$$D = \mathrm{DISSIM}(\boldsymbol{X}_1,\ \boldsymbol{X}_2) = \frac{4}{M}\sum_{m=1}^{M}(\lambda_m^i - 0.5)^2 \tag{17.14}$$

The dissimilarity index D varies from 0 to 1. When $\boldsymbol{X}_1$ and $\boldsymbol{X}_2$ are quite similar, eigenvalues of $\boldsymbol{S}_1$ and $\boldsymbol{S}_2$ are close to 0.5, D tends to be near 0. Otherwise, when $\boldsymbol{X}_1$ and $\boldsymbol{X}_2$ are different from each other, D should be close to 1.

17.2 Spatiotemporal synergetic operating performance assessment model

The accuracy and validity of measurement data in the process industries are the cornerstone of process monitoring and operating performance assessment. PHSMP is a typical production process with multi-correlated spatiotemporal-delay (MCSD) characteristic, which reflects the dynamic causality of the process. Therefore, if the MCSD issue in the PHSMP is not considered, the actual causality between process variables and key performance indicators (KPIs) will be disrupted. The primary work of this chapter is to fully mine the distribution law of the massive historical data, from which to estimate the MCSD parameters of process variables with respect to KPIs. In order to monitor the operating condition of the process, it is assumed that numerous sensors are used to collect real-time data, where y represents the KPI, x_i represents the i_{th} process variable, as shown in **Fig. 17. 2.** Assume that the samples are measured uniformly on the time axis, without considering the MCSD issue, the process variables and KPIs at time t are shown in **Fig. 17. 3** through the purple line. However, due to the differences in the spatial and temporal distribution of each rolling equipment, there are different delays between process variables and KPIs, namely, $\boldsymbol{\tau} = [\tau_1, \cdots, \tau_i, \cdots, \tau_M]$. By estimating the MCSD parameters based on MGRA, the corresponding relationship between process variables and KPIs is reconstructed, which is shown in **Fig. 17. 3**. And then, the reconstructed process data are utilized to train the DISSIM model, so as to realize the effective operating performance assessment for PHSMP.

17. 2. 1 MCSD parameter estimation based on RCM and mixed GRA

In order to analyze the MCSD characteristic between process variables and KPIs, it is necessary to summarize all process variables to be estimated for comprehensive consideration. In this chapter, the thickness of finishing mill is used as the zero-delay KPI, and process variables are reconstructed based on delay parameters to obtain a reconstructed correlation matrix.

Assume that the number of samples is N and the number of process variables is M, the data after standardized processing are as follows:

$$
\begin{aligned}
[\boldsymbol{X}, \boldsymbol{Y}] &= [\boldsymbol{x}_1, \cdots, \boldsymbol{x}_i, \cdots, \boldsymbol{x}_M, \boldsymbol{y}] \\
&= \begin{bmatrix}
x_1^1 & \cdots & x_i^1 & \cdots & x_M^1 & y^1 \\
\vdots & \ddots & \vdots & \ddots & \vdots & \vdots \\
x_1^j & \cdots & x_i^j & \cdots & x_M^j & y^j \\
\vdots & \ddots & \vdots & \ddots & \vdots & \vdots \\
x_1^N & \cdots & x_i^N & \cdots & x_M^N & y^N
\end{bmatrix}_{N\times(M+1)}
\end{aligned}
\tag{17.15}
$$

where the subscript of the entries in the matrix represents the process variable, the superscript represents the sampling time, and $\boldsymbol{y}$ is the time series data of KPI. The MCSD parameters between KPI and process variables can be expressed by delay sequence $\boldsymbol{\tau}$:

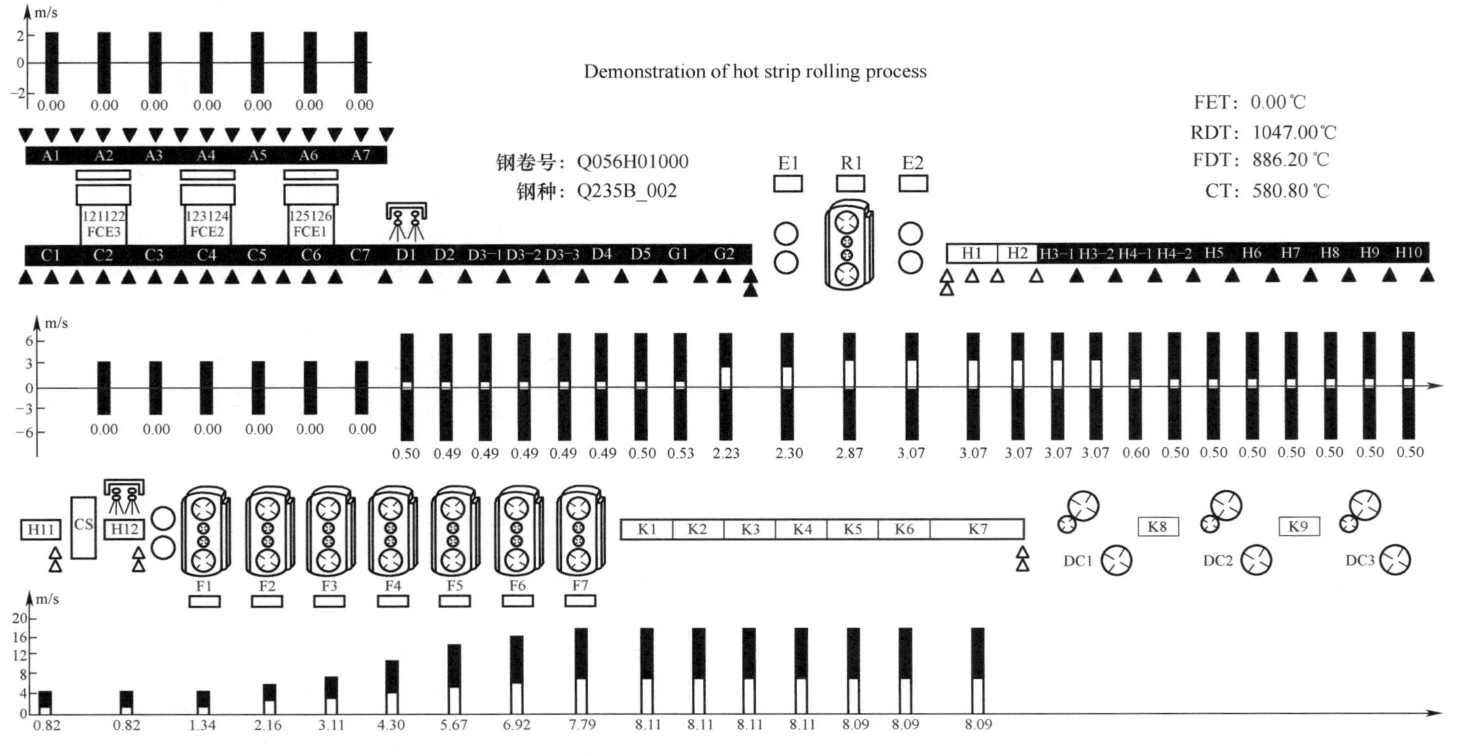

Fig.17.2 Schematic diagram of process variables and KPIs

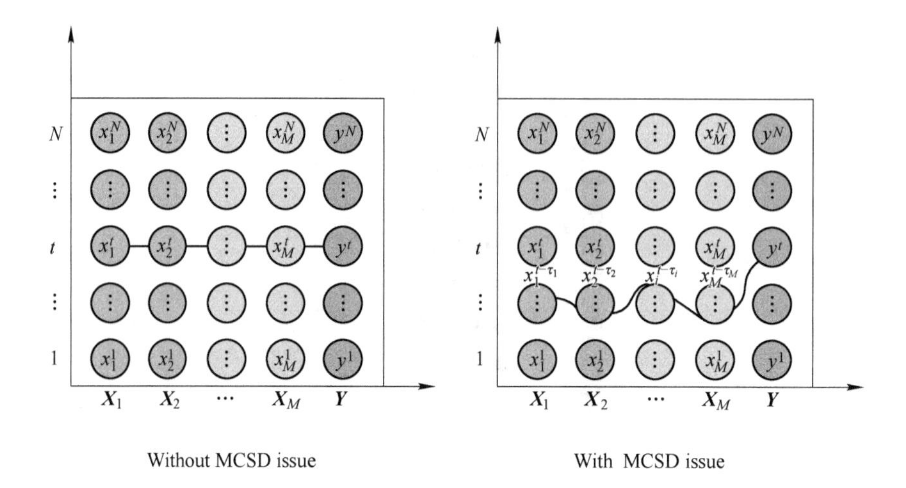

Fig. 17. 3 Illustration of MCSD characteristic in the PHSMP

$$\boldsymbol{\tau} = [\tau_1, \cdots, \tau_i, \cdots, \tau_M]$$
$$\tau_i = a_i T, \ i = 1, \cdots, M \tag{17.16}$$

where, a_i represents the time base of the i_{th} process variable, which is a dimensionless integer; T is the sampling period.

The MCSD parameter of the process variable is an integer multiple of the sampling period, which ensures the feasibility of reconstructing the process variable in time series. From time t, K samples of KPI are selected to form into a time series:

$$\boldsymbol{Y}(t) = [y^t, y^{t+T}, y^{t+2T}, \cdots, y^{t+(K-1)T}] \tag{17.17}$$

where, $KT \geqslant \max[\tau_1, \cdots, \tau_i, \cdots, \tau_M]$ makes the time series include at least one complete rolling cycle from the slab entering the 1_{st} stand to the exit of the 7_{th} stand in the finishing mill area (FMA).

Considering that the FMA is a tandem process consisting of seven stands, the measuring point of exit thickness is equivalent to the end of FMA in horizontal direction, and the measuring point of process variable $\boldsymbol{x}_i$ is located in front of the end of the FMA. According to the delay sequence defined above, the MCSD parameter of process variable $\boldsymbol{x}_i$ relative to thickness $\boldsymbol{y}$ is τ_i. Therefore, K samples of $\boldsymbol{x}_i$ are selected from time $t - \tau_i$ to generate the reconstructed correlation matrix as follows:

$$
\begin{aligned}
&[\boldsymbol{X}(t),\ \boldsymbol{Y}(t)] = [\boldsymbol{x}_1,\ \cdots,\ \boldsymbol{x}_i,\ \cdots,\ \boldsymbol{x}_M,\ \boldsymbol{y}] \\
&= \begin{bmatrix}
x_1^{t-d_1T} & \cdots & x_i^{t-a_iT} & \cdots & x_M^{t-a_MT} & y^t \\
x_1^{t-(a_1-1)T} & \ddots & x_i^{t-(a_i-1)T} & \ddots & x_M^{t-(a_M-1)T} & y^{t+T} \\
x_1^{t-(a_1-2)T} & \cdots & x_i^{t-(a_i-2)T} & \cdots & x_M^{t-(a_M-2)T} & y^{t+2T} \\
\vdots & \ddots & \vdots & \ddots & \vdots & \vdots \\
x_1^{t-(a_1-(K-1))T} & \cdots & x_i^{t-(a_i-(K-1))T} & \cdots & x_M^{t-(a_M-(K-1))T} & y^{t+(K-1)T}
\end{bmatrix}_{K\times(M+1)}
\end{aligned}
\tag{17.18}
$$

In order to effectively describe the correlations between process variables and KPI, the correlations between $\boldsymbol{x}_i$ and $\boldsymbol{y}$ must be quantified. GRA is a commonly used method in multivariate time series correlation analysis, which quantitatively describes the changing trend of multiple groups of variables through grey correlation degree. Aiming to improving similarity and proximity of GRA, the mixed GRA (MGRA) is proposed in this chapter. Let $\boldsymbol{y}$ be the reference sequence, and $\boldsymbol{x}_i$ be the compared sequence. Then, calculate the difference between the corresponding elements of the reference sequence $\boldsymbol{y}$ and the compared sequence $\boldsymbol{x}_i$, which can be represented by the deviation matrix $\boldsymbol{D}$:

$$
\boldsymbol{D} = \begin{bmatrix}
d_1^0 & \cdots & d_i^0 & \cdots & d_M^0 \\
\vdots & \ddots & \vdots & \ddots & \vdots \\
d_1^j & \cdots & d_i^j & \cdots & d_M^j \\
\vdots & \ddots & \vdots & \ddots & \vdots \\
d_1^{K-1} & \cdots & d_i^{K-1} & \cdots & d_M^{K-1}
\end{bmatrix}
\tag{17.19}
$$

where, $d_i^j = |y^{t+jT} - x_i^{t-(a_i-j)T}|$, $i = 1, 2, \cdots, M$, $j = 0, 1, \cdots, K-1$.

Considering the similarity and proximity of GRA, a new mixed grey relational coefficient (MGRC) is defined as follows:

$$
\gamma_i = \frac{\alpha}{K}\sum_{j=0}^{K-1}\frac{\min\{d_i^j\} + \zeta\max\{d_i^j\}}{d_i^j + \zeta\max\{d_i^j\}} + \frac{1-\alpha}{1 + \dfrac{1}{K}\sum\limits_{j=0}^{K-1} d_i^j + \dfrac{1}{K-1}\sum\limits_{j=0}^{K-2}|d_i^{j+1} - d_i^j| + \dfrac{1}{K-2}\sum\limits_{j=1}^{K-2}|d_i^{j+1} - 2d_i^j + d_i^{j-1}|}
\tag{17.20}
$$

where, ζ is the distinguishing coefficient. According to the previous works, ζ usually takes the value of 0.5, which can better reflect the actual correlation between multivariate variables[27]; $\alpha \in [0, 1]$ is the scalar factor, it is obvious that when $\alpha = 1$, MGRC degenerates into original GRC, while when $\alpha = 0$, MGRC degenerates into B-Mode GRC[28].

In order to quantitatively describe the multiple correlation degree between process variables and thickness in the reconstructed correlation matrix, the sum of GRC γ is calculated as follows:

$$\gamma = \sum_{i=1}^{M} \rho_i \gamma_i \tag{17.21}$$

where, $\rho_i (\sum_{i=1}^{M} \rho_i = 1)$ is the weight of process variable $\boldsymbol{x}_i$.

The larger the value of γ, the stronger the correlation between process variables and thickness. In this chapter, the predatory search-based genetic algorithm (PSGA) is used to optimize the MCSD parameters $[a_1, \cdots, a_i, \cdots, a_M]$, the scalar factor α, and the weights $[\rho_1, \cdots, \rho_i, \cdots, \rho_M]$ so that γ can be maximized[29-30]. The fitness function can be established based on equation (17.19) to equation (17.21). After that, real-valued coding is adopted to design the chromosome, which not only avoids the repetitive encoding and decoding operation of binary encoding scheme, but also avoids the impact of the limited length of binary string on the performance and solution accuracy of the algorithm. Each gene of the chromosome is designed to encode a parameter, which is illustrated in **Fig. 17.4**. The parameter optimization based on PSGA continues until a convergence criterion is achieved or a predetermined number of generations is reached. Then, according to the obtained MCSD parameters, the optimal reconstructed correlation matrix $[\boldsymbol{X}(t), \boldsymbol{Y}(t)]_{\mathrm{opt}}$ can be calculated to achieve data alignment between process variables and KPI.

Parameter	a_1	…	a_M	α	ρ_1	…	ρ_M
Chromosome	Value_1	…	Value_M	Value_{M+1}	Value_{M+2}	…	Value_{2M+1}

Fig. 17.4 Chromosome design in PSGA

17.2.2 SSOPA based on distributed DISSIM

From raw materials to final products, the PHSMP consists of many production units. Taking the FMA as an example, it consists of seven tandem finishing mill stands. The material and energy flows make each stand interconnected. In order to simplify the scale of operating performance assessment and make its model have a more reasonable strong physical significance, the FMA can be divided into upstream (F1-F2), midstream (F3-F4) and downstream subsystems (F5-F7), as shown in **Fig. 17.5**.

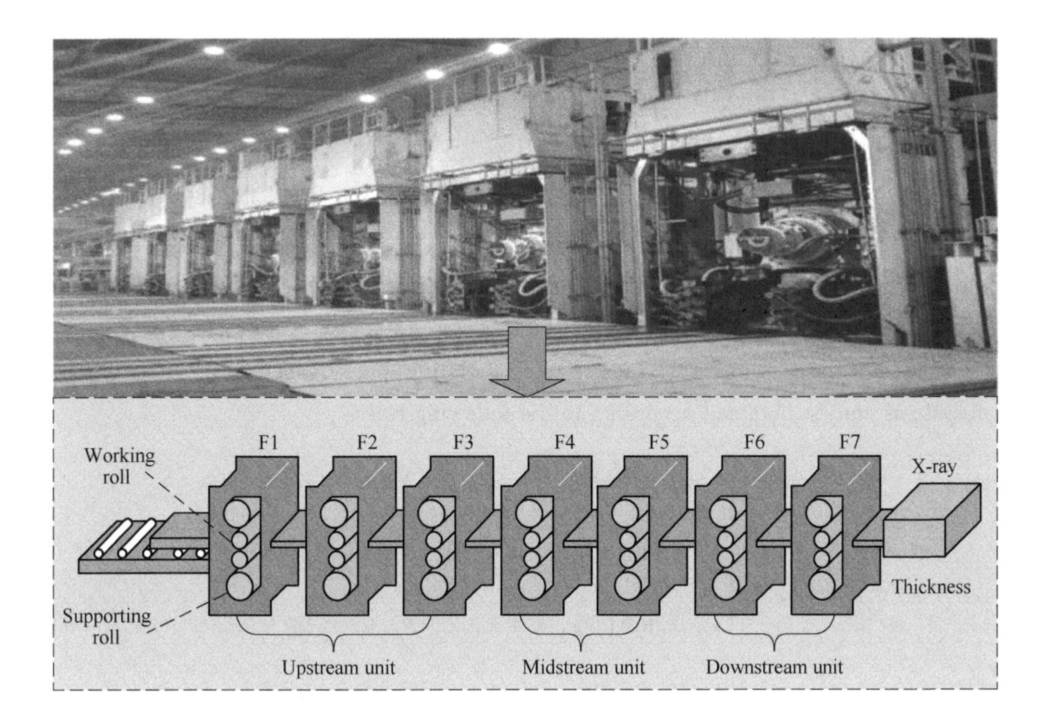

Fig. 17.5 The illustration of three-stream subsystem for FMA

In this chapter, operating performance levels for different products are divided according to the experience of field engineers and the Chinese National Standard GB/T 709—2006. For simplicity, the four levels (optimal, good, general and poor) are numbered as 1, 2, 3, and 4. Based on the estimated MCSD parameters, the training data of upstream, midstream, and downstream can be aligned as $\boldsymbol{Z}_{\mathrm{up}}^{c} = [\boldsymbol{X}_{\mathrm{up}}^{c}, \boldsymbol{Y}^{c}]$, $\boldsymbol{Z}_{\mathrm{mid}}^{c} = [\boldsymbol{X}_{\mathrm{mid}}^{c}, \boldsymbol{Y}^{c}]$ and $\boldsymbol{Z}_{\mathrm{down}}^{c} = [\boldsymbol{X}_{\mathrm{down}}^{c}, \boldsymbol{Y}^{c}]$, $c = 1, 2, 3, 4$.

For online assessment, the new testing data $\boldsymbol{Z}^{\mathrm{test}} = [\boldsymbol{Z}_{\mathrm{up}}^{\mathrm{test}}, \boldsymbol{Z}_{\mathrm{mid}}^{\mathrm{test}}, \boldsymbol{Z}_{\mathrm{down}}^{\mathrm{test}}]$ is updated along spatiotemporal direction, which represents the current multiple correlations of the process. Then it is normalized and aligned in the same way as the training data. The dissimilarity index of upstream, midstream, and downstream can be calculated as:

$$\begin{cases} D_{\text{up}}^c = \dfrac{4}{M_{\text{up}}} \sum_{m=1}^{M_{\text{up}}} (\lambda_{\text{up},m}^c - 0.5)^2 \\ D_{\text{mid}}^c = \dfrac{4}{M_{\text{mid}}} \sum_{m=1}^{M_{\text{mid}}} (\lambda_{\text{mid},m}^c - 0.5)^2 \\ D_{\text{down}}^c = \dfrac{4}{M_{\text{down}}} \sum_{m=1}^{M_{\text{down}}} (\lambda_{\text{down},m}^c - 0.5)^2 \end{cases}, \quad M_{\text{up}} + M_{\text{mid}} + M_{\text{down}} = M, \; c = 1, 2, 3, 4 \tag{17.22}$$

where, M_{up}, M_{mid}, and M_{down} are the number of process variables in upstream, midstream, and downstream, respectively; $\lambda_{\text{up},m}^c$, $\lambda_{\text{mid},m}^c$, $\lambda_{\text{down},m}^c$ are the extracted eigenvalues of corresponding covariance matrices (calculated by equation (17.13)).

Based on the dissimilarity index, the operating performance level of different subsystems can be obtained according to the following rules:

$$L_B = \begin{cases} 1, & \text{if } \min\{D_B^c,\ c = 1, 2, 3, 4\} = D_B^1 \\ 2, & \text{if } \min\{D_B^c,\ c = 1, 2, 3, 4\} = D_B^2 \\ 3, & \text{if } \min\{D_B^c,\ c = 1, 2, 3, 4\} = D_B^3 \\ 4, & \text{if } \min\{D_B^c,\ c = 1, 2, 3, 4\} = D_B^4 \end{cases}, \tag{17.23}$$

$$B = \{\text{up}, \text{mid}, \text{down}\}$$

On the basis of subsystem-level operating performance assessment, the global operating performance level can be evaluated as follows:

$$L_G = \text{Round}\Big(\sum_B \eta_B L_B\Big), \quad \sum_B \eta_B = 1 \tag{17.24}$$

where, η_B is the weight coefficient; Round(·) represents a rounding function.

Compared with upstream subsystem, midstream and downstream subsystems have greater influence on the thickness. Thus, η_{up}, η_{mid}, and η_{down} are set as 0.2, 0.3, and 0.5 in this chapter. In summary, the proposed spatiotemporal synergetic operating performance assessment framework are shown in **Fig. 17.6**.

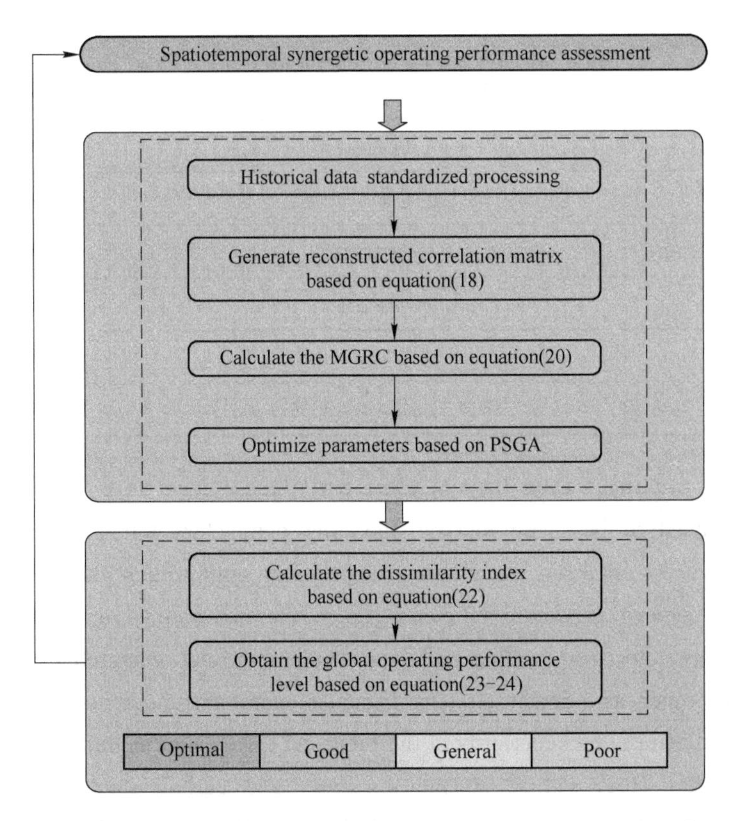

Fig. 17.6 Flowchart of the proposed operating performance assessment framework

17.3 Case study

17.3.1 Experimental setup

Modern PHSMP is a complex industrial process consisting of furnaces, transfer table, roughing mill area, finishing mill area (FMA), laminar cooling, and coiler in series. Among these production units, the FMA is the core unit. There are two supporting rolls and two working rolls in each finishing mill stand. Owing to the high temperature and speed, the gap of each stand is difficult to measure directly. Usually, the gap of each stand is estimated by the altitude difference between two supporting rolls. Any abnormal working condition of the current stand will affect operations of the subsequent stands[31]. Therefore, the operating performance assessment of FMA is very important. In this chapter, 27 process variables and the KPI are selected for experiments, as shown in **Table 17.1**.

In addition, readers can get more information about PHSMP in our previous work[32].

Table 17.1 Process variables and KPI in the FMA

Variable	Description	Unit
1-7	Average roll gap of the i_{th} stand, $i=1, \cdots, 7$	mm
8-14	Rolling force of the i_{th} stand, $i=1, \cdots, 7$	MN
15-21	Rolling speed of the i_{th} stand, $i=1, \cdots, 7$	mm/s
22-27	Bending force of the i_{th}, stand, $i=2, \cdots, 7$	MN
KPI	Finishing mill exit strip thickness	mm

Generally speaking, when the operating performance level is poor, corresponding adjustments need to be made for the operating parameters. When rolling in normal condition, process engineers should adopt appropriate optimization and decision-making strategies to achieve maximum efficiency. Since the non-optimal operating conditions could seriously affect the rolling efficiency and the final comprehensive economic benefits, a proper assessment strategy becomes more and more crucial in the real PHSMP. To illustrate the superiority of the proposed assessment method, two production scenarios (alloy steel with 2.70 mm-thickness and carbon steel with 3.95 mm-thickness) are also assessed by the MB-T-PLS[13], the IHLLE[14], and the SPSFA[15] for comparison. For each operating performance level, 2000 raw samples with MCSD characteristic are used for offline modeling. As the testing set, 800 aligned samples are generated from the raw process data, namely 200 poor samples, 200 general samples, 200 good samples, and 200 optimal samples. The sampling interval is 0.01 s. PSGA is used to tune the MCSD parameters, the scalar factor, and the weights to make γ achieve the maximum value. In this chapter, precision, recall, and F1-score are used to illustrate the performances of different assessment methods[17], which can be defined as:

$$\mathrm{Pre}_i = \frac{\mathrm{TP}_i}{\mathrm{TP}_i + \mathrm{FP}_i} \tag{17.25}$$

$$\mathrm{Rec}_i = \frac{\mathrm{TP}_i}{\mathrm{TP}_i + \mathrm{FN}_i} \tag{17.26}$$

$$\mathrm{F1\text{-}score}_i = \frac{2 \times \mathrm{Pre}_i \times \mathrm{Rec}_i}{\mathrm{Pre}_i + \mathrm{Rec}_i} \tag{17.27}$$

where, $i = 1, 2, 3, 4$ represents the operating performance level; TP, FP, and FN represent true positive, false positive, true negative, and false negative, respectively.

17.3.2 Results and discussion

The detailed analysis of each production scenario is as follows:

(1) **Scenario 1**: This case verifies the validity of the above assessment methods for the 2.70 mm-thickness alloy steel. For the first 200 samples, the process runs at optimal condition (Level 1). After that, the process degenerates to good condition (Level 2) starting with the 201_{st} sample. During the last 400 samples, the process runs at general condition (Level 3) and poor condition (Level 4), respectively. **Table 17.2** shows that the precision, recall, and F1-score of the aforementioned assessment methods. The highest value of each index is shown in bold type. Additionally, the assessment results and confusion matrices of above four assessment methods are shown in **Fig. 17.7**, respectively. Although IHLLE has better assessment results than MB-T-PLS and SPSFA, it is still inferior to the proposed method. Therefore, the proposed assessment method can effectively deal with the raw process data with MCSD characteristic in the modeling stage and accurately assess the operating performance of the PHSMP.

Table 17.2 Assessment indices of different methods in Scenario 1

Index	Methods	Optimal	Good	General	Poor	Average
Pre	MB-T-PLS	87.5%	86.5%	89.5%	85.5%	87.3%
	IHLLE	89.0%	87.5%	88.0%	87.0%	87.9%
	SPSFA	86.0%	83.0%	85.0%	84.5%	84.6%
	Proposed	**97.0%**	**97.0%**	**97.5%**	**96.0%**	**96.9%**
Rec	MB-T-PLS	90.2%	87.8%	85.2%	85.9%	87.3%
	IHLLE	87.3%	88.8%	87.1%	88.3%	87.9%
	SPSFA	86.9%	86.5%	82.1%	83.3%	84.7%
	Proposed	**97.0%**	**97.0%**	**97.0%**	**96.5%**	**96.9%**
F1 score	MB-T-PLS	88.8%	87.2%	87.3%	85.7%	87.3%
	IHLLE	88.1%	88.2%	87.6%	87.7%	87.9%
	SPSFA	86.4%	84.7%	83.5%	83.9%	84.7%
	Proposed	**97.0%**	**97.0%**	**97.3%**	**96.2%**	**96.9%**

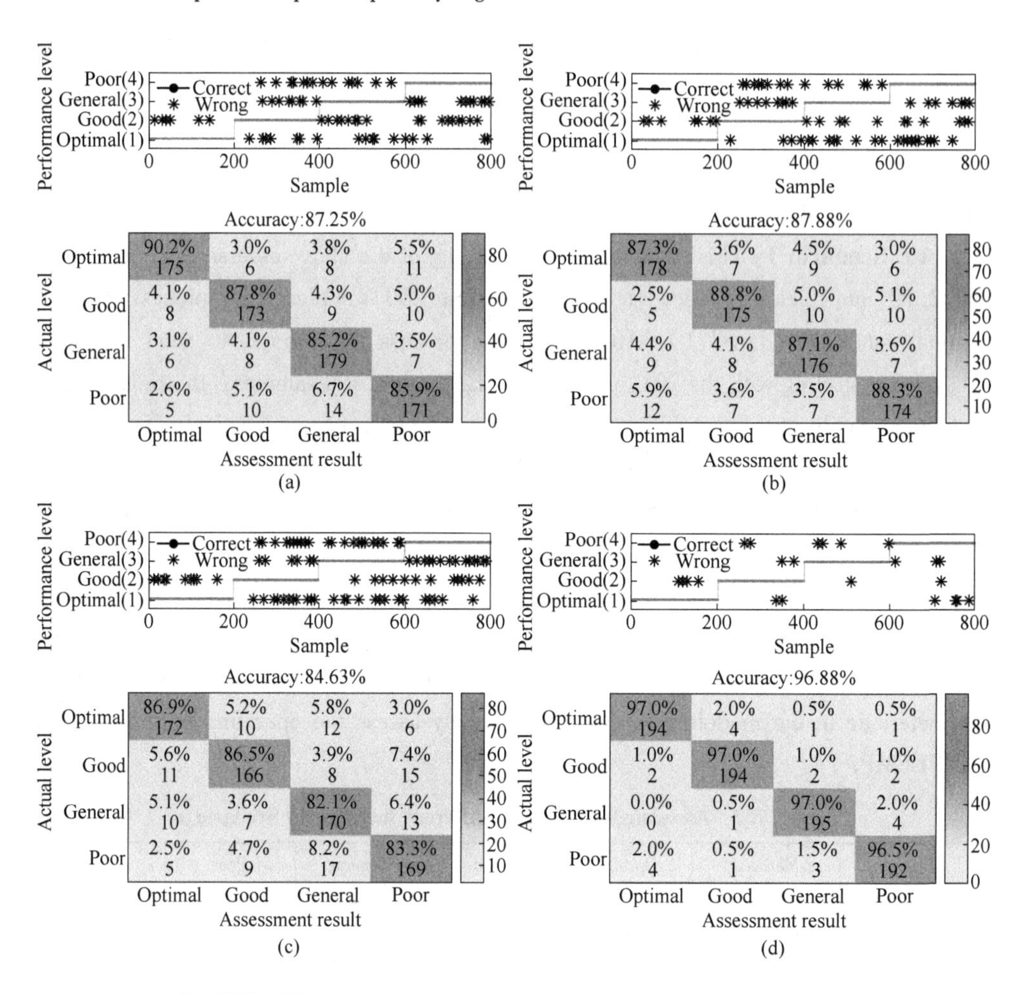

Fig. 17.7 The assessment results of different methods in Scenario 2

(a) MB-T-PLS; (b) IHLLE; (c) SPSFA; (d) Proposed

(Scan the QR code on the front of the book for color picture)

(2) **Scenario 2**: The PHSMP works under the conditions of producing 3.95 mm-thickness carbon steel. Compared to Scenario 1, the assessment indices for Scenario 2 are slightly lower to some extent, which is due to changes in the production process of different products. Detailed assessment indices are provided in **Table 17.3**. Compared with the MB-T-PLS algorithm, the IHLLE algorithm, and the SPSFA algorithm, it can be observed that the proposed method yields more accurate assessment results. The classification results of all models are shown in **Fig. 17.8**. Obviously, the proposed method has the fewest wrong assessed samples, indicating that the performance of the proposed model is the highest, which is consistent with the precision, recall, and F1-

score in **Table 17.3**. Moreover, **Fig. 17.8** gives the confusion matrices of different methods, which also indicates that the proposed method achieves ideal accuracy. Anyhow, the experimental results show that the proposed method is more reliable under the modeling conditions of MCSD characteristic, so it can improve the performance of the assessment stage. And the conclusion drawn from our assessment framework is consistent with the actual situation on-site.

Table 17.3 Assessment indices of different methods in Scenario 2

Index	Methods	Optimal	Good	General	Poor	Average
Pre	MB-T-PLS	85.5%	85.0%	84.0%	87.5%	85.5%
	IHLLE	84.0%	85.5%	85.0%	83.0%	84.4%
	SPSFA	82.5%	80.5%	81.5%	80.0%	81.1%
	Proposed	**95.5%**	**94.0%**	**94.5%**	**93.5%**	**94.4%**
Rec	MB-T-PLS	85.9%	87.6%	80.4%	88.4%	85.6%
	IHLLE	80.8%	87.2%	82.9%	86.9%	84.5%
	SPSFA	80.5%	81.3%	79.1%	83.8%	81.2%
	Proposed	**92.7%**	**94.0%**	**96.4%**	**94.4%**	**94.4%**
F1 score	MB-T-PLS	85.7%	86.3%	82.2%	87.9%	85.5%
	IHLLE	82.4%	86.4%	84.0%	84.9%	84.4%
	SPSFA	81.5%	80.9%	80.3%	81.8%	81.1%
	Proposed	**94.1%**	**94.0%**	**95.5%**	**94.0%**	**94.4%**

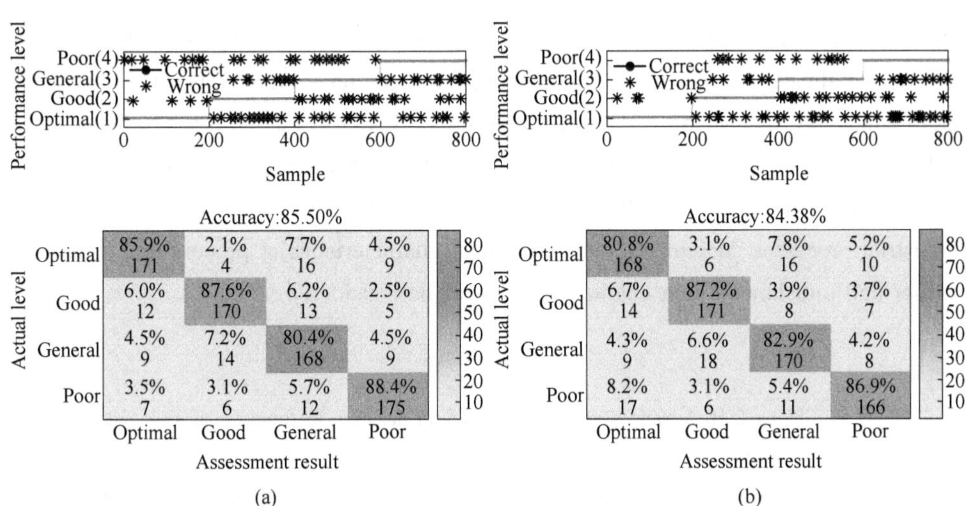

(a) (b)

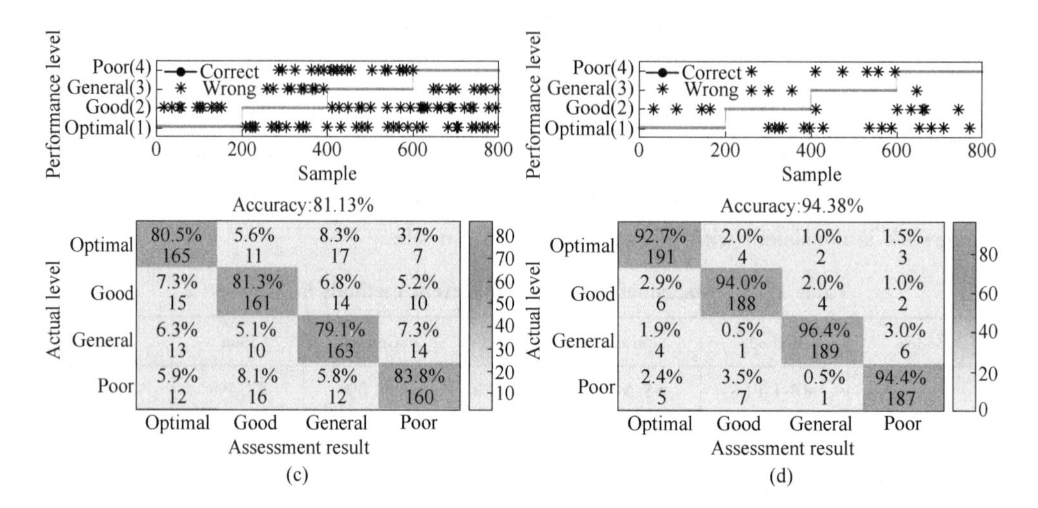

Fig. 17. 8 The assessment results of different methods in Scenario 2

(a) MB-T-PLS; (b) IHLLE; (c) SPSFA; (d) proposed

(Scan the QR code on the front of the book for color picture)

17. 4 Conclusions

In this chapter, a spatiotemporal synergetic operating performance assessment method has been proposed for the plant-wide hot strip mill process. Different from existing assessment methods based on aligned data modeling, the proposed method works under modeling condition of MCSD characteristic. First, the reconstructed correlation matrix of process variables and thickness are generated. Then, MCSD parameters are estimated by MGRA and PSGA. Finally, the aligned process data are used to train the distributed DISSIM model for operating performance assessment. The experiment results of two scenarios indicate that compared to other SOTA algorithms, the proposed method can more effectively assess the operating condition of PHSMP. For outlooks, this chapter only considers the relationship between one KPI (thickness) and process variables, future work will consider multiple KPIs, which will be more consistent with actual industrial processes. Moreover, other non-ideal characteristics of process data, such as outlier and unbalanced data characteristic, will be considered.

References

[1] Qin S J, Chiang L H. Advances and opportunities in machine learning for process data analytics [J]. Computers & Chemical Engineering, 2019, 126: 465-473.

[2] Yin S, Li X W, Gao H J, et al. Data-based techniques focused on modern industry: An

overview [J]. IEEE Transactions on Industrial Electronics, 2015, 62 (1): 657-667.

[3] Huang B, Kadali R. Dynamic Modeling, Predictive control and performance monitoring [J]. Springer-Verlag, London, 2008.

[4] Yu W, Zhao C. Online fault diagnosis for industrial processes with Bayesian network-based probabilistic ensemble learning strategy [J]. IEEE Transactions on Automation Science & Engineering, 2019, 16 (4): 1922-1932.

[5] Yu W, Zhao C, Huang B. MoniNet with concurrent analytics of temporal and spatial information for fault detection in industrial processes [J]. IEEE Transactions on Cybernetics, vol. 52, no. 8, pp. 8340-8351, 2022.

[6] Ge Z Q, Song Z H, Ding S X. Data mining and analytics in the process industry: The role of machine learning [J]. IEEE Access, 2017, 5 (99): 20590-20616.

[7] Ruiz-Cárcel C, Cao Y, Mba D, et al. Samuel. Statistical process monitoring of a multiphase flow facility [J]. Control Engineering Practice, 2015, 42, 74-88.

[8] Ding S X, Yin S, Peng K X, et al. A novel scheme for key performance indicator prediction and diagnosis with application to an industrial hot strip mill [J]. IEEE Transactions on Industrial Informatics, 2013, 9 (4): 2239-2247.

[9] Zou X, Zhao C. Meticulous assessment of operating performance for processes with a hybrid of stationary and nonstationary variables [J]. Industrial & Engineering Chemistry Research, 2019, 58 (3): 1341-1351.

[10] Wang H, Wang X, Wang Z. Performance assessment method of dynamic process based on SFA-GPR [J]. Journal of Process Control, 2022, 111: 27-34.

[11] Zhang H, Tang Z, Xie Y, et al. Siamese time series and difference networks for performance monitoring in the froth flotation process [J]. IEEE Transactions on Industrial Informatics, 2022, 18 (4): 2539-2549.

[12] Chu F, Mo S S, Hao L L, et al. Operating performance assessment of complex nonlinear industrial process based on kernel locally linear embedding PLS [J]. IEEE Transactions on Automation Science & Engineering, 2023: 1-13.

[13] Fan H P, Wu M, Lai X Z, et al. A decentralized operating performance assessment for geological drilling process via multi-block total projection to latent structures and Bayesian inference [J]. Journal of Process Control, 2022, 117: 26-39.

[14] Zhang H J, Zhang C, Dong J, et al. A new key performance indicator oriented industrial process monitoring and operating performance assessment method based on improved Hessian locally linear embedding [J]. International Journal of Systems Science, 2022, 53 (16): 3538-3555.

[15] Chu F, Hao L L, Shang C, et al. Assessment of process operating performance with supervised probabilistic slow feature analysis [J]. Journal of Process Control, 2023, 124: 152-165.

[16] Zhang C, Peng K, Dong J. A lifecycle operating performance assessment framework for hot strip mill process based on robust kernel canonical variable analysis [J]. Control Engineering

Practice, 2021, 107: 104698.

[17] Lu D, Wang F L, Wang S, et al. Operating performance assessment based on semi-supervised cluster generative adversarial networks for gold flotation process [J]. IEEE Transactions on Instrumentation & Measurement, 2022, 72: 1-12.

[18] Liu Y, Ma Z, Wang F L. Deep learning for operating performance assessment of industrial processes with layer attention-based stacked performance relevant denoising auto-encoders [J]. ACS Omega, 2023, 8 (16): 14583-14596.

[19] Jiang Z H, Jiang K, Xie Y F, et al. A cooperative silicon content dynamic prediction method with variable time delay estimation in the blast furnace ironmaking process [J]. IEEE Transactions on Industrial Informatics, 2023: 1-12.

[20] Saxen H, Gao C H, Gao Z W. Data-driven time discrete models for dynamic prediction of the hot metal silicon content in the blast furnace: A review [J]. IEEE Transactions on Industrial Informatics, 2013, 9 (4): 2213-2225.

[21] Ren H, Yang C H, Sun B, et al. Knowledge-data-based synchronization states analysis for process monitoring and its application to hydrometallurgical zinc purification process [J]. IEEE Transactions on Industrial Informatics, 2024, 20 (1): 546-559.

[22] Jiang K, Jiang Z H, Xie Y F, et al. Classification of silicon content variation trend based on fusion of multilevel features in blast furnace ironmaking [J]. Information Sciences, 2020, 521: 32-45.

[23] Deng J L. Introduction to grey system [J]. Journal of Grey System, 1989, 1 (1): 1-24.

[24] Wu H H. A comparative study of using grey relational analysis in multiple attribute decision making problems [J]. Quality Engineering, 2022, 159 (2): 209-217.

[25] Kano M, Hasebe S, Hashimoto I, et al. Statistical process monitoring based on dissimilarity of process data [J]. AIChE Journal, 2002, 48: 1231-1240.

[26] Chang T C, Lin S J. Grey relation analysis of carbon dioxide emissions from industrial production and energy uses in Taiwan [J]. Journal of Environmental Management, 1999, 56 (4): 247-257.

[27] Patel M T. Multi optimization of process parameters by using grey relation analysis: A review [J]. International Journal of Advanced Research in IT and Engineering, 2015, 4 (6): 1-15.

[28] Zhang W Q, Jiang S G, Wang L Y, et al. B-Mode grey relational analysis of surface functional groups change rules in coal spontaneous combustion [J]. Advanced Materials Research, 2011, 236-238: 762-766.

[29] Jiang Q C, Ding S X, Wang Y, et al. Data-driven distributed local fault detection for large-scale processes based on the GA-regularized canonical correlation analysis [J]. IEEE Transactions on Industrial Electronics, 2017, 64 (10): 8148-8157.

[30] Zhang D X, Guan Z H, Liu X Z. Genetic algorithm based on predatory search strategy [J]. Application Research of Computers, 2008, 25 (4): 1002-1007.

[31] Zhang X Y, Ma L, Peng K X, et al. A quality-related distributed fault detection method for

large-scale sequential processes [J]. Control Engineering Practice, 2022, 127: 105308.

[32] Zhang C F, Peng K X, Dong J, et al. KPI-related operating performance assessment based on distributed ImRMR-KOCTA for hot strip mill process [J]. Expert System with Application, 2022, 209: 118273.

Chapter 18 RUL Prediction for A Roller Based on Deep RNN

The roller is one of the most important tools in the HSMP, since its quality and service life directly affects the stability of strip production and can further influence the production efficiency. Once the roller is damaged, it will not only lead to the fracture of the strip but can cause a fire due to the high temperatures of more than 1000 ℃ during the rolling process. The roller is a consumable part, so it should be maintained every day. However, as quality requirements of products continue to increase, higher demands are placed on the use of rollers. As a consequence, maintenance strategy should shift from traditional breakdown maintenance and preventive time-based maintenance to condition-based maintenance (CBM), which is also called predictive maintenance (PdM) or prognostics and health management (PHM)[1].

PHM is an engineering process of failure prevention, and remaining useful life (RUL) prediction[2]. The core process of PHM is to make reliable predictions of the RUL[3-4]. The RUL of a roller refers to the lifetime left on the component from the current time to end-of-life (EoL). Accurate RUL prediction can provide helpful information for the management of rollers. It can not only guarantee the surface quality of the hot rolled products and avoid losses caused by accidental failure of the roller, but it also improves the operating rate of the rolling mill so as to increase production efficiency of the business. Therefore, accurate RUL prediction for rollers is of great significance for the production of hot strips, and special attention needs to be paid to the design of a reliable and accurate RUL prediction approach.

18.1 Review of DNN-based method

Recently, since the multilayer deep network architecture can fully capture the representative features from original measured data, deep learning methods have been widely used in many fields, such as speech recognition, image recognition, and fault diagnosis[5]. It has also been gradually applied to the field of PHM. In reference [6], an enhanced restricted Boltzmann machine was proposed to construct a heath indicator

(HI), and the RUL of the machines were estimated through a similarity-based method. The experimental results demonstrate the advantages of the proposed method. Li et al. proposed a data-driven method based on a deep convolution neural network for RUL prediction, where accurate prediction results can be obtained without prior knowledge and signal processing[7]. Compared to the previously mentioned deep network architecture, the recurrent neural network (RNN) is a more effective approach for processing time-series data since they have an internal state that can represent aging in formation[8]. Heimes et al. presented a RUL prediction model based on the RNN for turbofan engines, which won the 2008 PHM competition[9]. Guo et al. proposed a HI construction method based on RNN for predicting the RUL of bearings, where the experiment showed that the approach has better performance than some other methods[10]. These applications indicate that the RNN-based approach has enormous potential for PHM and RUL prediction.

In the hot strip production process, the roller is in direct contact with the metal to be rolled in order to plastically deform the metal. Due to the large rolling force of the hot strip mill (HSM), the friction between the work roller and the rolled metal is extremely high. Accordingly, wear becomes the major factor affecting the service life of the rollers[11]. With the increase of rolling time, the wear degree gradually increases, which will cause the surface of the roller to become rough, requiring maintenance of the component. The traditional theory of roller RUL prediction is based on the prediction of wear. However, there are still some lacunae which are worth discussing. Firstly, these theories are based on an empirical model and only consider central influencing factors such as rolling pressure, contact arc length, and rolling length, which is a simplification of the real situation. At present, there still does not exist an accurate prediction model based on the wear mechanism. Oike's equation is a widely used empirical model. However, it cannot predict the wear precisely when applied to different industrial sites[12]. In reference [13], a roller force model was integrated with Oike's equation to estimate the wear along the roller barrel, and the accuracy of prediction was improved. Secondly, the traditional model requires some important parameters, such as the coefficient of friction and the uneven friction coefficient of the steel plate, which are unmeasurable quantities in industrial cases. Although the empirical relationships between these coefficients and other measurable parameters, like rolled material, strip temperature, and lubrication, can be derived, these relationships are sometimes contradictory[14]. Thirdly, it is unreasonable that the RUL of the roller be determined only by the degree of wear. Roller deformation, roller fatigue, mechanical factors, and the processes of rolling also affect the service life of the roller. It is difficult to establish a mathematical model to describe the relationship between these factors and the RUL.

In order to overcome these shortcomings in predicting the RUL of rollers, a new method which can extract coarse-grained and fine-grained characteristics to estimate the health state and predict the RUL is proposed in this chapter. To form a feature set that contains various roller degradation characteristics for accurate RUL prediction, the coarse-grained and fine-grained characteristics are extracted from batch data based on the proposed deep RNN. Then, the features are used to construct a HI that is capable of indicating the health state of the roller. Following that, the RUL can be estimated by extrapolating the HI model, which is described by a double exponential function, to a predefined failure threshold (FT). Finally, the effectiveness of the proposed method is verified by the dataset collected from an industrial site. The main contributions and innovations are:

(1) The proposed deep RNN architecture can extract coarse-grained features from monitoring data and fine-grained features from maintenance data to develop a comprehensive HI that can reflect the health state of the roller. Furthermore, the model parameters can be obtained automatically during network training, rather than the model parameters of traditional methods are sometimes unmeasurable.

(2) The RUL of the roller is determined by the developed comprehensive HI rather than relying on just one factor as in the conventional method, which is more reasonable. In addition, the constructed HI has a value equal to one under fault state, so there is no need to artificially specify a FT.

(3) A RUL prediction and health state estimation framework based on deep RNN is proposed for the RUL estimation of the roller of HSMP, where many monitored variables related to the roller wear that are not considered by empirical models are considered in this chapter.

18.2 Deep RNN network architecture

18.2.1 Basic theory of RNN

As a kind of deep learning approach, RNN is well suited for dealing with sequential data. This can be attributed to the special network structure that remembers previous information and applies it to the calculation of the current output. As shown in **Fig. 18.1**, at time t, the input of the hidden layer comes not only from the input layer

x_t, but also from its own output at the previous moment h_{t-1}. As a result, the output y_t is determined by both the present input information and the information at time $t-1$.

The RNN is generally trained using a back-propagation through time (BPTT) algorithm. However, as the time information stored in the RNN increases, the gradients tend to vanish or explode[15]. In order to overcome this problem, a long short-term memory (LSTM) network architecture including memory cells is proposed[16]. The memory cell is introduced to replace the hidden neuron in the hidden layer of a traditional RNN. It records the information for a long time through a forget gate, an input gate, and an output gate. There is a initial cell state in the memory cell, which is assigned with certain unit weight. All components of the LSTM are shown in **Fig. 18.2**.

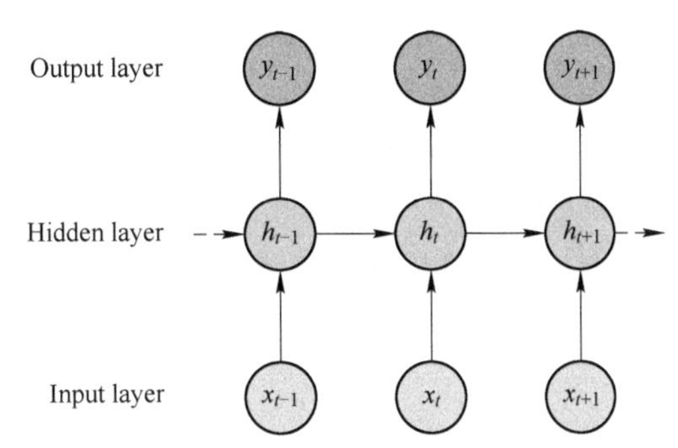

Fig. 18.1 Structural diagram of the RNN

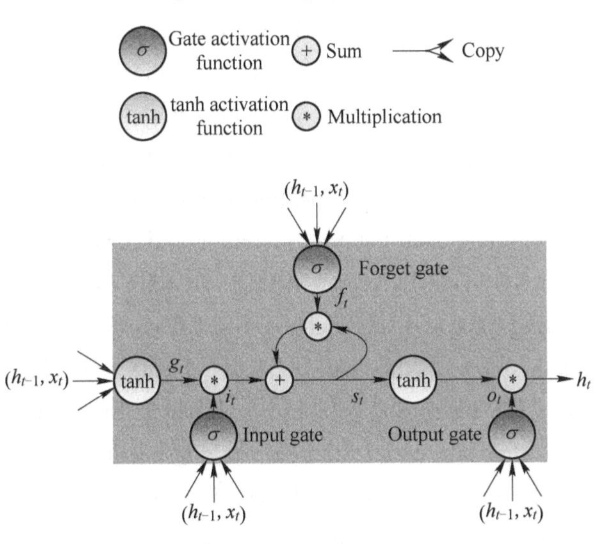

Fig. 18.2 The network architecture of the LSTM memory cell

(1) Forget gate: The forget gate f_t is used to decide how much information will be discarded. With a sigmoid layer, it outputs a number between 0 and 1, where 1 indicates the cell state value is fully reserved and 0 represents the value is completely forgotten. The forget gate can be calculated as:

$$f_t = \sigma(w_{fx}x_t + w_{fh}h_{t-1} + b_f) \tag{18.1}$$

(2) Input gate: The input gate i_t is used to determine what new information is to be stored in the cell state and to update the cell state. This step consists of two parts. First, a sigmoid layer is used to update the input value. After that, a tanh layer g_t is used to create a candidate state, which can be multiplied by the input value to update the cell state. These can be implemented as:

$$i_t = \sigma(w_{ix}x_t + w_{ih}h_{t-1} + b_i) \tag{18.2}$$

$$g_t = \tanh(w_{gx}x_t + w_{gh}h_{t-1} + b_g) \tag{18.3}$$

With equation (18.1)-equation (18.3), the current cell state can be updated as:

$$s_t = f_t^* s_{t-1} + i_t^* g_t \tag{18.4}$$

(3) Output gate: The output gate o_t is introduced to decide which information will be output. First, a sigmoid layer is used to determine which part of the cell state will be output. Then, the remaining state value can be obtained by multiplying the outputs of a tanh layer and the sigmoid layer. This can be calculated as:

$$o_t = \sigma(w_{ox}x_t + w_{oh}h_{t-1} + b_o) \tag{18.5}$$

$$h_t = \tanh(s_t) * o_t \tag{18.6}$$

where, w_{fx}, w_{ix}, w_{gx}, and w_{ox} are the weights of the input layer to the hidden layer; w_{fh}, w_{ih}, w_{gh}, and w_{oh} are hidden layer weights between time t and $t-1$; b_f, b_i, b_g, and b_o are the biases.

18.2.2 Novel deep RNN architecture

The hot rolling process is a typical batch process, which produces strip steel coil-by-coil and each coil represents a batch[17]. Considering that the HSM production process has its own unique characteristics, it is necessary to design a targeted modeling method. A HSM can be used to produce a wide range of products in different materials, shapes, and sizes, and different products inevitably lead to different wear rates. The HSM processes a large number of products by continuously repeating the rolling operation. Thus, a certain periodicity may exist between each batch of the monitored data. In view of the above characteristics, a deep RNN model is proposed that can fully extract the fine-grained features from the monitoring data and the coarse-grained features from the maintenance data. The monitoring data refers to the data collected during the rolling process, which contains approximately 180 measuring points in each batch. Thus, fine-grained features can be used to model wear rate under different working conditions. The maintenance data refers to the data collected after rolling a batch of steel, where each batch of steel contains only one measurement point. Thus, coarse-grained features can be used to model the periodicity of rolling. The network architecture consists of three sub-structures, i. e. a multilayer LSTM for extracting fined-grained features which is represented as LSTM1, a multilayer LSTM for extracting coarse-grained features which is represented as LSTM2, and a fully connected layer for regression. **Fig. 18.3** shows the network architecture of the proposed approach.

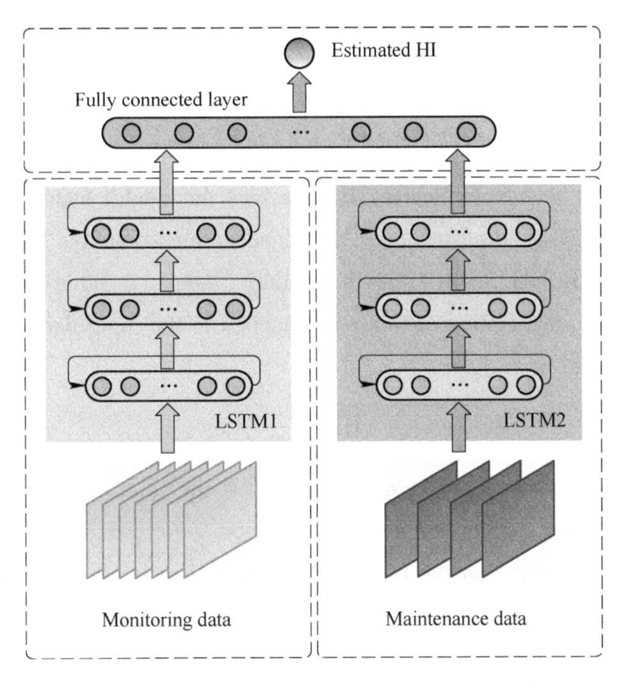

Fig. 18.3 The architecture of deep RNN

First, the collected data is prepared as input data for the LSTM in 3-dimensional (3D) format for the convenience of training. The dimension of the data is $N_s \times N_{tw} \times N_{id}$, where N_s is the number of samples, N_{tw} indicates the time window, and N_{id} denotes the number of selected input dimension.

Afterwards, the multilayer LSTM1 is developed to extract fine-grained features from the monitoring data, and the multilayer LSTM2 is constructed to extract coarse-grained features from the maintenance data, which can both learn the most suitable feature representations from each raw input data. After model convergence, each vector of the outputs in the two multilayer LSTMs naturally represents the fine-grained and the coarse-grained features, respectively.

Finally, all the extracted features are connected to a fully connected layer, and one neuron as a representative of the outputs is added at the top of the network for HI estimation. To further improve the performance of the proposed deep RNN, the Adam algorithm is used to optimize the neural network[18]. In addition, the dropout technique is used to regularize the hidden layers, which is able to alleviate overfitting and improve generalization ability[19]. The parameters of the deep RNN model are optimized to minimize the error between the actual HI and the estimated HI.

18.3 RUL prediction based on deep RNN

18.3.1 Health indicator construction

In order to develop a comprehensive HI to assess the health state of the roller, the proposed deep RNN is used to fuse the fined-grained features and coarse-grained features as the HI. Specifically, in the training step, a label set indicating the percentage of roller degradation at time t is attached to the input data set. Thus, the HI can be expressed as:

$$\mathrm{HI}_t = \frac{t}{T} \tag{18.7}$$

where, HI_t represents the label at moment t; T is the number of steel batches rolled during the last maintenance to the next maintenance.

For example, assuming that the roller maintenance time is after rolling 380 batches of steel, and the current monitoring time point is the 228_{th} batch of steel, then the label HI is 0.6. Accordingly, the deep RNN model is trained to minimize the cost function as follow:

$$J = \frac{1}{2}\sum_{t=0}^{T} \| \mathrm{HI}_t - \hat{\mathrm{HI}}_t \|_2^2 \tag{18.8}$$

where, $\hat{\mathrm{HI}}_t$ is the output of the deep RNN model.

In the testing step, the monitored data are directly input into the trained deep RNN model for HI estimation. As can be derived from equation (18.7) that the FT of the constructed HI is expected to equal to one. Thus, the RUL of roller can be predicted based on the fixed FT value.

18.3.2 Remaining useful life

For the purpose of estimating the RUL, the double exponential model is introduced to describe the HI model, which has been proved to be an effective model for RUL prediction. Thus, the state-space model can be built as follow:

$$\begin{cases} z_{1,k} = z_{1,k-1} + N(0, \sigma_1^2) \\ z_{2,k} = z_{2,k-1} + N(0, \sigma_2^2) \\ z_{3,k} = z_{3,k-1} + N(0, \sigma_3^2) \\ z_{4,k} = z_{4,k-1} + N(0, \sigma_4^2) \\ HI_k = z_{1,k} \cdot \exp(z_{2,k} \cdot k) + z_{3,k} \cdot \exp(z_{4,k} \cdot k) + \eta_k \end{cases} \tag{18.9}$$

where, $z_k = [z_{1,k}, z_{2,k}, z_{3,k}, z_{4,k}]$ is the state variable at time k; $N(0, \sigma^2)$ is the

Gaussian noise with mean zero and variance σ^2; η_k is the independent measurement noise.

The particle filter (PF) algorithm is used to predict the future HI based on this state-space model. The main idea of PF is to sample a large number of random particles using the Monte Carlo method, and give each particle an importance weight to represent the posterior probability density. More details about PF can be found in reference [20]. The PF algorithm includes the following four steps:

(1) Particle Initialization: at time $k = 0$, the initial particles $\{z_0^i\}_{i=1}^N$ are produced from a priori probability density function (PDF) $p(z_0)$, and the weights are assigned as $w_o^i = 1/N$.

(2) Particle Update: weights are calculated at time k as:

$$w_k^i = w_{k-1}^i \frac{p(\mathrm{HI}_k \mid z_k^i)p(z_k^i \mid z_{k-1}^i)}{q(z_k^i \mid z_{k-1}^i, \mathrm{HI}_{1:k})} \tag{18.10}$$

where, $q(z_k^i \mid z_{k-1}^i, \mathrm{HI}_{1:k})$ is the important density function which is selected as the prior probability density function in this paper, and normalized as:

$$w_k^i = w_k^i \Big/ \sum_{j=1}^{N} w_k^j \tag{18.11}$$

(3) Particle Resampling: the new particle set $\{\hat{z}_k^i\}_{i=1}^N$ is generated by duplicating the particles with large weights and assigned with equal weight $1/N$.

(4) State Estimation: the new state is calculated by using new particles and weights as:

$$\hat{z}_k^i = \frac{1}{N}\sum_{i=1}^{N} \hat{z}_k^i \tag{18.12}$$

Once the t measurements of the HI have been obtained, the posterior PDF of the state variable at time t can be derived with the particle states as:

$$p(z_t \mid \mathrm{HI}_{0:t}) \approx \sum_{i=1}^{N} w_t^i \delta(z_t - z_t^i) \tag{18.13}$$

where, N is the number of particles; w_t^i denotes the importance weight of particle; z_t^i is the state variable estimated by the i_{th} particle; and $\delta(\cdot)$ is the Dirac delta function.

Based on equation (18.9), the k-step ahead prediction of the corresponding PDF is calculated using

$$p(z_{t+k} \mid \mathrm{HI}_{0:t}) \approx \sum_{i=1}^{N} w_t^i \delta(z_{t+k} - z_{t+k}^i) \tag{18.14}$$

and the prediction value of the state is

$$\tilde{z}_{t+k} \approx \sum_{i=1}^{N} w_t^i z_{t+k}^i \tag{18.15}$$

Accordingly, the HI can be predicted by substituting equation (18.15) into equation (18.9). Suppose that L_t^i is the failure time that is estimated by the i_{th} particle, which can be calculated by extrapolating the state-space model until the HI exceeds the predefined FT for the first time. Thus, the RUL associated with the i_{th} particle is obtained as follows:

$$\mathrm{RUL}_t{}^i = \{L_t^i - t\}_{i=1}^N \tag{18.16}$$

and the PDF of the RUL prediction is further can be estimated by

$$p(\mathrm{RUL}_t{}^i \mid \mathrm{HI}_{o:t}) \approx \sum_{i=1}^{N} w_t^i \delta(\mathrm{RUL}_t - \mathrm{RUL}_t{}^i) \tag{18.17}$$

The whole process of the proposed RUL prediction framework is shown in **Fig. 18.4**.

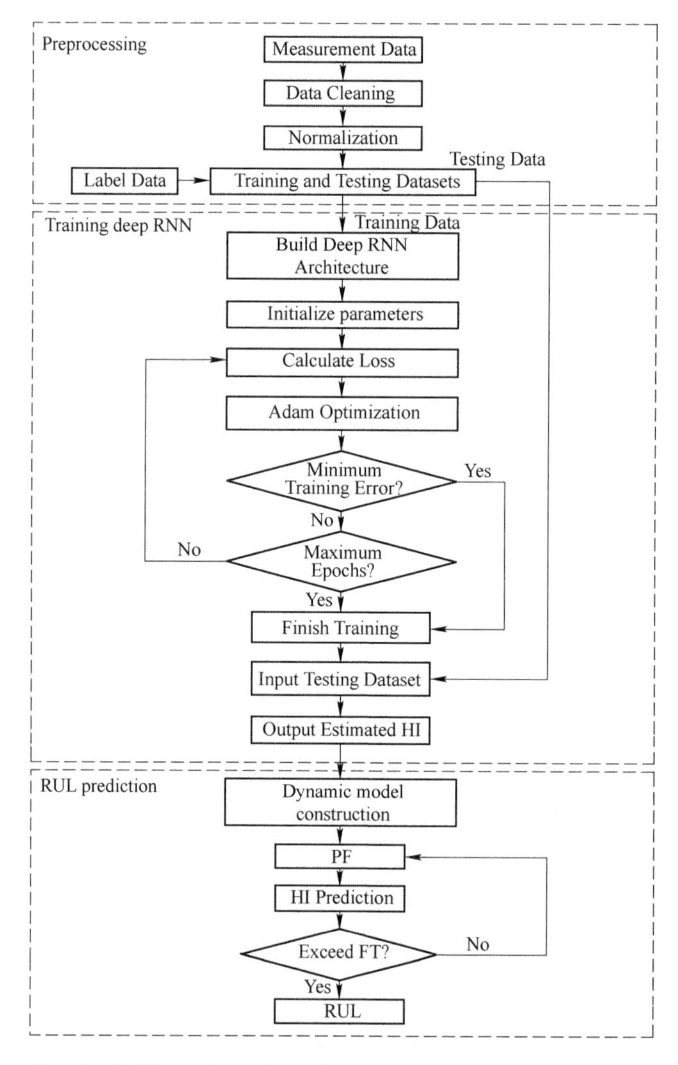

Fig. 18.4 Flowchart of proposed RUL prediction framework

18.4 Case study

The experimental data set was collected from the HSMP of a 1700-mm hot rolling line in March 2019. It is generally believed that the rollers have reached the end of their service life and need to be replaced after rolling a large number of batches of steel. The work roller maintenance data of the last stand were collected during 12 cases of maintenance, and the corresponding rolling monitored data were also collected. Among them, 10 of the collected data sets were treated as the training set, and the other two were used to test the performance of the proposed method.

18.4.1 Data preprocessing

Fifteen variables were used in total. Ten of them were input to LSTM1 as shown in **Table 18.1**, which are collected during the rolling process. The other five variables collected after rolling a batch of steel were used as the input for LSTM2, and the detailed descriptions are shown in **Table 18.2**. Due to the difference in rolling time for each batch, as shown in **Fig. 18.5**, the monitoring data for LSTM1 should be trimmed to the same length. Since the data lengths have little difference, the shortest batch data could be found out from all modeling data, and then the data of remaining batches were cut to this length. In addition, the value of each variable was normalized using min-max normalization to eliminate the dimensional effect between the variables.

Table 18.1 Detailed descriptions of variables for LSTM1

Variable	Unit	Variable	Unit
Strip thickness	mm	Rolling force	kN
Strip width	mm	Bending force	kN
Temperature	℃	Rolling speed	m/s
Crown	mm	Roller gap	mm
Flatness	I	Power	kW

Table 18.2 Detailed descriptions of variables for LSTM2

Variable	Unit	Variable	Unit
Cumulative rolling length	km	Target thickness	mm
Weight of the steel	kg	Target width	mm
Carbon content	%		

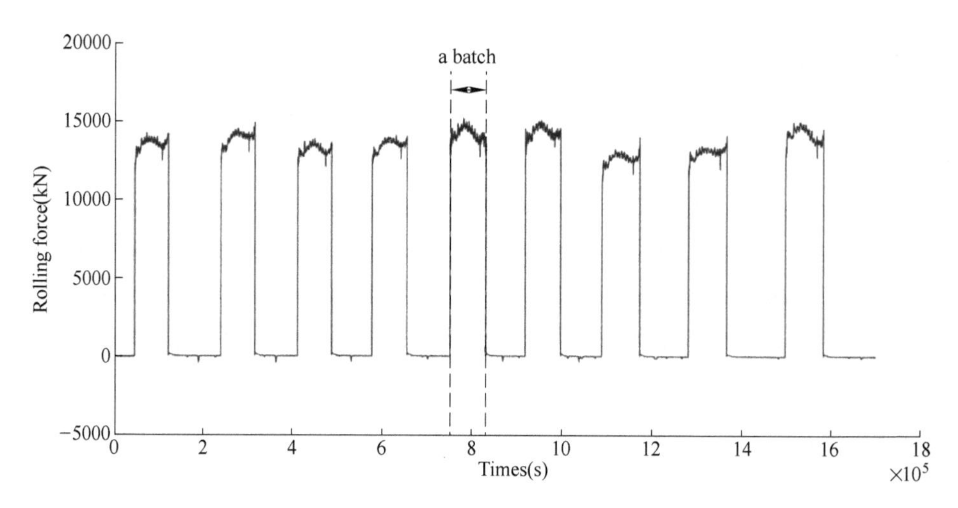

Fig. 18. 5 An example of a monitoring sample of the rolling force

18. 4. 2 Experimental setup

In the proposed deep RNN, the network structure parameters strongly depend on the complexity of the problem. Considering the complexity of the research problem and the number of available training samples, the LSTM1 for fine-grained feature extraction was set to have five LSTM layers which contain 128, 200, 256, 360, and 512 hidden nodes, respectively, and the LSTM2 for coarse-grained feature extraction was set to three LSTM layers with 32, 64, and 128 hidden nodes, respectively. The truncated BPTT with the minibatch stochastic gradient descent method was used to update the weights during training, and Adam optimization algorithm was used to optimize the neural network. In the testing phase, once the collected test data set is input into the deep RNN, the network outputs a HI corresponding to the health of the roller as the recognized result. A detailed configuration of this deep RNN is summarized in **Table 18. 3**, and all the parameters were selected based on a large number of trials.

Table 18. 3 Details of deep RNN

Setting items	Detail
Learning rate adjustment	Adam
Number of LSTM layers for LSTM1	5
Number of LSTM layers for LSTM2	3
Number of hidden nodes for LSTM1	128; 200; 256; 360; 512
Number of hidden nodes for LSTM2	32; 64; 128

Continued Table 18.3

Setting items	Detail
Output dimension of LSTM1	4
Output dimension of LSTM2	2
Time window of LSTM1	6
Time window of LSTM2	12
Mini-batch size	64
Dropout rate	0.5

18.4.3 Results and discussion

18.4.3.1 HI construction results

In order to quantitatively evaluate the constructed HI, two metrics of monotonicity (Mon) and correlation (Corr) are considered[21]. The monotonicity metric is used to evaluate the tendency of an HI, and a higher score means a better performance in monotonicity, which can be calculated by:

$$\mathrm{Mon(HI)} = \left| \frac{\text{Num of } d\mathrm{HI} > 0}{T-1} - \frac{\text{Num of } d\mathrm{HI} < 0}{T-1} \right| \tag{18.18}$$

where, T represents the length of the HI during the lifetime; $d/d\mathrm{HI} = \mathrm{HI}_{T+1} - \mathrm{HI}_T$ is the difference of the HI sequence; Num of $d/d\mathrm{HI} > 0$ and Num of $d/d\mathrm{HI} < 0$ are the number of the positive differences and the negative differences, respectively.

The correlation metric is able to reflect the correlation property between HI and operating time, which is denoted as:

$$\mathrm{Corr(HI)} = \frac{\left| \sum_{t=1}^{T} (\mathrm{HI}_t - \overline{\mathrm{HI}})(l_t - \bar{l}) \right|}{\sqrt{\sum_{t=1}^{T} (\mathrm{HI}_t - \overline{\mathrm{HI}})^2 \sum_{t=1}^{T} (l_t - \bar{l})^2}} \tag{18.19}$$

where, HI_t and l_t denote the HI and the time value at moment t; $\overline{\mathrm{HI}}$ and $\bar{l}$ are the means of HI_t and l_t, respectively.

According to equation (18.18) and equation (18.19), the Mon and Corr of the proposed approach are obtained and listed in **Table 18.4**. To demonstrate the advantages of the proposed HI construction method, an RNN was modeled for comparison using the multilayer LSTM1 for fine-grained feature extraction, which is denoted as RNN1. Another RNN with multilayer LSTM2 for coarse-grained feature extraction was also used to construct a HI for comparison, and marked as RNN2.

Table 18.4 HI construction results

Approach	Mon	Corr
RNN1	0.4342	0.9628
RNN2	0.3249	0.9137
Deep RNN	0.5973	0.9762

It can be seen from Table IV that the Mon and Corr values of deep RNN and RNN1 are both larger than RNN2, and the deep RNN achieves the highest values. The large values imply that the HI established by deep RNN has a better performance in monotonicity and correlation. Through the comparison, it is demonstrated that the proposed deep RNN network structure which can extract fine-grained and coarse-grained features has better performance than other RNN-based methods.

18.4.3.2 RUL prediction results

To verify the effectiveness of the proposed RUL prediction approach based on deep RNN, the two collected test sets, represented as test01 and test02, were used for prediction and analysis. Due to the statistical properties of the PF algorithm, the estimation results are in distribution forms approximated by 5000 samples. The RUL of test01 was estimated on the 161_{st} and 242_{nd} batch, and the estimation and prediction results are shown in **Fig. 18.6**. The mean of the HI prediction is indicated by the blue line, and the 95% confidence interval (CI) for HI estimation and prediction are also given. **Fig. 18.6** (a) shows the EoL predicted at the 161_{st} batch, which means that the data from the first 161 batches are used to update the model. Thus, the RUL and 95% CI can be derived based on equation (18.16) and equation (18.17). From **Fig. 18.6** (c), it can be seen that the median value and 95% CI of RUL are 37 batches and [31, 47], respectively, which indicates that roller maintenance is required after rolling 37 further batches of steel. As shown in **Fig. 18.6** (b), the HI estimated for test01 at 242_{nd} batch exceeded the FT at the 252_{nd} batch. The RUL prediction results are shown in **Fig. 18.6** (d), and the 5_{th}, 50_{th}, and the 95_{th} percentiles of the RUL are 5, 10, and 17, respectively.

The estimation and prediction results for test02 are shown in **Fig. 18.7**, which are estimated on the 195_{th} and 263_{rd} batch. As shown in **Fig. 18.7** (a), the estimated HI

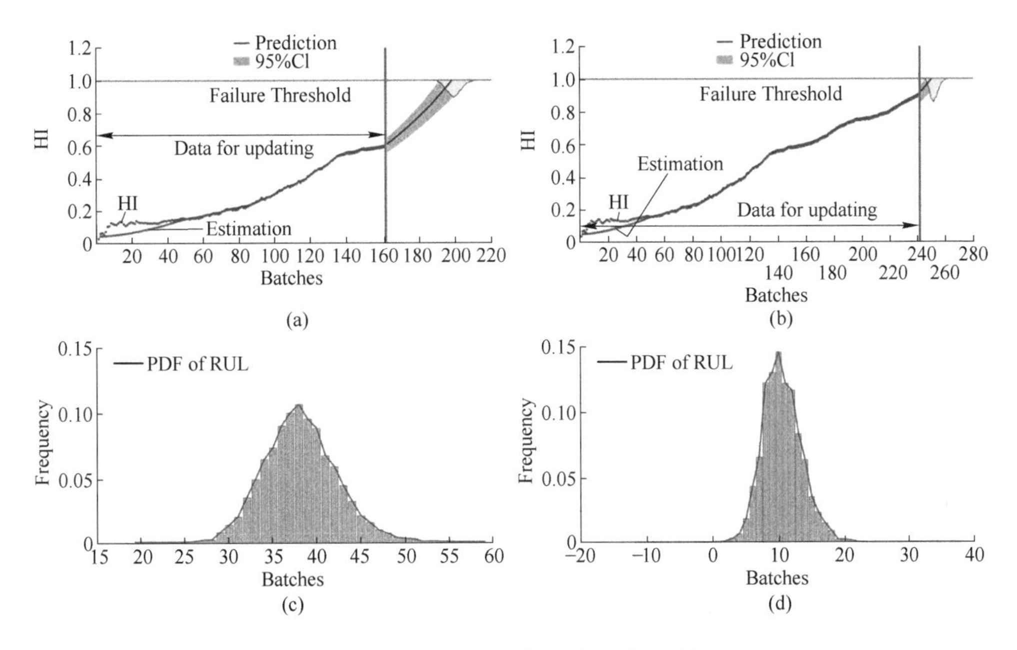

Fig. 18.6 Predicted results of test01

(a) Prediction results for test01 at 161 batches; (b) Prediction results for test01 at 242 batches; (c) Estimated RUL for test01 at 161 batches; (d) Estimated RUL for test01 at 242 batches

at the 195_{th} batch is higher than FT for the first time at 223rd batches, indicating that the EoL are 223 batches. The PDF of the RUL is shown in **Fig. 18.7** (c), the median value and 95% CI of RUL are 28 batches and [22, 35], respectively. As can be seen from **Fig. 18.7** (b) that the EoL predicted at 263_{rd} batch is 277 batches, and the PDF of the RUL is plotted in **Fig. 18.7** (d). The 5_{th}, 50_{th}, and the 95_{th} percentiles of the RUL are 7, 14, and 21, respectively, which indicates that it is necessary to maintain the rollers after rolling 14 more batches of steel.

The comprehensive results of the prediction performance are listed in **Table 18.5**. It can be seen that the RUL prediction results with long available batch data for both test01 and test02 are more accurate than those with shorter ones. This can be attributed to the limitation of the Bayesian method. If there is no new measurement information to update the model, the prediction accuracy will decrease as the prediction window increases. On the other hand, the uncertainty of the rolling schedule and the uncertainty in the actual rolling process can also lead to the inaccuracy of the initial predictions. However, once new measurement data are acquired, the parameters of state-space model are automatically updated to improve the prediction results. Although the prediction at the beginning is imprecise, the prediction results become more and more accurate as the measured data increases.

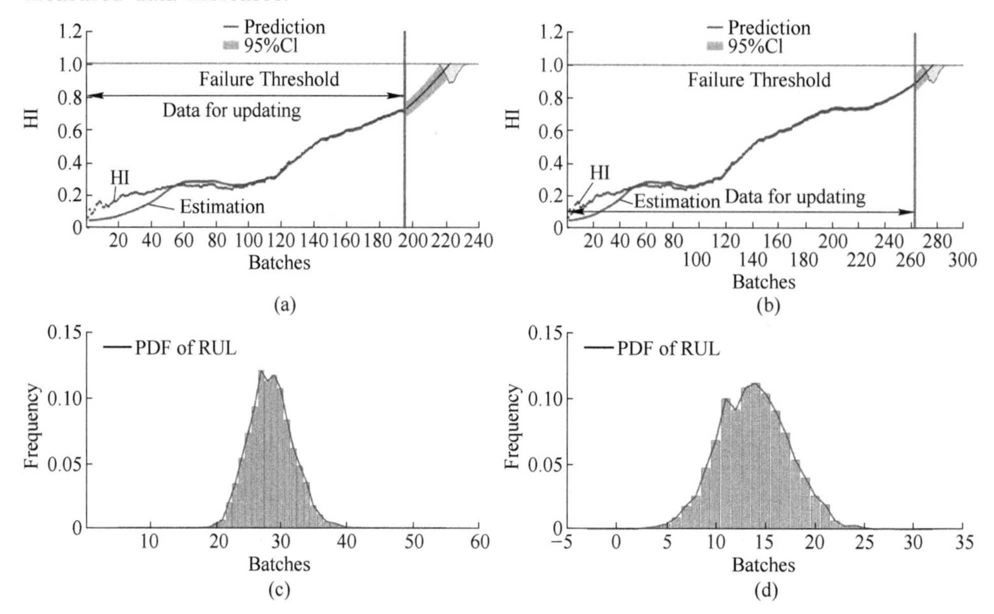

Fig. 18.7 Predicted results of test02

(a) Prediction results for test02 at 195 batches; (b) Prediction results for test02 at 263 batches; (c) Estimated RUL for test02 at 195 batches; (d) Estimated RUL for test02 at 263 batches

Table 18.5 RUL prediction results

Testing data	Current time	Estimated EoL	Predicted RUL	95% CI for RUL	Actual EoL
Test01	161	198	37	[31, 47]	256
	242	252	10	[5, 17]	256
Test02	195	223	28	[22, 35]	278
	263	277	14	[7, 21]	278

In order to further demonstrate the advantages of the proposed deep network architecture, convolutional neural network (CNN) and deep belief network (DBN) are used for comparison. The CNN has five convolution layers, and a fully connected layer is also added at the end so as to have the same architecture as deep RNN. The DBN is composed of four restricted Boltzmann machines, and the basic neural network is attached at the top layer. The approaches are carried out on the two test datasets, and root mean square error (RMSE) and mean absolute error (MAE) are used to assess the performance, which are defined as:

$$\mathrm{RMSE} = \sqrt{\frac{1}{N}\sum_{i=1}^{N}(\mathrm{RUL}_a - \mathrm{RUL}_e)^2} \tag{18.20}$$

$$\mathrm{MAE}=\frac{1}{N}\sum_{i=1}^{N}\left|\mathrm{RUL_a}-\mathrm{RUL_e}\right| \tag{18.21}$$

where, $\mathrm{RUL_a}$ is the actual RUL value; $\mathrm{RUL_e}$ is the estimated value.

The comparison results with respect to the RMSE and MAE are summarized in **Table 18.6**. RNN2 achieves the highest RMSE value, which indicates its prediction error is the largest. The RMSE values of RNN1, CNN, and DBN are 17.67, 18.97, and 15.23, respectively, which are higher than the deep RNN. Deep RNN achieves relatively low MAE value as 9.85, while the MAE values of RNN1, RNN2, CNN, and DBN are 15.42, 26.25, 15.74, and 14.37, respectively. From the results, it is observed that the deep RNN has the smallest prediction error. This demonstrates that the deep RNN network structure proposed in this study is able to extract coarse-grained and fine-grained features to construct the HI, and can further accurately predict the RUL for rollers of HSM. Furthermore, the prediction result based on the proposed RUL framework include a time range of possible failures, which can provide more useful information for roller management decision.

Table 18.6 Comparison of the prediction results

Approach	RMSE	MAE
RNN1	17.67	15.42
RNN2	31.26	26.25
CNN	18.97	15.74
DBN	16.23	14.37
Deep RNN	11.05	9.85

18.5 Conclusions

Rollers plays a significant role in the hot rolling process. With their wide and frequent application in steel production, the RUL prediction of rollers has become a crucial and challenging issue. To solve this problem, a deep learning architecture based on RNN was developed to estimate the RUL in this chapter. Firstly, a deep RNN model was established to extract the coarse-grained and the fine-grained features to construct the HI. Afterwards, the PF algorithm was introduced to iteratively update the parameters of a constructed state-space model, and the PDF of RUL could be estimated by extrapolating the dynamic model to the FT. In the end, the proposed methods were applied to the HSM to predict the RUL of the rollers. The proposed RUL prediction framework is helpful for the management of the roller, since it can effectively reduce the production cost and improve the operating rate of the HSM so as to improve the production efficiency of the business.

The following conclusions can be drawn from the experimental results. First, the proposed deep RNN network structure is able to extract coarse-grained and fine-grained features to create the HI. Second, the developed comprehensive HI is able to reflect the health of the roller, and it gradually increases with the deterioration of health state. Third, compared with some other popular deep learning methods, the proposed RUL prediction framework is able to accurately predict the RUL for rollers of HSM since it obtained relatively low RMSE and MAE values.

Despite the good experimental results achieved by the proposed approach, it still needs further optimization in the future. On one hand, the parameters in the deep network structure can be further optimized through some optimization algorithms, such as [22] and [23]. On the other hand, some efforts should be made to improve the PF so as to obtain higher prediction accuracy.

References

[1] Azadeh A, Asadzadeh S M, Salehi N, et al. Condition-based maintenance effectiveness for series-parallel power generation system-a combined Markovian simulation model [J]. Reliability Engineering System Safety, 2015, 142: 357-368.

[2] Sutharssan T, Stoyanov S, Bailey C, et al. Prognostic and health management for engineering systems: a review of the data-driven approach and algorithms [J]. The Journal of Engineering, 2015, 7: 215-222.

[3] Jardine A K S, Lin D, Banjevic D. A review on machinery diagnostics and prognostics implementing condition-based maintenance [J]. Mechanical Systems & Signal Processing, 2006, 20 (7): 1483-1510.

[4] Pecht M, Jaai R. A prognostics and health management for information and electronics-rich systems [J]. Microelectronics Reliability, 2010, 50 (3): 317-323.

[5] Liu W, Wang Z, Liu X, et al. A survey of deep neural network architectures and their applications [J]. Neurocomputing, 2017, 234 (19): 11-26.

[6] Liao L, Jin W, Pavel R. Enhanced restricted Boltzmann machine with prognosabillity regularization for prognostics and health assessment [J]. IEEE Transactions on Industrial Electronics, 2016, 63 (11): 7076-7083.

[7] Li X, Ding Q, Sun J Q. Remaining useful life estimation in prognostics using deep convolution neural networks [J]. Reliability Engineering System Safety, 2018, 172: 1-11.

[8] Zhang Y, Xiong R, He H, et al. Long short-term memory recurrent neural network for remaining

useful life prediction of lithium-ion batteries [J]. IEEE Transactions on Vehicular Technology, 2018, 67 (7): 5695-5705.

[9] Heimes F O. Recurrent neural networks for remaining useful life estimation [C]//2008 International Conference on prognostics and health management, 2008: 1-6.

[10] Guo L, Li N, Jia F, et al. A recurrent neural network-based health indicator for remaining useful life prediction of bearings [J]. Neurocomputing, 2017, 240: 98-109.

[11] Schey J A. Tribology in Metalworking: Friction, Lubrication, and Wear [J]. Journal of Applied Metalworking, 1984, 3 (2): 173.

[12] Puchhala S, Franzke M, Hirt G. Interaction effects between strip and work roll during flat rolling process [J]. Process Machine Interactions, 2013, 439-458.

[13] John S, Sikdar S, Mukhopadhyay A, et al. Roll wear prediction model for finishing stands of hot strip mill [J]. Ironmaking & Steelmaking, 2013, 33 (2): 169-175.

[14] Rath S, Singh A P, Bhaskar U, et al. Artificial neural network modeling for prediction of roll force during plate rolling process [J]. Materials & Manufacturing Processes, 2010, 25 (1/2/3): 149-153.

[15] Pascanu R, Mikolov T, Bengio Y. On the difficulty of training recurrent neural networks [J]. International Conference on machine learning, 2013, 1310-1318.

[16] Hochreiter S, Schmidhuber J. Long short-term memory [J]. Neural Computing, 1997, 9 (8): 1735-1780.

[17] Peng K, Zhang K, You B, et al. A quality-based nonlinear fault diagnosis framework focusing on industrial multimode batch processes [J]. IEEE Transactions on Industrial Electronics, 2016, 63 (4): 2615-2624.

[18] Kingma D P, Ba J. Adam: A method for stochastic optimization [J]. Int. Conf. Learn. Representat. (ICLR), 2015: 1-15.

[19] Srivastava N, Hinton G, Krizhevsky A, et al. Dropout: A simple way to prevent neural networks from overfitting [J]. J. Mach. Learn. Res, 2014, 15 (1): 1929-1958.

[20] Liu Z, Sun G, Bu S, et al. Particle learning framework for estimating the remaining useful life of lithium-ion batteries [J]. IEEE Transactions on Instrument Measurement, 2016, 66 (2): 280-293.

[21] Lei Y, Li N, Guo L, et al. Machinery health prognostics: A systematic review from data acquisition to RUL prediction [J]. Mechanical Systems & Signal Processing, 2018, 104: 799-834.

[22] Liu W, Wang Z, Yuan Y, et al. A novel Sigmoid-function-based adaptive weighted particle swarm optimizer [J]. IEEE Transactions on Cybernetics, 2021, 51 (2): 1085-1093.

[23] Rahman I U, Wang Z, Liu W, et al. An N-state Markovian jumping particle swarm optimization algorithm [J]. IEEE Transactions on Systems, Man, and Cybernetics: Systems, 2021, 51 (11): 6626-6638.

Chapter 19 Cloud-Edge-End Based HSMP Lifecycle Health Management Prototype System

In this chapter, we build HSMP real-time simulation environment by using embedded real-time processors such as Power PC and Compact RIO, read real-time control data of digital thread, and complete the simulation verification of real-time control at the unit level at the end side by using embedded real-time operating system such as VxWorks and Linux. Build a process level edge server and database, complete the reading and analysis of relevant data in the digital thread, build a manufacturing process cloud server and database, and achieve the reading, analysis, and visualization of cloud side data. Finally, use cloud open Software architecture, Python, STK, Matlab Simulink, Simul8 and other simulation software to realize the closed-loop optimization simulation of "cloud side" collaborative operation and maintenance integration, and convert it into executable program through compiler to complete the simulation of optimization and control. The prototype system structure diagram is shown in **Fig. 19. 1**.

19. 1 Detailed design of the prototype system

The overall system design of the prototype system includes hardware (network diagram, configuration) and software design. **Fig. 19. 2** shows the schematic diagram of the prototype system hardware and network configuration.

19. 1. 1 System design

19. 1. 1. 1 Digital thread platform

Digital thread platform uses a ServMAX G404 data server. The data server is a 4U rack server, configured with 12×3/2. 5 SATA/SAS/NVMe U. 2 hot swappable hard disk (4 ×NVMe U. 2 native) and 2×NVMe M. 2 (2242/2260/2280/22110) Solid-state drive, which supports a maximum of 128TB of data storage capacity and a storage speed of 320MB/S. At the same time, the server supports 12 hot swappable hard drives, which can scale up to 128TB of data storage capacity. Supports the PCIe 4. 0 protocol,

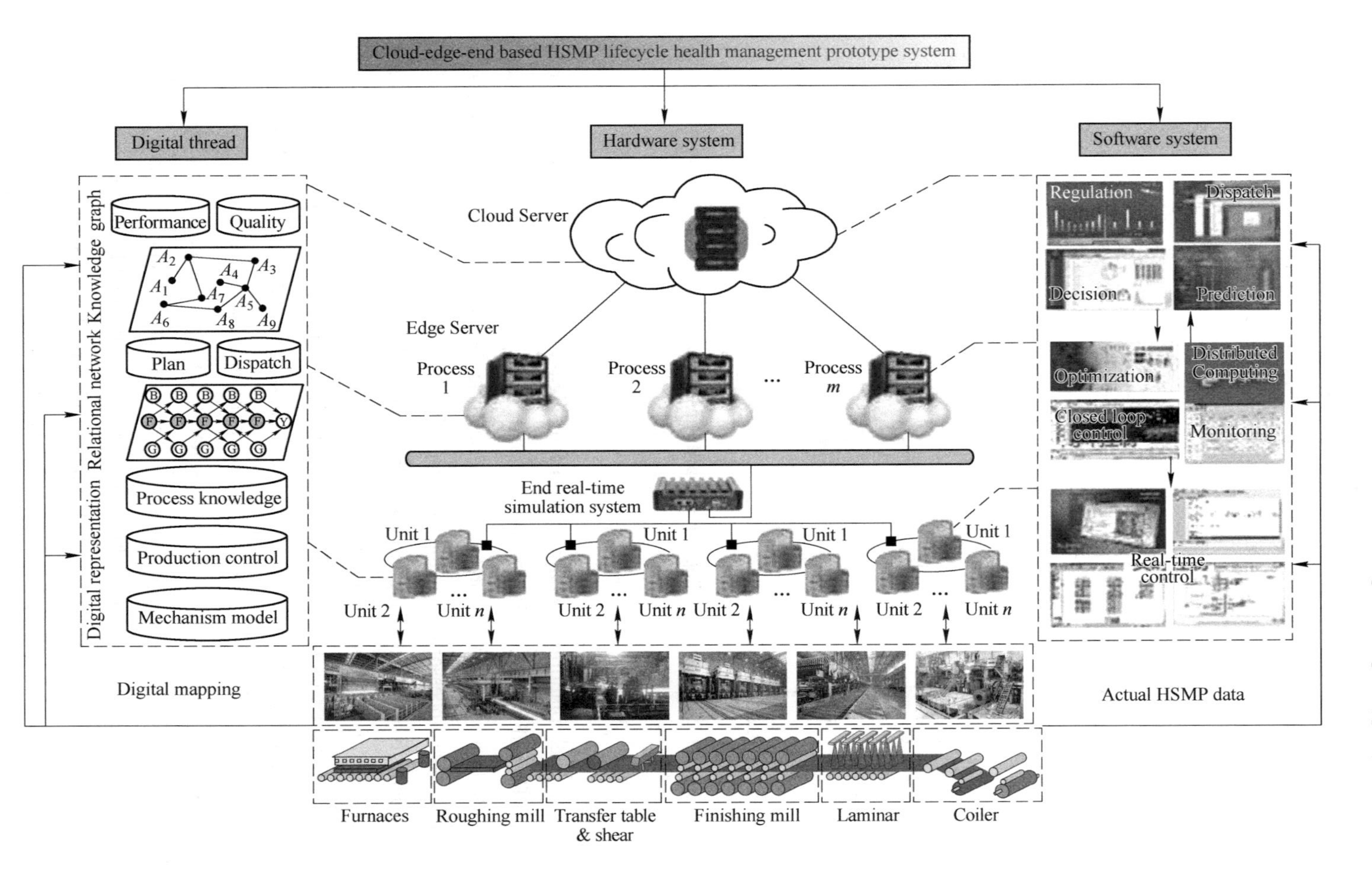

Fig.19.1 Structure diagram of HSMP prototype system

equipped with 6 × PCIe 4.0 ×16 slots and 1 × PCIe 4.0 ×8 slots, with a bidirectional bandwidth of 64GB/s for ×16, allowing for faster bus data transmission. The server is configured with a 2 × 10GbE RJ45 network interface that supports 10Gb/s network transmission speed.

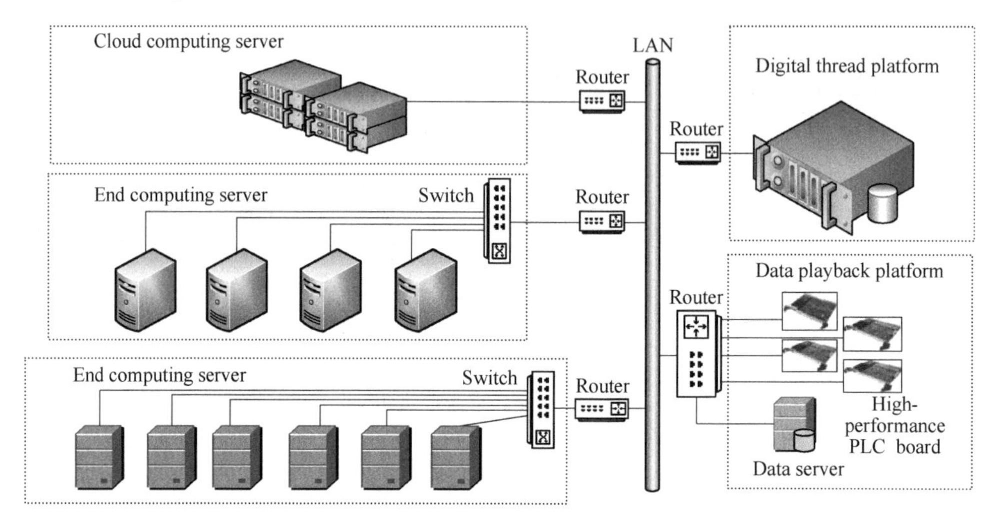

Fig. 19.2 Schematic diagram of hardware and network configuration

19.1.1.2 Data playback platform

The data playback platform is equipped with four high-performance PLC cards based on the PowerPC architecture MPC85 series, and the backboards between the cards communicate through the VME protocol. This board is designed for embedded industrial applications. The processor core is the Freescale P1013 core, equipped with 2GB DDR3 memory space, 64MB Boot/Program Flash for loading BootLoader, saving environment variables or solidifying user programs, 128 KB of nonvolatile memory FRAM for power-off saving of certain variables, and 1MB of dual port memory, which can be directly accessed by the local CPU, It can also serve as a slave for other CPUs (Masters) to access through the VME bus, achieving memory sharing. The fastest control cycle of the CPU can be less than 1ms, making it suitable for application in fast systems of distributed production processes. Supports mass storage, with 1 microSD and 1 mSATA interface on the board. There are 2 Ethernet ports that support Ethernet communication (1Gbit/s) for network startup or shell interaction, etc. There is one serial port that supports serial communication, mainly used for configuring environment variables and obtaining printing information. Both the network port and serial port are RJ45 interfaces.

19. 1. 1. 3 Cloud computing server

The server on this side is a ServMAX G404-X3 high-performance computing server. This server adopts a dual third-generation Intel Xeon scalable series processor, with a single processor up to 40 cores and supporting 4 dual width GPUs for parallel computing. It has a thermal design power (TDP) of up to 270W, and the motherboard is built using the Intel C621A chipset. It supports up to 4TB of 3DS LRDIMM/LRDIMM/3DS RDIMM/RDIMM DDR4 ECC memory, with a speed of 3200/2933/22666 MHz among 16 DIMMs and up to 4TB of Intel Optane Persistent Memory (PMem) 200 series, with a maximum speed of 3200 MHz. It also provides state-of-the-art data protection features, including TPM (Trusted Platform Module) and RoT (Trusted Root) support. Based on the functions and capabilities of the third-generation Intel Xeon scalable processor (Socket P+) and Intel C621A chipset, the motherboard improves energy efficiency and system performance for many applications such as High-performance computing, artificial intelligence (AI), deep learning (DL), Big data and enterprise applications.

19. 1. 1. 4 Edge and End computing servers

Edge and End computing servers are equipped with Lenovo ThinkStation P350 workstations. The workstation adopts Intel Core i7-11700, with an architecture of Rocket Lake-S. It has 8 cores, 16 threads, and a thermal design power (TDP) of 65W, with a maximum core frequency of 5. 2GHz; The motherboard adopts AIoT0-W580, with 4 DDR4 U-DIMM slots supporting 128GB, and 6 PCIe expansion slots capable of expanding up to 22 Gigabit Ethernet ports. The machine has built-in Lenovo 32G memory, Lenovo 2TSSD Solid-state drive, Nvidia T400 2G unique display, 17L tower case and high expansion power supply, which can support multi-core high frequency CPU and professional graphics card with stronger performance; The SSD using NVMe transmission protocol (PCle 4. 0 channel) has greatly increased the data transmission efficiency, and can achieve SSD performance several times that of AHCI transmission protocol (SATA channel). It is an ideal choice for big data and enterprise applications, and can fully meet the performance calculations such as edge computing/IoT, industrial integration, and human intelligence (AI).

19. 1. 1. 5 Database software

The operating system of the prototype system digital thread platform is Windows Server, and the system is equipped with pSpace v7. 1 real-time historical database, which is used to write real-time data, transfer real-time and historical data, and analyze and call historical data. Force Control pSpace v7. 1 is a high-performance, high-throughput, reliable, and cross platform real-time historical database system that can be used for automatic collection, compression, storage, statistical analysis, and monitoring of factory process data. It can convert massive data generated during the production process into information that meets business needs, and assist managers in making accurate decisions on safety production, reducing energy consumption, and optimizing production processes. The single node throughput rate of pSpace v7. 1 is not less than 1. 5 million records per second. This database is an open architecture that provides multiple secondary development interfaces, achieving various functions such as measurement point management, security management, data management, and alarm processing. Provide API interfaces, including various secondary development interfaces such as C API/. NET API/Java API/Nodejs API/Python API; Provide SQL access interfaces, including ODBC, JDBC, and OLEDB interfaces; OPC DA/HDA Server&OPC UA Server provide data using standard OPC protocols; Support MQTT interface, provide transfer and interaction interfaces with relational databases, such as ODBCRouter/SQLRouter.

The digital thread platform has installed a relational database MySQL v8. 0. 33, and its database management software is Navicat Premium v15. 0. 17. This database stores form information composed of manufacturing process multi process process parameters, product performance parameters, enterprise order and scheduling information, and other data. MySQL database is an Open-source model relational Database management system, which can use the most commonly used structured Query language for database operations. Because of its open source characteristics, it can be downloaded under the license of the General Public License and modified according to personalized needs. The platform is also equipped with a timing database, InfluxDB v1. 8. 4, for storing manufacturing process timing history data. The management software is InfluxDB Studio V0. 2. InfluxDB is an open-source distributed temporal, event, and metric database written in Go language, focusing on high-performance reading, writing, efficient storage, and real-time analysis of massive temporal data. It is widely used in scenarios such as DevOps monitoring, IoT monitoring, and real-time analysis. The digital thread platform is equipped with a Graph database Neo4j Desktop v1. 4. 13, which is used to store the Knowledge graph and associated network extracted and constructed from each link of the manufacturing process. Neo4j is a high-performance NoSQL graphical database that stores structured data on a network (mathematically called a graph) rather than in a table. Neo4j can also be seen as a high-performance graph engine that has all the characteristics of a mature database.

In order to better interpret the real data recorded during the operation and maintenance process of the manufacturing process, the digital thread is also equipped with the data interpretation and analysis software ibaAnalyzer v7.3.6. IbaAnalyzer is a powerful data analysis software used to analyze *.dat files generated by ibaPDA, ibaQDR, and ibaLogic, as well as query data files and alarm events in ibaHD Server. It has been widely used in process data collection and analysis systems in metallurgical enterprises.

19.1.1.6 Cloud Edge Computing Server Software Configuration

The operating system of the "cloud" side computing server is Windows Server, and the operating system of the "edge" and "end" side workstations is Windows 10 Professional Edition. The "Cloud Edge" side is equipped with pSpace v7.1, which can be used in a cascading mode to achieve collaborative computing between servers. Its OPC UA communication function can be used to read real-time data replayed by the digital thread in real time. At the same time, each server is also equipped with the FSmartWorx v2.1 platform for developing front-end display interfaces. This platform is based on web visualization and model driven concepts, combined with cloud native and multi terminal experience technology, and adopts a micro front-end framework to gather core functions such as data collection, real-time monitoring, remote control, data analysis, strategy optimization, and 3D systems. The server is also equipped with FinforWorx 3.0 as an integrated data management platform, which enables different types of data interaction with digital threads.

The algorithm development tool for the prototype system includes Python v3.8 is a newer version of the Python language that is suitable for various tasks such as scripting, automation, machine learning, and web development. Compared to previous versions, Python 3.8 has brought many syntax changes, memory sharing, more efficient serialization and deserialization, improved dictionaries, and more new features. The Python language can connect multiple mainstream databases through the standard database interface Python DB-API. At the same time, Python is at the forefront in solving data science tasks and challenges, and some convenient and easy-to-use libraries can help developers develop efficiently. Python provides modules such as sockets for programming transport layer protocols. Python Socket UDP communication is a network communication method based on the UDP protocol, which can achieve point-to-point data transmission and has the advantages of efficiency, speed, and simplicity.

19. 1. 1. 7 Data playback platform software configuration

The development tool for high-performance PLC boards on the data playback platform is CodeSys v3. 5 SP11. CodeSys programming software provides an environment for programming, compiling, and debugging PLC applications. The software is based on the IEC61131-3 standard and supports five commonly used PLC programming languages: Instruction List (IL), Function Block Diagram (FBD), Ladder Diagram (LD), Structured Text (ST), and Sequential Function Diagram (SFC); In addition, CodeSys also supports the 6th programming language continuous function diagram (CFC) that is converted from function block diagrams. Compared with other soft PLC products such as SMATIC WinAC from SIEMENS and Soft-PLC from SOFTPLC, CodeSys's biggest advantage lies in its integration of logic control, motion control, and visualization. Visualization can be easily achieved without the need for other configuration software. **Table 19. 1** lists the basic configuration of the prototype system software.

Table 19. 1 Basic configuration of prototype system software

Server name	Operating system	Basic software and version	Development environment
Digital thread platform	Windows Server	pSpace v7. 1 MySQLv8. 0. 33 InfluxDB v1. 8. 4 Neo4j Desktop v1. 4. 13 ibaAnalyzer v7. 3. 6	Python v3. 8
Cloud computing server	Windows Server	pSpace v7. 1 FSmartWorx v2. 1 FinforWorx3. 0	Python v3. 8
Edge computing server	Windows 10 Professional Edition	pSpace v7. 1 FSmartWorx v2. 1 FinforWorx3. 0	Python v3. 8
Edge computing server	Windows 10 Professional Edition	pSpace v7. 1 FSmartWorx v2. 1	Python v3. 8

Continued Table 19. 1

Server name	Operating system	Basic software and version	Development environment
Data playback platform	Windows 10 Professional Edition	ibaAnalyzer v7. 3. 6 CodeSys v3. 5 SP11	Python v3. 8

19. 1. 2 Function design

19. 1. 2. 1 Digital thread

Aiming at the characteristics of large data volume, multiple sources, high real-time and reliability requirements throughout the entire lifecycle of complex products, data management systems are established for each level of the manufacturing process system. Combining transmission technologies such as field bus, Industrial Ethernet, wireless network, etc., and comprehensively considering transmission efficiency, energy balance, reliability, delay and other performance indicators, build a "cloud side" real-time data transmission network, and use OPC UA architecture to achieve cross system, cross hierarchy interconnection and compatibility and conversion of multiple communication protocols. Based on the database cascade architecture, a multi copy distributed storage architecture is used to build a Distributed database, giving consideration to efficient massive data throughput and low latency random access capabilities, and realizing data organization, storage, analysis, calculation, and query.

19. 1. 2. 2 Correlation analysis

Based on the different dimensions, computing power, and real-time requirements of cloud, edge, and end data, performance monitoring tasks are divided into different levels of monitoring tasks such as units, processes, and systems. Monitoring indicators are constructed based on corresponding subgraphs in the association model, which are used for dynamic monitoring of multiple performance degradation at different levels. Considering the problems such as insufficient sample number and unbalanced labels on the side of equipment, the shortcomings of long training time and poor real-time performance on the cloud side, and different models have different advantages in monitoring performance, research the fusion mechanism of collaborative management and monitoring information of different models, conduct model training based on meta learning and Transfer learning, and fuse monitoring information based on Ensemble learning and Bayesian model averaging. For faults detected in performance monitoring, we extend from simple fault classification to cross process fault tracing. Under the constraints of control flow, information flow, material flow, etc. before and after the fault occurs, we use hybrid Bayesian network inference and other algorithms to trace and locate the fault cause to a more accurate unit, process, or operation level based on heterogeneous dynamic graphs that depict correlation relationships.

19. 1. 2. 3 Closed loop scheduling and optimization

Based on the characteristics of dynamic adjustment strategies for scheduling schemes that can be distributed and deployed, the concept of "cloud edge" collaboration is applied. Through the linkage of "cloud" side servers and "edge" side servers, a production scheduling scheme is formulated to efficiently respond to uncertain disturbances during the production process. On the "cloud" side, resource and energy allocation and multi-level and multi timescale forecasting and scheduling methods are deployed to form a system level forecasting and scheduling module, and high-performance computing resources are used to develop a multi-level and multi timescale linkage forecasting and scheduling scheme. Issue scheduling plans on the "edge" side and monitor real-time data information during the manufacturing process. Based on dynamic event perturbation, a device level reactive adjustment scheme is developed in the dynamic scheduling module of the "edge" side server, achieving fast response to dynamic disturbances without changing the global scheduling strategy. When the local adjustment on the "edge" side cannot effectively cope with uncertain disturbances, the disturbance information is fed back to the "cloud" side, and the system level reactive scheduling module is used to dynamically adjust the system level scheduling plan based on global information, and further guide the "edge" side to execute device level dynamic scheduling, achieving a predictive reactive closed-loop scheduling for "cloud edge" collaboration.

19. 1. 2. 4 Closed-loop optimization

In response to the situation of abnormal quality, low efficiency, and cost during process operation caused by abnormal sensors, actuators, and equipment themselves, based on the constructed correlation coupling network and the "edge end" collaborative framework, research is conducted on edge end collaborative optimization and self-healing control technology to achieve "edge end" collaborative regulation and intelligent optimization for rapid recovery of performance anomalies under abnormal performance. Based on the complex multi flow correlation network formed by the operational information in different processes/systems during the manufacturing process, this paper analyzes the genetic coupling mechanism of product multi process based on the historical data, process information, process path and other heterogeneous data information of the manufacturing process and studies the multi-objective optimization compensation and control strategy between multi process and multi system. In the "cloud" environment, construct an objective function for the "manufacturing energy quality" collaboration, establish coupling constraints for multiple information transmission, and achieve multi-objective comprehensive performance optimization of the "cloud edge" . Aiming at the problems of complex manufacturing process coupling relationship, huge amount of data, limited computing capacity, and high demand for security timeliness, the information Physical system architecture of the "cloud edge" collaborative environment is constructed, and the integrated closed-loop framework of dynamic production multi-objective/multitask real-time optimal operation and control decision-making is designed.

19. 2 Cloud-edge-end collaborative environment of prototype system

The real-time data collaboration diagram of the prototype system is shown in the figure. Real time data is sent to the real-time database of the digital mainline through the high-performance PLC board of the data playback platform. The real-time database, as the server of OPC UA, forwards real-time data to the real-time database on the computing server side, and the real-time database on the "cloud edge" side receives the required data as the client of OPC UA. According to the functional deployment of the "cloud side" side, the "end" side real-time database can read real-time data through OPC according to Functional requirement, the "edge" real-time database can read the real data of the corresponding process, and the "cloud" real-time database can read the real-time data of the whole process. **Fig. 19. 3** is the collaboration diagram of real-time data.

The construction of a cloud edge collaborative environment is shown in the figure. Taking cloud edge collaboration as an example, the cloud side server has strong computing power, and the algorithm execution framework performs more model training tasks. During the training process, it needs to receive large-scale, batch real-time, historical data, and pay more attention to the iteration speed and convergence rate of

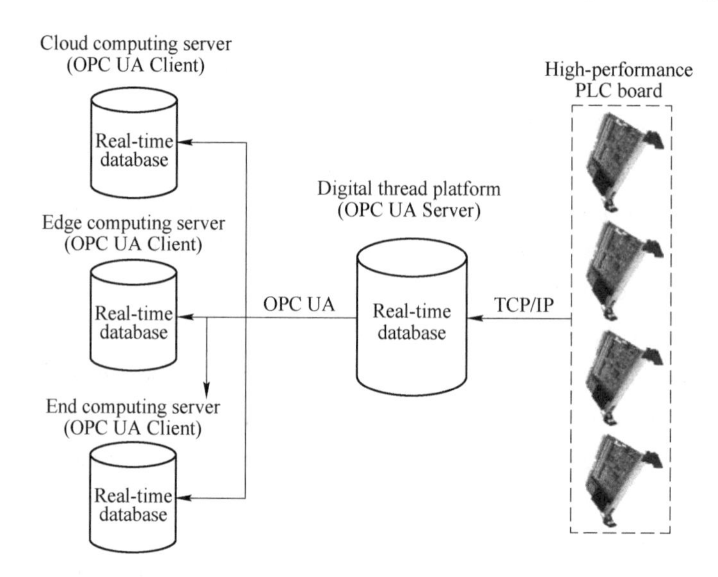

Fig. 19. 3 Schematic diagram of hardware and network configuration

training. The trained models or parameters on the "cloud" side can be updated in real-time to the real-time database on the "cloud" side. The real-time database on the "cloud" side can serve as the algorithm parameter center shared by the cloud side and form a cascading mode with the real-time database configured on the "edge" side. In this mode, the "edge" side real-time library can update the parameters of the "cloud" side real-time library selection cascade in real time, and the parameters of the "cloud" side can be sent to the "edge" side for execution or called. The cascading mode only requires configuring the IP addresses and corresponding cascading properties of the real-time libraries on both sides. At the same time, the parameters calculated by a certain "edge" server can be updated in real-time to the proprietary area opened up by the real-time database on the "cloud" side. The updated parameters can be distributed to other "edge" servers through a cascade mode between cloud edges, thereby achieving edge collaboration. Similarly, the collaborative operating environment on the "edge" and "end" sides can be configured using the same method to achieve collaborative computing on the edge. **Fig. 19. 4** shows the schematic diagram of building a cloud edge collaborative environment for the prototype system.

19. 3 Integration of cloud-edge-end data management platform

In response to the characteristics of large volume, multiple sources, high real-time and reliability requirements of HSMP lifecycle data, a "cloud edge" data management

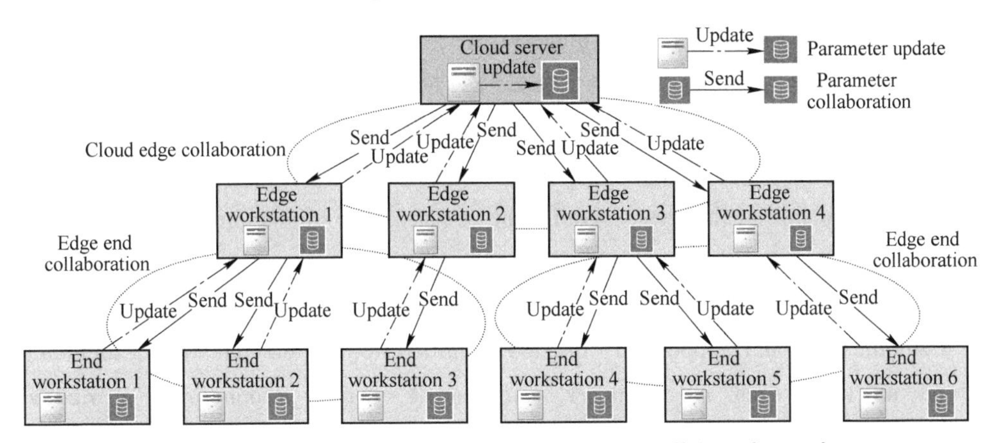

Fig. 19.4 Schematic diagram of building a cloud edge collaborative environment

platform has been established based on various levels of manufacturing process systems and deployed within the digital mainline. The FinforWorx integrated control platform is based on online low code development, providing a visual development environment, low code drag and drop configuration development, and simple operation. The FinforWorx platform can achieve cloud edge data access, multi database integration, basic platform services, unified data access, and platform functional applications.

Cloud edge data access: used to collect real-time data from different data sources. The collection driver can collect different data sources based on APIs, and supports standard format industrial messages, custom format development, message data docking, and database docking. Data conversion is achieved through data processing services. It can face massive cloud edge data access, meet the requirements of large-scale industrial on-site data, and do not limit the number of data access. Real time subscribed data can be pushed to the front-end through a unified interface, and data can also be permanently stored. At the same time, the client supports receiving and processing APIs, providing personalized data processing functions for development.

Support the integration of multiple types of databases, comprehensive data management, and unified docking and cross database association queries of different types of databases such as pSpace real-time historical database, InfluxDB temporal database, MySQL relational database, and Neo4j graphic database. Manage platform basic data, business data, process data, statistical data, and alarm data, and achieve the sharing and exchange of data resources, can provide a collaborative information environment for further digital mainline platform construction in the future.

Basic services: realize the management of various services of the basic platform based on Microservices technology, including system services, authentication services, data services, workflow services, report services, alarm services and other application function development.

Unified data access interface: through standardized and open access to data services, based on standardized interfaces, realize the encapsulation of tag, index and other data query APIs, realize real-time interaction with multiple systems, reflect the value of data, reduce Data redundancy, and ensure data security and consistency.

Platform functional application: Provide a unified application framework, including basic functions such as multiuser management, service monitoring, distributed management, load balancing, dictionary management, and log management.

19.4 Cloud edge collaborative closed-loop scheduling system

On the basis of the above software and hardware configurations and cloud edge collaborative environment, the prototype system deploys a predictive reactive closed-loop scheduling method for multi device collaboration to "cloud" and "edge" servers, initially forming a closed-loop scheduling mode for "cloud" and "edge" collaboration, and successfully deploying a predictive reactive closed-loop scheduling method for multi device collaboration. And it is verified by the application of Angang Steel's HSMP production line. The detailed functional description is as follows:

Fig. 19.5 shows the interface for process base table maintenance and historical data

tqsbmv1__薄板钢级_炼钢工艺数据表

记录创建责任者 | 00065353 | (全选)

开始筛选 | 查询

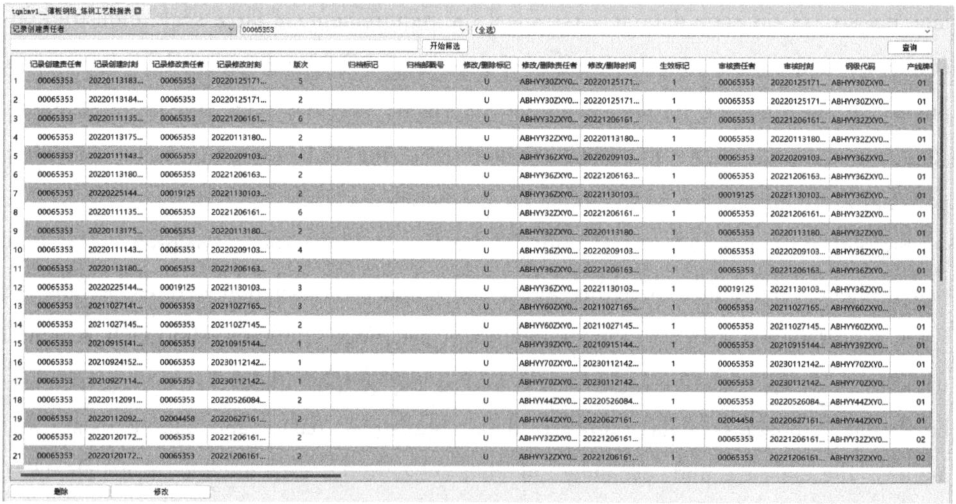

	记录创建责任者	记录创建时刻	记录修改责任者	记录修改时刻	版次	归档标记	归档邮戳号	修改/删除标记	修改/删除责任者	修改/删除时间	生效标记	审核责任者	审核时刻	钢级代码	产线牌
1	00065353	20220113183...	00065353	20220125171...	5			U	ABHYY30ZXY0...	20220125171...	1	00065353	20220125171...	ABHYY30ZXY0...	01
2	00065353	20220113184...	00065353	20220125171...	2			U	ABHYY30ZXY0...	20220125171...	1	00065353	20220125171...	ABHYY30ZXY0...	01
3	00065353	20220111135...	00065353	20221206161...	6			U	ABHYY32ZXY0...	20221206161...	1	00065353	20221206161...	ABHYY32ZXY0...	01
4	00065353	20220113175...	00065353	20220113180...	2			U	ABHYY32ZXY0...	20220113180...	1	00065353	20220113180...	ABHYY32ZXY0...	01
5	00065353	20220111143...	00065353	20220209103...	4			U	ABHYY36ZXY0...	20220209103...	1	00065353	20220209103...	ABHYY36ZXY0...	01
6	00065353	20220113180...	00065353	20221206163...	2			U	ABHYY36ZXY0...	20221206163...	1	00065353	20221206163...	ABHYY36ZXY0...	01
7	00065353	20220225144...	00019125	20221130103...	2			U	ABHYY36ZXY0...	20221130103...	1	00019125	20221130103...	ABHYY36ZXY0...	01
8	00065353	20220111135...	00065353	20221206161...	6			U	ABHYY32ZXY0...	20221206161...	1	00065353	20221206161...	ABHYY32ZXY0...	01
9	00065353	20220113175...	00065353	20220113180...	2			U	ABHYY32ZXY0...	20220113180...	1	00065353	20220113180...	ABHYY32ZXY0...	01
10	00065353	20220111143...	00065353	20220209103...	4			U	ABHYY36ZXY0...	20220209103...	1	00065353	20220209103...	ABHYY36ZXY0...	01
11	00065353	20220113180...	00065353	20221206163...	2			U	ABHYY36ZXY0...	20221206163...	1	00065353	20221206163...	ABHYY36ZXY0...	01
12	00065353	20220225144...	00019125	20221130103...	3			U	ABHYY36ZXY0...	20221130103...	1	00019125	20221130103...	ABHYY36ZXY0...	01
13	00065353	20211027141...	00065353	20211027165...	3			U	ABHYY60ZXY0...	20211027165...	1	00065353	20211027165...	ABHYY60ZXY0...	01
14	00065353	20211027145...	00065353	20211027145...	2			U	ABHYY60ZXY0...	20211027145...	1	00065353	20211027145...	ABHYY60ZXY0...	01
15	00065353	20210915141...	00065353	20210915144...	1			U	ABHYY39ZXY0...	20210915144...	1	00065353	20210915144...	ABHYY39ZXY0...	01
16	00065353	20210924152...	00065353	20230112142...	1			U	ABHYY70ZXY0...	20230112142...	1	00065353	20230112142...	ABHYY70ZXY0...	01
17	00065353	20210927114...	00065353	20230112142...	1			U	ABHYY70ZXY0...	20230112142...	1	00065353	20230112142...	ABHYY70ZXY0...	01
18	00065353	20220112091...	00065353	20220526084...	2			U	ABHYY44ZXY0...	20220526084...	1	00065353	20220526084...	ABHYY44ZXY0...	01
19	00065353	20220112092...	02004458	20220627161...	2			U	ABHYY44ZXY0...	20220627161...	1	02004458	20220627161...	ABHYY44ZXY0...	01
20	00065353	20220120172...	00065353	20221206161...	2			U	ABHYY32ZXY0...	20221206161...	1	00065353	20221206161...	ABHYY32ZXY0...	02
21	00065353	20220120172...	00065353	20221206161...	2			U	ABHYY32ZXY0...	20221206161...	1	00065353	20221206161...	ABHYY32ZXY0...	02

删除 | 修改

Fig. 19.5 Sketch map of process base table maintenance and historical data management interface

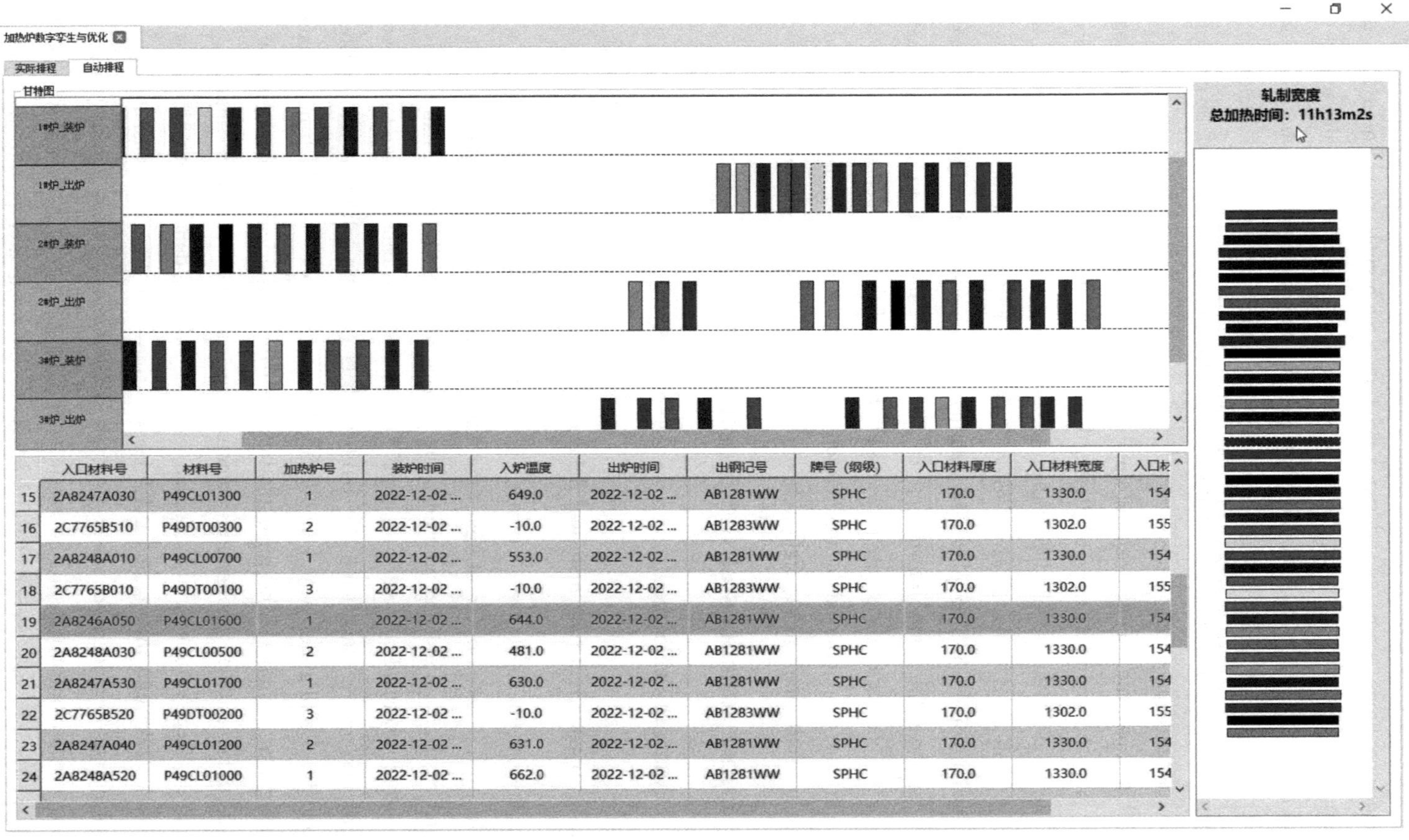

	入口材料号	材料号	加热炉号	装炉时间	入炉温度	出炉时间	出钢记号	牌号（钢级）	入口材料厚度	入口材料宽度	入口
15	2A8247A030	P49CL01300	1	2022-12-02 ...	649.0	2022-12-02 ...	AB1281WW	SPHC	170.0	1330.0	154
16	2C7765B510	P49DT00300	2	2022-12-02 ...	-10.0	2022-12-02 ...	AB1283WW	SPHC	170.0	1302.0	155
17	2A8248A010	P49CL00700	1	2022-12-02 ...	553.0	2022-12-02 ...	AB1281WW	SPHC	170.0	1330.0	154
18	2C7765B010	P49DT00100	3	2022-12-02 ...	-10.0	2022-12-02 ...	AB1283WW	SPHC	170.0	1302.0	155
19	2A8246A050	P49CL01600	1	2022-12-02 ...	644.0	2022-12-02 ...	AB1281WW	SPHC	170.0	1330.0	154
20	2A8248A030	P49CL00500	2	2022-12-02 ...	481.0	2022-12-02 ...	AB1281WW	SPHC	170.0	1330.0	154
21	2A8247A530	P49CL01700	1	2022-12-02 ...	630.0	2022-12-02 ...	AB1281WW	SPHC	170.0	1330.0	154
22	2C7765B520	P49DT00200	3	2022-12-02 ...	-10.0	2022-12-02 ...	AB1283WW	SPHC	170.0	1302.0	155
23	2A8247A040	P49CL01200	2	2022-12-02 ...	631.0	2022-12-02 ...	AB1281WW	SPHC	170.0	1330.0	154
24	2A8248A520	P49CL01000	1	2022-12-02 ...	662.0	2022-12-02 ...	AB1281WW	SPHC	170.0	1330.0	154

Fig.19.6 Digital twinning and optimization interface of heating furnace

(Scan the QR code on the front of the book for color picture)

management. The main function of this interface is to perform maintenance and managementoperations such as adding, deleting, modifying, and querying data. Support custom field display function. By selecting the desired field in the upper right corner of the interface and clicking the query button, the display of custom fields can be achieved. Supports all data queries and custom field filtering. The " Delete " and "Modify" buttons in the lower left corner of the interface can be used to delete and modify the original data.

Fig. 19. 6 shows the digital twin and optimization scheduling interface for the heating furnace, with the main functions of playing back historical data of the heating furnace and displaying algorithm rescheduling results. The interface is mainly composed of three parts: the Gantt chart of the heating furnace, the data sheet of the heating furnace and the result chart of the rolling plan. The heating furnace Gantt chart on the top of the interface displays the time information of slab entry and exit in different heating furnaces, the heating furnace data table on the bottom of the interface displays the basic information and production data information of the current batch of slabs, and the rolling plan result chart on the right side of the interface displays the order of slab exit.

19. 5 Other functions of the prototype system

19. 5. 1 Database deployment

19. 5. 1. 1 pSpace real-time historical database

The pSpace real-time historical database completes the writing of real-time data in the HSMP lifecycle, the transfer of real-time and historical data, and the analysis and invocation of historical data. In addition, the pSpace real-time historical database is also used to store intermediate data such as the results of various development model processing and model parameters. Its built-in tools such as psConfig and psMonitor enable database configuration and monitoring functions. pSpace is a high-performance, high-throughput, reliable, and cross platform real-time historical database system that can be used for automatic collection, compression, storage, statistical analysis, and monitoring of factory process data. It can convert massive data generated during the production process into information that meets business needs, and assist managers in making accurate decisions on safety production, reducing energy consumption, and optimizing production processes. The single node throughput rate shall not be less than 1. 5 million records per second. This database is an open architecture that provides multiple secondary development interfaces, achieving various functions such as measurement point management, security management, data management, and alarm processing. Provide API interfaces, including various secondary development interfaces such as C API/. NET API/Java API/Nodejs API/Python API; provide SQL access interfaces, including ODBC, JDBC, and OLEDB interfaces; OPC DA/HDA Server&OPC UA Server provide data using standard OPC protocols; support MQTT interface, provide transfer and interaction interfaces with relational databases, such as ODBCRouter/SQLRouter. At present, more than 10000 variables (including 2100 variables in the heating furnace process, 1400 variables in the rough rolling process, 4600 variables in the finishing rolling process, and 2200 variables in the laminar cooling and coiling process) have been simultaneously stored in a real-time database, with an upload cycle of 100ms. **Fig. 19. 7** to **Fig. 19. 10** show the relevant figures of the pSpace real-time historical database. In the digital mainline, according to the functional deployment of the "cloud edge" side, the "edge" real-time database can read the real data of the corresponding process through OPC, and the "end" real-time database can also complete the corresponding real-time data reading.

19. 5. 1. 2 MySQL relational database

The MySQL relational database mainly stores mechanism model setting information, production scheduling information, and performance prediction related information during the manufacturing process. In the interactive integration with the pSpace real-time historical database, the data is saved in different tables, and these tables are placed in different databases for quick access and management through SQL structured query statements. Choose the mainstream MySQL database management tool Navicat for database creation, management, and maintenance.

The MySQL relational database consists of six sub databases: rsu model, fsu model, ctc model, performance, rolling plan, and scheduling, which store information on rough rolling model setting, finish rolling model setting, laminar cooling model setting, performance prediction related indicators, rolling plan, and scheduling and scheduling related information, as shown in **Fig. 19. 11**. The green icon represents the overall directory of the established database.

Rsu model, fsu model, and ctc model specifically store model related parameters such as slab zero setting, primary setting, and secondary setting, involving key setting information such as temperature, roll gap, thickness, and speed during the rolling process, corresponding to the secondary model setting parameters during the rolling

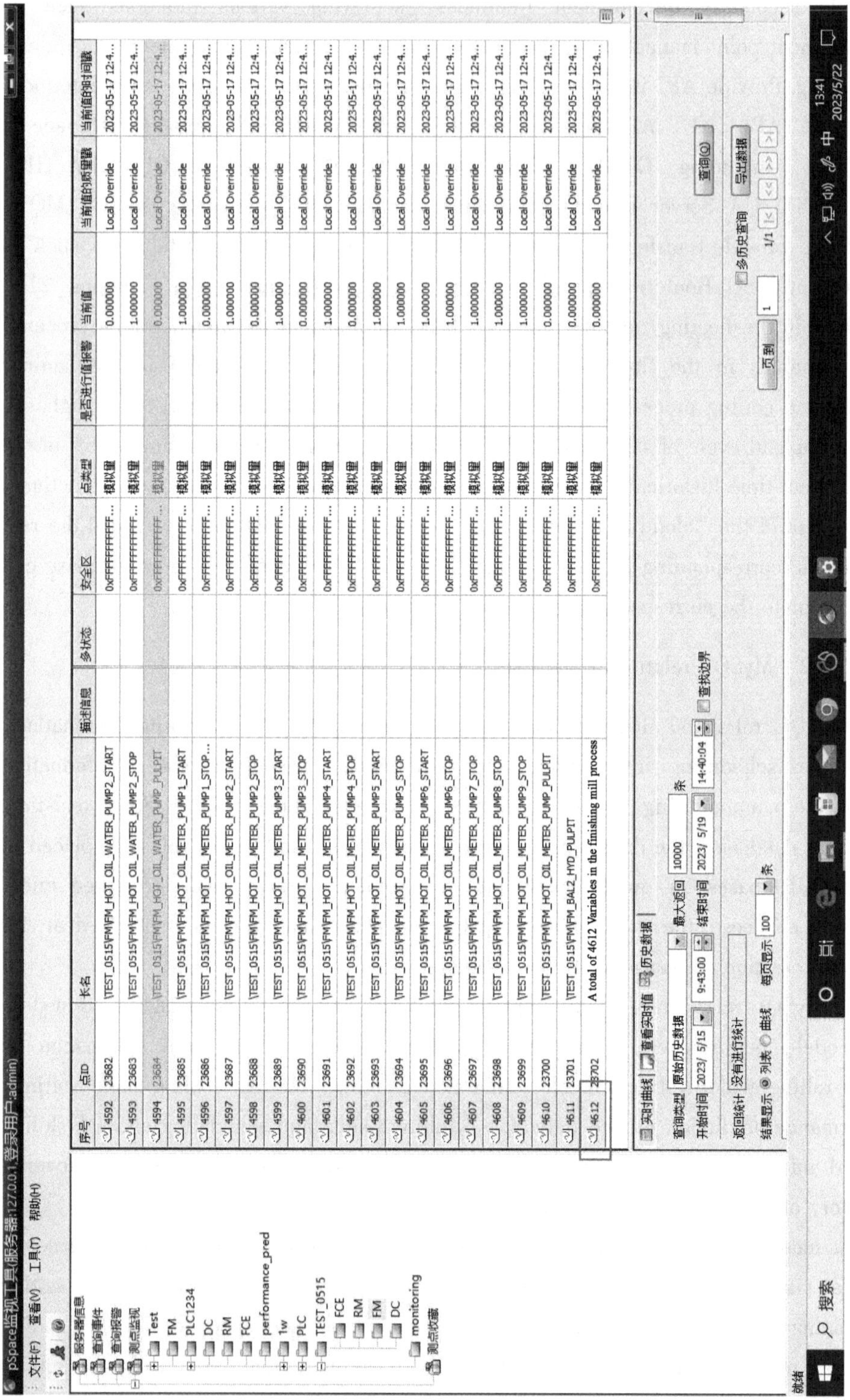

Fig.19.7 pSpace real-time historical database interface

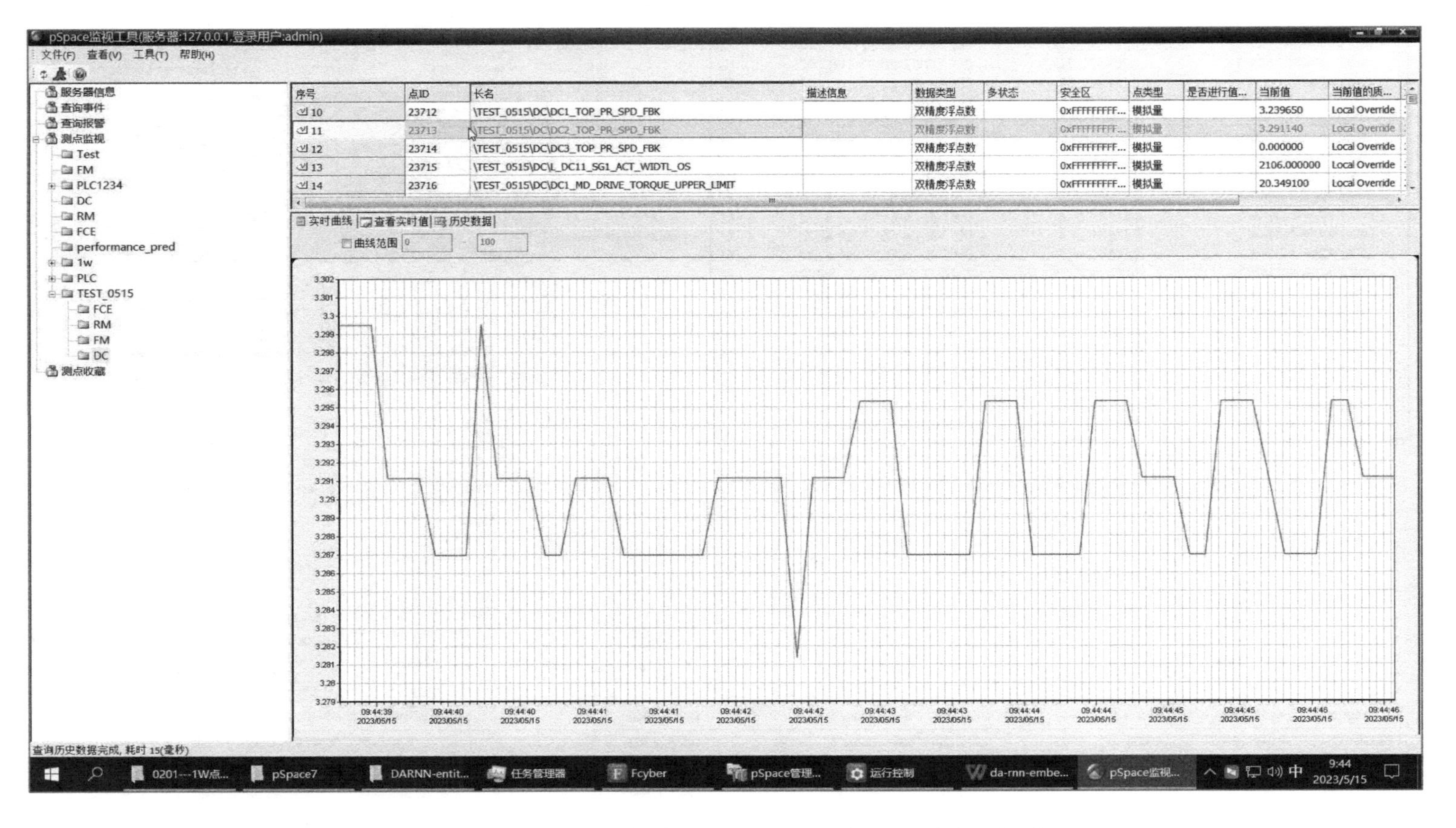

Fig.19.8 Sketch map of 10000 points upload to pSpace real-time database

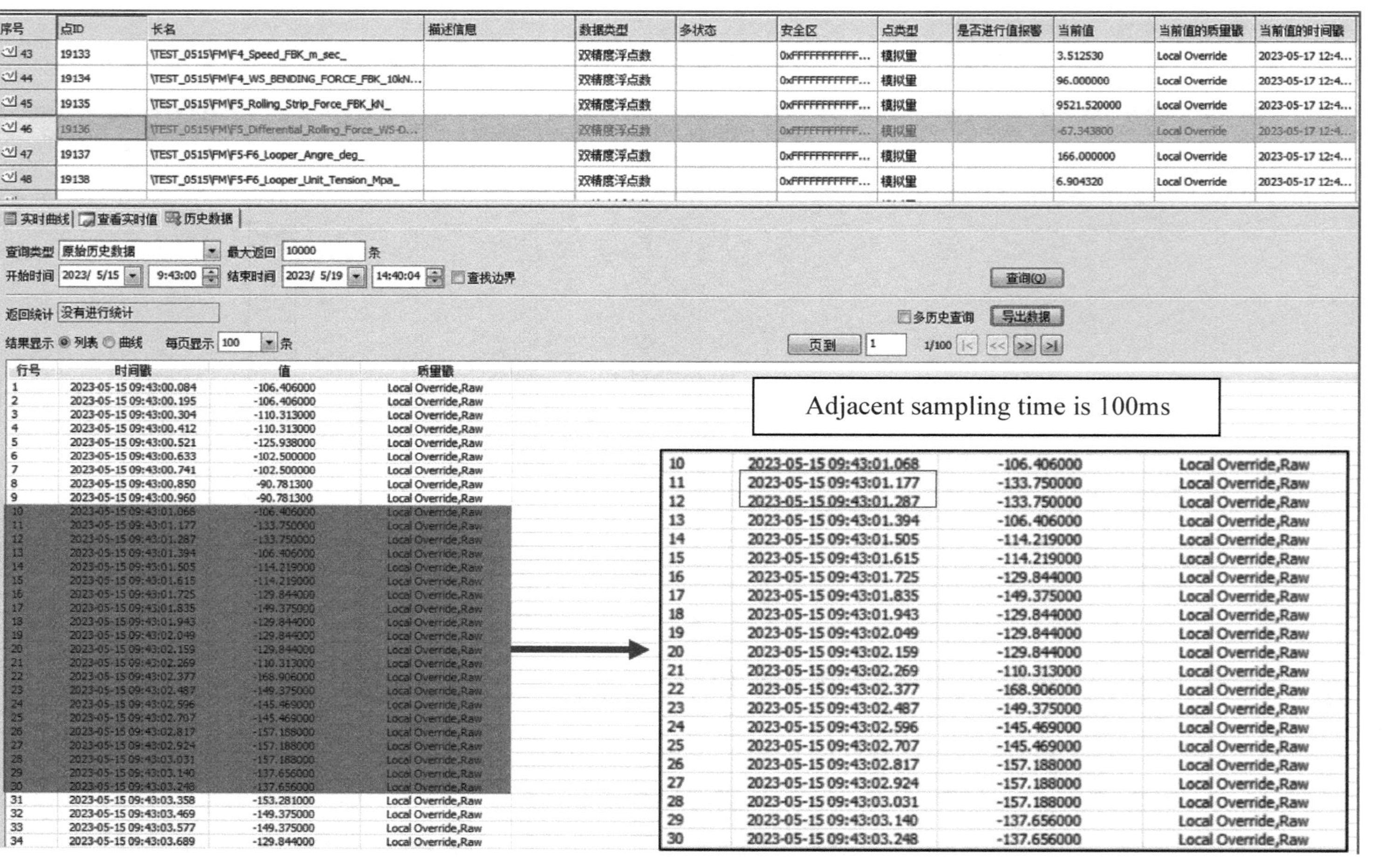

Fig.19.9　pSpace real-time historical database upload time results

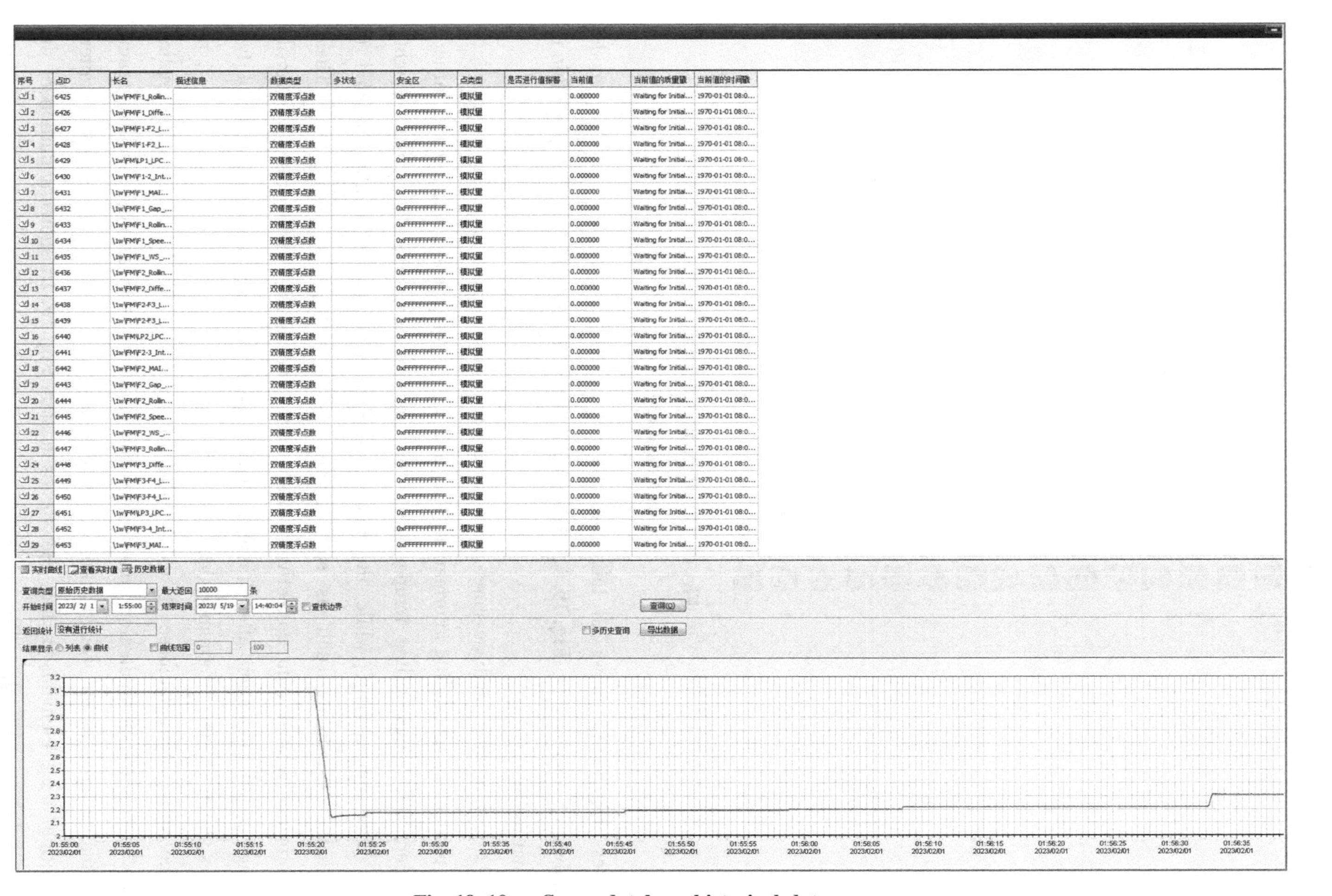

Fig. 19. 10 pSpace database historical data

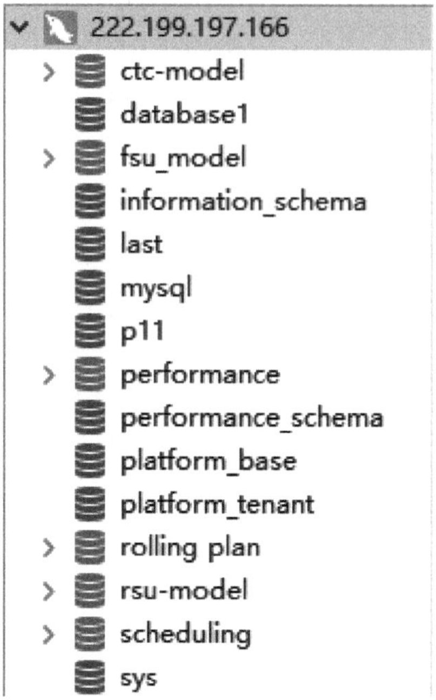

Fig. 19. 11 MySQL database overall directory

process. **Fig. 19. 12** shows all tables in the rsu model, fsu model, and ctc model databases, respectively. **Fig. 19. 13** shows some of the contents of the rsuc0 gap table in the rsu model.

The performance database stores selected measured data related to quality performance and production scheduling information. The former involves primary data such as furnace temperature, rough rolling temperature, finishing rolling temperature, slab weight, and rolled out weight, while the latter includes secondary information related to production scheduling such as steel grade, rolling date, and rolling shift. These performance related data collectively affect three indicators: yield strength, tensile strength, and elongation after fracture. **Fig. 19. 14** shows the partial contents of two tables stored in the performance database.

The rolling plan database stores the slab rolling plan, which includes information such as the rolling production time, steel coil number, slab number, rolling team, nominal weight, actual weight, nominal thickness, actual thickness, and the content of various trace elements in the steel for each slab. **Fig. 19. 15** shows a portion of the rolling plan stored in the rolling plan.

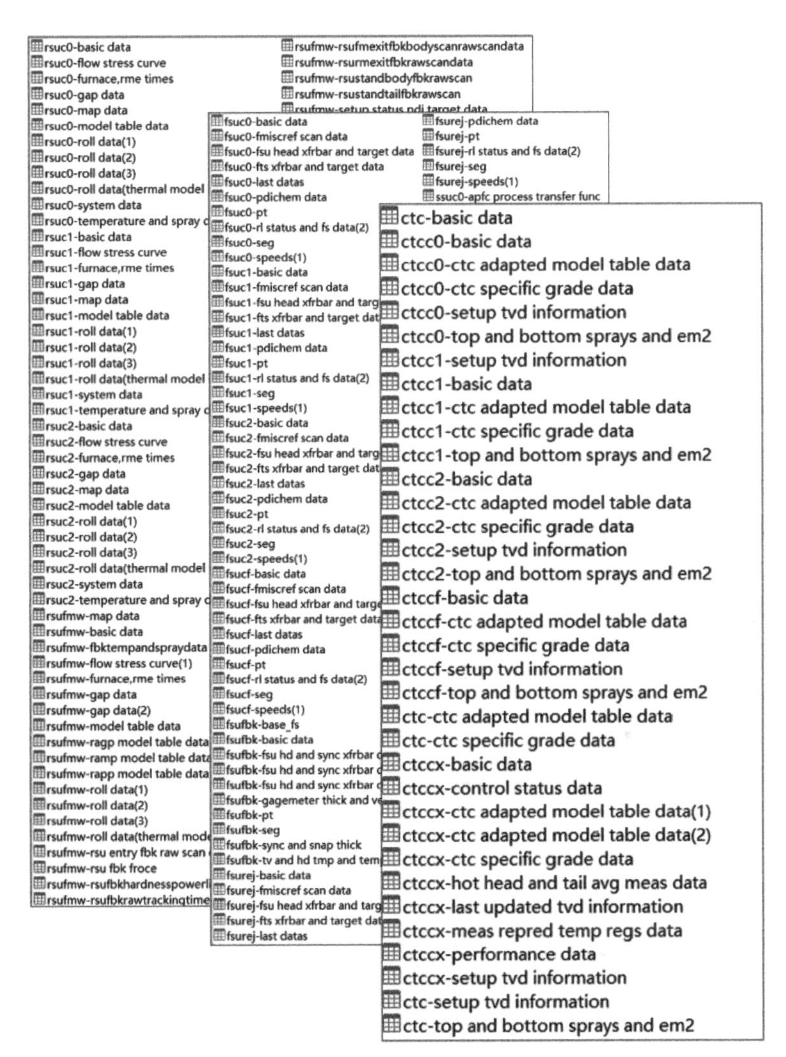

Fig. 19.12 Database all forms

The scheduling database and rolling plan database store slab rolling plans, including the rolling production time, steel coil number, slab number, rolling team, nominal weight, actual weight, nominal thickness, actual thickness, and the content of various trace elements in the steel for each slab. **Fig. 19.16** shows all the tables in the scheduling database. **Fig. 19.17** shows some of the contents of the table in scheduling.

ps[1]	obj[1]	gap(1)[1]	gap(2)[1]	strch[1]	gapof[1]	gap(3)[1]	gap(4)[1]	DFP[1]	ps[2]	obj[2]	gap(1)[2]	gap(2)[2]	strch[2]	gapof[2]	gap(3)[2]	gap(4)[2]	DFP[2]	ps[3]	obj[3]	gap(1)[3]
1	e2	1498.9	1747.9	0.81	-0.00	(Null)	(Null)	FFT	1	r2	109.5	1624.5	0.38	1.13	1615.4	1686.4	FFF	2	r2	69.8
1	e2	1498.5	1747.9	0.84	-0.00	(Null)	(Null)	FFT	1	r2	109.3	1624.5	0.54	1.09	1614.9	1686.0	FFF	2	r2	69.7
1	e2	1496.9	1747.9	0.84	-0.00	(Null)	(Null)	FFT	1	r2	108.7	1624.5	0.41	1.14	1615.2	1685.5	FFF	2	r2	68.2
1	e2	1501.1	1747.9	0.78	-0.00	(Null)	(Null)	FFT	1	r2	108.8	1624.5	0.29	1.14	1615.6	1687.8	FFF	2	r2	68.3
1	e2	1501.8	1747.9	0.76	-0.00	(Null)	(Null)	FFT	1	r2	108.9	1624.5	0.22	1.13	1615.9	1688.3	FFF	2	r2	68.4
1	e2	1503.2	1747.9	0.75	-0.00	(Null)	(Null)	FFT	1	r2	109.0	1624.5	0.13	1.12	1616.1	1689.1	FFF	2	r2	68.4
1	e2	1500.9	1747.9	0.77	-0.00	(Null)	(Null)	FFT	1	r2	109.0	1624.5	0.14	1.13	1616.2	1688.0	FFF	2	r2	68.4
1	e2	1499.9	1747.9	0.78	-0.00	(Null)	(Null)	FFT	1	r2	109.0	1624.5	0.12	1.13	1616.2	1687.5	FFF	2	r2	68.4
1	e2	1502.3	1747.9	0.75	-0.00	(Null)	(Null)	FFT	1	r2	109.1	1624.5	0.11	1.15	1616.9	1689.0	FFF	2	r2	68.6
1	e2	1500.9	1747.9	0.78	-0.00	(Null)	(Null)	FFT	1	r2	108.9	1624.5	0.26	1.17	1615.7	1687.8	FFF	2	r2	68.3
1	e2	1501.7	1747.9	0.76	-0.00	(Null)	(Null)	FFT	1	r2	109.1	1624.5	0.14	1.13	1616.8	1688.7	FFF	2	r2	68.5
1	e2	1500.7	1747.9	0.78	-0.00	(Null)	(Null)	FFT	1	r2	108.8	1624.5	0.29	1.16	1615.7	1687.6	FFF	2	r2	68.3
1	e2	1499.5	1747.9	0.77	-0.00	(Null)	(Null)	FFT	1	r2	109.1	1624.5	0.16	1.17	1616.8	1687.6	FFF	2	r2	68.6
1	e2	1500.4	1747.9	0.79	-0.00	(Null)	(Null)	FFT	1	r2	108.8	1624.5	0.31	1.14	1615.7	1687.5	FFF	2	r2	68.2
1	e2	1501.7	1747.9	0.76	-0.00	(Null)	(Null)	FFT	1	r2	109.0	1624.5	0.18	1.14	1616.8	1688.7	FFF	2	r2	68.5
1	e2	1500.5	1747.9	0.78	-0.00	(Null)	(Null)	FFT	1	r2	108.8	1624.5	0.32	1.17	1615.7	1687.6	FFF	2	r2	68.2
1	e2	1500.8	1747.9	0.78	-0.00	(Null)	(Null)	FFT	1	r2	108.9	1624.5	0.29	1.17	1615.8	1687.8	FFF	2	r2	68.2
1	e2	1501.6	1747.9	0.76	-0.00	(Null)	(Null)	FFT	1	r2	109.0	1624.5	0.18	1.15	1616.8	1688.6	FFF	2	r2	68.5
1	e2	1500.3	1747.9	0.77	-0.00	(Null)	(Null)	FFT	1	r2	109.1	1624.5	0.17	1.18	1616.7	1688.0	FFF	2	r2	68.5
1	e2	1499.8	1747.9	0.77	-0.00	(Null)	(Null)	FFT	1	r2	109.1	1624.5	0.17	1.20	1616.6	1687.7	FFF	2	r2	68.6
1	e2	1499.5	1747.9	0.78	-0.00	(Null)	(Null)	FFT	1	r2	108.9	1624.5	0.25	1.20	1615.9	1687.1	FFF	2	r2	68.3
1	e2	1501.7	1747.9	0.76	-0.00	(Null)	(Null)	FFT	1	r2	109.1	1624.5	0.19	1.20	1616.6	1688.6	FFF	2	r2	68.5
1	e2	1499.1	1747.9	0.79	-0.00	(Null)	(Null)	FFT	1	r2	108.9	1624.5	0.26	1.20	1615.9	1687.0	FFF	2	r2	68.3
1	e2	1500.9	1747.9	0.79	-0.00	(Null)	(Null)	FFT	1	r2	108.8	1624.5	0.49	1.33	1615.5	1687.7	FFF	2	r2	68.4
1	e2	1499.6	1747.9	0.84	-0.00	(Null)	(Null)	FFT	1	r2	108.6	1624.5	0.64	1.31	1614.6	1686.6	FFF	2	r2	68.3
1	e2	1499.6	1747.9	0.78	-0.00	(Null)	(Null)	FFT	1	r2	109.0	1624.5	0.15	1.14	1616.0	1687.3	FFF	2	r2	68.4
1	e2	1501.6	1747.9	0.76	-0.00	(Null)	(Null)	FFT	1	r2	109.1	1624.5	0.10	1.14	1616.5	1688.5	FFF	2	r2	68.5
1	e2	1503.3	1747.9	0.75	-0.00	(Null)	(Null)	FFT	1	r2	109.0	1624.5	0.12	1.13	1616.1	1689.2	FFF	2	r2	68.4
1	e2	1504.4	1747.9	0.74	-0.00	(Null)	(Null)	FFT	1	r2	109.1	1624.5	0.09	1.13	1616.7	1690.0	FFF	2	r2	68.6
1	e2	1505.0	1747.9	0.73	-0.00	(Null)	(Null)	FFT	1	r2	109.1	1624.5	0.10	1.15	1616.8	1690.4	FFF	2	r2	68.6
1	e2	1501.9	1747.9	0.76	-0.00	(Null)	(Null)	FFT	1	r2	109.0	1624.5	0.16	1.16	1616.0	1688.4	FFF	2	r2	68.4
1	e2	1504.6	1747.9	0.74	-0.00	(Null)	(Null)	FFT	1	r2	109.0	1624.5	0.17	1.18	1615.9	1689.8	FFF	2	r2	68.4
1	e2	1498.3	1747.9	0.83	-0.00	(Null)	(Null)	FFT	1	r2	108.7	1624.5	0.54	1.29	1615.2	1686.3	FFF	2	r2	68.4
1	e2	1498.1	1747.9	0.85	-0.00	(Null)	(Null)	FFT	1	r2	108.6	1624.5	0.60	1.27	1614.8	1685.9	FFF	2	r2	68.3
1	e2	1231.7	1747.9	0.89	-0.00	(Null)	(Null)	FFT	1	r2	108.8	1624.5	0.47	0.71	1342.5	1415.0	FFF	2	r2	69.2
1	e2	1498.9	1747.9	0.79	-0.00	(Null)	(Null)	FFT	1	r2	109.5	1624.5	0.37	1.05	1615.8	1686.6	FFF	2	r2	69.8
1	e2	1491.2	1747.9	0.92	-0.00	(Null)	(Null)	FFT	1	r2	108.8	1624.5	0.69	0.75	1614.8	1682.1	FFF	2	r2	69.1
1	e2	1493.0	1747.9	0.89	-0.00	(Null)	(Null)	FFT	1	r2	108.8	1624.5	0.70	0.79	1614.9	1683.2	FFF	2	r2	69.1
1	e2	1494.9	1747.9	0.89	-0.00	(Null)	(Null)	FFT	1	r2	108.8	1624.5	0.71	0.81	1614.5	1684.0	FFF	2	r2	69.1
1	e2	1495.8	1747.9	0.85	-0.00	(Null)	(Null)	FFT	1	r2	108.9	1624.5	0.66	0.82	1615.2	1684.7	FFF	2	r2	69.2
1	e2	1494.6	1747.9	0.89	-0.00	(Null)	(Null)	FFT	1	r2	108.8	1624.5	0.74	0.88	1614.5	1683.8	FFF	2	r2	69.2

Fig.19.13 Sketch map of partial content in the rsu model table

基地区分	机组代码	牌号（钢级）	标准	合同号	出钢记号	材料目标厚度	材料目标宽度	材料目标长度	材料实际厚度	材料实际宽度	材料实际长度	品名中文	产品规范码	最终用户中文名称
鞍山本部	A400	SPHC	Q/ASB 212-2021	B2310059001	AB2682WW	3.5	1230	0	3.497	1244	651252	热轧直发卷	BBQ2121BBM00000(	普通用户
鞍山本部	A400	SPHC	Q/ASB 212-2021	B2310316001	AB2682WW	3.5	1230	0	3.497	1244	648323	热轧直发卷	BBQ2121BBM00000(	普通用户
鞍山本部	A400	SPHC	Q/ASB 212-2021	B2310316001	AB2682WW	3.5	1230	0	3.499	1242	648117	热轧直发卷	BBQ2121BBM00000(	普通用户
鞍山本部	A400	SPHC	Q/ASB 212-2021	B2310316001	AB2682WW	3.5	1230	0	3.498	1241	645890	热轧直发卷	BBQ2121BBM00000(	普通用户
鞍山本部	A400	SPHC	Q/ASB 212-2021	B2310316001	AB2682WW	3.5	1230	0	3.498	1242	646629	热轧直发卷	BBQ2121BBM00000(	普通用户
鞍山本部	A400	SPHC	Q/ASB 212-2021	B2310316001	AB2682WW	3.5	1230	0	3.497	1237	648458	热轧直发卷	BBQ2121BBM00000(	普通用户
鞍山本部	A400	SPHC	Q/ASB 212-2021	B2310316001	AB2682WW	3.5	1230	0	3.494	1241	646923	热轧直发卷	BBQ2121BBM00000(	普通用户
鞍山本部	A400	SPHC	Q/ASB 212-2021	B2310316001	AB2682WW	3.5	1230	0	3.495	1242	437561	热轧直发卷	BBQ2121BBM00000(	普通用户
鞍山本部	A400	SPHC	Q/AS											
鞍山本部	A400	SPHC	Q/AS											
鞍山本部	A400	SPHC	Q/AS											
鞍山本部	A400	SPHC	Q/AS											
鞍山本部	A400	SPHC	Q/AS											
鞍山本部	A400	SPHC	Q/AS											
鞍山本部	A400	SPHC	Q/AS											
鞍山本部	A400	SPHC	Q/AS											
鞍山本部	A400	SPHC	Q/AS											
鞍山本部	A400	SPHC	Q/AS											
鞍山本部	A400	SPHC	Q/AS											
鞍山本部	A400	SPHC	Q/AS											
鞍山本部	A400	SPHC	Q/AS											
鞍山本部	A400	SPHC	Q/AS											
鞍山本部	A400	SPHC	Q/AS											
鞍山本部	A400	SPHC	Q/AS											
鞍山本部	A400	SPHC	Q/AS											
鞍山本部	A400	SPHC	Q/AS											
鞍山本部	A400	SPHC	Q/AS											
鞍山本部	A400	SPHC	Q/AS											
鞍山本部	A400	SPHC	Q/AS											
鞍山本部	A400	SPHC	Q/AS											
鞍山本部	A400	SPHC	Q/AS											
鞍山本部	A400	SPHC	Q/AS											
鞍山本部	A400	SPHC	Q/AS											
鞍山本部	A400	SPHC	Q/AS											
鞍山本部	A400	SPHC	Q/AS											
鞍山本部	A400	SPHC	Q/AS											
鞍山本部	A400	SPHC	Q/AS											
鞍山本部	A400	SPHC	Q/AS											
鞍山本部	A400	Q235B	GB/T											
鞍山本部	A400	SPHC	Q/AS											
鞍山本部	A400	Q235B	GB/T											
鞍山本部	A400	SPHC	Q/AS											
鞍山本部	A400	SPHC	Q/AS											
鞍山本部	A400	SPHC	Q/AS											

f1	炉温	粗轧温度	精轧温度	订货厚度	订货宽度	实际重量	卷取温度	板坯重量	轧出重量	AS_ACT	C_ACT	SN_ACT	ALS_ACT	SI_ACT	V_ACT	CR_ACT	CU_ACT	TI_ACT	P_ACT	CA_ACT	MO_ACT
0	1217.48	1078.25	854.35	2.5	1500	27410	638.21	28016	27450	0.00342	0.16	0.00168	0.029	0.1	0.0031	0.02	0.024	0.0017	0.01	0.0015	0.002
1	1217.14	1079.48	850.01	2.5	1500	27360	637.93	28028	27400	0.00276	0.17	0.00158	0.011	0.1	0.0023	0.024	0.024	0.0025	0.012	0.0011	0.002
2	1262	1088	886	2.5	1500	27380	641	27996	27420	0.00248	0.16	0.00149	0.016	0.09	0.0048	0.044	0.024	0.0021	0.023	0.0013	0.002
3	1238	1091	886	3.75	1500	27260	641	28001	27260	0.00248	0.16	0.00149	0.016	0.09	0.0048	0.044	0.024	0.0021	0.023	0.0013	0.002
4	1243	1102	895	3.75	1500	27260	641	28023	27260	0.00276	0.17	0.00158	0.011	0.1	0.0023	0.024	0.024	0.0025	0.012	0.0011	0.002
5	1265	1098	883	3	1500	27160	639	28018	27210	0.00294	0.16	0.00164	0.011	0.09	0.003	0.035	0.021	0.0028	0.015	0.001	0.002
6	1270	1106	887	3	1500	27260	640	28005	27280	0.00248	0.16	0.00149	0.016	0.09	0.0048	0.044	0.024	0.0021	0.023	0.0013	0.002
7	1270	1112	893	2.75	1500	27370	643	28005	27400	0.00294	0.16	0.00164	0.011	0.09	0.003	0.035	0.021	0.0028	0.015	0.001	0.002
8	1270	1112	893	2.75	1500	27260	643	28023	27300	0.00294	0.16	0.00164	0.011	0.09	0.003	0.035	0.021	0.0028	0.015	0.001	0.002
9	1272	1118	895	2.5	1500	27200	639	28000	27320	0.0035	0.16	0.00182	0.022	0.1	0.0037	0.045	0.026	0.0029	0.016	0.0011	0.002
10	1262	1088	886	2.5	1500	27280	641	27989	27370	0.00248	0.16	0.00149	0.016	0.09	0.0048	0.044	0.024	0.0021	0.023	0.0013	0.002
11	1271	1116	896	2.3	1500	27400	642	28037	27500	0.00294	0.16	0.00164	0.011	0.09	0.003	0.035	0.021	0.0028	0.015	0.001	0.002
12	1277	1126	898	2.3	1500	27320	645	27998	27350	0.0035	0.16	0.00182	0.022	0.1	0.0037	0.045	0.026	0.0029	0.016	0.0011	0.002
13	1271	1117	894	2.3	1500	27510	639	28028	27520	0.00338	0.17	0.00134	0.017	0.1	0.002	0.049	0.019	0.0023	0.018	0.0015	0.002
14	1271	1117	894	2.3	1500	27420	639	28027	27450	0.00338	0.17	0.00134	0.017	0.1	0.002	0.049	0.019	0.0023	0.018	0.0015	0.002
15	1277	1126	898	2.3	1500	27310	645	28023	27360	0.0035	0.16	0.00182	0.022	0.1	0.0037	0.045	0.026	0.0029	0.016	0.0011	0.002
16	1279	1134	897	2.3	1500	27180	640	28018	27270	0.00288	0.16	0.00174	0.01	0.13	0.0035	0.04	0.024	0.0033	0.019	0.001	0.002
17	1271	1117	894	2.3	1500	27450	639	28009	27480	0.00338	0.17	0.00134	0.017	0.1	0.002	0.049	0.019	0.0023	0.018	0.0015	0.002
18	1279	1135	896	2.3	1500	27450	638	28007	27460	0.00286	0.18	0.00078	0.013	0.08	0.0029	0.042	0.018	0.0021	0.018	0.0016	0.001
19	1279	1134	897	2.3	1500	27350	640	28019	27360	0.00288	0.16	0.00174	0.01	0.13	0.0035	0.04	0.024	0.0033	0.019	0.001	0.002
20	1279	1135	896	2.3	1500	27410	638	28021	27520	0.00286	0.18	0.00078	0.013	0.08	0.0029	0.042	0.018	0.0021	0.018	0.0016	0.001
21	1277	1126	898	2.3	1500	27400	645	28019	27410	0.0035	0.16	0.00182	0.022	0.1	0.0037	0.045	0.026	0.0029	0.016	0.0011	0.002
22	1278	1133	895	2.5	1500	27280	639	28016	27300	0.00286	0.18	0.00078	0.013	0.08	0.0029	0.042	0.018	0.0021	0.018	0.0016	0.001
23	1268	1108	885	2.5	1500	27500	642	28027	27520	0.00288	0.16	0.00174	0.01	0.13	0.0035	0.04	0.024	0.0033	0.019	0.001	0.002
24	1279	1133	894	2.5	1500	27520	641	28034	27520	0.00342	0.16	0.00172	0.018	0.11	0.0051	0.045	0.022	0.0027	0.021	0.0015	0.002
25	1268	1108	885	2.5	1500	27340	642	28019	27350	0.00288	0.16	0.00174	0.01	0.13	0.0035	0.04	0.024	0.0033	0.019	0.001	0.002
26	1268	1108	885	2.5	1500	27580	642	28023	27580	0.00288	0.16	0.00174	0.01	0.13	0.0035	0.04	0.024	0.0033	0.019	0.001	0.002
27	1279	1133	894	2.5	1500	27350	641	28014	27360	0.00342	0.16	0.00172	0.018	0.11	0.0051	0.045	0.022	0.0027	0.021	0.0015	0.002
28	1278	1131	892	2.5	1500	27400	641	28027	27420	0.00303	0.15	0.00171	0.014	0.14	0.0034	0.036	0.021	0.0027	0.016	0.0011	0.002
29	1278	1131	892	2.5	1500	27440	641	28018	27460	0.00303	0.15	0.00171	0.014	0.14	0.0034	0.036	0.021	0.0027	0.016	0.0011	0.002
30	1279	1133	894	2.5	1500	27560	641	28028	27570	0.00342	0.16	0.00172	0.018	0.11	0.0051	0.045	0.022	0.0027	0.021	0.0015	0.002
31	1278	1131	892	2.5	1500	27540	641	28054	27550	0.00303	0.15	0.00171	0.014	0.14	0.0034	0.036	0.021	0.0027	0.016	0.0011	0.002
32	1279	1133	894	2.5	1500	27580	641	28030	27600	0.00342	0.16	0.00172	0.018	0.11	0.0051	0.045	0.022	0.0027	0.021	0.0015	0.002
33	1245	1115	897	4.5	1500	27330	643	28021	27330	0.00342	0.16	0.00172	0.018	0.11	0.0051	0.045	0.022	0.0027	0.021	0.0015	0.002
34	1250	1127	902	4.5	1500	27430	639	28018	27430	0.00303	0.15	0.00171	0.014	0.14	0.0034	0.036	0.021	0.0027	0.016	0.0011	0.002
35	1233	1113	893	5.5	1500	27260	634	28027	27390	0.00296	0.15	0.00158	0.023	0.1	0.0026	0.041	0.019	0.0017	0.019	0.001	0.002
36	1235.98	1117	895	5.5	1500	27450	629	28012	27450	0.00271	0.14	0.0017	0.014	0.11	0.004	0.039	0.02	0.0024	0.014	0.0014	0.002
37	1230	1108	889	7	1500	27450	630	28028	27530	0.00271	0.14	0.0017	0.014	0.11	0.004	0.039	0.02	0.0024	0.014	0.0014	0.002
38	1233	1111	889	7.5	1500	27490	631	28014	27490	0.00295	0.16	0.00167	0.015	0.09	0.0042	0.04	0.022	0.0018	0.018	0.001	0.002
39	1228	1102	881	7.5	1500	27410	632	28000	27410	0.00296	0.15	0.00158	0.023	0.1	0.0026	0.041	0.019	0.0017	0.019	0.001	0.002
40	1228	1096	876	7.5	1500	27410	632	28021	27490	0.00271	0.14	0.0017	0.014	0.11	0.004	0.039	0.02	0.0024	0.014	0.0014	0.002
41	1231	1103	881	7.5	1500	27480	630	28032	27480	0.00308	0.16	0.00193	0.014	0.1	0.0031	0.04	0.019	0.002	0.018	0.0009	0.002
42	1228	1096	876	7.5	1500	27470	632	28036	27470	0.00271	0.14	0.0017	0.014	0.11	0.004	0.039	0.02	0.0024	0.014	0.0014	0.002
43	1231	1103	881	7.5	1500	27560	630	28052	27560	0.00308	0.16	0.00193	0.014	0.1	0.0031	0.04	0.019	0.002	0.018	0.0009	0.002

Fig.19.14 Sketch map of part of the content in the performance table

生产时刻	钢卷号	合同号	成品牌号	名义厚度	名义宽度	实际重量	实际厚度	实际宽度	钢卷长度	钢卷内径	钢卷外径	板坯号	出钢记号	坯厚	坯宽	坯长	坯重	加热炉号	总在炉时间
2023-01-01 00:0	P53BR00900	E22C0842005	St12	3.5	1517	27.17	3.499	1523	649.495	762	1864	2A8884B010	AB1281WW	170	1535	13668	27.998	3	172
2023-01-01 00:0	P53CN00500	E2310902051	M3A42	4.25	1674	26.96	4.25	1691	477.878	762	1779	2B8554C520	KB01T1WW	200	1706	10827	27.548	2	182
2023-01-01 00:0	P53CH00300	E2310902103	M3A42	5.25	1648	27.1	5.251	1663	395.335	762	1795	2B8555C540	KB01T1WW	200	1658	11140	27.543	3	167
2023-01-01 00:0	P53CC00500		M3A42	3.75	1634	0	3.75	1639	589	762	1842	2B8556C540	KB01T1WW	200	1658	11140	27.543	0	0
2023-01-01 00:1	P53CH00500	E2310902049	M3A42	3.75	1621	27.09	3.75	1631	564.227	762	1810	2B8556C030	KB01T1WW	200	1658	11140	27.543	2	178
2023-01-01 00:4	P53CC00600	E2310902023	M3A42	5.25	1631	26.77	5.251	1648	394.075	762	1793	2B8557C010	KB01T1WW	200	1658	11140	27.543	2	192
2023-01-01 00:5	P53CP00200	G22C0079006	DX51D+Z	5	1424	27.13	5	1443	479.009	762	1905	2A8614A010	AB1281WW	170	1450	13440	26.007	2	186
2023-01-01 00:5	P53CP00500	E22C0650001	SAE1008	4	1265	25.14	4.002	1273	628.622	762	1945	2B8204A510	AB2783WW	170	1305	14642	25.499	2	183
2023-01-01 01:0	P53CP00100	E22C0543005	SAE1008	4	1212	25.32	4	1225	658.26	762	1983	2A8586A030	AB2785WW	170	1250	15286	25.499	2	183
2023-01-01 01:0	P53C005700	B2310590004	SPHC-JG2	5	1750	29.59	5.001	1764	427.375	762	1820	2B8538C020	FA0581WW	200	1809	10562	29.997	3	172
2023-01-01 01:0	P53C006000	B2310591002	SPHC-JG2	5	1750	29.9	5	1765	431.605	762	1824	2B8538C510	FA0581WW	200	1809	10562	29.997	3	172
2023-01-01 01:1	P53C005600	B2310590004	SPHC-JG2	5	1750	29.9	4.999	1766	431.447	762	1824	2B8538C010	FA0581WW	200	1809	10562	29.997	3	169
2023-01-01 01:1	P53CC00900	E2310902050	M3A42	3.75	1638	27.03	3.751	1653	555.336	762	1798	2B8557C510	KB01T1WW	200	1658	11140	27.543	1	231
2023-01-01 01:1	P53CC01000	E2310902050	M3A42	3.75	1638	26.91	3.756	1653	552.135	762	1795	2B8557C520	KB01T1WW	200	1658	11140	27.543	2	171
2023-01-01 01:2	P53CC00100	E2310902023	M3A42	5.25	1631	26.94	5.251	1650	396.097	762	1797	2B8556C010	KB01T1WW	200	1658	11140	27.543	2	161
2023-01-01 01:2	P53CC00200	E22C1020182	M3A42	3.75	1639	27.06	3.752	1653	555.805	762	1799	2B8556C040	KB01T1WW	200	1658	11140	27.543	1	227
2023-01-01 01:2	P53CC00800	E2310929092	M3A42	3.75	1629	26.98	3.746	1646	557.409	762	1800	2B8557C030	KB01T1WW	200	1658	11140	27.543	3	182
2023-01-01 01:3	P53CC00400	E2310902103	M3A42	5.25	1648	26.97	5.25	1663	393.513	762	1792	2B8556C530	KB01T1WW	200	1658	11140	27.543	1	226
2023-01-01 01:3	P53C004200	E2310929067	M3A42	5.5	1714	27.5	5.499	1723	369.737	762	1780	2A8873C020	KB01T1WW	200	1753	10536	27.55	2	160
2023-01-01 01:3	P53CC00300	E2310340156	M3A42	3.75	1634	26.89	3.752	1652	552.647	762	1795	2B8556C510	KB01T1WW	200	1658	11140	27.543	3	185
2023-01-01 01:3	P53CC00700	B2310578033	IF-2	3.75	1638	26.73	3.752	1648	550.692	762	1792	2B8557C020	KB01T1WW	200	1658	11140	27.543	1	230
2023-01-01 01:3	P53C003800	E2310902054	M3A42	3.75	1730	27.01	3.753	1741	526.596	762	1760	2A8872C020	KB01T1WW	200	1753	10536	27.55	2	158
2023-01-01 01:4	P53C004100	E2310929067	M3A42	5.5	1714	26.95	5.5	1730	360.811	762	1763	2A8873C010	KB01T1WW	200	1753	10536	28.272	3	182
2023-01-01 01:4	P53C006100	B2310590004	SPHC-JG2	5	1750	29.51	5	1767	425.493	762	1814	2B8538C520	FA0581WW	200	1809	10562	29.997	1	213
2023-01-01 01:4	P53C003500	E2310902102	M3A42	5	1718	26.96	4.997	1730	397.278	762	1763	2A8871C510	KB01T1WW	200	1753	10536	27.55	2	147
2023-01-01 01:4	P53C004400	E2310902054	M3A42	3.75	1730	26.97	3.751	1748	523.989	762	1756	2A8873C510	KB01T1WW	200	1753	10536	27.55	3	188
2023-01-01 01:5	P53C005100	E22C1019068	M3A21	3.25	1673	28.29	3.254	1690	655.328	762	1815	2B8471C020	KF37T1WW	200	1700	11240	27.385	1	192
2023-01-01 01:5	P53C700300	E2310902051	M3A42	4.25	1674	26.97	4.249	1692	477.885	762	1779	2B8554C530	KB01T1WW	200	1706	10827	27.548	2	150
2023-01-01 01:5	P53C004000	E2310902054	M3A42	3.75	1730	26.88	3.751	1746	522.839	762	1754	2A8872C520	KB01T1WW	200	1753	10536	27.55	3	174
2023-01-01 01:5	P53C004500	E2310340208	M3A42	3.75	1730	27.05	3.749	1747	526.125	762	1758	2A8873C520	KB01T1WW	200	1753	10536	27.55	1	193
2023-01-01 02:0	P53C003900	E2310902054	M3A42	3.75	1730	26.74	3.75	1746	520.254	762	1751	2A8872C510	KB01T1WW	200	1753	10536	27.55	3	176
2023-01-01 02:0	P53AX00200	B22C0577005	Q235B	5.75	1800	29.27	5.749	1804	359.52	762	1792	2B8565A510	GC3182WW	170	1830	12202	29.799	2	149
2023-01-01 02:0	P53AX00400	B2310092003	Q235B	5.75	1800	29.57	5.748	1812	361.665	762	1797	2B8565A010	GC3182WW	170	1830	12202	29.799	2	148
2023-01-01 02:1	P53AX00600	B2310029008	Q235B	5.75	1800	29.44	5.749	1812	360.012	762	1793	2B8565A020	GC3182WW	170	1830	12202	29.799	2	143
2023-01-01 02:1	P53C004300	E2310929067	M3A42	5.5	1714	27.14	5.498	1730	363.487	762	1768	2A8873C030	KB01T1WW	200	1753	10536	27.55	1	205
2023-01-01 02:1	P53C003100	E2310902102	M3A42	5	1718	27.03	4.998	1734	397.311	762	1763	2A8870C040	KB01T1WW	200	1753	10536	27.55	3	186
2023-01-01 02:1	P53CN00200		M3A21	3.25	1673	0	3.25	1678	638	762	1795	2B8457C530	KF37T1WW	200	1700	10444	26.943	0	0
2023-01-01 02:1	P53CN00300	E22C1020105	M3A05	3.25	1659	28.31	3.254	1675	661.664	762	1823	2B8468C050	CB04T1WW	200	1680	11373	28.993	1	191
2023-01-01 02:4	P53C003400	E2310902036	M3A42	5.5	1704	26.93	5.501	1720	362.573	762	1766	2A8871C020	KB01T1WW	200	1753	10536	27.55	1	207
2023-01-01 02:5	P53C700100	E22C1020146	M3A42	5.25	1664	27.05	5.249	1682	390.296	762	1786	2B8554C040	KB01T1WW	200	1706	10827	27.548	3	214
2023-01-01 02:5	P53AX00100	B22C0582014	Q235B	5.75	1800	29.28	5.75	1810	358.389	762	1790	2A8885A540	GC3182WW	170	1830	12202	29.799	1	210
2023-01-01 02:5	P53CS00100	B2310043004	Q235B	6	1500	23.89	5.997	1508	336.52	762	1775	2A8827A090	GC3182WW	170	1530	11725	23.94	3	202
2023-01-01 02:5	P53AX00900	B2310552001	Q235B	5.75	1800	29.45	5.75	1817	359.081	762	1792	2B8563A540	GC3182WW	170	1830	12202	29.799	2	185
2023-01-01 03:0	P53AX00700	B2310029008	Q235B	5.75	1800	29.41	5.747	1818	358.583	762	1790	2B8565A530	GC3182WW	170	1830	12202	29.799	3	196

Fig.19.15 Sketch map of part of the content in the rolling-plan table

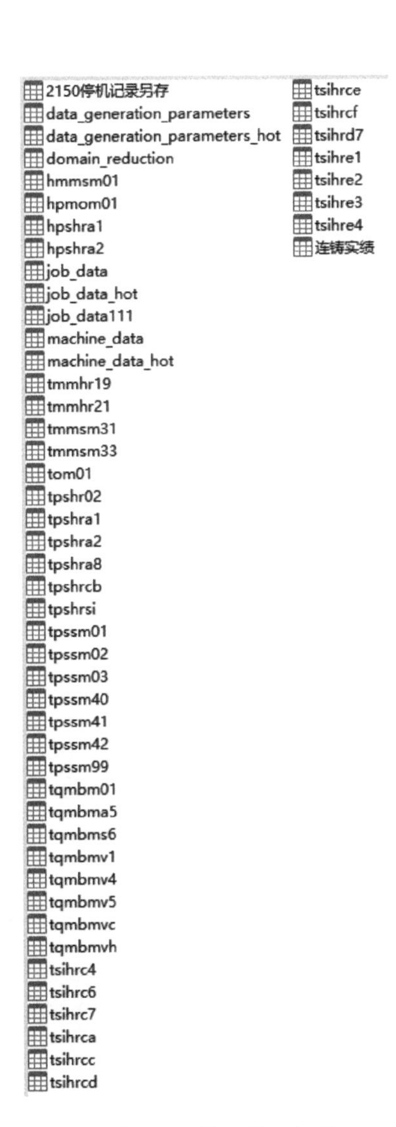

Fig. 19.16 All tables in the scheduling database

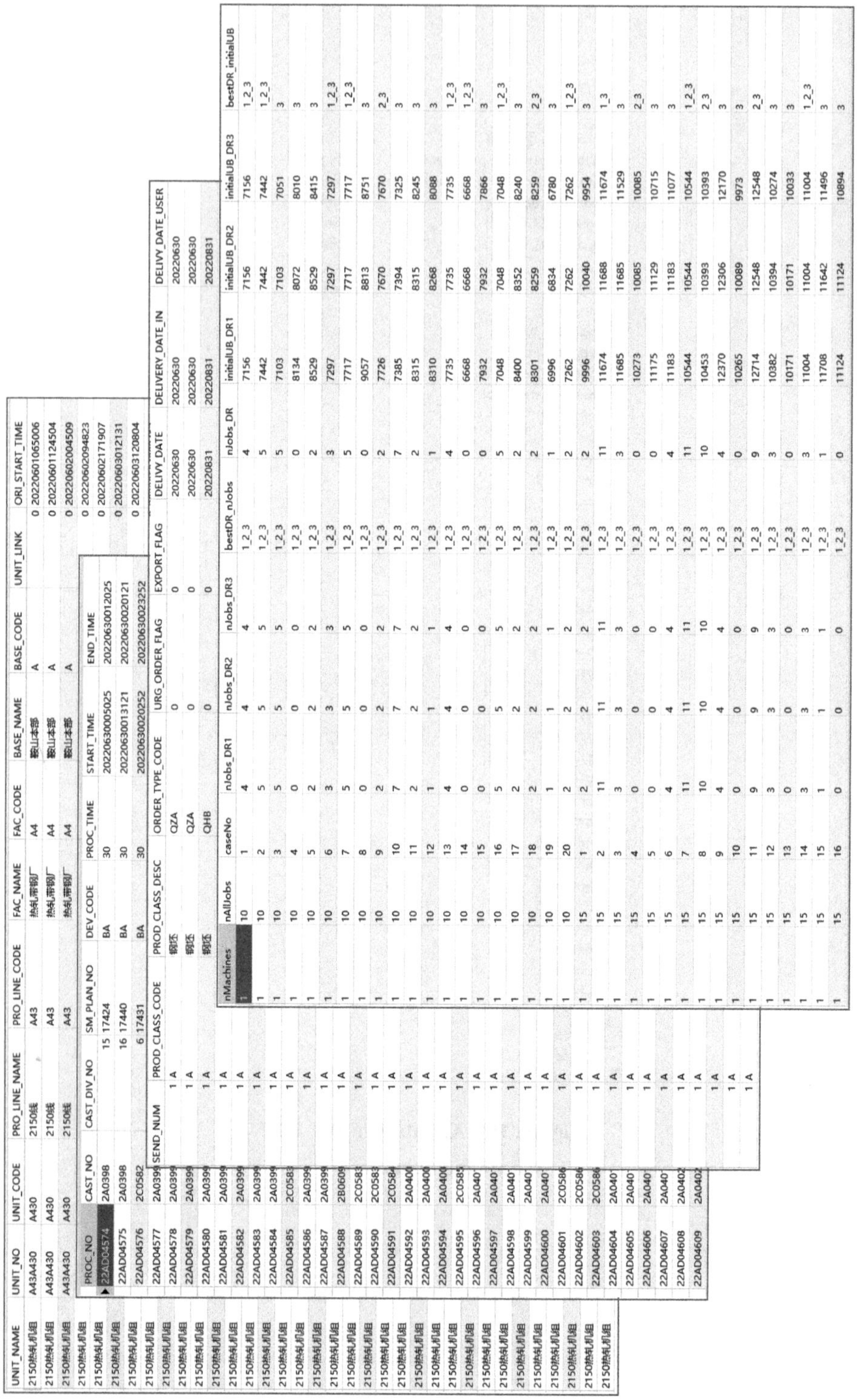

UNIT_NAME	UNIT_NO	UNIT_CODE	PRO_LINE_NAME	PRO_LINE_CODE	FAC_NAME	FAC_CODE	BASE_NAME	BASE_CODE	UNIT_LINK	ORI_START_TIME
2150热轧机组	A43A430	A430	2150线	A43	热轧带钢厂	A4	鞍山本部	A	0	20220601065006
2150热轧机组	A43A430	A430	2150线	A43	热轧带钢厂	A4	鞍山本部	A	0	20220601124504
2150热轧机组	A43A430	A430	2150线	A43	热轧带钢厂	A4	鞍山本部	A	0	20220602004509
2150热轧机组									0	20220602094823
2150热轧机组									0	20220602171907
2150热轧机组									0	20220603012131
2150热轧机组									0	20220603120804
2150热轧机组										
2150热轧机组										
2150热轧机组										
2150热轧机组										
2150热轧机组										
2150热轧机组										
2150热轧机组										
2150热轧机组										
2150热轧机组										
2150热轧机组										
2150热轧机组										
2150热轧机组										
2150热轧机组										
2150热轧机组										
2150热轧机组										
2150热轧机组										
2150热轧机组										
2150热轧机组										
2150热轧机组										
2150热轧机组										
2150热轧机组										
2150热轧机组										
2150热轧机组										
2150热轧机组										
2150热轧机组										
2150热轧机组										
2150热轧机组										
2150热轧机组										

PROC_NO	CAST_NO	CAST_DIV_NO	SM_PLAN_NO	DEV_CODE	PROC_TIME	START_TIME	END_TIME
22AD04574	2A0398	15	17424	BA	30	20220630005025	20220630012025
22AD04575	2A0398	16	17440	BA	30	20220630013121	20220630020121
22AD04576	2C0582	6	17431	BA	30	20220630020252	20220630023252
22AD04577	2A0399						
22AD04578	2A0399						
22AD04579	2A0399						
22AD04580	2A0399						
22AD04581	2A0399						
22AD04582	2A0399						
22AD04583	2A0399						
22AD04584	2A0399						
22AD04585	2C0583						
22AD04586	2A0399						
22AD04587	2A0399						
22AD04588	2B0609						
22AD04589	2C0583						
22AD04590	2C0583						
22AD04591	2C0584						
22AD04592	2A0400						
22AD04593	2A0400						
22AD04594	2A0400						
22AD04595	2C0585						
22AD04596	2A0401						
22AD04597	2A0401						
22AD04598	2A0401						
22AD04599	2A0401						
22AD04600	2A0401						
22AD04601	2C0586						
22AD04602	2C0586						
22AD04603	2C0586						
22AD04604	2A0401						
22AD04605	2A0401						
22AD04606	2A0401						
22AD04607	2A0401						
22AD04608	2A0402						
22AD04609	2A0402						

SEND_NUM	PROD_CLASS_CODE	PROD_CLASS_DESC	ORDER_TYPE_CODE	URG_ORDER_FLAG	EXPORT_FLAG	DELIVY_DATE	DELIVERY_DATE_IN	DELIVY_DATE_USER
1	A	钢坯	QZA	0	0	20220630	20220630	20220630
1	A	钢坯	QZA	0	0	20220630	20220630	20220630
1	A	钢坯	QHB	0	0	20220831	20220831	20220831
1	A							
1	A							
1	A							
1	A							
1	A							
1	A							
1	A							
1	A							
1	A							
1	A							
1	A							
1	A							
1	A							
1	A							
1	A							
1	A							
1	A							
1	A							
1	A							
1	A							
1	A							
1	A							
1	A							
1	A							
1	A							
1	A							
1	A							
1	A							
1	A							
1	A							
1	A							
1	A							

nMachines	nAllJobs	caseNo	nJobs_DR1	nJobs_DR2	nJobs_DR3	bestDR_nJobs	nJobs_DR	initialUB_DR1	initialUB_DR2	initialUB_DR3	bestDR_initialUB
1	10	1	4	4	4	1_2_3	4	7156	7156	7156	1_2_3
1	10	2	5	5	5	1_2_3	5	7442	7442	7442	1_2_3
1	10	3	5	5	5	1_2_3	5	7103	7103	7051	3
1	10	4	0	0	0	1_2_3	0	8134	8072	8010	3
1	10	5	2	2	2	1_2_3	2	8529	8529	8415	3
1	10	6	3	3	3	1_2_3	3	7297	7297	7297	1_2_3
1	10	7	5	5	5	1_2_3	5	7717	7717	7717	1_2_3
1	10	8	0	0	0	1_2_3	0	9057	8813	8751	3
1	10	9	2	2	2	1_2_3	2	7726	7670	7670	2_3
1	10	10	7	7	7	1_2_3	7	7385	7394	7325	3
1	10	11	2	2	2	1_2_3	2	8315	8315	8245	3
1	10	12	1	1	1	1_2_3	1	8310	8268	8088	3
1	10	13	4	4	4	1_2_3	4	7735	7735	7735	1_2_3
1	10	14	0	0	0	1_2_3	0	6668	6668	6668	1_2_3
1	10	15	0	0	0	1_2_3	0	7932	7932	7866	3
1	10	16	5	5	5	1_2_3	5	7048	7048	7048	1_2_3
1	10	17	2	2	2	1_2_3	2	8400	8352	8240	3
1	10	18	2	2	2	1_2_3	2	8301	8259	8259	2_3
1	10	19	1	1	1	1_2_3	1	6996	6834	6780	3
1	10	20	2	2	2	1_2_3	2	7262	7262	7262	1_2_3
1	15	1	2	2	2	1_2_3	2	9996	10040	9954	3
1	15	2	11	11	11	1_2_3	11	11674	11688	11674	1_3
1	15	3	3	3	3	1_2_3	3	11685	11685	11529	3
1	15	4	0	0	0	1_2_3	0	10273	10085	10085	2_3
1	15	5	0	0	0	1_2_3	0	11175	11129	10715	3
1	15	6	4	4	4	1_2_3	4	11183	11183	11077	3
1	15	7	11	11	11	1_2_3	11	10544	10544	10544	1_2_3
1	15	8	10	10	10	1_2_3	10	10453	10393	10393	2_3
1	15	9	4	4	4	1_2_3	4	12370	12306	12170	3
1	15	10	0	0	0	1_2_3	0	10265	10089	9973	3
1	15	11	9	9	9	1_2_3	9	12714	12548	12548	2_3
1	15	12	3	3	3	1_2_3	3	10382	10394	10274	3
1	15	13	0	0	0	1_2_3	0	10171	10171	10033	3
1	15	14	3	3	3	1_2_3	3	11004	11004	11004	1_2_3
1	15	15	1	1	1	1_2_3	1	11708	11642	11496	3
1	15	16	0	0	0	1_2_3	0	11124	11124	10894	3

Fig.19.17 Sketch map of part of the content in the scheduling

19.5.1.3 InfluxDB temporal database

TheInfluxDB database is a temporal database, which focuses more on the temporal characteristics of the data compared to relational databases. The InfluxDB database stores the data required for the full process demonstration of hot strip rolling, such as the sensor values for coiling machine startup, rolling force of the first to seventh finishing rolling stands, roller speed, etc., and outputs the data in a continuous time series to the Django program through the InfluxDB database, used to demonstrate the entire process of hot rolling of strip steel on the front-end page, **Fig. 19.18** shows some of the content in the InfluxDB database.

Settings Help

SELECT * FROM "FDA01_TEST"

Time stamp

Roll-call

#	time	DC1_ON	DC2_ON	DC3_ON	FIX_IN	FIX_OFF	F1_Rolling_Force	F2_Rolling_Force	F3_Rolling_Force	F4_Rolling_Force	F5_Rolling_Force	F6_Rolling_Force	F7_Rolling_Force	M12_ON	MMD410A_ON	MMD411A_ON	MMD412A_ON	MMD413A_ON	L_E6_EXIST	L_X7_EX
1	2023/2/2 8:05:00	0	1	0	0	0	1.94861	16.2321	181.172	67.4083	-132.416	-70.9608	-55.9795	0	0	0	0	0	1	1
2	2023/2/2 8:05:00	0	1	0	0	0	36.9396	10.4553	199.685	108.672	-94.2188	-167.554	-63.5938	0	0	0	0	0	1	1
3	2023/2/2 8:05:00	0	1	0	0	0	76.0021	39.8348	212.422	92.8816	-22.5956	-136.292	-61.4063	0	0	0	0	0	1	1
4	2023/2/2 8:05:00	0	1	0	0	0	64.2833	-1.26343	212.422	59.4305	-94.0089	-132.805	-70.0639	0	0	0	0	0	1	1
5	2023/2/2 8:05:00	0	1	0	0	0	34.9038	-9.07593	196.797	28.5111	-200.998	-102.866	-56.8447	0	0	0	0	0	1	1
6	2023/2/2 8:05:00	0	1	0	0	0	81.6271	-24.7009	188.954	83.3639	-180.156	-133.906	-42.7402	0	0	0	0	0	1	1
7	2023/2/2 8:05:00	0	1	0	0	0	60.5424	-9.07593	178.284	48.0424	-76.6277	-101.136	-41.875	0	0	0	0	0	1	1
8	2023/2/2 8:05:00	0	1	0	0	0	42.8816	0.607056	165.547	63.6674	-129.375	-92.0126	-52.7286	0	0	0	0	0	1	1
9	2023/2/2 8:05:00	0	1	0	0	0	121.172	10.4553	181.172	83.9978	-67.5303	-112.199	-48.8223	0	0	0	0	0	1	1
10	2023/2/2 8:05:00	0	1	0	0	0	58.5066	-3.13391	169.453	36.1583	-140.229	-109.604	-80.0723	0	0	0	0	0	1	1
11	2023/2/2 8:05:01	0	1	0	0	0	38.8101	0.607056	171.324	18.4976	-107.013	-107.638	-92.6563	0	0	0	0	0	1	1
12	2023/2/2 8:05:01	0	1	0	0	0	25.3861	26.2455	181.172	36.3594	-25.6367	-125.438	-65.9795	0	0	0	0	0	1	1
13	2023/2/2 8:05:01	0	1	0	0	0	25.3861	4.67859	208.516	139.922	-48.6544	-125.884	-71.6045	0	0	0	0	0	1	1
14	2023/2/2 8:05:01	0	1	0	0	0	27.4219	18.2678	186.949	108.507	-67.9758	-122.397	-48.167	0	0	0	0	0	1	1
15	2023/2/2 8:05:01	0	1	0	0	0	88.3549	26.0803	215.31	116.65	-93.3536	-121.632	-37.1036	0	0	0	0	0	1	1
16	2023/2/2 8:05:01	0	1	0	0	0	38.9753	65.1428	211.404	98.8236	-117.866	-106.353	-64.4473	0	0	0	0	0	1	1
17	2023/2/2 8:05:01	0	1	0	0	0	43.0469	22.1741	200.516	49.5823	-86.4063	-152.572	-45.7813	0	0	0	0	0	1	1
18	2023/2/2 8:05:01	0	1	0	0	0	56.8014	10.29	192.891	42.1003	-83.8107	-128.889	-49.6875	0	0	0	0	0	1	1
19	2023/2/2 8:05:01	0	1	0	0	0	46.7878	26.0803	173.359	46.1719	-118.312	-133.906	-80.9375	0	0	0	0	0	1	1
20	2023/2/2 8:05:01	0	1	0	0	0	17.4083	24.2098	184.06	34.2878	-171.898	-133.041	-35.7929	0	0	0	0	0	1	1
21	2023/2/2 8:05:02	0	1	0	0	0	31.1628	22.1741	185.078	59.7611	-96.3348	-98.9598	-35.583	0	0	0	0	0	1	1
22	2023/2/2 8:05:02	0	1	0	0	0	89.5913	20.3036	186.096	87.9041	-87.7169	-104.177	-54.2491	0	0	0	0	0	1	1
23	2023/2/2 8:05:02	0	1	0	0	0	42.8816	10.4553	187.966	73.1851	-153.913	-107.873	-47.3018	0	0	0	0	0	1	1
24	2023/2/2 8:05:02	0	1	0	0	0	44.5868	4.67859	186.949	63.5021	-111.364	-150.842	-56.6348	0	0	0	0	0	1	1
25	2023/2/2 8:05:02	0	1	0	0	0	40.6805	8.41956	156.716	59.9264	-107.668	-121.532	-51.208	0	0	0	0	0	1	1
26	2023/2/2 8:05:02	0	1	0	0	0	54.4351	-5.16968	185.078	42.1003	-40.1866	-63.7803	-57.5	0	0	0	0	0	1	1
27	2023/2/2 8:05:02	0	1	0	0	0	42.8816	-7.20544	165.547	88.9753	-107.248	-151.917	-48.167	0	0	0	0	0	1	1
28	2023/2/2 8:05:02	0	1	0	0	0	78.0378	-5.16968	182.19	73.6808	-151.947	-107.638	-41.0098	0	1	0	0	0	0	0
29	2023/2/2 8:05:02	0	1	0	0	0	72.2611	6.54907	187.966	59.5958	-122.638	-108.083	-62.0616	0	1	0	0	0	0	0
30	2023/2/2 8:05:02	0	1	0	0	0	33.0333	-5.16968	180.154	92.551	-114.615	-137.603	-46.4366	0	1	0	0	0	0	0
31	2023/2/2 8:05:03	0	1	0	0	0	38.9753	-1.26343	161.641	59.7611	-72.9572	-110.679	-65.9678	0	1	0	0	0	0	0
32	2023/2/2 8:05:03	0	1	0	0	0	46.9531	16.2321	196.797	71.6451	-148.487	-85.5107	-64.4473	0	1	0	0	0	0	0
33	2023/2/2 8:05:03	0	1	0	0	0	62.5781	10.4553	214.292	53.8191	-64.2794	-145.415	-49.6875	0	1	0	0	0	0	0
34	2023/2/2 8:05:03	0	1	0	0	0	-3.99341	-1.26343	195.779	81.7969	-113.095	-128.27	-45.7813	0	1	0	0	0	0	0
35	2023/2/2 8:05:03	0	1	0	0	0	38.9753	2.64282	208.516	86.0201	-98.7803	-121.532	-59.0205	0	1	0	0	0	0	0
36	2023/2/2 8:05:03	0	1	0	0	0	60.5424	0.772339	196.797	83.0333	-81.8837	-135.637	-52.7286	0	1	0	0	0	0	0
37	2023/2/2 8:05:03	0	1	0	0	0	31.3281	18.433	169.453	45.8413	-150.427	-93.1134	-50.3428	0	1	0	0	0	0	0
38	2023/2/2 8:05:03	0	1	0	0	0	38.9753	22.1741	188.984	36.4889	-122.638	-104.177	-65.9678	0	1	0	0	0	0	0

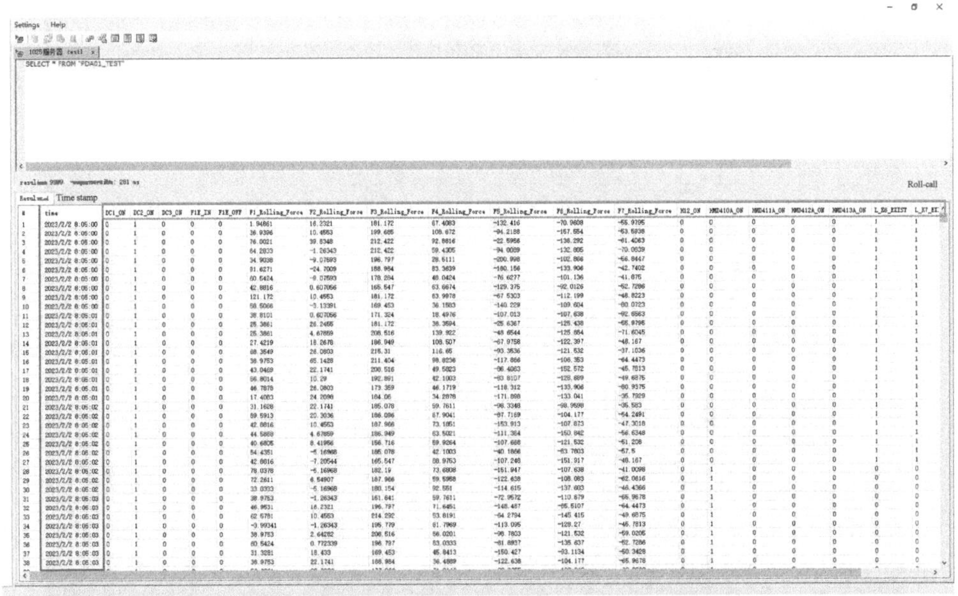

Fig. 19.18 Sketch map of partial content in the InfluxDB database

19.5.1.4 Neo4j Graph database

Neo4j is a high-performance NoSQL database based on graph theory, mainly used for storing, querying, and processing highly correlated data with complex structures. The graph in the Neo4j data model consists of nodes (entities) and edges (relationships), each of which can have any number of attributes. The database has its own Query language, Cypher, which is a highly optimized declarative graph Query language, and can effectively match and query complex graph structure data. In addition, Neo4j's transaction management ensures full ACID (Atomicity, Consistency, Isolation,

Durability) characteristics, emphasizing data consistency and reliability, especially when performing complex operations to ensure data integrity. Neo4j also has scalability, which is reflected in its backup solution and high availability services, enabling it to meet the needs of large-scale graphic data processing.

The rolling procedure Knowledge graph built by this project is stored in Neo4j, which contains 1672 entities and 2465 triplets, and two entity types of equipment and attribute are built. **Fig. 19.19** shows the overall overview of the Knowledge graph, in which the green entity is the equipment entity, the yellow entity is the attribute entity, and the edge is the relationship between entities. Neo4j has powerful query functions, such as querying all device entities through the Cypher statement "match (n: Device) return n".

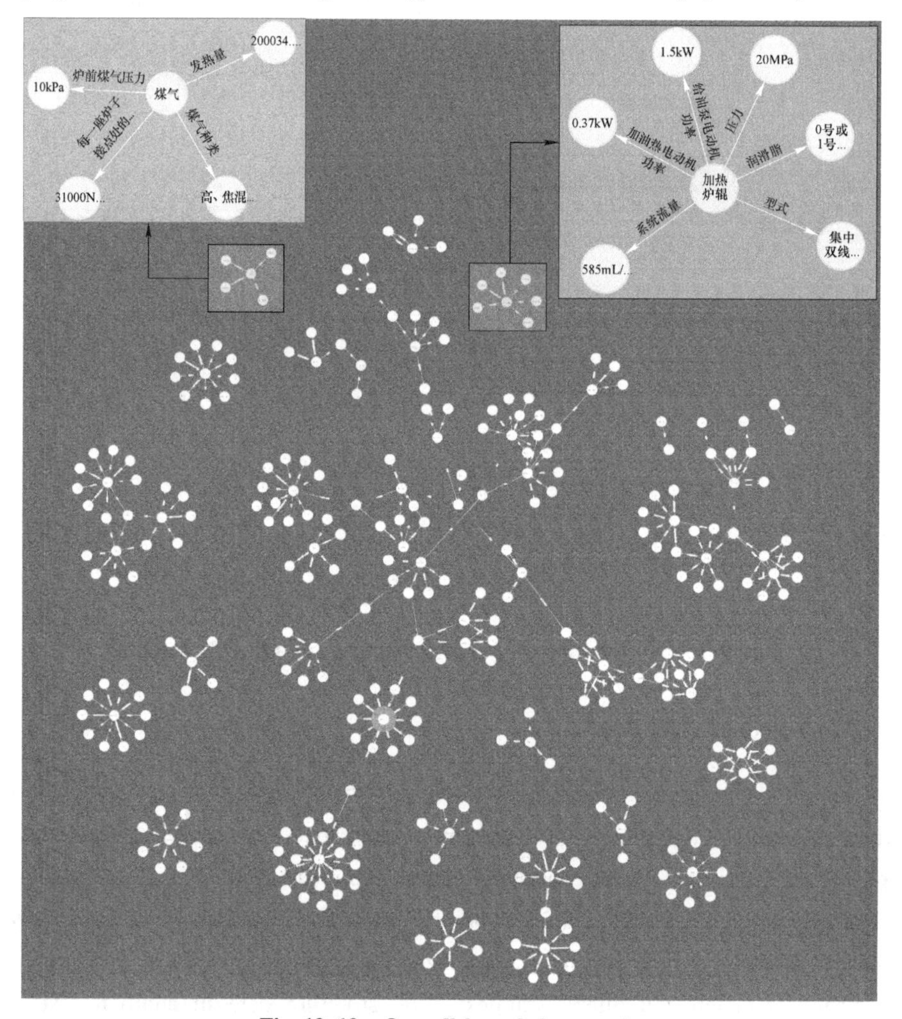

Fig. 19.19 Overall knowledge graph

19.5.2 Real-time data playback

The hardware platform of the HSMP data playback system is shown in **Fig. 19.20**. A total of four PLCs are installed in the VME chassis, simulating the four main processes of finishing mill, roughing mill, coiling, and heating furnace in the industrial hot strip rolling process. When the four processes work normally, CodeSys and Python establish a connection through TCP/IP, and the data stored in the Global variable GVL. wrbuffer in CodeSys is sent to the Socket client established in Python in real time, and then the data is immediately uploaded from Python to the pSpace real-time database in real time.

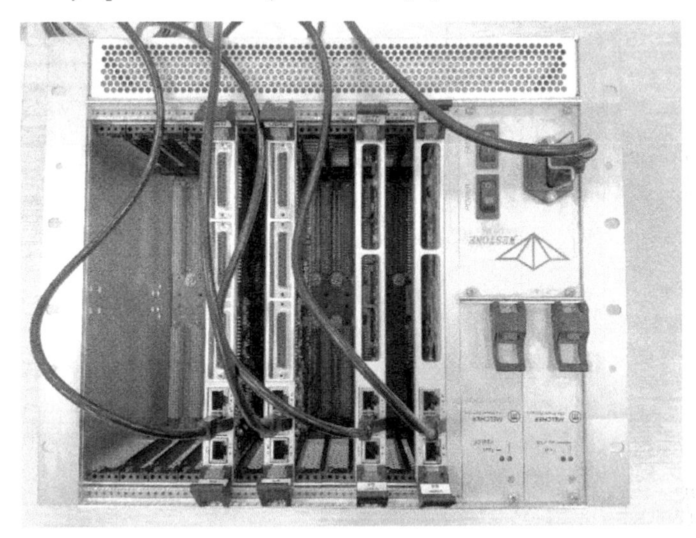

Fig. 19.20 Overall knowledge graph

When fault injection is carried out for the four processes, the normal data enters the fault injection system of the hot continuous rolling process based on Python to get the fault data, then Python and CodeSys establish socket communication, and send the fault data to the PLC through TCP/IP communication protocol. At the same time, CodeSys software stores the received data into the Global variable GVL. rdbuffer. When the PLC writes a signal, CodeSys transfers the data into the Global variable GVL. wrbuffer (as shown in the figure), and then sends it to the Socket client established in Python. Python immediately uploads the data to the pSpace real-time database (as shown in **Fig. 19.21** and **Fig. 19.22**) after receiving the data.

Based on the real-time data in thepSpace real-time database, the full process demonstration interface of University of Science and Technology Beijing's hot strip rolling process can read historical data and play back the full process of hot strip

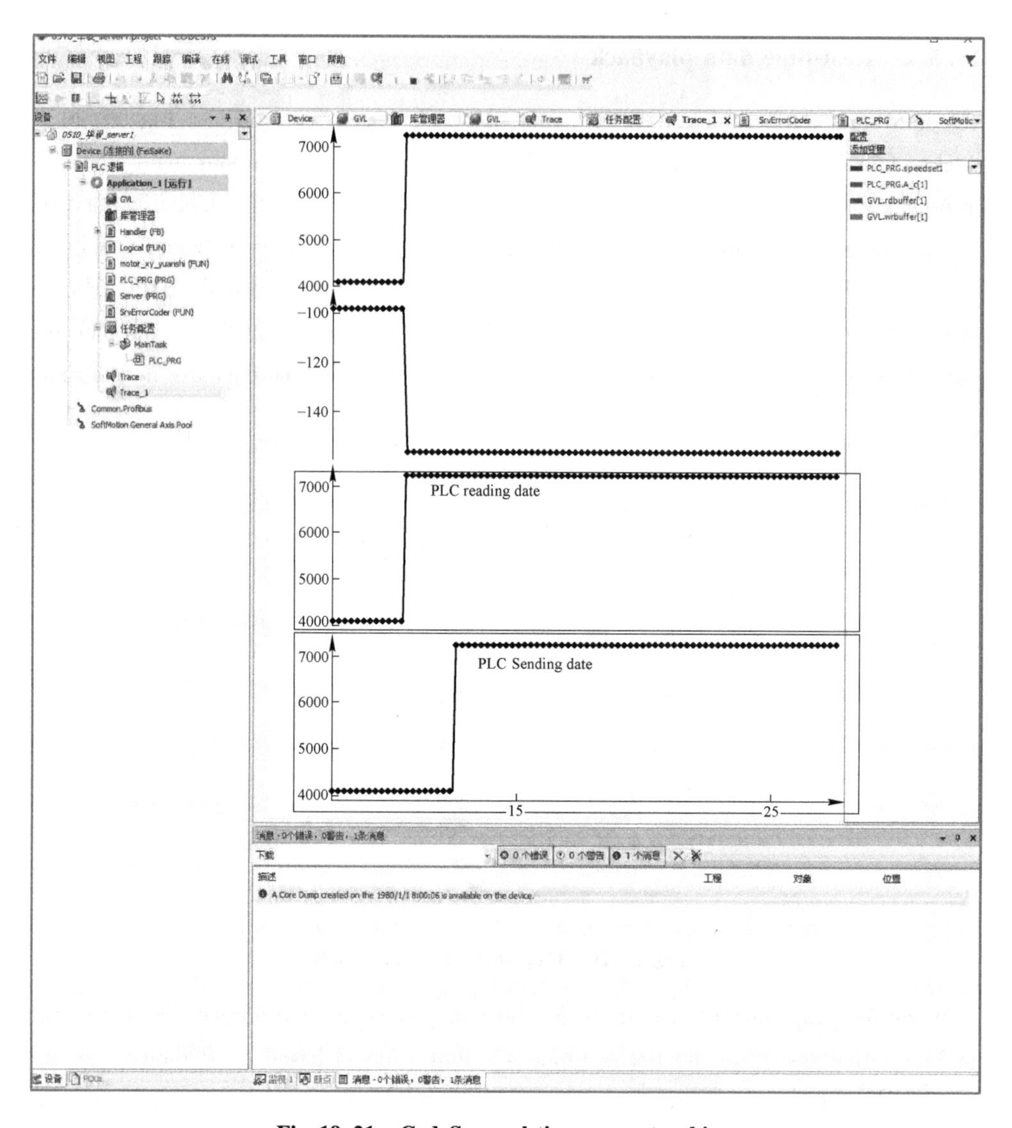

Fig. 19. 21　CodeSys real-time curve tracking

rolling. **Fig. 19. 23** shows the roller tables of each part in rolling and marks their running speed below each roller table. At the top right corner and middle position of the page, the exit temperature of finishing rolling, the entrance temperature of finishing rolling, the exit temperature of rough rolling, the laminar cooling temperature Information such as steel coil number and steel grade. Through this page, you can have a clear and intuitive understanding of the complete rolling process of the slab from the heating furnace to the roughing mill stand, then to the finishing mill stand, followed by laminar cooling and final coiler.

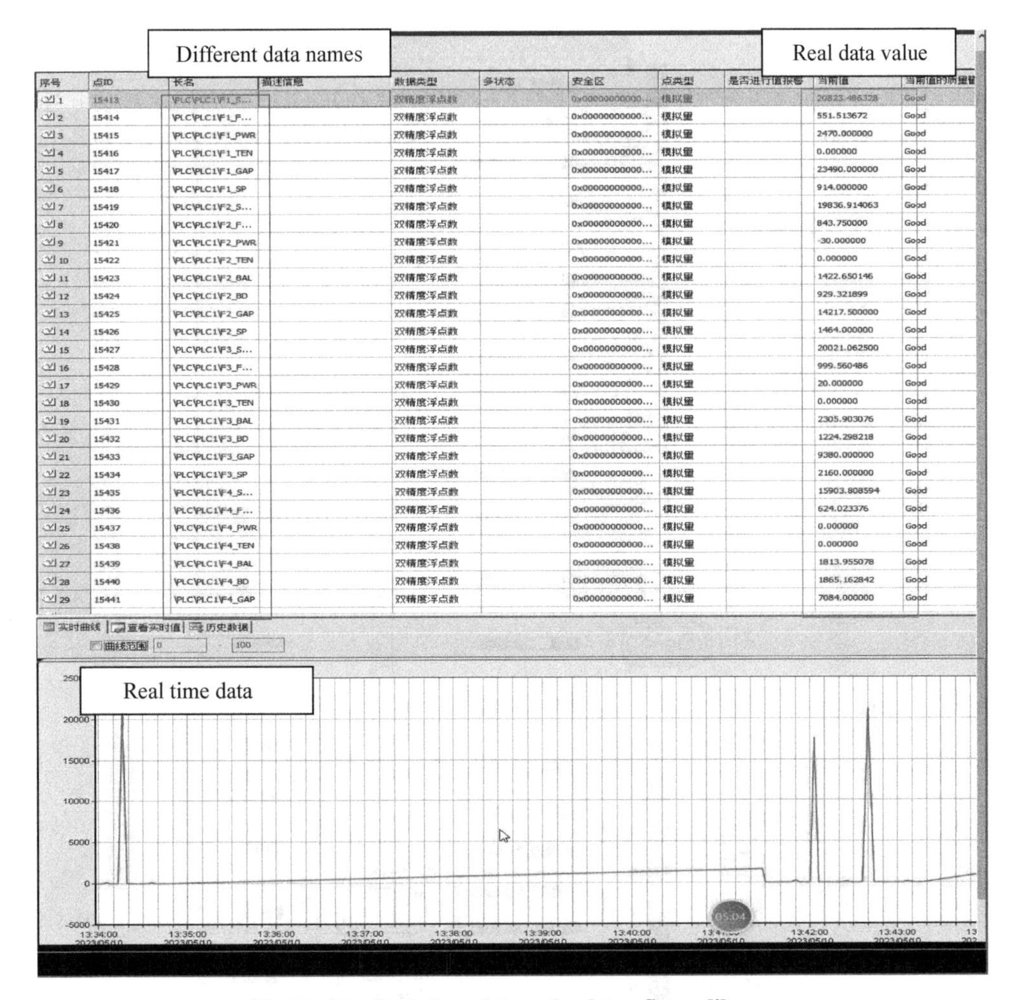

Fig. 19.22 Real time data upload to pSpace library

19.5.3 Process monitoring interface

The process monitoring method of "edge to end" collaboration, the "edge" side data management and calculation platform, and the "end" side data collection and preprocessing platform. The "end" side is responsible for collecting a large amount of real-time data in a distributed manner and preprocessing the original data. The "edge" side corresponds to the corresponding processes of the industrial process, with a large number of processes monitoring algorithms deployed on top. After receiving data from the "end" side, the sub processes of the process are monitored based on their own algorithms, facilitating the integration of "edge" side monitoring results on the "cloud" side and achieving real-time monitoring of the entire industrial process.

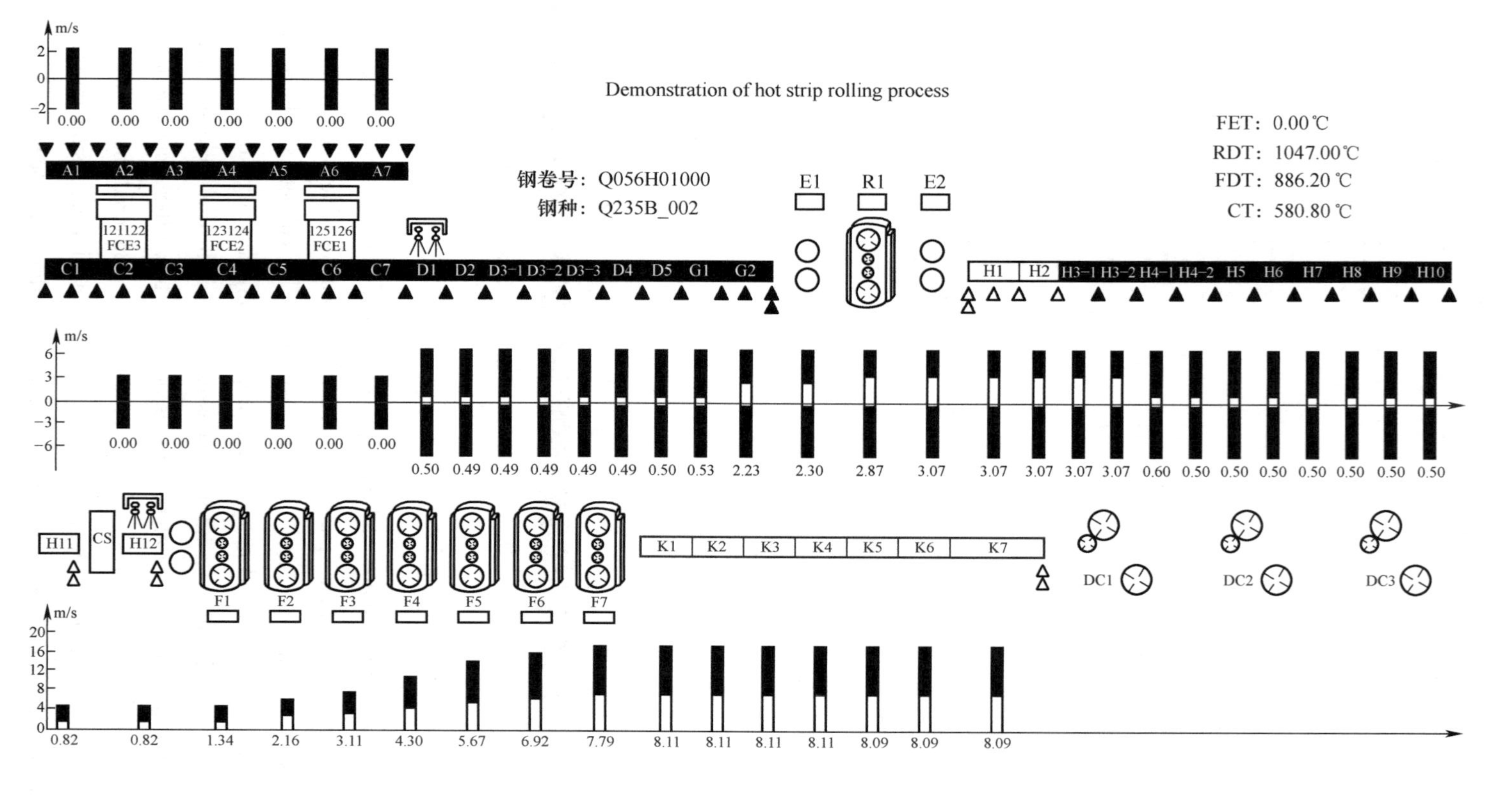

Fig.19.23 HSMP flow demonstration interface

The "end" side is a data collection and preprocessing platform. Due to the diverse forms of data transmission protocols and complex communication methods used in the hot rolling process of strip steel, specific data acquisition equipment is required to meet practical needs. At the same time, this also makes the original data collected at the "end" side have strong multi-source heterogeneous characteristics. Before uploading the data to the "edge" side to build the monitoring model, data pre-processing is an essential Committed step. The purpose of this step is to improve the data quality and transform the original data into a more convenient form. For example, for non-ideal data, Outlier need to be deleted from the original dataset, and missing data needs to be filled or estimated. Afterwards, training datasets are often standardized, with one method being to automatically scale each variable to zero mean and unit variance. In this case, the scale differences between process variables can be eliminated, so the data model will not lean towards variables with larger scalars. In addition, with the introduction of various multi-sampling data processing strategies, it is also necessary to synchronize the sampling rates of different process variables through upsampling or downsampling.

The "edge" side is a data management and computing platform. After the preprocessed data on the "end" side is transmitted to the "edge" platform, it can be visually displayed through a web interface, making it easy for operation and maintenance management personnel to monitor services in real time. The "edge" side integrates real-time data collected from the "end" side, which is the foundation for storing data, and constructs a large-scale industrial process dataset to provide sufficient data support for service monitoring model training. The "edge" side further processes data, trains offline to build an efficient process monitoring model, extracts potential information through the deep representation structure of the model, monitors the operation status of the production process, and provides guarantees for the sustained and safe operation of production equipment. **Fig. 19.24** shows the overall schematic diagram of "edge-end" collaboration, and **Fig. 19.25** shows the actual process monitoring results.

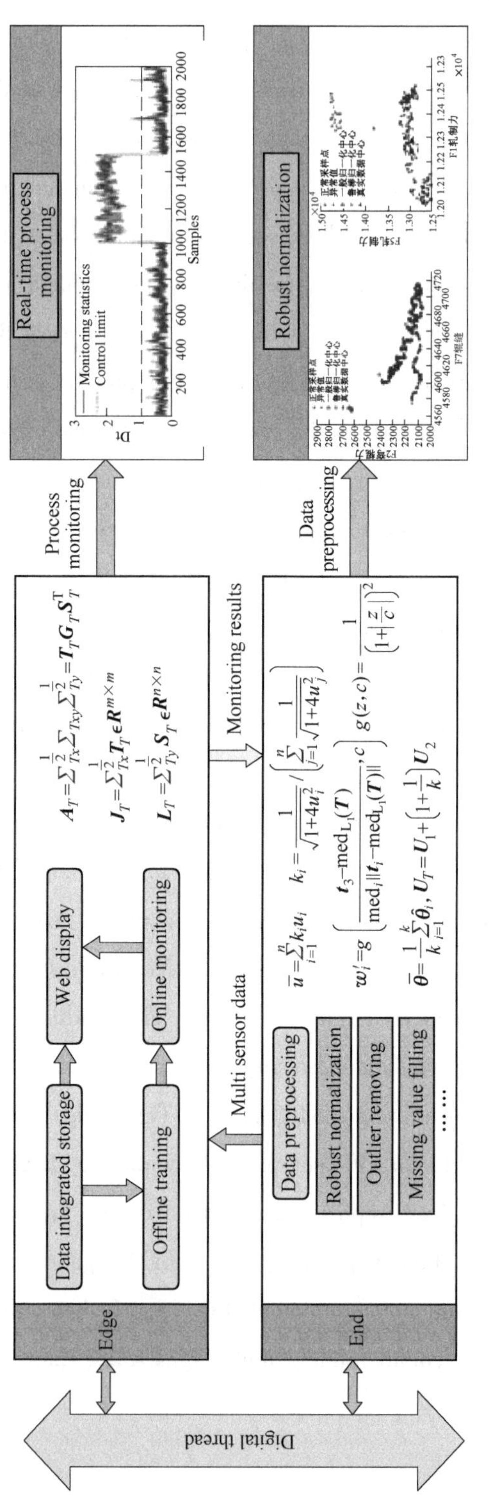

Fig.19.24 Schematic diagram of "edge-end" collaboration

Digital thread project

Monitoring display

2023年05月23日
星期二
16:31:08

T²统计量:

88.38
80
60
40
20
0
T²控制限
16:29:28 16:29:34 16:29:40 16:29:46 16:29:52 16:29:58 16:30:04 16:30:10 16:30:16 16:30:22 16:30:28 16:30:34 16:30:40 16:30:46 16:30:52 16:30:58 16:31:04

时间	来源	统计量	控制限
2023/05/23 16:29:21	精轧过程	45.7	44.19
2023/05/23 16:29:21	精轧过程	50.55	44.19
null	精轧过程	null	44.19
null	精轧过程	null	44.19
null	精轧过程	null	44.19

SPE统计量:

17.94
15
12
9
6
3
0
SPE控制限
16:29:28 16:29:34 16:29:40 16:29:46 16:29:52 16:29:58 16:30:04 16:30:10 16:30:16 16:30:22 16:30:28 16:30:34 16:30:40 16:30:46 16:30:52 16:30:58 16:31:04

时间	来源	统计量	控制限
2023/05/23 16:29:25	精轧过程	9.44	8.97
2023/05/23 16:29:23	精轧过程	14.61	8.97
null	精轧过程	null	8.97
null	精轧过程	null	8.97
null	精轧过程	null	8.97

Fig.19.25 Process monitoring results of "edge to end" collaboration